Microorganisms for Sustainability

Volume 22

Series Editor

Naveen Kumar Arora, Environmental Microbiology, School for Environmental Science, Babasaheb Bhimrao Ambedkar University, Lucknow, Uttar Pradesh, India

More information about this series at http://www.springer.com/series/14379

Pankaj Kumar Arora
Editor

Microbial Technology for Health and Environment

Springer

Editor
Pankaj Kumar Arora
Department of Microbiology
Babasaheb Bhimrao Ambedkar University
Lucknow, Uttar Pradesh, India

ISSN 2512-1901 ISSN 2512-1898 (electronic)
Microorganisms for Sustainability
ISBN 978-981-15-2678-7 ISBN 978-981-15-2679-4 (eBook)
https://doi.org/10.1007/978-981-15-2679-4

This Springer imprint is published by the registered company Springer Nature Singapore Pte Ltd.
The registered company address is: 152 Beach Road, #21-01/04 Gateway East, Singapore 189721, Singapore

Preface

Pollution is creating a critical situation for health and environment due to its hazardous nature and has become a worldwide issue for environmentalists. A number of carcinogenic compounds are daily discharged into soil and water by various means. The presence of these compounds in water and soil not only disturbs the ecosystem of our earth but also creates a risk for health of living beings. Several technologies have been developed to remove these compounds from soil and water. Microbial technology has gained attention as microbes are able to degrade these compounds without harming the environment.

The book focuses on various aspects of microbial technology in health, environment, and agriculture. This book consists of 14 chapters related to microbial peroxidases, wastewater treatment, solid waste management, quorum quenching, antitumor products, microbe-assisted phytoremediation, microbial endophytes, role of microbes in agriculture, microbial degradation of nitroaromatics, and organophosphates.

This book describes the recent advances in the field of microbial degradation and microbial remediation of various xenobiotic compounds in soil and wastewater. It also explains various modern microbial technologies for biodegradation and wastewater treatment. It covers various microbial technologies for wastewater treatment, biodegradation, bioremediation, and solid waste management. Contributions from authoritative experts in the world are compiled. It focuses on the current scenario of industrial wastewater treatment and its biodegradation.

The book is meant for researchers in wastewater industry, students of environmental sciences, environmental microbiologist, and practitioners in water pollution abatement.

I acknowledge the Department of Biotechnology, India, for awarding me DBT-Ramalingaswami Re-entry Fellowship.

Lucknow, India
Pankaj Kumar Arora

Contents

About the Series Editor

Naveen Kumar Arora, PhD, Microbiology, is a Professor in the Department of Environmental Science at Babasaheb Bhimrao Ambedkar University (A Central University), Lucknow, Uttar Pradesh, India. He is a renowned researcher in the field of environmental microbiology and biotechnology. His specific area of research is rhizosphere biology and PGPRs. He has more than 50 research papers published in premium international journals and several articles published in magazines and dailies. He is an editor of 10 books, published by Springer. He is a member of several national and international societies, Secretary General of Society for Environmental Sustainability, in editorial board of 4 journals, and reviewer of several international journals. He is also the editor in chief of the journal *Environmental Sustainability* published by Springer Nature. He has delivered lectures in conferences and seminars around the globe. He has a long-standing interest in teaching at the PG level and is involved in taking courses in bacteriology, microbial physiology, environmental microbiology, agriculture microbiology, and industrial microbiology. He has been advisor to 118 postgraduate and 8 doctoral students. Recently, he was awarded for excellence in research by the Honorable Governor of Uttar Pradesh. Although an academician and researcher by profession he has a huge obsession for the wildlife and its conservation and has authored a book, *Splendid Wilds*. He is the President of Society for Conservation of Wildlife and has a dedicated website www.naveenarora.co.in for the cause of wildlife and environment conservation.

About the Editor

Pankaj Kumar Arora is currently an Assistant Professor and DBT-Ramalingaswami Re-entry Fellow in the Department of Microbiology at Babasaheb Bhimrao Ambedkar University in Lucknow, India. Dr. Arora is also an editorial board member for the journal *Scientific Reports*, an associate editor for the journal *Frontiers in Microbiology*, and an academic editor for *PLOS ONE*. He is the recipient of several national awards and fellowships including a Young Botanist Award and Dr. Y. S. Murty Medal from the Indian Botanical Society. His major focus area is environmental microbiology and is currently investigating the biodegradation and bioremediation of various xenobiotic compounds including nitrophenols, chlorinated nitrophenols, and indole. He has published a total of 36 papers in reputed journals, two edited books, and has eight years of teaching and research experience at national and international institutes.

Microbial Peroxidases and Their Unique Catalytic Potentialities to Degrade Environmentally Related Pollutants

1

Muhammad Bilal and Hafiz M. N. Iqbal

Abstract

Industrial sectors play an imperative role in the economic growth and development of any nation. Nevertheless, the discharge of industrial wastewater polluted by various textile dyes, pharmaceuticals, recalcitrant organic compounds, hormones, xenobiotic compounds (i.e., insecticides, pesticides, plastics, fertilizers, and hydrocarbons), and personal care products into the receiving water bodies seriously threatened the natural ecosystem owing to their extremely toxic consequences. This problem is pervasively increasing due to the lack of efficient waste management procedures for the proper disposal and treatment of waste. Considering the diverse nature of wastewater from industrial processes, designing a cost-competitive, efficient, and eco-friendlier technology with stable remediation performance has become a challenging task for the research investigators and environmental engineers. In the past couple of decades, environmental biotechnology has witnessed a tremendous upsurge in exploring some judicious substitutes to the existing technologies for waste management. Conventionally, in practice, approaches dealing with wastewater remediation such as chemical, physical, and biological methods are either inefficient or restrictive due to techno-economic constraints. In this perspective, enzyme-assisted treatment is a rapid, easy, eco-sustainable approach and therefore has been keenly explored to degrade and mineralize an array of xenobiotic and recalcitrant organic contaminants. Peroxidases isolated and characterized by different microbial or plant-based natural resources have demonstrated great bioremediation potential.

M. Bilal (✉)
School of Life Science and Food Engineering, Huaiyin Institute of Technology, Huaian, China

H. M. N. Iqbal
Tecnologico de Monterrey, School of Engineering and Sciences, Campus Monterrey, Monterrey, NL, Mexico

P. K. Arora (ed.), *Microbial Technology for Health and Environment*, Microorganisms for Sustainability 22,
https://doi.org/10.1007/978-981-15-2679-4_1

Genetic engineering and enzyme immobilization approaches have made it possible to produce a significant amount of recombinant enzymes and upgrade the half-life, catalytic stability, and activity of the biocatalyst, respectively. Moreover, the development of nanozymes might display the potential remediation capability toward a wide variety of toxic pollutants. In this chapter, we have presented a comprehensive overview of the peroxidases and advanced enzyme tools and technologies, i.e., immobilized enzyme-based constructs, nanozymes as robust catalytic tools, and genetic engineering along with their use in the degradation and detoxification of toxic substances, human-health related hazardous compounds, carcinogenic and mutagenic entities, and environmentally related contaminants of high concern.

Keywords
Green biotechnology · Industrial waste · Immobilization · Enzymatic remediation · Genetic engineering · Nanozymes

1.1 Introduction

The functioning of society and its future developments relied on the availability of bio-renewable energy resources and the capability to transform environmentally polluting production processes with green and sustainable bioprocesses. Water contamination by various recalcitrant organic compounds, textile dye pollutants, pharmaceutically active compounds, hormones, xenobiotic compounds (i.e., insecticides, pesticides, plastics, fertilizers, and hydrocarbons), and personal care products is a high concern environmental issue because of their toxic, carcinogenic, mutagenic, and teratogenic consequences. These facts garnered a growing concern among the scientific community, in particular environmental engineers, to develop a greener and sustainable relevance between the Earth and its biological resources (Marco-Urrea et al. 2009). Undoubtedly, a range of various chemical, physical, and biological methods have been developed and attempted to degrade and eliminate these environmentally related chemicals from the contaminated water bodies. Biological strategies, explicitly the use of enzymes, have received considerable attention for eliminating industrial and environmental contaminants owing to their pronounced catalytic efficacy, selectivity, and eco-friendlier processes. Among the enzymes ascertained for biodegradation purpose, fungal peroxidases, for instance, horseradish peroxidase (HRP), manganese-dependent peroxidase (MnP), lignin peroxidase (LiP), and others, are gaining a lot of attention as promising biocatalysts for the biodegradation of refractory xenobiotic and removal of harmful compounds from wastewater. Peroxidases are a large group of enzymes, which catalyze the reduction of peroxides (i.e., H_2O_2) and oxidation of a large range of inorganic and organic compounds. These enzymes exhibit considerable potential to abate environmental contamination by bioremediating wastewater polluted with phenols, cresols, and chlorinated phenols, degradation of industrial or textile dye pollutants, and the elimination of peroxide from foodstuffs and industrial wastes. Processed water from *fabrics*, garment, or textile units often characterizes an intense pigmentation owing to the occurrence of different dyes, which are unaffected to classical decolorizing treatment technologies and could be potentially decolorized by the use of peroxidases (Huber and Carré 2012). Peroxidases from different sources are also able to oxidize dimethoxybenzene, amines, aromatic alcohols, lignin dimers, dyes, and phenolic and nonphenolic substrates in the absence of Mn(II). In this chapter, we provided an updated overview of the peroxidases and advanced enzyme tools i.e., immobilized enzyme-based constructs, nanozymes as robust catalytic tools, and genetic engineering along with their use in the degradation and detoxification of toxic substances, human health-related hazardous compounds, carcinogenic and mutagenic entities, and environmentally related contaminants of high concern.

1.2 Peroxidases: Potential Sources

Peroxidases are widely spread, structurally diverse and abundant enzymes produced by bacteria, fungi, algae, animals, and plants (Battistuzzi et al. 2010). These enzymes carry out depolymerization of lignin and other aromatic pollutants in an H_2O_2-dependent reaction using a suitable redox mediator. Peroxidases are characterized as heme and nonheme peroxidases. The heme-containing peroxidases can be classified into two groups, where enzymes in group 1 are mainly present in animals and those found in bacteria, plants, and fungi constitute group 2. Peroxidases found in bacteria, plants, and fungi are additionally categorized into three classes: class 1 contains yeast-secreted cytochrome C peroxidase, bacterial-originated catalases, and ascorbate peroxidase secreted by some species of plants (Dunford 1999; Smulevich et al. 2006). Extracellularly secreted fungal enzymes explicitly MnP and LiP are included in class 2, whereas class 3 encompasses plant-derived peroxidase such as HRP (Veitch and Smith 2001; Battistuzzi et al. 2010). Nonheme peroxidases, on the other hand, constitute five diverse independent families, including alkyl hydro peroxidase, NADH peroxidase, thiol peroxidase, manganese catalase, and haloperoxidase (Koua et al. 2009). Among all the above-stated peroxidases, LiP, MnP, and HRP are the most studied enzymes owing to their superior catalytic performance for decontamination of toxic compounds and environmental pollutants.

1.2.1 Physiochemical and Catalytic Properties of LiP (EC 1.11.1.14)

LiP (diaryl propane oxygenase) is a robust heme-containing peroxidase that executes H_2O_2-driven oxidative depolymerization of the lignin molecule (Fig. 1.1). After the discovery and isolation in ligninolytic extract of *P. chrysosporium*, various isozymes of LiP were identified in different microbial isolates such as *T. versicolor*, *P. chrysosporium*, *P. sordida*, and *Phlebia radiata* (Tien and Kirk 1983; Farrell et al.

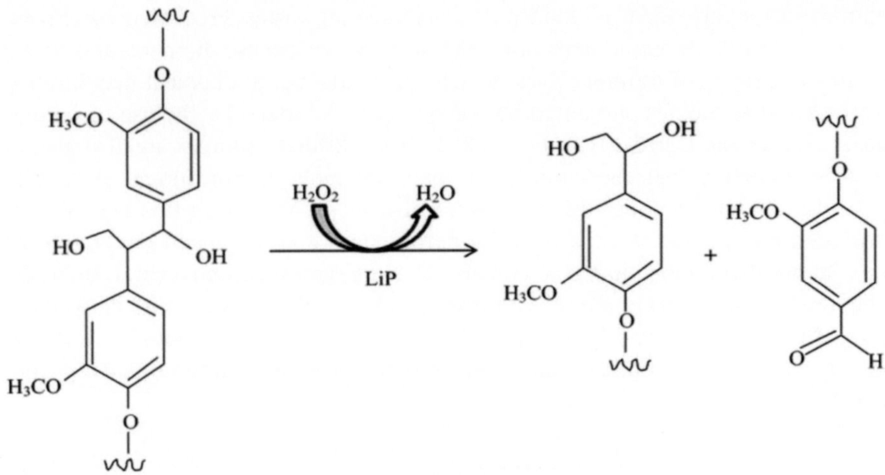

Fig. 1.1 LiP-assisted cleavage of β-1 linkage in lignin moiety. (Adapted from Falade et al. 2017)

1989; Johansson et al. 1993; Sugiura et al. 2009). For example, Farrell et al. (1989) identified the presence of six LiP isozymes designating H1, H2, H6, H7, H8, and H10 in the culture broth of *P. chrysosporium* BKM-F-1767. Similarly, Glumoff et al. (1990) characterized five LiP isozymes also from *P. chrysosporium* and found that the purified isozymes exhibited different substrate specificity, stability, sugar content, and isoelectric points. The different N-terminal amino acid sequences revealed that different genes encoded them. The gene sequence study of *P. chrysosporium* strain RP78 showed almost 10 *lip* genes that further confirmed the presence of LiP isozymes (Martinez et al. 2004). In consistence with earlier reports, Morgenstern et al. (2008) described that the genome of *P. chrysosporium* harbor 10 LiP genes designating LiPA–LiPJ that were responsible for encoding various LiP isoforms.

LiP is a monomeric hemoprotein with an isoelectric point and molecular weight of 38–43 kDa and 3.3–4.7, respectively (Kirk et al. 1986). LiP uses VA as a substrate and functions typically at an acidic pH of about 3.0 (Tien and Kirk 1988). Optimal activity at a very low pH optimum differentiates LiP from many other types of peroxidases. LiP displays high redox potential for oxidizing nonphenolic compounds constituting more than 90% of recalcitrant lignin molecules. In addition to nonphenolic moieties, LiP also exhibits an additional aptitude to oxidizing a diversity of phenolic compounds such as catechol, acetosyringone, guaiacol, syringic acid, and vanillyl alcohol (Wong 2009). The presence of tryptophan residue (Trp171) on the surface of the enzyme produces a tryptophanyl radical by transferring an electron to the heme molecule. Thus, it is involved in the oxidative potential of LiP to degrade lignin moieties and related substrates with greater redox potential. Notably, variable tryptophan microenvironment modulates the substrate specificity, enzyme activity, and catalytic stability (Ivancich et al. 2001).

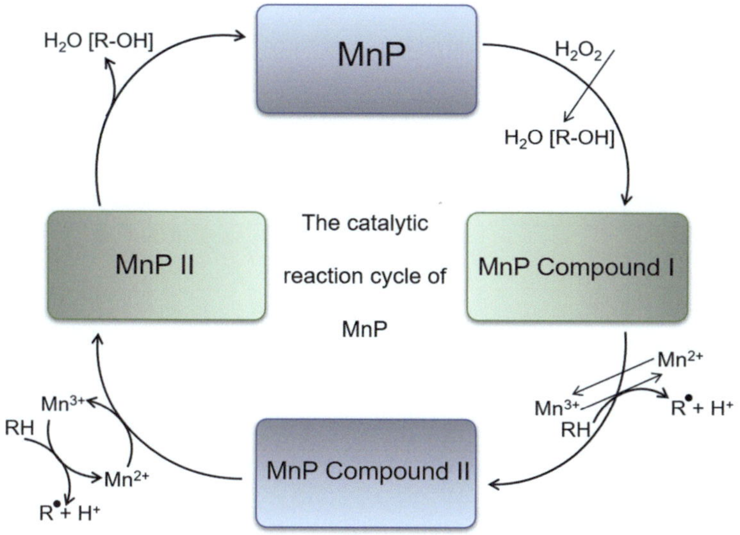

Fig. 1.2 A stepwise illustration of MnP catalytic cycle

1.2.2 Physiochemical and Catalytic Properties of MnP

MnP from *P. chrysosporium* was first discovered by Kuwahara and coworkers, in 1984. This enzyme has also been demonstrated as one of the most common ligninolytic enzymes produced by WRF. Though its participation in lignin modification has been well-studied and described in fungi (Hofrichter 2002), limited reports exist on MnP-producing bacteria. Wood-rotting *Basidiomycota* commonly secretes several forms of MnP in their culture extract with molecular masses varying from 40 to 50 kDa. Up to 11 various MnP isoforms have been reported in the culture extract of *Ceriporiopsis subvermispora* (Lobos et al. 1994). MnP follows a catalytic cycle similar to other heme-containing peroxidases, such as LiP or HRP, except utilizing Mn^{2+}as a typical electron donor (Fig. 1.2). From the reaction chemistry viewpoint, MnP has enormous potential to turn phenolic molecules into phenoxy radicals. The whole reaction assists under the Mn^{2+}oxidation to reactive Mn^{3+} in the presence of H_2O_2, which consequently can decompose an array of phenolic structures such as phenols, amines, aryl diamines, dyes, and phenolic lignin structures (Tuor et al. 1992). However, nonphenolic structures can also be degraded by MnP in the presence of some suitable low-molecular-weight mediators. A literature survey demonstrated numerous reports on the MnP-assisted oxidative depolymerization of natural and synthetic lignin structures as well as a range of many other refractory compounds (Hofrichter et al. 2001, 2010; Hofrichter 2002).

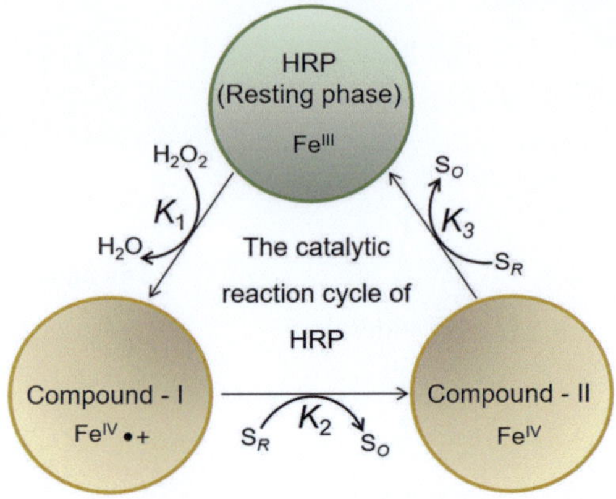

Fig. 1.3 A stepwise illustration of HRP catalytic cycle. Abbreviations: S_R, substrate in reduced phase; S_O, substrate in oxidized phase under respective reaction environment

1.2.3 Physiochemical and Catalytic Properties of HRP

HRP is an imperative heme-containing enzyme and subject of substantial research for more than a century. Though the HRP term is used rather broadly, the roots of the horseradish (*Armoracia rusticana*) herb comprises a large variety of characteristic peroxidase with HRP-C as the most abundant form (Veitch 2004). HRP-C comprises two different metal centers, one heme group, and two calcium atoms. Both metal centers play a crucial role in the structure and functional stability and integrity of the enzyme. The catalytic mechanism of HRP, particularly, the C isozyme, has been widely studied (Veitch and Smith 2001). Figure 1.3 depicts the catalytic cycle of HRP using ferulic acid as a reducing substrate. The formation of radical species in the two one-electron reduction steps leads to an intricate profile of reaction products containing dimers, trimers, and oligomers that might function as reducing substrates in succeeding turnovers (Veitch 2004). In recent years, HRP has gained remarkable interest in researchers, scientists, and biotechnology-related communities, around the globe, and thousands of research articles have appeared in the literature.

1.3 New or Advanced Enzyme-Based Techniques

The real-time practical implementation of enzyme biocatalysts for environmental remediation displays many inadequacies such as productivity, catalytic activity, and marginal stability. Since enzymes are biomacromolecules, any physical and chemical change in their structural conformation results in the loss of the enzymatic catalytic activity. Microbial strains are incompetent to secrete sufficient enzyme quantities under natural environmental conditions. In this context, scientists and environmental engineers are continuously striving to extract/isolate and identify new and unique enzyme-producing microbial strains with considerable ability to entirely biotransform or biodegrade the toxic environmental pollutants into harmless substances.

Some of the sophisticated and state-of-the-art techniques, namely, immobilization and genetic engineering, can address the above-stated issues.

1.3.1 Genetic Engineering

The overall productivity and titer of enzymes from their inherent producers are often very low under the natural process conditions. The use of genetic engineering approaches provides an efficient way to improve the biosynthesis of an array of enzymes by the identification, isolation, and cloning of coding genes into an appropriate expression candidate. In addition to cost-efficiency for larger-scale production of enzymes, overexpressing enzymes by recombinant DNA technology also enhances the catalytic features, such as activity, specificity, selectivity, and stability of the enzymes. Moreover, the isolation and purification of enzyme-based catalytic cues by recombinant technology are also easier than that of the original strain. The pH, substrate scope, temperature steadiness, and shelf life of the recombinant enzymes can be substantially upgraded using genetic engineering technologies. Genetically engineered enzymes display a greater capacity to decompose or decontaminate the environmental pollutants under defined reaction conditions. For instance, the peroxidase enzyme system of *Thanatephorus cucumeris* strain Dec 1 has been exploited for the degradation or decolorization of synthetic dye-based contaminants. Furthermore, the extracted peroxidase enzyme system was also expressed in *Aspergillus oryzae* RD 005. Recombinant approaches such as site-directed mutagenesis, cassette mutagenesis, error-prone PCR, staggered extension protocol, and DNA shuffling were employed to archive the enzyme with expanded substrate range and transformation of a wide spectrum of environmental pollutants (Dua et al. 2002).

1.3.2 Enzyme Engineering

Enzyme engineering relates to the utilization of recombinant DNA technology causing alteration in the amino acid sequence of the enzymes to boost up their features such as catalytic activity, pH and temperature stability, stress tolerance, etc. (Singh et al. 2013). Figure 1.4 illustrates various features of a biocatalyst that could be improved by protein engineering. At contemporary, enzyme engineering strategies have been applied for highly efficient, selective, and hyperactive catalytic constructs to efficiently degrade or remove radionuclides and heavy metals (Dhanya 2014). For instance, nitrobenzene was effectively transformed into nitrite and catechol by an engineered nitrite nitrobenzene 1,2-dioxygenase. However, modification in the amino acid residues near its active site (at the position 293) by site-specific mutagenesis resulted in a 2.5-fold greater oxidation efficiency toward 2,6-dinitrotoluene (Ju and Parales 2006). The same site-directed mutagenesis approach was also employed to 2-nitrotoluene dioxygenase that catalyzes the oxidative transformation of nitrotoluene to 3-methyl catechol and nitrite (Haigler and Spain 1993). The amino acid residues immediate to catalytic regions were selected

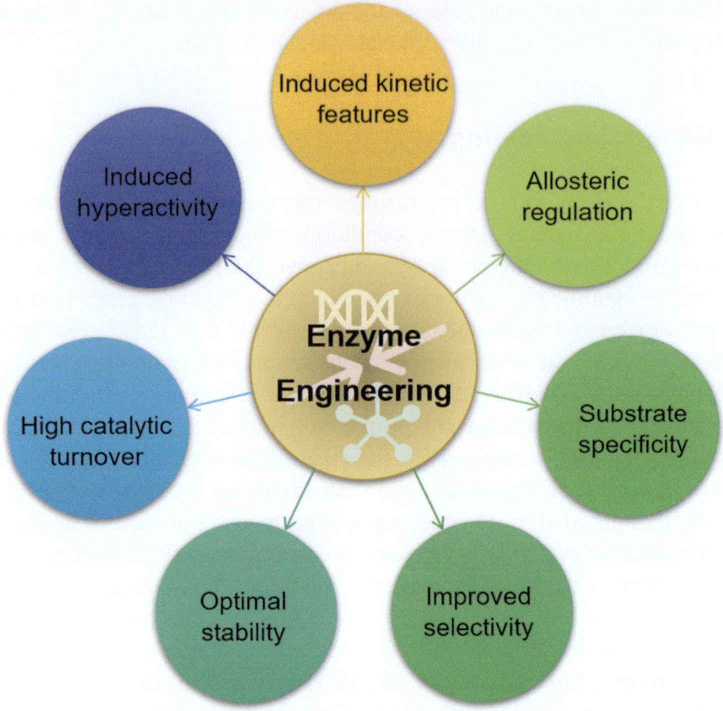

Fig. 1.4 Improved catalytic properties of enzymes by enzyme engineering approach for environmental remediation prospects

for mutagenesis, where the enantio specificity of the enzyme was changed by the replacement at the position 258 (Lee et al. 2005).

1.3.3 Nanozymes

Over the past several years, exhaustive research has been directed to designing many artificial enzyme biocatalysts with a variety of smart materials to mimic the structures and functions of various naturally occurring enzymes. In recent times, nanozymes, so-called nanoconstructs, with enzyme-like properties or next-generation artificial enzymes have gained great researchers' interest as novel and unique artificial enzymes with low cost and high stability. Owing to their exceptional features than natural and classic synthetic enzymes, significant improvements have been made in the field of nanozymes, and consequently several nano-based materials or nanoparticles have been pursued to mimic a great variety of naturally occurring enzymes, i.e., catalase, esterase, ferroxidase, peroxidase, oxidase, protease, phosphatase, and superoxide dismutase, for broad-spectrum applications (Wang et al. 2016). Generally, nanozymes lacking a suitable active site, and thus only a particular

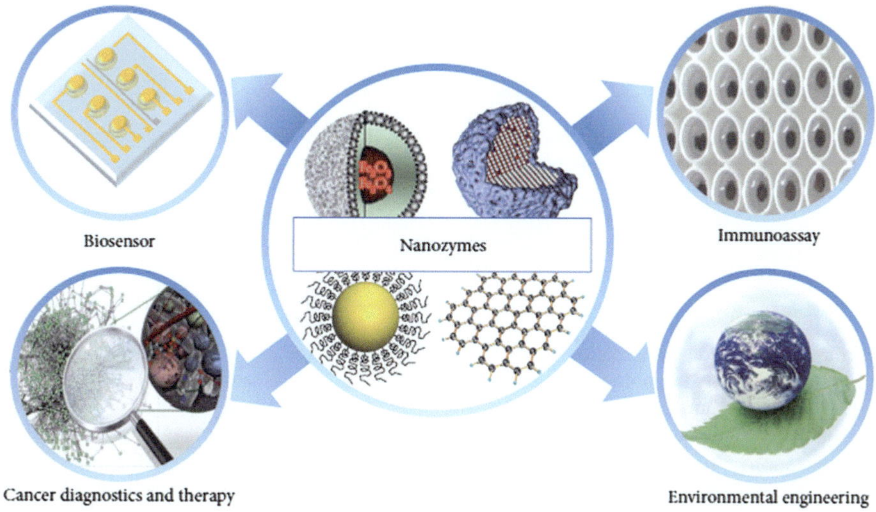

Fig. 1.5 Application perspective of nanozymes in different fields (Adpated from Shin et al. 2015)

substrate, could attach and undergo a chemical reaction. These nanoconstructs can catalyze the transformation of substrates to products and follow the identical mechanisms and reaction kinetics as of native enzymes in the physiological environment (Gao and Yan 2017). Nanozymes have shown potential applications in the development of novel biosensor, immunoassay, therapeutics, cancer diagnostics, and environmental remediation for the sensing and removal of dyes, organic compounds, and industrial wastes, (Liang et al. 2017). Peroxidases mimicking magnetic nanoparticles (Fe_3O_4-MNPs) have been found to be an incredibly simple and powerful method for mineralization and decomposition of many organic contaminants, such as phenol, rhodamine, and methylene blue from the aqueous media (Wu et al. 2015). Figure 1.5 illustrates the application perspective of nanozymes in different fields.

1.3.4 Immobilized Peroxidases and Properties

Immobilization of enzyme is an association of native or soluble enzyme onto several different kinds of insoluble materials (inert, organic, or inorganic) to hold in suitable reactor geometry for improved catalytic efficiency, stability, and economic recycling of enzymes even under unfavorable reaction conditions (Bilal et al. 2017a, c). Several immobilization procedures such as adsorption on alginate matrix, encapsulation, entrapment in polymer networks, covalent coupling to various insoluble support matrices (i.e. silica gel), and bifunctional reagents-assisted cross-linking have been developed and reported to attach a broad variety of industrially relevant enzymes (Table 1.1). Notably, the supporting matrices selected for enzyme attachment should exhibit a larger surface area, be less expensive, and be circumventing the diffusional limitations of substrate/product for enzyme catalytic reactions. With

Table 1.1 Commonly used immobilization strategies and their merits and demerits

Immobilization technique	Type of interactions involved	Strength of interactions	Merits	Demerits
Adsorption	Ionic interactions,hydrogen bonds, hydrophobic interactions	Weak	Simple, inexpensive No conformation modification High enzyme activity Support reusability	Low stability Biocatalyst desorption Loss of biomolecules
Entrapment	Ionic interactions, hydrophobic interactions, covalent bonds	Weak/strong	No modification in enzyme structure	Low enzyme loading
Encapsulation	Ionic and hydrophobic interactions	Weak	No enzyme alteration Protection of the enzyme	Mass transfer limitations
Cross-linking	Covalent bonds	Strong	High strength of interactions No support required Decrease desorption Prevent leakage	Diffusion limitations Loss of enzyme activity Alteration in active site
Covalent binding	Covalent bonds	Strong	Robust and stable interactions Strong biocatalyst binding High stability Reduction in enzyme leaking	Decreased enzyme activity due to limited enzyme mobility Support materials are not renewable

reference to the free enzyme, the insolubilized enzymes have retained their initial activity for a longer duration and offer simple recovery for recycling in multiple consecutive cycles (Bilal et al. 2017a, c). Moreover, the peroxidases in immobilized form may exhibit long-term economic and ecological merits for the mineralization of recalcitrant xenobiotics because of their repeatability and excellent durability. They also present increased thermal tolerance and remarkable activities in varying environmental conditions of extreme reaction temperature and pH. The hyperactivity and stability of intracellular enzymes, which are not active in a cell-free system, can also be enhanced by deploying immobilization and encapsulation strategies (Skoronski et al. 2017).

1.4 LiP-Assisted Degradation of Hazardous Contaminants

The textile industries consume various chemicals and a large number of pigments and synthetic dyes such as anthraquinone, azo, diazo, disperse, and reactive dyes that are the main sources of environmental contamination. It is reported that about 10–15% of these dyes find their way to effluents during the textile dyeing process, thus posing a danger to public health due to profound toxicity, carcinogenicity, or teratogenicity effects (Bilal and Asgher 2015). Immobilized peroxidases (microbial and plant) have shown superior performance to degrade these dye-related pollutants owing to their distinctive oxidative potential, low specificity, and minimum associated limitations. Table 1.2 illustrates the application of immobilized LiP to biotransform and biodegrade a variety of various environmental pollutants.

For instance, Ferreira-Leitao et al. (2007) assessed the efficacy of fungal LiP and compared it with a plant-based HRP for the decolorization and degradation of methylene blue (MB) dye and its demethylated products. Though both peroxidases were capable of oxidizing MB and its demethylated derivatives efficiently, the oxidative capacity of fungal LiP was found to be double in contrast to the HRP. Moreover, HRP lacks the capability to accomplish aromatic ring cleavage indicating LiP as the preferred candidate for degrading phenothiazine dyes and eliminating color from waste streams.

In another study, Qiu et al. (2009) used nanoporous gold (NPG) of pore diameter 40–50 nm as a bolster supporting matrix for the covalent attachment of LiP. The NPG-immobilized LiP showed an optimal temperature of 10 °C higher compared to the free counterpart. After an incubation period of 120 min at 45 °C, the immobilized LiP maintained about 55% of its initial activity, while the native LiP was completely denatured under the identical process conditions. Finally, the industrial applicability of the NPG-immobilized LiP was tested for decolorizing three structurally different dyes including pyrogallol red, fuchsine, and rhodamine B by two H_2O_2-mediated approaches. Notably, a high biotransformation efficiency of 84.6%, 75.5%, and 87.2% was recorded for fuchsine, rhodamine B, and pyrogallol red dyes, respectively, indicating a potential catalytic and dye-decolorizing efficacy of the immobilized LiP. Recently, LiP from *Ganoderma lucidum* IBL-05 was optimally entrapped in good quality Ca-alginate microspheres using sodium alginate (4.0% w/v), calcium chloride (0.2 M), and glutaraldehyde (0.02%) as a cross-linking agent. The pH and temperature optima for Ca-alginate-entrapped LiP were improved in comparison with free LiP. After Ca-alginate entrapment, the catalytic behavior, as well as the temperature and pH stability of LiP, was considerably enhanced. The free LiP exhibited 48%, 40%, 52%, 59%, and 66% color removal efficiency for S.F Golden yellow CRL, S.F Black CKF, S.F Foron Blue E2BLN, S.F Turq Blue GWF, and S.F Red C4BLN dyes, respectively, and were meaningfully improved to 80%, 70%, 89%, 83%, and 93% by the Ca-alginate-immobilized LiP using VA as a redox mediator. In addition, a noticeable reduction in water quality parameters including biological oxygen demand (66.44–98.22%), chemical oxygen demand (81.34–98.82%), and total organic content (80.21–97.77%) values were also recorded. The toxicity tested (brine shrimp lethality and hemolytic test) further substantiated the

Table 1.2 Degradation efficiency of immobilized lignin peroxidase for various environmental pollutants

Enzyme source	Immobilization method	Immobilization matrix	Pollutant	Degradation efficiency (%)	References
P. ostreatus	Covalent attachment	Carbon nanotubes	Remazol Brilliant Blue R	≥50	Oliveira et al. (2018)
G. lucidum	Entrapment	Ca-alginate beads	Sandal-fix Red C4BLN	93	Shaheen et al. (2017)
G. lucidum	Entrapment	Ca-alginate beads	Sandal-fix Turq Blue GWF	83	Shaheen et al. (2017)
G. lucidum	Entrapment	Ca-alginate beads	Sandal-fix Foron Blue E2BLN	89	Shaheen et al. (2017)
G. lucidum	Entrapment	Ca-alginate beads	Sandal-fix Black CKF	70	Shaheen et al. (2017)
G. lucidum	Entrapment	Ca-alginate beads	Sandal-fix Golden yellow CRL	80	Shaheen et al. (2017)
Schizophyllum commune	Cross-linking	Chitosan beads	Sandal Fix Foron Blue E2BLN	89.71	Sofia et al. (2016)
S. commune	Cross-linking	Chitosan beads	Sandal Fix Red C4BLN	87.55	Sofia et al. (2016)
S. *commune*	Cross-linking	Chitosan beads	Sandal Fix Turq Blue GWWF 165%	95.43	Sofia et al. (2016)
S. commune	Cross-linking	Chitosan beads	Sandal Fix Golden Yellow CRL	83.87	Sofia et al. (2016)
S. commune	Cross-linking	Chitosan beads	Sandal Fix Black CKF	63.76	Sofia et al. (2016)
S. commune	Cross-linking	Chitosan beads	Reactive T Blue GWF	69.86	Sofia et al. (2016)
Coriolusversicolor	Covalent immobilization	Chitosan microspheres	Molasses wastewater	80	Ran et al. (2012)
P. chrysosporium	Encapsulation	Polyacrylamide hydrogel	Bisphenol A	90	Gassara et al. (2013)

P. chrysosporium	Covalent immobilization	Mesoporous silica	Acid Orange II	77	Hu et al. (2013)
P. chrysosporium	Adsorption immobilization	Mesoporous 2D silica	Phenol	60	Xu et al. (2010)
P. *chrysosporium*	Adsorption immobilization	Nanoporous gold	Fuchsine	85	Qiu et al. (2009)
P. chrysosporium	Adsorption immobilization	Nanoporous gold	Rhodamine B	75	Qiu et al. (2009)
P. chrysosporium	Adsorption immobilization	Nanoporous gold	Pyrogallol Red	87	Qiu et al. (2009)

efficiency of Ca-alginate-encapsulated LiP treatment for reducing the cytotoxicity of dye solutions (Shaheen et al. 2017). Oliveira et al. (2018) used carbon nanotubes (CNTs) to immobilize crude extract containing LiP produced from the lignocellulosic residue (*Jatropha curcas* seed cake) by two fungal strains such as *P. ostreatus* (PLO9) and *G. lucidum* (GRM117). In comparison with native enzymes, the CNT-immobilized LiPs revealed improved catalytic efficiency, greater specific activities, and higher substrate affinities. Furthermore, it also showed an efficient decolorization of RBBR dye in several consecutive decolorization cycles. It can be concluded that LiP in immobilized state appears as a good biocatalytic system for the treatment of textile and dyeing industrial effluents loaded with a variety of dyes.

1.5 MnP-Assisted Degradation of Hazardous Contaminants

MnP is conceived as one of the earliest peroxidases with an ability to be utilized for the decolorization and degradation of environmental pollutants. Interestingly, Yao et al. (2013) reported MnPs to be much more efficient in degrading dyes or other toxic compounds as compared to LiPs and laccase. Table 1.3 portrays the potential of immobilized MnP to degrade and remove a variety of various environmental pollutants. For example, a purified MnP isolated from the culture extract of *G. lucidum* fungus was entrapped in the agar-agar matrix. The resultant entrapped MnP showed better tolerance to varying temperature and pH conditions with reference to the free counterpart. Thermal stability was also considerably enhanced following immobilization on agar-agar support. The insolubilized biocatalyst preserved more than up to 70% and 60% of its preliminary activity after incubating for 5 days at 30 °C and 40 °C, respectively. Moreover, MnP treatment results in the complete degradation and detoxification of three textile dyes, viz., reactive blue 21, reactive red 195A, and reactive yellow 145A (Bilal et al. 2016). Encapsulated MnP on agarose beads demonstrated wider pH and thermal resistance, as well as enhanced thermal and storage stability relative to the soluble enzyme. Additionally, it has shown potential to degrade different industrial effluents to varying extents in a packed bed reactor in several successive cycles (Bilal et al. 2017b). In another study, Bilal et al. (2017e) prepared cross-linked enzyme aggregates of *G. lucidum* MnP utilizing a range of aggregating agents including ammonium sulfate, ethanol, acetone, and tert-butanol and 2-propanol and GA as a cross-linker. The biocatalytic capacity of as-prepared MnP-CLEAs was investigated by reacting with two potential endocrine disrupters (triclosan and nonylphenol) and different textile dyes in an enzyme-loaded packed bed bioreactor. Results revealed that MnP-CLEAs were efficient in the transformation of both the tested endocrine disrupters and achieved 84.2%, 88.0%, 95.5%, and 100% degradation of Nishat textile-, K&N textile-, Crescent textile-, and Sitara textile-based wastewater effluents, respectively.

PAHs are widely distributed organic compounds in aquatic and terrestrial atmospheres, as products of the incomplete fossil fuel incineration. Several of these contaminants pose toxic, mutagenic, and highly carcinogenic effects to animal and human health (Johnsen et al. 2005). They are highly recalcitrant to nucleophilic and

Table 1.3 Degradation efficiency of immobilized manganese peroxidase for various environmental pollutants

Enzyme source	Immobilization method	Immobilization matrix	Target pollutant	Degradation efficiency (%)	References
G. lucidum	Entrapment	Agarose beads	Kalash textile effluent	94.5	Bilal et al. (2017b)
P. chrysosporium	Cross-linking	Chitosan beads	Crescent Textile industry	67.73	Bilal et al. (2016a)
G. lucidum	Encapsulation	Sol-gel matrix	Arzoo textile effluent	92.36 in 5 h	Bilal and Asgher (2016)
G. lucidum	Encapsulation	Sol-gel matrix	Ayesha textile effluent	86.42 in 5 h	Bilal and Asgher (2016)
G. lucidum	Encapsulation	Sol-gel matrix	Crescent textile effluent	82.01 in 5 h	Bilal and Asgher (2016)
G. lucidum	Encapsulation	Sol-gel matrix	Itmad textile effluent	93.92 in 5 h	Bilal and Asgher (2016)
G. lucidum	Encapsulation	Sol-gel matrix	Kalash textile effluent	92.28 in 5 h	Bilal and Asgher (2016)
G. lucidum	Encapsulation	Gelatin Hydrogel	Reactive Red 195A	More than 90 in 5 h	Bilal et al. (2016b)
G. lucidum	Entrapment	Ca-alginate beads	Sandal-fix Red C4BLN	87.5	Bilal and Asgher (2015)
G. lucidum	Entrapment	Ca-alginate beads	Sandal-fix Turq Blue GWF	82.1	Bilal and Asgher (2015)
G. lucidum	Entrapment	Ca-alginate beads	Sandal-fix Foron Blue E2BLN	89.4	Bilal and Asgher (2015)
G. lucidum	Entrapment	Ca-alginate beads	Sandal-fix Black CKF	95.7	Bilal and Asgher (2015)
G. lucidum	Entrapment	Ca-alginate beads	Sandal-fix Golden Yellow CRL	83	Bilal and Asgher (2015)
P. Chrysosporium	Covalent immobilization	Chitosan/biomimetic silica nanoparticles	2,6-dimethoxyphenol	95	Luan et al. (2014)

(continued)

Table 1.3 (continued)

Enzyme source	Immobilization method	Immobilization matrix	Target pollutant	Degradation efficiency (%)	References
Anthracophyllum discolor	Adsorption immobilization	Vulcanicnanoclay	Pyrene	86	Acevedo et al. (2010)
A. discolor	Adsorption Immobilization	Vulcanicnanoclay	Anthracene	65	Acevedo et al. (2010)
A. discolor	Adsorption Immobilization	Vulcanicnanoclay	Fluoranthene	15	Acevedo et al. (2010)
A. discolor	Adsorption Immobilization	Vulcanicnanoclay	Phenanthrene	10	Acevedo et al. (2010)

microbial attack because of the presence of fused aromatic rings in their structure with high biochemical persistence (Johnsen et al. 2005; Nikiforova et al. 2009). Among the group of ligninolytic consortium secreted by WRF, LiP, MnP, and laccase were demonstrated to exhibit a crucial role in the decomposition of these toxic PAHs. Acevedo et al. (2010) immobilized *Anthracophyllum discolor* MnP on naturally derived nanoclay support by simple adsorption to evaluate its ability for PAH decomposition. In comparison with the free enzyme, the insolubilized biocatalyst achieved a >65% and >86% degradation rate toward anthracene and pyrene, respectively. Nevertheless, phenanthrene and fluoranthene hydrocarbons were comparatively less degraded by the nanoclay-adsorbed enzyme with a transformation efficiency of <8.6% and <15.2%, respectively.

1.6 HRP-Assisted Degradation of Hazardous Contaminants

Peroxidase extracted and purified from horseradish roots have also been assessed in environmental applications. In recent years, HRPs have been employed in enormous biotechnological processes, such as lignocellulosic biomass delignification and chemical synthesis; nevertheless, environmental bioremediation, treatment of the phenolic compounds containing wastewaters, and removal of toxic compounds such as xenobiotics, dye pollutants from industrial and drinking water, and microbial-resistant pharmaceuticals and hazardous pollutants are the major practical applications of plant-based HRPs (Bilal et al. 2017a, b, c). Table 1.4 summarizes the recent biodegradation studies of the immobilized HRP for a variety of various environmental pollutants.

Zinc oxide (ZnO) nanocrystals in the form of nanodiscs, nanoflowers, and nanorods were synthesized to immobilize HRP and employed for the degradation of phenol. Notably, high removal efficiency of 86.09%, 79.46%, and 77.03% was achieved by the HRP immobilized on nanodiscs, nanoflowers, and nanorods, respectively, in comparison with the equivalent non-immobilized fraction exhibiting 61.52%. Also, the phenol degradative activity of ZnO-insolubilized HRP was tested with five additional model phenolic pollutants. The ZnO-coupled enzyme displayed superior catalytic efficiency for removing all these phenolic compounds from the aqueous media than that to the native HRP. Results evidenced that immobilization technology exhibited a noticeable shielding effect on the HRP enzyme against the inhibition/inactivation factors, and as a result, immobilized enzyme promoted the oxidation and elimination of phenol from the solution (Zhang et al. 2016). In a recent study, Zhang and Cai (2019) immobilized HRP onto newly fabricated Fe_3O_4-incorporated nanotubes used for the degradation of phenol. The synthesized nanobiocatalytic system removed about 74.1% of total phenol at a final concentration of 50 mg/L under the optimal processing conditions and preserved over 60% of its starting activity even after six consecutive cycles. HRP covalently coupled to silica-coated iron nanoparticles functionalized by 3-aminopropyltriethoxysilane results in effective degradation of 2,4-dichlorophenol with an utmost removal rate of 80% after incubating for 3 h (Chang and Tang 2014). Of most recent, Vineh et al. (2018)

Table 1.4 Degradation efficiency of immobilized horseradish peroxidase for various environmental pollutants

Enzyme source	Immobilization method	Immobilization matrix	Target pollutant	Degradation efficiency (%)	References
	In situ cross-linking	ZnO nanowires/macroporous silicon dioxide composite	Reactive Blue 19	100 within 2–3 h	Sun et al. (2018)
	In situ cross-linking	ZnO nanowires/macroporous silicon dioxide composite	Acid Violet 109	100 within 2–3 h	Sun et al. (2018)
Horseradish roots	Cross-linking	Polyacrylamide Gel	Methyl orange	More than 90	Bilal et al. (2018)
Merck company	Covalent attachment	Functionalized reduced graphene oxide	Phenol	100	Vineh et al. (2018)
Source leaves Biotechnology Co. (Shanghai, China)	Covalent Immobilization	Macroporous SiO_2/ZnO nanowires	Acid Blue 113	95	Sun et al. (2017)
Source leaves Biotechnology Co. (Shanghai, China)	Covalent Immobilization	Macroporous SiO_2/ZnO nanowires	Acid Black 10 BX	90	Sun et al. (2017)
Aladdin Industry (Shanghai, China)	Covalent Immobilization	Carbon nanospheres	2,4 Dichlorophenol,	95	Lu et al. (2017)
Aladdin Industry (Shanghai, China)	Covalent Immobilization	Carbon nanospheres	4-Methoxyphenol	99	Lu et al. (2017)
Aladdin Industry (Shanghai, China)	Covalent Immobilization	Carbon nanospheres	Bisphenol A	52	Lu et al. (2017)
Horseradish roots	Entrapment	Chitosan beads	Remazol Brilliant Blue R	82.17	Bilal et al. (2017c)
Horseradish roots	Entrapment	Chitosan beads	Reactive Black 5	97.82	Bilal et al. (2017c)
Horseradish roots	Entrapment	Chitosan beads	Congo Red	94.35	Bilal et al. (2017c)
Horseradish roots	Entrapment	Chitosan beads	Crystal Violet	87.43	Bilal et al. (2017c)

Enzyme source	Immobilization method	Immobilization matrix	Target pollutant	Degradation efficiency (%)	References
Horseradish roots	Covalent attachment	Polyvinyl alcohol-alginate beads	Methyl orange	100	Bilal et al. (2017d)
Sinopharm Chemical Reagent Co., Ltd. (Beijing, China)	Covalent Immobilization	Multi-walled carbon nanotube/ cordierite	4-Aminoantipyrine	96	Li et al. (2017)
HRP	Covalent attachment	Nanodiscs	Phenol	86.09	Zhang et al. (2016)
HRP	Covalent attachment	Nanoflowers	Phenol	79.46	Zhang et al. (2016)
HRP	Covalent attachment	Nanorods	Phenol	77.03	Zhang et al. (2016)
Horseradish roots	Cross-linking	Calcium-alginate support	Reactive Red 120	72.39	Bilal et al. (2016c)
Horseradish Roots	Cross-linking	Calcium-alginate support	Reactive Blue 4	87.23	Bilal et al. (2016c)
Horseradish roots	Cross-linking	Calcium-alginate support	Reactive Orange 16	79.57	Bilal et al. (2016c)
Horseradish roots	Adsorption	Kaolin	Pyrogallol	70	Šekuljica et al. (2016)
Horseradish roots	Adsorption	Kaolin	Acid Violet 109	87	Šekuljica et al. (2016)
Tianyuan Biologic Engineering Corp. (China)	Covalent Immobilization	Magnetic nanoparticles/graphene oxide nanocomposite	2-Chlorophenol	82	Chang et al. (2015)
Tianyuan Biologic Engineering Corp. (China)	Covalent Immobilization	Magnetic nanoparticles/graphene oxide nanocomposite	4-Chlorophenol	52	Chang et al. (2015)
Tianyuan Biologic Engineering Corp. (China)	Covalent immobilization	Magnetic nanoparticles/graphene oxide nanocomposite	2,4-Dichlorophenol	33	Chang et al. (2015)
Majorbio Biotech Company, USA	Covalent Immobilization	Zinc oxide nanocrystals	4-Aminoantipyrine	<95	Zhang et al. (2011)

recorded a complete removal of phenol (2.5 mg/mL) by plant HRP covalently immobilized onto modified graphene oxide using GA as a cross-linker, as compared to free enzyme that achieved only 55% degradation under the same conditions. HRP immobilized on GA-modified carbon nanospheres demonstrated twofold increased efficiency for the biotransformation of different phenolic compounds from aqueous media, in particular 4-methoxyphenyl, bisphenol A, and chlorophenols, as compared to the free form of the enzyme. The immobilized nanobiocatalyst eliminated 51.75%, 95.4%, 43.1%, 100%, 95%, 55.4%, and 99.3% for bisphenol A, paracetamol, phenol, catechol, *p*-chlorophenol, 4-methoxyphenol, and 2,4-dichlorophenol contaminants, respectively (Lu et al. 2017). The findings might represent a high potential to develop a cost-efficient, robust, and eco-sustainable biocatalytic system for wide environmental applications and wastewater remediation.

1.7 Conclusion

With the rapid urbanization and growth in population size and industrial sectors over the past few years, the accumulation of contaminants into the natural environment reached an alarming stage. Enzyme-based treatment appears to be a green, viable, and eco-friendly solution to overcome this dilemma. A broad family of peroxidases isolated from different microbial and plant sources has demonstrated the ability to biodegrade various pollutants. Initially, peroxidase-mediated treatment was not proven as an efficient remediation approach because of the insufficient production of enzymes by microorganisms under the physiological conditions. However, the production and titer of these enzymes can be substantially increased by the development of recombinant DNA technology together with providing optimal growth requirements to native producers. Furthermore, the shelf life, catalytic performance, and the stability of enzymes under stress environment can be boosted up to an incredible level by novel immobilization and enzyme engineering approaches. Nanozymes also holds a significant promise to clean up environmental pollutants up to remarkable levels because of their distinctive catalytic properties. In addition to pollutant detection and degradation, nanozymes can increase the bioavailability by acting on the distantly occurring substrates.

Acknowledgments Both authors are highly obliged to their respective departments and universities for providing the literature services.

Conflict of Interest Both authors declare no conflicting interests related to this work.

References

Acevedo F, Pizzul L, González ME, Cea M, Gianfreda L, Diez MC (2010) Degradation of polycyclic aromatic hydrocarbons by free and nanoclay-immobilized manganese peroxidase from *Anthracophyllum discolor*. Chemosphere 80(3):271–278

Battistuzzi G, Bellei M, Bortolotti CA, Sola M (2010) Redox properties of heme peroxidases. Arch Biochem Biophys 500:21–36

Bilal M, Asgher M (2015) Dye decolorization and detoxification potential of Ca-alginate beads immobilized manganese peroxidase. BMC Biotechnol 15(1):111

Bilal M, Asgher M (2016) Enhanced catalytic potentiality of *Ganoderma lucidum* IBL-05 manganese peroxidase immobilized on sol-gel matrix. J Mol Catal B Enzym 128:82–93

Bilal M, Asgher M, Shahid M, Bhatti HN (2016) Characteristic features and dye degrading capability of agar-agar gel immobilized manganese peroxidase. Int J Biol Macromol 86:728–740

Bilal M, Asgher M, Iqbal M, Hu H, Zhang X (2016a) Chitosan beads immobilized manganese peroxidase catalytic potential for detoxification and decolorization of textile effluent. Int J Biol Macromol 89:181–189

Bilal M, Asgher M, Hu H, Zhang X (2016b) Kinetic characterization, thermo-stability and Reactive Red 195A dye detoxifying properties of manganese peroxidase-coupled gelatin hydrogel. Water Sci Technol 74(8):1809–1820

Bilal M, Iqbal HM, Shah SZH, Hu H, Wang W, Zhang X (2016c) Horseradish peroxidase-assisted approach to decolorize and detoxify dye pollutants in a packed bed bioreactor. J Environ Manag 183:836–842

Bilal M, Asgher M, Parra-Saldivar R, Hu H, Wang W, Zhang X, Iqbal HMN (2017a) Immobilized ligninolytic enzymes: an innovative and environmental responsive technology to tackle dye-based industrial pollutants–a review. Sci Total Environ 576:646–659

Bilal M, Asgher M, Iqbal HM, Hu H, Wang W, Zhang X (2017b) Bio-catalytic performance and dye-based industrial pollutants degradation potential of agarose-immobilized MnP using a Packed Bed Reactor System. Int J Biol Macromol 102:582–590

Bilal M, Iqbal HM, Hu H, Wang W, Zhang X (2017c) Enhanced bio-catalytic performance and dye degradation potential of chitosan-encapsulated horseradish peroxidase in a packed bed reactor system. Sci Total Environ 575:1352–1360

Bilal M, Rasheed T, Iqbal HM, Hu H, Wang W, Zhang X (2017d) Novel characteristics of horseradish peroxidase immobilized onto the polyvinyl alcohol-alginate beads and its methyl orange degradation potential. Int J Biol Macromol 105:328–335

Bilal M, Asgher M, Iqbal HM, Hu H, Zhang X (2017e) Bio-based degradation of emerging endocrine-disrupting and dye-based pollutants using cross-linked enzyme aggregates. Environ Sci Pollut Res 24(8):7035–7041

Bilal M, Rasheed T, Zhao Y, Iqbal HM, Cui J (2018) "Smart" chemistry and its application in peroxidase immobilization using different support materials. Int J Biol Macromol 119(2018):278–290

Chang Q, Tang H (2014) Immobilization of horseradish peroxidase on NH2-modified magnetic Fe3O4/SiO2 particles and its application in removal of 2, 4-dichlorophenol. Molecules 19(10):15768–15782

Chang Q, Jiang G, Tang H, Li N, Huang J, Wu L (2015) Enzymatic removal of chlorophenols using horseradish peroxidase immobilized on superparamagnetic Fe3O4/graphene oxide nanocomposite. Chin J Catal 36(7):961–968

Dhanya MS (2014) Advances in microbial biodegradation of chlorpyrifos. J Environ Res Develop 9(1):232

Dua M, Singh A, Sethunathan N, Johri A (2002) Biotechnology and bioremediation: successes and limitations. Appl Microbiol Biotechnol 59(2–3):143–152

Dunford HB (1999) Heme peroxidases. Wiley, Chichester, p 507

Falade AO, Nwodo UU, Iweriebor BC, Green E, Mabinya LV, Okoh AI (2017) Lignin peroxidase functionalities and prospective applications. Microbiol Open 6(1):e00394

Farrell RL, Murtagh KE, Tien M, Mozuch MD, Kirk TK (1989) Characterization of lignin peroxidase isoenzymes. Enyzm Microb Technol 11:322–328

Ferreira-Leitao VS, de Carvalho MEA, Bon EP (2007) Lignin peroxidase efficiency for methylene blue decolouration: comparison to reported methods. Dyes Pigments 74(1):230–236

Gao L, Yan X (2017) Nanozymes: an emerging field bridging nanotechnology and biology. Sci China Life Sci 59:400–402

Gassara F, Brar SK, Verma M, Tyagi RD (2013) Bisphenol A degradation in water by ligninolytic enzymes. Chemosphere 92(10):1356–1360

Glumoff T, Harvey P, Molinari S, Goble M, Frank G, Palmer JM et al (1990) Lignin peroxidase from *Phanerochaete chrysosporium* molecular and kinetic characterization of isozymes. Eur J Biochem 187:515–520

Haigler BE, Spain JC (1993) Biodegradation of 4-nitrotoluene by Pseudomonas sp. strain 4NT. Appl Environ Microbiol 59:2239–2243

Hofrichter M (2002) Review: lignin conversion by manganese peroxidase (MnP). Enzym Microb Technol 30:454–466

Hofrichter M, Steffen K, Hatakka A (2001) Decomposition of Humic substances by Ligninolytic Fungi. 5th Finnish Conference on Environmental Science and Proceedings, Turku, Finland, pp 56–60

Hofrichter M, Ullrich R, Pecyna MJ, Liers C, Lundell T (2010) New and classic families of secreted fungal heme peroxidases. Appl Microbiol Biotechnol 87:871–897

Hu Z, Xu L, Wen X (2013) Mesoporous silicas synthesis and application for lignin peroxidase. Chem Mater 21(6):1144–1153

Huber P, Carré B (2012) Decolorization of process waters in deinking mills and similar applications: a review. BioRes 7(1):1366–1382

Ivancich A, Mazza G, Desbois A (2001) Comparative electron paramagnetic resonance study of radical intermediates in turnip peroxidase isozymes. Biochemistry 40:6860–6866

Johansson T, Welinder KG, Nyman PO (1993) Isozymes of lignin peroxidase and manganese (II) peroxidase from the white-rot basidiomycete *Trametes versicolor*. II. Partial sequences, peptide maps, and amino acid and carbohydrate compositions. Arch Biochem Biophys 300:57–62

Johnsen AR, Wick LY, Harms H (2005) Principles of microbial PAH-degradation in soil. Environ Pollut 133:71–84

Ju KS, Parales RE (2006) Control of substrate specificity by active-site residues in nitrobenzene dioxygenase. Appl Environ Microbiol 72(3):1817–1824

Kirk TK, Croan S, Tien M, Murtagh KE, Farrell RL (1986) Production of multiple ligninases by *Phanerochaete chrysosporium*: effect of selected growth conditions and use of a mutant strain. Enyzm Microb Technol 8:27–32

Koua D, Cerutti L, Falquet L et al (2009) PeroxiBase: a database with new tools for peroxidase family classification. Nucleic Acids Res 37:D261–D266

Kuwahara M, Glenn JK, Morgan MA, Gold MH (1984) Separation and characterization of two extracellular H2O2-dependent oxidases from ligninolytic cultures of *Phanerochaete chrysosporium*. FEBS Lett 169:247–250

Lee KS, Parales JV, Friemann R, Parales RE (2005) Active site residues controlling substrate specificity in 2-nitrotoluene dioxygenase from Acid *ovorax sp*. strain JS42. J Ind Microbiol Biotechnol 32:465–473

Li ZL, Cheng L, Zhang LW, Liu W, Ma WQ, Liu L (2017) Preparation of a novel multi-walled-carbon-nanotube/cordierite composite support and its immobilization effect on horseradish peroxidase. Process Saf Environ Prot 107:463–467

Liang H, Lin F, Zhang Z, Liu B, Jiang S, Yuan Q, Liu J (2017) Multicopper laccase mimicking nanozymes with nucleotides as ligands. ACS Appl Mater Interfaces 9(2):1352–1360

Lobos S, Larraín J, Salas L, Cullen D, Vicuña, R (1994) Isoenzymes of manganese-dependent peroxidase and laccase produced by the lignin-degrading basidiomycete *Ceriporiopsis subvermispora*. Microbiol 140(10):2691–2698

Lu YM, Yang QY, Wang LM, Zhang MZ, Guo WQ, Cai ZN et al (2017) Enhanced activity of immobilized horseradish peroxidase by carbon nanospheres for phenols removal. CLEAN–Soil, Air, Water 45(2):1600077

Luan PP, Jiang YJ, Zhang SP, Gao J, Su ZG, Ma GH, Zhang YF (2014) Chitosan-mediated formation of biomimetic silica nanoparticles: an effective method for manganese peroxidase immobilization and stabilization. J Biosci Bioeng 118(5):575–582

Marco-Urrea E, Aranda E, Caminal G, Guillén F (2009) Induction of hydroxyl radical production in *Trametes versicolor* to degrade recalcitrant chlorinated hydrocarbons. Bioresour Technol 100(23):5757–5762

Martinez D, Larrondo LF, Putnam N, Gelpke MD, Huang K, Chapman J (2004) Genome sequence of the lignocellulose degrading fungus *Phanerochaete chrysosporium* strain RP78. Nat Biotechnol 22:695–700

Morgenstern I, Klopman S, Hibbett DS (2008) Molecular evolution and diversity of lignin degrading heme peroxidases in the *Agarico mycetes*. J Mol Evol 66:243–257

Nikiforova SV, Pozdnyakova NN, Turkovskay OV (2009) Emulsifying agent production during PAHs degradation by the white rot fungus *Pleurotus ostreatus* D1. Curr Microbiol 58:554–558

Oliveira SF, da Luz JMR, Kasuya MCM, Ladeira LO, Junior AC (2018) Enzymatic extract containing lignin peroxidase immobilized on carbon nanotubes: potential biocatalyst in dye decolourization. Saudi J Biol Sci 25:651–659

Qiu H, Li Y, Ji G, Zhou G, Huang X, Qu Y, Gao P (2009) Immobilization of lignin peroxidase on nanoporous gold: enzymatic properties and in situ release of H_2O_2 by co-immobilized glucose oxidase. Bioresour Technol 100(17):3837–3842

Ran YH, Che ZF, Chen WQ (2012) Co-immobilized lignin peroxidase and manganese peroxidase from *Coriolus versicolor* capable of decolorizing molasses wastewater. Appl Mech Mater 138:1067–1071. Trans Tech Publications.

Šekuljica NŽ, Prlainović NŽ, Jovanović JR, Stefanović AB, Djokić VR, Mijin DŽ, Knežević-Jugović ZD (2016) Immobilization of horseradish peroxidase onto kaolin. Bioprocess Biosyst Eng 39(3):461–472

Shaheen R, Asgher M, Hussain F, Bhatti HN (2017) Immobilized lignin peroxidase from *Ganoderma lucidum* IBL-05 with improved dye decolorization and cytotoxicity reduction properties. Int J Biol Macromol 103:57–64

Shin HY, Park TJ, Kim MI (2015) Recent research trends and future prospects in nanozymes. J Nanomater 2015:7

Singh RK, Tiwari MK, Singh R, Lee JK (2013) From protein engineering to immobilization: promising strategies for the upgrade of industrial enzymes. Int J Mol Sci 14(1):1232–1277

Skoronski E, Souza DH, Ely C, Broilo F, Fernandes M, Junior AF, Ghislandi MG (2017) Immobilization of laccase from *Aspergillus oryzae* on graphene nanosheets. Int J Biol Macromol 99:121–127

Smulevich G, Jakopitsch C, Droghetti E, Obinger C (2006) Probing the structure and bifunctionality of catalase-peroxidase (KatG). J Inorg Biochem 100:568–585

Sofia P, Asgher M, Shahid M, Randhawa MA (2016) Chitosan beads immobilized schizophyllum commune ibl-06 lignin peroxidase with novel thermo stability, catalytic and dye removal properties. J Animal Plant Sci 26(5)

Sugiura T, Yamagishi K, Kimura T, Nishida T, Kawagishi H, Hirai H (2009) Cloning and homologous expression of novel lignin peroxidase genes in the white-rot fungus *Phanerochaete sordida* YK-624. Biosci Biotechnol Biochem 73:1793–1798

Sun H, Jin X, Long N, Zhang R (2017) Improved biodegradation of synthetic azo dye by horseradish peroxidase cross-linked on nano-composite support. Int J Biol Macromol 95:1049–1055

Sun H, Jin X, Jiang F, Zhang R (2018) Immobilization of horseradish peroxidase on ZnO nanowires/macroporous SiO2 composites for the complete decolorization of anthraquinone dyes. Biotechnol Appl Biochem 65(2):220–229

Tien M, Kirk K (1983) Lignin-degrading enzyme from the hymenomycete *Phanerochaete chrysosporium*. Science 221:661–663

Tien M, Kirk TK (1988) Biomass, part B: lignin, pectin and chitin. In: Wood WA, Kellog SC (eds) Methods in enzymology. Academic Press, San Diego, California, USA, pp 238–249
Tuor U, Wariishi H, Schoemaker HE, Gold MH (1992) Oxidation of phenolic arylglycerol β-aryl ether lignin model compounds by manganese peroxidase from *Phanerochaete chrysosporium*: oxidative cleavage of an α-carbonyl model compound. Biochemistry 31:4986–4995
Veitch NC (2004) Horseradish peroxidase: a modern view of a classic enzyme. Phytochemistry 65(3):249–259
Veitch NC, Smith AT (2001) Horseradish peroxidase. Adv Inorg Chem 51:107–101
Vineh MB, Saboury AA, Poostchi AA, Rashidi AM, Parivar K (2018) Stability and activity improvement of horseradish peroxidase by covalent immobilization on functionalized reduced graphene oxide and biodegradation of high phenol concentration. Int J Biol Macromol 106:1314–1322
Wang X, Hu Y, Wei H (2016) Nanozymes in bionanotechnology: from sensing to therapeutics and beyond. Inorgan Chem Front 3(1):41–60
Wong DWS (2009) Structure and action mechanism of ligninolytic enzymes. Appl Biochem Biotechnol 157:174–209
Wu J, Xiao D, Zhao H, He H, Peng J, Wang C et al (2015) A nanocomposite consisting of graphene oxide and Fe_3O_4 magnetic nanoparticles for the extraction of flavonoids from tea, wine and urine samples. Microchim Acta 182(13–14):2299–2306
Xu LQ, Wen XH, Ding HJ (2010) Immobilization of lignin peroxidase on spherical mesoporous material. Huanjingkexue 31(10):2493–2499
Yao J, Jia R, Zheng L et al (2013) Rapid decolorization of azo dyes by crude manganese peroxidase from *Schizophyllum sp.* F17 in solid-state fermentation. Biotechnol Bioproc E 18:868–877
Zhang C, Cai X (2019) Immobilization of horseradish peroxidase on Fe_3O_4/nanotubes composites for Biocatalysis-degradation of phenol. Compos Interfaces 26(5):379–396
Zhang Y, Wu H, Huang X, Zhang J, Guo S (2011) Effect of substrate (ZnO) morphology on enzyme immobilization and its catalytic activity. Nanoscale Res Lett 6(1):450
Zhang F, Zhang W, Zhao L, Liu H (2016) Degradation of phenol with Horseradish Peroxidase immobilized on ZnO nanocrystals under combined irradiation of microwaves and ultrasound. Desalin Water Treat 57(51):24406–24416

Microalgal Technology: A Promising Tool for Wastewater Remediation

2

Meenu Thakur, Sakshi Bajaal, Neha Rana, and Madan L. Verma

Abstract

Many species of microalgae have excellent ability to remove nitrogen, phosphorus, heavy metals, pesticides, organic and inorganic compounds, and pathogens from wastewater. Microalgae species grow well in wastewater and may be used for treatment of municipal, industrial, agro-industrial, and livestock wastewaters. Furthermore, microalgae biomass is an excellent source of production of various valuable products. In this chapter, applications of microalgae for treatment of wastewater and production of valuable products are discussed.

Keywords

Phycoremediation · Microalgae · Wastewater · Bioenergy · Metabolites

2.1 Introduction

Pollution is the major threat to our environment which has resulted from increased mushrooming of industries and more urbanization. It affects our ecosystems, flora, fauna, and human health worldwide by contaminating soil, water, and air. Pollution arises due to increased concentrations of unwanted and harmful substances as the results of anthropogenic activities. Most of organic and inorganic constituents have

M. Thakur · S. Bajaal · N. Rana
Shoolini Institute of Life Sciences and Business Management, Affiliated to Himachal Pradesh University, Solan, Himachal Pradesh, India

M. L. Verma (✉)
Centre for Chemistry and Biotechnology, Deakin University, Victoria, Australia

Department of Biotechnology, School of Basic Sciences, Indian Institute of Information Technology, Una, Himachal Pradesh, India

P. K. Arora (ed.), *Microbial Technology for Health and Environment*, Microorganisms for Sustainability 22,
https://doi.org/10.1007/978-981-15-2679-4_2

been released in water bodies due to household, agriculture, and industries which have led to organic and inorganic pollution (Mouchet 1986; Lim et al. 2010). This pollution has greatly affected the availability and quality of water resources around the globe (Abdel-Raouf et al. 2012). Moreover, there are common incidences of discharging the wastewater into water bodies without proper treatment. There is no parallel connection between the planning and implementation of such discharges in municipal plans which are posing serious problems to public health. The wastewater contains nitrogen, phosphorus, heavy metals, pesticides, organic and inorganic toxins, and pathogens. The major reason for water pollution is the discharge of industrial waste and sewage without proper treatment. Several wastewater treatment plants discharge water that contains significant amounts of toxic metals and organic and inorganic compounds. Therefore, it is a major challenge to develop efficient wastewater treatment technologies.

Agriculture pollution is also a source of water contamination. Agrochemical residues containing high concentrations of insecticides/pesticides/fertilizers pose serious threat to aquatic ecosystems. Nitrate is identified as one of the most common sources of agriculture pollution that causes eutrophication (Abdel-Raouf et al. 2012).

Wastewater can be mainly categorized into household and industrial wastewater (Chiu et al. 2015). In the present world, one of the major challenges is the availability of clean and potable water for drinking and household. However, to meet this challenge, there is a need to develop different new methods for wastewater treatment (Bansal et al. 2018). One of the major resolutions can be phycoremediation which efficiently uses algae for treatment of wastewater (Bansal et al. 2018). Algae are eukaryotic organisms with great variety ranging from single cell to highly differentiated plants. Algae are efficient carbon fixer as it can utilize carbon to release oxygen into atmosphere (Rehnstam Holm and Godhe 2003). More than 50% of total photosynthetic activity can be attributed to algae, and it significantly affects the food chain (Day et al. 2017). Moreover, algae can be used to convert carbon dioxide to oxygen by utilization of carbon for its own growth. Thus, algal cells can be efficiently used for wastewater treatment because they can remove the organic compounds, metals, and nutrients left out in wastewater (Laurens et al. 2017). Heavy metals have been detected in industrial wastewater. Microalgae can efficiently remove/remediate heavy metal ions from the effluents. Kumar et al. (2015) have reviewed various biochemical mechanisms present in microalgal cells for removal of heavy metals. Algae can be used for production of value-added products along with safe cleaning of wastewater such as algal char to replace coal and production of effective biofuel, lipids, and active metabolites which can be used as colorants, preservatives, and medicines. Microalgal cells can be used to produce alternative bioenergy sources. It has been observed that microalgae due to high amount of polyunsaturated fatty acids (PUFA) have enormous potential to be used as biodiesel (VenkataMohan et al. 2015). They can be harnessed as valuable biodiesel sources due to their high cell densities and accumulation of large quantities of triacylglycerols. But some of the major problems using algal technology for wastewater treatment, including availability of space, sunlight, contamination, resilience time, etc., need to be managed before employing phycotechnology for wastewater treatment.

Some biotechnological strategies such as hyperconcentrated cultures, immobilized cell system, photobioreactors, and genetic engineering can be used to improve phycotechnology. Photobioreactors can be used to improve phycoremediation. A microalgae photobioreactor has been developed by Marbelia et al. (2014). In this bioreactor, they used lab effluent as input as growth media for *Chlorella vulgaris*. Photobioreactors can be used efficiently for growing high density of algae due to less washout problem, and dilutions can be maintained at optimal levels. Moreover, it has helped to achieve higher cell density and enabled high waste removal.

This chapter attempts to discuss the role of microalgae in phycoremediation of wastewater, current technologies used, and future technologies to improve the process further.

2.2 Adverse Effects of Wastewater on the Environment

The composition of wastewater reflects the lifestyles and technologies practiced in producing society (Gray 1989). It is a mixture of organic and inorganic materials and xenobiotic compounds. Major portions of sewage are carbohydrates, fats, proteins, volatile acids, etc. Major constituents of inorganic pollutants include various ions and heavy metals, viz., sodium, calcium, magnesium, chlorine, bicarbonate, and ammonium salts, which are among the causative agents of water pollution (Tebbutt 1983; Horan 1990; Lim et al. 2010). Various pollutant sources include untreated direct discharge of human wastes from household, municipal wastes, and agricultural leach outs including high concentrations of insecticides and pesticides. It includes the industrial drains containing higher concentrations of heavy metals (Horan 1990). Pollutants can be classified into two categories depending on their sources in biological and chemical wastes. Chemical wastes include various inorganic ions, heavy metals from industries, detergents from household, and agricultural leach outs containing insecticides and pesticides (Akpor 2011). Other than these different sources, pathogenic microorganisms such as bacteria, viruses, and protozoans are common problems which affect the quality of drinking water (Akpor 2011). Moreover, the largest contributor of pollution is discharge of effluents from wastewater treatment. There are a number of previous studies on the negative impact of these effluents, which may result into death of aquatic life, algal blooms, habitat destruction from sedimentation, debris, and toxicity from chemical contaminants and even can interfere with food chain (Canada Gazzette 2010).

The adverse effects of wastewater effluent on environment can be classified into two, that is, ecological and health impact (Akpor 2011). Wastewater includes a number of different inorganic pollutants such as nitrogen, phosphorus, and heavy metals (Larsdotter 2006). Major forms of nitrogen can be ammonium ions and nitrite and nitrate ions (Hurse and Connor 1999). Nitrogen in untreated wastewaters may be organic and inorganic (Sabalowsky 1999). Nitrate in water causes methemoglobinemia, which is the most significant health problem. Blood contains hemoglobin (iron-based compound), but in the presence of nitrite, it gets converted to methemoglobin that does not carry oxygen. Likewise, nitrogen is toxic to fish in its ammoniacal form and adds on to oxygen demand (Jenkins et al. 2003). However, it involves increased concentration of phosphorus in water which is an important constituent of living beings. But after assimilation high concentration of phosphorus in organisms poses major problem (Rybicki 1997). Increased concentration of phosphorus leads to eutrophication, a phenomenon which gives rise to algal bloom. This algal bloom has negative effect on wastewater treatment and can hinder the potability of drinking water. Thus, this type of major limitation can be removed by controlling the concentration of phosphate in wastewater discharge (Vanlarsdrecht 2005).

Human and animal wastes are major sources for pathogenic organisms which are released in the wastewater, leading it to become a major reason for public health hazards. Most commonly occurring pathogenic microorganisms involve bacteria, viruses, and protozoa which contaminates the water resources (Kris 2007). These microorganisms are responsible for major waterborne diseases. Such severe waterborne diseases involve higher risk to human health. Some of the names include typhoid fever, shigellosis, salmonellosis, campylobacteriosis, and giardiasis. Hepatitis A is one of the virus-borne diseases due to drinking of contaminated water (WHO 2004), whereas other pathogens may cause critical diseases which have costly treatment such as stomach ulcers, etc. Viruses can be the most dangerous and harming pollutants present in water. Reasons for this may be high pathogenicity, difficulty in diagnosis, and high dose of antiviral compound requirement (Okoh et al. 2007). Bacteria are another form of pathogens by causing various types of health hazards associated with digestive systems and skin such as diarrhea, dysentery, and skin and tissue infections. Major disease-causing bacteria found in wastewater are different types of bacteria, such as *E. coli* O157:H7, *Listeria*, *Salmonella*, and *Leptospirosis*. *Giardia* and *Cryptosporidium* are among other protozoans causing serious diseases. More concentrations of nitrates can cause methemoglobinemia whose permissible limit has been set as 10 mg/mL by the US Environmental Protection Agency (EPA 2002). Nitrite can further interact with amine to form nitrosamines which are potent carcinogens. Thus, inorganic constituents such as nitrogen and phosphorus cause most favorable conditions for growth of such pathogenic organisms. The microbial toxins cause acute problems ranging from gastroenteritis to nervous system impairment. According to a health report from the World Health Organization, these pathogenic organisms can be a cause of liver cancer in humans.

2.3 Newer Approaches Over Conventional Wastewater

There are several approaches for cleaning wastewater, viz., conventional and advanced treatment (Fig. 2.1). Conventional approaches include various physical as well as chemical treatments. Chemical treatment is one of the most effective treatments for wastewater. The general purposes of the chemical treatment are to change the properties of water such as removal of suspended solids (turbidity) from the water, pH adjustment, and removal of dissolved material in the water, thus improving water quality. The prevalent methods in chemical treatment can be coagulation/flocculation, chlorination, chloramination, ozonation, and ultraviolet light (UV) (Gray 2002).

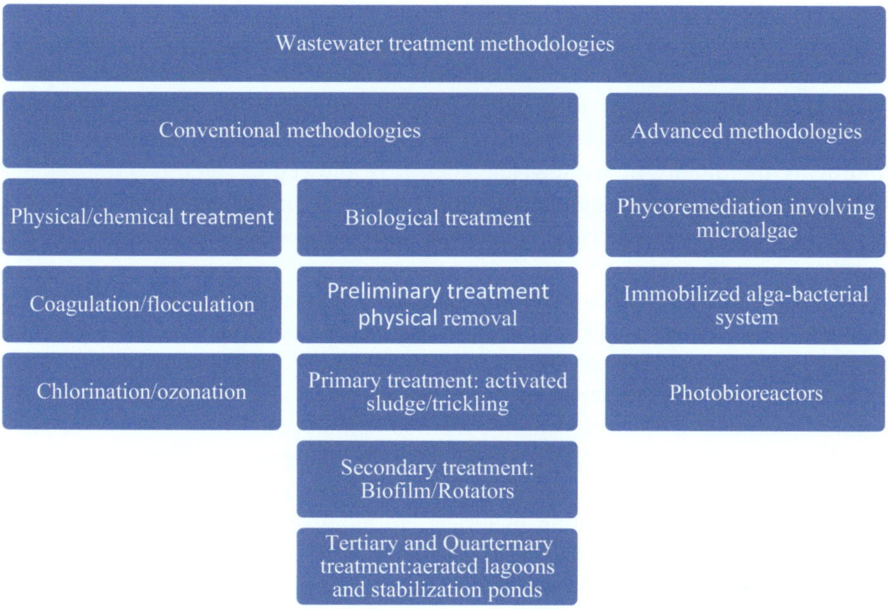

Fig. 2.1 Diagrammatic representation showing conventional and advanced methodologies for wastewater treatment

Flocculation aids in chemical and thermal destruction of pathogens and ultimate killing. Flocculation can contribute to the removal of various heavy metals and pathogenic organisms including *E. coli* (60–98%), viruses (60–90%), and *Giardia* (60–98%) (Tchobanoglous et al. 2003). The addition of alum (coagulating agent) increases the rate at which the suspended particles settle out flocculation. In bulk water treatment, the alum dose can be optimized (Gomez et al. 2006). Chlorination is the best known method of disinfection. It needs more contact time due to its high oxidation potential. Chlorine can react with NH^{4+}, NO^{2-}, H_2S, Fe^{2+}, and other organic compounds and leads to the formation of compound called trihalomethanes, always leading to toxicity problems. It is still the most commonly employed method in the treatment of wastewater with high organic compound concentration (Tchobanoglous et al. 2003).

Ozonation is used mainly in secondary wastewater treatment that has antimicrobial activity. Major drawbacks of ozonation include high cost involved and lack of maintenance. Moreover, there is always the possibility of microbial regrowth. The efficiency of UV and chlorine method has been found to be optimal for disinfection in wastewater. Treatment with UV light results in no toxicity but involves various limitations such as high cost, increased volume of sludge, and dewatering capability. The additional advantages of chemical processes are mineralization of nonbiodegradable components and reduction in size of reactor (Tchobanoglous et al. 2003).

Biological wastewater treatment can be classified as on-site and off-site treatment systems. Both treatments involve different conditions to be fulfilled before treatment. Moreover, both have different impacts on public health as well as environment (Akpor 2011). All biological methods use the metabolic activities of microorganisms to utilize the contaminants of wastewater, thereby reducing the BOD and COD values of wastewater effluent. Almost every biological method involves utilization of microorganisms for digesting inorganic constituents and improving the quality of wastewater. Then it is allowed to settle down as sludge which can be separated, and eventually BOD value can be lowered down. Moreover, the pathogenic organisms can be removed, and wastewater can be recycled efficiently (Abraham et al. 1997).

No single treatment is available for efficient treatment of wastewater. Primary, secondary, and tertiary treatment methods can be used for removal of contaminants and pathogens. Sometimes preliminary treatment can be used prior to primary treatment to increase the efficiency of the treatment methods. Moreover, most of the time, a combination of more than one method is required (Horan 1990). The stepwise biological treatment involves various steps. Preliminary treatment removes the coarse solid materials which will otherwise cause blockage and equipment damage. Very large particles and floating materials are removed by coarse removal which involves using bars (Tebbutt 1983). The next step is primary treatment which includes sedimentation processes. It enhances the settling of solids under gravity. Sedimentation tanks can contribute significantly by lowering the BOD value by 40% (Horan 1990). However, secondary treatment involves the reduction in organic matter. This involves the action of microorganisms such as heterotrophic bacteria and their digestive enzymes on utilization of organic matter for energy and growth. All processes can be divided into fixed film reactors and dispersed growth processes for the removal of the microbial population. Tertiary treatment is used to lower the concentration of organic ions. Biological and chemical methods can be used for tertiary treatment. Tertiary treatment is responsible for deciding factor of overall cost involved because it itself costs four times more than primary treatment (De la Noüe et al. 1992). Sometimes, the next step in quaternary treatment is also intended for removal of heavy metals, organic compounds, and soluble fractions.

2.4 Microalgal Species Involved in the Wastewater Treatments

Several species of microalgae have remarkable ability to remove nitrogen, phosphorus, heavy metals, pesticides, organic and inorganic toxins, and pathogens from wastewater. Example of these microalgae includes *Chlorella, Scenedesmus, Phormidium, Botryococcus, Chlamydomonas*, and *Arthrospira* (Abdel-Raouf et al. 2003; Rawat et al. 2013; Molazadeh et al. 2019).

Many microalgae species including *Chlorella, Scenedesmus, Euglena, Chlamydomonas, Oscillatoria,* and *Ankistrodesmus* have been demonstrated to grow efficiently in wastewater. Few microbial algae are known for heavy metal removal (Molazadeh et al. 2019). Examples are *Oscillatoria* spp. (for chromium removal), *Chlorella vulgaris (for* cadmium, copper, and zinc removal), *Chlamydomonas* spp.

Table 2.1 Algae used in the remediation of the pollutants present in the wastewater

Pollutants	Algae used for bioremediation	References
Oil effluents	*Scenedesmus obliques* *Prototheca zopfii* *Ankistrodesmus* and *Scenedesmus quadricauda*	Rajasulochana et al. (2009) Walker et al. (1975) Abeliovich (1986); Pinto et al. (2003)
Textile waste effluents	*Chlorella vulgaris* *Chlorella pyrenoidosa*	El-kassas and Mohamed 2014; Jinqi and Houtian (1992)
Phenolics compounds	*Chlorella vulgaris, Spirulina*	Ismail et al. (2013)
Nitrate, organic, and inorganic phosphorous	*Oscillatoria, Synechococcus, Nostoc, Spirulina platensis* *Chlorella sorokiniana*	Dubey et al. (2011); Laliberte et al. (1997); Sawayama et al. (1992); Sawayama et al. (1998) and Ogbonna et al. (2000)
Copper- and iron-containing effluent	*Botryococcus braunii and Anabaena doliolum*	Rai and Mallich (1992)

(lead removal), and *Scenedesmus chlorelloides (*molybdenum removal). It is also noticed that tolerance to organic pollutants in wastewater varies from species to species (Molazadeh et al. 2019). *Euglena, Oscillatoria, Chlamydomonas, Scenedesmus, Chlorella, Nitzschia, Navicula*, and *Stigeoclonium* have been described as the most resistant genera to organic pollutants (Palmer (1974). Table 2.1 summarizes utilization of various pollutants of wastewater by microalgae.

Microalgae regulate eutrophication process by removing phosphorous and nitrogen components. Microalgae can be effective substitute for biological treatment which converts the organic as well as inorganic unwanted constituents to valuable biomass. Microalgal species have been widely used for treatment of municipal, industrial, agro-industrial, and livestock wastewaters. Algae harvested from the treatment pond may be a source of food and valuable products. Algae are able to accumulate toxic compounds including selenium, zinc, and arsenic in their cells and eliminate them from the aquatic environments. A variety of physical, chemical, and biological methods can be used at different stages of primary, secondary, or tertiary levels for wastewater treatment.

2.5 Factors Affecting the Wastewater Remediation

2.5.1 Carbon

Carbon is the most important constituent for microalgal growth and cell growth as it follows the autotrophic mode of nutrition. However, in other two modes, that is, heterotrophic and mixotrophic, it behaves as organic carbon. In *Chlorella protothecoides*, biomass as well as lipid content varied depending on the modes of nutrition. Heterotrophic mode resulted in 3.4 times more biomass and 4.2 times more lipid content than autotrophic mode (Yanna and Hyde 2002). Microalgae convert carbon dioxide into inorganic carbon source, and water acts as electron donor for production of glucose which gives rise to complex carbohydrates. Moreover, these microalgal cells are efficient for utilization of Na_2CO_3 and $NaHCO_3$. Other important constituents than carbon involve nitrogen and phosphorus which plays a significant role in growth and development of microalgae. There can be so many sources of nitrogen such as detergents, ammonium ions, nitrates, and nitrites present in wastewater which can be efficiently utilized for growth of microalgae.

2.5.2 pH

pH is an important abiotic factor that affects the microalgal growth. In the cultivation of the microalgae, the pH value increases due to the photosynthetic assimilations of the CO_2. pH is another factor responsible for availability of carbon (Azov 1982). Moreover, absorption of nitrogen increases pH of the medium. Mechanism involves reduction of nitrate to ammonia ions which produces hydroxyl ion (Xu et al. 2006). Increased pH induces precipitation of phosphate in the medium. However, this incidence can be lowered by process of respirations. pH is known to influence the growth rate of microalgae. The microalgae use the inorganic carbon and HNO_3 for the growth of the cell productivity. Depending on these parameters, pH value may vary from low to high in the alkaline region. pH from 7 to 9 is optimum for the algal growth. Carbon dioxide acts as buffer system in bicarbonate-carbonate for photosynthesis,

2.5.3 Salinity

Marine phytoplankton is tolerant to changes in salinity. The best algal growing conditions for most species are salinity levels that are lower than that of their native habitat. Lipids can perform both structural and storage functions as they can be used in synthesis of the cell membrane and storage products. The lipid contents and the composition of microalgae have been shown to change the responses to the environmental variables such as the light, temperature, and salinity. Microalgae have greater impact on lipid content due to salinity (Asulabh et al. 2012). However, increased salinity can negatively affect the photosynthetic activity which may be due to hindrance in electron transport chain (Zhang et al. 2012).

2.5.4 Temperature

It is also a very important factor for the growth of the microalgae. Usually microalgae grow profusely in elevated temperature, and growth stops beyond a critical temperature (Ras et al. 2013). Most common temperature range is from 16 °C to 27 °C. More heat along with humidity can result in less growth of microalga. After a critical temperature, growth rate decreases; however, effect of temperature may vary depending on the species.

2.5.5 Light

Microalgae are phototroph which means they obtain energy from the light. Some of the microalgae are capable of growing in dark conditions using simple organic compounds as the energy and the carbon sources. Light conditions directly affect the growth and the photosynthesis process of the microalgae. It was also reported that the conversion efficiency of the sunlight energy into chemical energy is 2% (Fontes et al. 1987). Direct sunlight can often be too intense and cause photoinhibition at the surface. At the same time, algal cells deeper down may suffer from photo-deprivation, as the radiation has been absorbed or reflected by cells closer to the surface. To deal with these challenges, cultivations must be designed with a large surface-to-volume ratio and adequate mixing of the algal mass to make sure all cells are illuminated for an appropriate amount of time (Christenson and Sims 2011). Algal cultures prevent from the light limitation to decrease the depth of the culture vessel. Moreover, depth also affects the productivity in light through inverse relationship.

2.5.6 Inhibitory Substance

Many substances act as inhibitory for cell growth as well as photosynthetic efficiency or process. Inhibitory substances include phenolics, heavy metals, herbicides, pesticides, substances in detergents, some microbes, household cleaning products, and personal care products. Ammonia is one of the inhibitors which reduces the microalgal growth in high temperature and pH (Aharon and Yosef 1976). Mechanism of toxicity by organic compounds is associated with inhibition of nutrient uptake, ultimately leading to permanent damage of cell membrane.

2.6 Problems Encountered During Wastewater Remediation

One of the important drawbacks of wastewater remediation by microalgae is that it requires spacious system for the growth of the algae and a good operation speed which is not fulfilled by present-day phycoremediation technologies. Downstream equipment used for the wastewater remediation is failing due to a build of large solid hairs and fibers during the primary treatments. The treated effluents are not giving the total nitrogen and phosphorous targets. Ammonia removal is a strictly aerobic process. If more ammonia is released into the wastewater remediation, then it results into more retention time and low food-to-microorganism ratio and affects pH buffering. Algae-treated wastewater is not meeting biochemical oxygen demand target due to organic overloading, low oxygen concentrations, and sludge accumulation and old sludge to the effluents. Loss of opportunity to maintain the fertility of the soil is achieved through wastewater rescue. This leads to the need to purchase the organic fossil fertilizers. The downstream processing parameters are very expensive for the harvesting and the recovery of the secondary metabolites. For short-term treatment processes, algal pond treatment can be better alternative for bioflocculation of the algae. Algal growth for the clearing of the wastewater requires the large amount of algae to grow in the water bodies and destabilize the ecosystem if the animals feed the large amount of the algae growing in the water bodies and leads to the death of the animals.

2.7 Mechanism of Action of Microalgae During Wastewater Treatment

Microalgae are a type of microscopic photosynthetic organisms usually found in marine as well as freshwater environments. They possess a photosynthetic mechanism which is somewhat similar to land plants. Such photosynthetic capabilities of microalgae make them significant for treatments with microbial aids where higher concentration of nitrogen and phosphorus can be utilized in conversion of solar energy into biomass. General mechanism adopted by microalgae to treat wastewater includes assimilation, precipitation, biosorption, and bioaccumulation.

2.7.1 Assimilation

Wastewater contains phosphorous in organic as well as inorganic form. Most common forms of phosphorus in aqueous solutions are orthophosphates and polyphosphates which can be utilized by organism for production of biomass. In the next level, polyphosphates can be converted into orthophosphates. This process is usually quite slow. The removal of phosphorous from wastewater in the biological system comprises of the treatment of the influent wastewater which is incorporated into cell biomass and further involves cleaning with sludge wasting. Microalgal cells need phosphorus for metabolic processes such as ATP production, phospholipids, and nucleic acids. Algae can assimilate orthophosphates as inorganic ions with the aid of energy (Becker 1994). Microalgal cells can store the excess phosphorus in its storage (volutin) granules. These reserves can be used for prolonged growth of microalgal cells (Fogg 1975; Oliver and Ganf 2000). Therefore, it may be concluded that phosphorus is not associated with immediate effects on microalgal growth as compared to temperature and pH (Mostert and Grobbelaar 1987). Moreover, concentration of phosphorus may vary in wastewater ranging from 1 mg phosphorus per g dry mass. It has been reported that average concentration of phosphorus in algal cell is 13 mg phosphorus per g dry weight (Oswald 1988). Higher concentrations of phosphorus may not necessarily result in higher growth, whereas various different conditions can be optimized to increase the assimilation efficiency of microalgal cells. For instance, microalgal cells deficient in nutrients can result in better uptake and thus assimilation of phosphorus in less time span. It results in more efficient bioremediation. In turn, phosphorus assimilation depends on fixed carbon in algae. One of the various optimization strategies involves the starving conditions in bioreactor for enhancing the assimilation of other pollutants.

2.7.2 Precipitation

Carbon species are one of important constituents among others. Microalgae take up inorganic carbon in the form of carbon dioxide and bicarbonate ions during photosynthesis (Oswald 1988; Borowitzka 1988), which can be subsequently converted into carbon dioxide using carbonic anhydrase. When bicarbonate is used as carbon source, the pH in the medium increases. This pH increase, which can elevate the pH in algal cultures to values above 11, strongly affects the water chemistry. Phosphorus may as a result precipitate with available cations to form metal phosphates, where calcium phosphates are the most common. Besides being promoted by high pH values, precipitation reactions can be enhanced by higher concentrations of calcium and phosphorus along with high temperature (Song et al. 2002). Precipitation usually results in neutral pH and concentrations of phosphate and calcium to be 50 mg and 100 mg, respectively (Carlsson et al. 1997). In soft water which is usually with less concentration of 50 mg, raised levels of phosphate concentration can be used to induce precipitation. Carbonate enhances the production of amorphous calcium phosphate and promotes calcite formation from calcium at pH above 8.0. Various calcium phosphates can be present in the wastewater which may lie in molar ratios between 1 and 1.67. Some salts like amorphous calcium phosphate and octacalcium phosphate can act as precursor to hydroxyapatite (Arvin 1983). However, hydroxyapatite formation is inhibited by different ion concentrations such as magnesium, carbonate, and pyrophosphates ($P_2O_7^{4-}$) (Fergusson et al. 1973; Arvin 1983). The effect of magnesium is pronounced when the Mg/Ca ratio exceeds 0.45. At pH levels above 10.5, magnesium forms precipitates with hydroxide ions and loses its adverse effect on phosphorus solubility and in turn its utilization (Jenkins et al. 1971). Moreover, carbonate reduces the crystalline nature of calcium phosphate resulting in formation of amorphous calcium phosphate (Fergusson and McCarty 1971; Arvin 1983). It can be concluded that phosphorus precipitation is inversely correlated with carbonate concentration (Fergusson et al. 1973).

Chemical precipitation contributes significantly to phosphorus uptake by algal wastewater treatment and thus bioremediation (Doran and Boyle 1979; Moutin et al. 1992; Proulx and LessardP 1994; Mesple et al. 1996; Tam and Wong 2000). Particularly in areas with hard water, i.e., water with high concentrations of calcium and magnesium, this effect may be pronounced. One of the major effects is chemical stripping of phosphorus which can be specifically beneficial for algal growth with enhanced phosphorus removal as a result. It makes chemical sludge harvesting easier as compared to free-floating cells.

2.7.3 Biosorption

Metabolism-independent binding or adsorption of heavy metals to living or dead cells, extracellular polysaccharides, capsules, and slime layers is referred to as "biosorption." Walls and envelopes of algae are very efficient in biosorption due to the charged groups present in them (Table 2.2). Algae can be immobilized in polyacrylamide gel and

Table 2.2 Parameters used for the algal cultivations

Operational parameters	Description	References
Inorganic carbon effect	CO_2 and HCO_3^-act as the inorganic source of carbon for the microalgae	De Morias and Costa (2007)
Salinity	High evaporation causes the salinity effect. High salinity can cause the cellular ionic stress and osmotic stress due to the selective ion permeability of the cell wall	Moheimani (2005); Salama et al. (2014)
Light	Strong illumination can inhibit the photosynthesis process	Kim et al. (2015)
pH effect	Higher photosynthetic activity can increase the pH. Neutral pH is favorable but pH as high as 10 and low as 4 are tolerate by some species	Moheimani (2005)

packed into columns or used in fluidized beds for the considerable binding of heavy metals such as zinc, cadmium, copper, lead, gold, and uranium from wastewater (Chojnacka et al. 2005).

Microalgae have enormous potential in cleansing water as they show a strong affinity toward polyvalent metal and dissolved metal ions in wastewater (De Bashan and Bashan 2010), for example, *Chlorella* and *Scenedesmus*.

Biosorption of metals by microorganisms proceeds through a two-stage pathway (Fig. 2.2):

1. An initial rapid, reversible, and passive adsorption onto the cell surface (where metal ions adsorb via electrostatic interactions to cell wall functional groups).
2. A less speedy, irreversible active process which involves the transport of metal cations across the cell membrane. The first stage occurs in both living and non-living cells, whereas the second one takes place only in living ones (Jjemba 2004; Sud et al. 2008).

Garnham et al. (1992) focused on removal of three metals (Zn, Co, and Mn) by *Chlorella salina* and showed that their uptake was essentially biphasic. The initial phase of biosorption is not dependent on physicochemical conditions such as light, temperature, and metabolic inhibitors. A slower phase of uptake followed that was instead dependent on metabolism and other abiotic factors. For those three metals, cellular compartment analysis indicated that large amounts were bound to intracellular components and to the cell wall itself. A higher concentration of each metal in the vacuole than in the cytosol was also observed, thus unfolding a possible mechanism of regulation of the free metal ion and detoxification. The capacity of biomaterials to adsorb metals depends on the composition of their cellular surface and is promoted by the presence of negatively charged functional groups, coupled with chemical composition of the outer solution undergoing treatment (Monteiro et al. 2011). This is especially true with regard to competing anionic groups and pH, which affect protolysis and consequently drive such changes. Microalgae are suitable chiefly as biosorbents owing to their natural abundance in seas and oceans

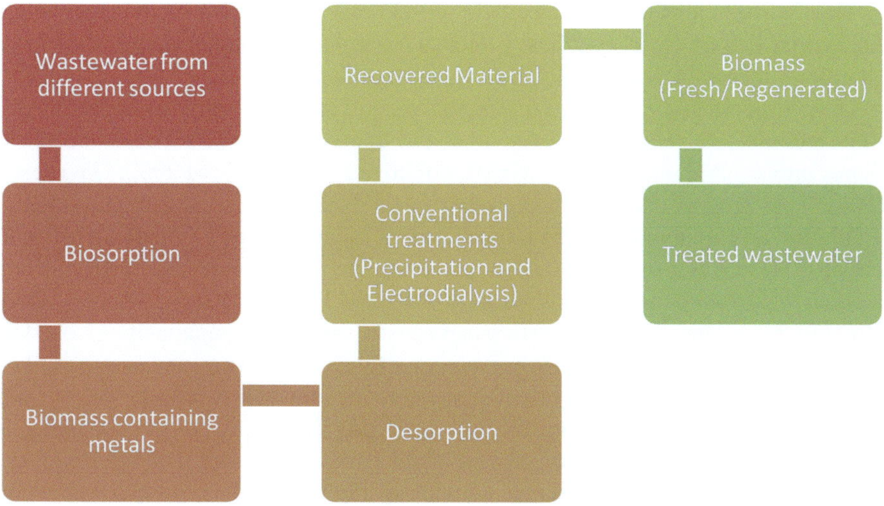

Fig. 2.2 Flowchart showing steps in biosorption

(which allows harvesting and culturing at relatively low cost) and their high sorption capacity when compared with other biological or physicochemical sorbents. Such a capacity spans the range of 25.4–389 mg/g under acidic/neutral pH values (from as low as 2 up to 7) while withstanding concentrations from 20 up to as high as 20,000 mg/L. These features are, on average, better than those claimed for other sorbents from biological or physicochemical origins. Therefore, microalgal biomass might be an economically feasible (besides technologically efficient) alternative to existing physicochemical methods of metal removal and recovery from wastewaters (Mehta and Gaur 2005; Romera et al. 2006) even though actual engineering/cost analyses have not been carried out in full. Note, however, that the yield of microalgal biomass on the original volume of (sea or fresh) water is relatively poor, unlike happens with, e.g., macroalgae, and harvesting is in addition difficult to achieve.

2.7.4 Bioaccumulation

Bioaccumulation is defined as intracellular accumulation of unwanted substances, which occurs in two stages: the first is similar to biosorption involving attachment of potentially toxic elements to the surface, and the second is active transport of metal ions into cells. Bioaccumulation is nonequilibrium process (Aksu and Dönmez 2000). The process is more complex than biosorption itself and requires metabolic activity of cells. It has been reported that metabolic processes support the bioaccumulation process and the following reaction takes place in the cell:

$$HCO_3 - \rightarrow CO_2 + OH$$

The hydroxyl ions thus produced are present on cell surface which can scavenge the hydroxides of toxic and heavy metals by enhancing precipitation reactions. Therefore, these reactions aid in wastewater treatment processes.

Other processes such as biotransformations and biomineralization involve metabolic processes to remove the toxic ions and convert insoluble sulfides and phosphates into soluble ions so as to achieve effective removal (Lloyd 2002; Gavrilescu 2004). This property is used in the removal of ions of iron, manganese, and lead (Loukidou and Zouboulis 2005). This aspect of bioconversions includes batch systems for wastewater treatment (Aksu and Dönmez 2000). Bioaccumulation offers another important benefit that is separate; biomass harvesting step is not required, and both the treatment and harvesting can be performed simultaneously. Also, additional unit processes are reduced: harvesting, drying, processing, and storage (Aksu and Dönmez 2005). Bioremoval of pollutants present in wastewater has greater impact by operational conditions maintained during the process, as some of the pollutants have negative impact on growth of microalgal cells and thus the treatment process. Generally, the wastewater containing high load of pollutants cannot be treated by bioaccumulation which poses a major limitation to biological method of remediation. Moreover, energy source such as sucrose has to be supplied for providing energy to the growing cells for effective growth and removal (Aksu and Dönmez 2005). By employing effective methods of strain selection, those strains can be selected which can utilize the organic and inorganic pollutants and leads to bioaccumulation. Present-day practical application of bioaccumulation is that the majority of conventional municipal wastewater treatment plants based on living organisms have a significant contribution of bioaccumulation itself. However, much work is needed to confirm the significant contribution of bioaccumulation in municipal wastewater treatment plants (Aksu and Dönmez 2000).

In bioaccumulation, pollutants are transported across cell wall and membrane. Inside the cells are bound to intracellular structures (Kujan et al. 1995). It has been studied in previous work that bioaccumulation involves redox reactions to scavenge the unwanted constituents present in wastewater (Yilmazer and Saracoglu 2009). These metabolic processes are complex in nature, and different conditions such as pH, temperature, growth inhibitors, etc., affect the metabolic processes (Kujan et al. 1995). Metal ions cause toxicity by complexation with lipid content of cell membrane which causes damage in integrity (Yilmazer and Saracoglu 2009). It has been concluded that with increased concentration of pollutants, accumulation adversely affects the cell morphology and physiology (DeSiloniz et al. 2002). Major route for mechanism of action is through sulfhydryl groups of enzymes which can be easily attacked by metal ions, thus causing toxicity in the cell. Another route is by synthesis of low molecular weight proteins rich in thiol groups which can be synthesized in response to complexation of metal ions with these pollutants (Martin-Gonzalez et al. 2006). There are few reports on some adapted microorganisms which are better suited for bioaccumulation than non-adapted ones (Kocberber and Donmez 2007; DeSiloniz et al. 2002). Effective biotreatment results have been obtained by using the enriched cultures isolated from polluted environments (Kocberber and Donmez 2007). A study on bioaccumulation in which plasmid of *Escherichia coli* from pea with genes of metallothionein has been used showed improved accumulation of mercury (Deng and Wilson 2001).

2.8 Biotechnological Strategies for Improvement of Phycoremediation of Wastewater

All the problems faced during the wastewater treatment can be improvised by adopting various biotechnological strategies (Table 2.3) which can be explained in the following sections.

2.8.1 Immobilized Cell System

One of the major problems in the utilization of microalgae for the biological tertiary treatment of wastewater is their recovery from the treated effluent (Chevalier and De la Noüe 1985a; b). Among the ways of solving this problem which have been recently studied are immobilization techniques (De la Noüe and Proulx 1988). Immobilization of the cells provides better utilization as well as stability to the cells as compared to free cells. There are several reports on using immobilized cells in both batch and continuous culture systems (Hall and Rao 1989). Chevalier and De la Noüe (1985a, 1985b) discovered that *Scenedesmus* cells when immobilized using k-carrageenan were capable of bioaccumulating at same rates as that of free microalgal cells. There are numerous advantages related to using the immobilized living cells as compared to suspended cells. Immobilized microalgal cells on suitable

Table 2.3 Different biotechnological strategies used for improvement of phycoremediation

Biotechnological strategy	Microalgal sp. used	Approach used	References
Immobilized microalgal cells	*Phormidium laminosum*	Polymer foam was used as matrix for immobilization of microalgal cells	Sawayama et al. (1998)
Hyperconcentrated cultures	*Scenedesmus obliquus*	Algal biomass greater than 1.5 g/L. more biomass can sequester more carbon and thus result in energy-generating process along with wastewater treatment	Chevalier and De La Noüe (1985a, b)
Microalgal fixed biofilm	*Enterobacter cloacae* DT-1	Natural biofilm # 52 was used as feedstock for bioenergy production	Miranda et al. (2017)
Bacterial-algal ggconsortium	*Chlorella* sp. and bacterial sp.	Bacteria is known for efficient remediation of wastewater, and algal cells can be used for production of value-added compounds and energy production	Foladori et al. (2018)
Photobioreactors	*Dunaliella salina* and *Chlorella* sp.	Photosequencing batch reactor has been developed	

support are advantageous as cell retention time increases in the reactor (Travieso et al. 1992). Polymer coated form of *Phormidium laminosum* removes nitrate components up to 90% in a continuous-flow reactor (De la Noüe et al. 1990; Garbisu et al. 1991; Travieso et al. 1992; Sawayama et al. 1998). Sawayama et al. (1998) have reported that hollow fiber-immobilized cyanobacterial systems are easy to construct and immobilization does not take a long time. Markov et al. (1995) have observed that removal of inorganic nutrients from wastewater can be improved by immobilizing cyanobacteria on hollow fibers and hydrogen production was increased. In a similar study on direct generation of electricity using cyanobacterial species *Mastigocladus* and *Phormidium* immobilized on suitable matrix which has improved the process.

2.8.2 Hyperconcentrated Cultures

Hyperconcentrated cultures have also been employed for wastewater effluent remediation in which high algal biomass >1.5 g/L. Chitosan has been used for immobilizing algae using flocculent (Lavoie and De la Noüe 1983; Morales et al. 1985), whereas cell concentrations of up to 1.9 g/L dry weight have been obtained for *Oscillatoria* sp. grown on sewage sludge (Hashimoto and Furukawa 1989). Studies on hyperconcentrated cultures are very limited, and more work should be done to strengthen the work.

2.8.3 Genetic Engineering

Microalgae consist of characteristics which are helpful for using it for phytoremediation of wastewater, but still no one microalga is the most efficient one. Biotechnology can help in improving the bioremediation efficiency of microalgal cells by inserting the gene of interest in target cells (Guihéneuf et al. 2016). Mutagenesis has also been used for improving the microalgal cells. Selective mutagenesis can be performed using various physical and chemical mutagens. The genetically modified microalgal cells can be used to enhance the production of valuable products (Hlavova et al. 2015). Moreover, for the development of suitable genetically engineered algal strains capable of effective degradation of nitrogen and phosphorus, genome databases like NCBI and GenBank can be used for selection of suitable genes and data mining approach (NCBI directory 1995).

2.9 Microalgal-Bacterial Aggregate System for Wastewater Treatment

Different microorganisms can be complexed and used as aggregates for wastewater treatment. For example, microalgal-bacterial aggregates have been employed for wastewater treatment. Microalgal-bacterial consortium can be used due to synergistic effect in which microalga provides oxygen for the process and bacteria utilizes nitrogen due to nitrification-denitrification. Major problems associated with such aggregates are poor settlement of algal biomass and harvesting problem (Bansal et al. 2018). However, different conditions should be optimized and evaluated for effective wastewater treatment, and economic feasibility of the process can be determined (Quijano et al. 2017). In a similar study, Filadori and coworkers (2018) found enhanced nitrogen removal using energy efficient microalgal-bacterial consortium on real municipal wastewater. Photosequencing batch reactor (PSBR) has been developed for the removal of nitrogen. However, various kinetic characteristics should be evaluated, and mass balance analysis should be performed to improve the process further.

2.10 Development of Photobioreactors

Microalgal cells can be produced on a large scale for employing them in different applications such as bioremediation processes and bioconversion of biomass into valuable products and bioactive compounds (Gupta et al. 2015). A large number of efficient photobioreactors have been proposed that are very advantageous for mass cultivation of algae (Ugwu et al. 2008). Photobioreactor is a reactor with facility for light so as to grow photosynthetic microorganisms such as microalgae. Microalgal cultivation needs photobioreactor for different purposes. For generating high-value-added products, axenic cultivation of microalgae is needed. Until now, different types of PBRs have been invented and produced for algae cultivation during the past decades, and some of them have achieved large-scale commercial production (Singh and Sharma 2012; Fernández et al. 2013; Gupta et al. 2015). Both types of photobioreactors (open and closed ones) can be used for production of valuable products. Open bioreactors have been preferred due to limited control of physical and chemical conditions such as water, temperature, light, and pH. However, closed bioreactor can be used for large-scale production, but major limitation is less light and photosynthetic activity (Bansal et al. 2018).

Microalgal cultivation has been studied for over 70 years. Large-scale microalgal cultivation was firstly raised by the research of Carnegie Institute in 1952 (Burlew 1953). To deal with problems encountered in open system, closed vessels have been developed to achieve a better yield of microalgae biomass, which does not allow direct mass transfer between culture media and the atmosphere and is able to provide a controllable environment such as light, CO_2, temperature, and nutrients (Vasumathi et al. 2012; Wang et al. 2014). Closed photobioreactor can be used for production of various valuable products which can be used in biopharmaceuticals, cosmetics, human health, and biofuels which are produced from microalgae that became more and more important; therefore, the development of suitable and sustainable closed PBRs has a great potential. The current common closed PBRs generally include flat panel, vertical tube (bubble column and airlift), horizontal tube, stirred tank, and their modified configurations (Han et al. 2017). Most commonly used closed bioreactors for phycoremediation of wastewater are suspended system and fixed systems (Hoffman 2002).

Membrane bioreactors (MBR) are the most popular and an effective wastewater treatment technology used in the water treatment area. For the traditional PBR such as the flat bioreactor, microalgae can be easily washed out of the bioreactor. The membrane has a well-known function of excellent micro-size particle separation. Hence, applying membranes containing microalgae to treat wastewater allows decoupling the dilution rate (related to HRT) and biomass retention time (MRT). MBR was applied in the microalgae wastewater treatment processing, called microalgae MBR (MMBR) (Han et al. 2017).

2.11 Applications of Phycoremediation in Wastewater Treatment

Algae can be used in wastewater treatment for a range of purposes (Fig. 2.3) which are beneficial for the environment as well as for producing valuable products. Some of the applications of wastewater treatment by microalgae are given below:

2.11.1 Microalgae in Wastewater Treatment

2.11.1.1 Removal of Nutrients

Microalgae can be efficiently used to remediate the nutrients present in wastewater. Major nutrient is nitrogen which contributes to 10% of biomass and is the second most important nutrient to microalgal cells (Becker 1994). Nitrogen can be mainly present as ammonium ions and nitrate ions which can be easily accumulated by microalgal cells (Oliver and Ganf 2000). However, cyanobacteria can assimilate various amino acids such as arginine, glutamine, and asparagine and thus can fix nitrogen (Bhaya et al. 2000). Several species of microalgae can utilize excess of nitrogen. Phosphorus is the next significant macronutrient which can be utilized by algae in form of orthophosphate. Phosphates can be utilized by phosphatases and stored within the microalgal cells in polyphosphate granules which is known as assimilation. Thus, wastewater containing high amounts of phosphorous can be treated by using microalgal cultures (Fogg 1975; Oliver and Ganf 2000).

Moreover, microalgae can be efficiently used for tertiary and quaternary treatments as it can easily take up major nutrients for its growth.

2.11.1.2 Reduction of Biological and Chemical Oxygen Demand (BOD/COD)

Algae are more efficient carbon fixer as well as scavenge excess nutrients at effective cost. It can relieve the biological oxygen demand of wastewater by oxygen produced by photosynthetic process (Laliberte et al. 1994). Microalgae are considered to be better for remediation process because it results in less toxic waste as well as are nonpathogenic. Algae can remediate the wastewater by the use of enzymes for conversion and degradation of pollutants (Oswald 1988). More recalcitrant metals and xenobiotic compounds can be remediated by algal metabolism. Many

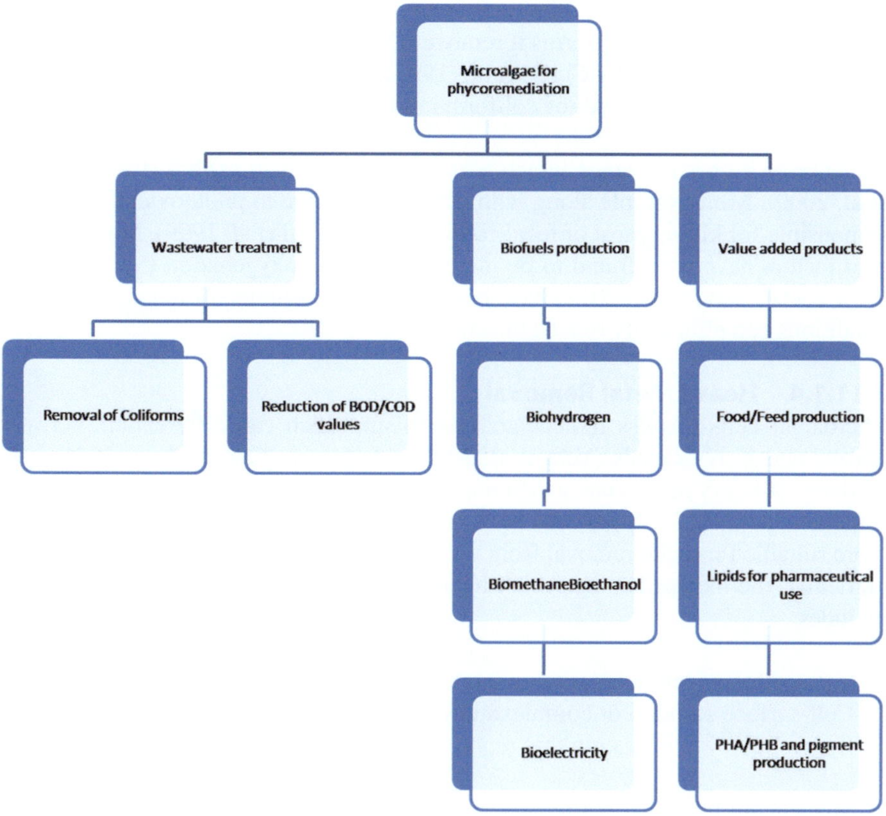

Fig. 2.3 Applications of microalgae for wastewater treatment and production of valuable compounds

researchers have studied microalgae as possible solution for environmental problems (Yoshida et al. 2006). The growth of algal cells in natural waters makes it fit for human consumption. Important algae species which can be employed for wastewater treatment are *Chlorella, Scenedesmus, Synechocystis, Gloeocapsa, Chroococcus, Anabaena, Lyngbya, Oscillatoria,* and *Spirulina* (Palmer 1974).

However, due to excess BOD values of wastewater, the dissolved oxygen can be depleted which results in anaerobiosis and thus has adverse effect on aquatic life. Hence, its removal is necessary. Colak and Kaya (1988) have found that possibilities of biological wastewater treatment by algae eliminated BOD and COD by 68.4% and 67.2%, respectively.

2.11.1.3 Removal of Coliform Bacteria

Microalgae can result in scanty growth of fecal coliforms (FC) due to less availability of dissolved oxygen in wastewater. This affects the growth of coliforms. Most of the coliform bacteria are responsible for pathogenic outbreaks which make water unfit for drinking purposes. The efficiency of microalgae for bioremediation is decided on the basis of coliforms it removes (Sebastian and Nair 1984).

In a study carried out by Curtis et al. (1992), different necessary conditions have been checked out for growth of coliforms in the wastewater. Sunlight can damage fecal coliform by depleting the oxygen concentration. This depletion occurs in the presence of higher pH range which is usually greater than pH 8.5 (Colley Davies et al. 2000). Moreover, pH along with oxygen gives rise to photooxidation process responsible for killing most of fecal coliforms (Maynard et al. 1999). Major significant factors have been found to be increased pH and oxygenation (Van der Steen et al. 2000). Ansa et al. (2011) have observed that simulation of algal cells in lake conditions can effectively reduce the coliforms.

2.11.1.4 Heavy Metal Removal

Microalgal cells possess molecular mechanism which can differentiate between normal and heavy metals (Vela et al. 2006). Moreover, they can be used for easy recovery of heavy metals using different desorption chemicals (Figueira et al. 1999; Rajamani et al. 2007). Algal cells have higher affinity for metals which makes them more suitable for metal removal from wastewater (De Bashan and Bashan 2010). In particular, the mechanism by which microorganism removes metals from solution includes:

1. Extracellular accumulation/precipitation.
2. Cell-surface sorption or complexation by live or dead biomass.
3. Intracellular accumulation that requires microbial activity (Cossich et al. 2002).

Mechanism for biosorption involves the entrapment of heavy metals in the cellular structure and subsequent absorption on binding sites of cells. This method is also termed as "biosorption" or "passive uptake" (Malik 2004). Heavy metals can interfere with the metabolic processes of cell and thus cause bioaccumulation or active uptake.

2.12 Formation of Valuable Products

Extensive use of fossil fuels due to rapid industrialization has resulted in increase in global warming, thereby changing climatic conditions. There is aroused interest in utilizing microalgal cells for efficient alternative energy sources known as biofuels. Microalgae are thus extensively used for the production of biomass which in turn can be used for biofuel production. Microalgae have several advantages for biofuel production as these cells have short generation time, rapid growth, high lipid content, and minimal land requirements. Moreover, wastewater stream can be utilized as nutrient feed for these microalgal cells for conversion of biomass to bioenergy.

2.12.1 Biomass Production

Microalgal cells can be easily grown using the wastewater stream as they contain various constituents that serve as nutrients for the cells. Biomass thus produced can be used in a number of beneficial ways. As biomass can be used for accumulation of heavy metals that are toxic and after extraction of lipid left out, biomass can be used for animal feed or for increasing fertility of soil (Pittman et al. 2011). The biomass can be utilized for production of biofuels and can be converted into many valuable products such as biofertilizers, biofilms, and biopolymers, and recent report is on using them for production of electricity (Gouveia et al. 2016).

2.12.2 Bioethanol Production

Generally, two methods are employed for the production of bioethanol from microalgal biomass, namely, fermentation (biochemical process) and gasification (Singh and Gu 2010). Microalgae are rich in carbohydrates and proteins which can be used as carbon sources for fermentation, so microalgae can replace the requirement of food crops. This provides the scope for utilization of microalgae in the third-generation biofuel production, as using the food crops will cause the scarcity of the same. Moreover, there is a temporary prohibition on the use of food crops for the production of bioethanol due to food security and availability of agricultural land which can be easily resolved using microalgal cells. Therefore, microalgae are generating a lot of interest as biomass feedstock for bioethanol production (Harun et al. 2010). Fermentation of the microalgal biomass is catalyzed by microbes such as bacteria, yeast, and fungi, and the main by-products are CO_2 and water. The spent biomass after fermentation is used for anaerobic digestion process for methane production so in essence all the organic matter is accounted for (Singh and Gu 2010; Harun et al. 2010). Bioethanol production using microalgal cells is still in initial stage, and more studies are required to evaluate the utilization of microalgae for conversion of biomass to bioethanol (Harun et al. 2010).

2.12.3 Biomethane Production

The interest in biomethane production emanates from the fact that biomethane fermentation technology produces valuable products such as biogas (Singh and Gu 2010; Harun et al. 2010). Biogas is a mixture of methane (55–75%) and CO_2 (25–45%) and is produced via anaerobic digestion of microalgal biomass by anaerobic microorganisms (Singh and Gu 2010; Harun et al. 2010). Biomethane can be used as fuel gas and can also be used to generate electricity, while the spent biomass is used to make biofertilizers (Singh and Gu 2010). Biogas production in turn depends on several factors such as temperature, pH, organic load, and retention time in reactors (Harun et al. 2010). Though microalgae hold enormous potential for biogas production, more studies are needed to strengthen the possibilities of using them on commercial scale for biogas production (Singh and Gu 2010).

2.12.4 Biochar Production

Biomass obtained from microalgal cells can be used as important alternative of energy due to its high efficiency to fix carbon dioxide and easy availability that can be very useful in present scenario of energy crisis around the globe. Algal biomass can be converted into biochar that involves slow pyrolysis to prepare a carbon-rich product used for increasing alkalinity of acidic soils. Moreover, high concentrations of nitrogen and phosphorus give additional advantage to increase fertility of soil for agriculture (Chaiwong et al. 2013). In some previous studies, this biochar due to the presence of functional groups and inorganic elements helps it to be used as absorbent for wastewater remediation (Yu et al. 2017). Functional groups present on microalgal biochar are responsible for better biosorption for various organic molecules. Biochar has potential to be used for increasing fertility and alkalinity of soil for agricultural processes.

2.12.5 Microalgae in Bioelectrochemical Systems

Due to huge energy demands, there is a constant need to find alternative energy resources which are clean, renewable, and cost-effective as well as environmentally friendly. Microbial fuel cells (MFCs) are such energy-generating systems that fulfill all the above characteristics. Microbial fuel cells are based on important property of microalgal cells to fix atmospheric carbon dioxide and produce oxygen by photosynthesis that can enhance the cathodic reaction (Saba et al. 2017). These microalgal cells can also act as efficient electron acceptors and can behave as electron acceptors at cathodic end and electron donors at anodic end. They can be used for removal of various organic and inorganic constituents from wastewater (Gude et al. 2013; Wu et al. 2014; Commault et al. 2014). Baicha et al. (2016) reviewed microalgal cells as MFCs for bioproduction of electricity and concluded that carbon dioxide can play a significant role in biomass cultivation. Along with MFCs, microalgal cells can be also employed as microbial desalination cells (MDCs) and bioelectrochemical systems (BES). In another study, Saba et al. (2017) have reviewed the effect of several parameters on energy production from MFCs.

The major limitation in MFC is the low current flow; however, considerable amount of energy is generated (Otadi et al. 2011). Photosynthetic activity of microalgal cells can be used to generate electricity like bioelectrochemical system. Thus, solar energy can be converted to electricity using microalgal cells. These play a significant role in power generation and can consume the light at night generated in daylight (Soni et al. 2016).

2.12.6 Microalgal Biofilms

Biofilms are produced when microalgal cells are covered by surface molecules such as exopolysaccharides. Algal biofilms are films which are generated by colonization of microalgal cells on illuminated surfaces in the presence of humid conditions and nutrients (Leadbeater and Callow 1992; Jarvie et al. 2002). Algal biofilms can adapt the change in environment and survive all the adverse conditions as single cell or in clumps (Menicucci 2010). Microalgal cells can be used in a wide number of applications ranging from agriculture, alternative energy sources, and personal care products and nutraceuticals (Pulz and Gross 2004; Mata et al. 2010). Algal biofilm is used for removing water impurities; thus, algae are significant due to nutrient removal from wastewater due to their enhanced nitrogen (N) and phosphorous (P) metabolism ability. Algal biofilms can be employed in wastewater treatment as well. It avoids expensive harvesting techniques used in suspension cultivation like centrifugation and flocculation (Gross et al. 2016).

2.13 Other Applications of Microalgae

2.13.1 Production of Secondary Metabolites

Microalgae share some of the common properties like plants, and thus they can be used for production of some important secondary metabolites, viz., carotenoids, sterols, proteins, lectins, oils, unsaturated fatty acids, antioxidants, fibers, and amino acids. Their potential can be explored for commercial production (Cardozo et al. 2007a, b; Rosenberg et al. 2008; Ioannou and Roussis 2009; Ibañez et al. 2012).

2.13.2 Sulfated Polysaccharides

Marine algae can be used as source for sulfated polysaccharides (SPs) with so many structural variants (Wijesekara et al. 2011; Zhang et al. 2012). But most common sulfated polysaccharides are fucoidan and laminarins derived from brown algae, carrageenan from red algae, and ulvan obtained from green algae (Li and Kim 2011). Some of previous studies have been carried out with objective of using these sulfated polysaccharides in food, feed, pharmaceutical, and beauty industry (Li and Kim 2011). Some of the studies have confirmed the role of sulfated polysaccharides as antiviral compounds against enveloped viruses (Baba et al. 1998; Zhu et al. 2003). In a similar study, anti-HIV activity has been reported in microalgal and cyanobacterial extracts. Moreover, the antiviral activity depends on the molecular weight as well as grade of sulfation in the compounds (Witvrouw and De Clercq 1997). Antiviral compounds were extracted with anti-HIV activities from *Fucus vesiculosus*. It showed water solubility and high anti-HIV activity (Béress et al. 1993). Likewise, different algae Phaeophyta, Rhodophyta, and Chlorophyta have been explored for antiviral activity (Zhu et al. 2003).

2.13.3 Proteins and Amino Acids

Rhodophyta (red algae) have enormous amount of proteins as compared to other types of microalgae (Mendis and Kim 2011). For instance, phycobiliproteins (PBPs) are generally present in red algae and cyanobacteria (Sekar and Chandramohan 2008). PBPs possess high solubility, stability, and fluorescent properties (Su et al. 2010). These can perform several important functions such as light harvesting reactions in cyanobacteria and formation of cryptomonads and cyanelles (Glazer 1994). There are extensive reports of previous studies on PBPs, and they are known to contain medicinal properties such as antitumor, anti-inflammatory, serum lipid-reducing properties, and antiviral properties. Moreover, these can be utilized for adsorption of pollutants from the body (Romay et al. 2003; Sekar and Chandramohan 2008).

2.13.4 PHA and PHB Production

Novel compounds such as sustainable polymers like PHAs (polyhydroxyalkanoates) and PHBs (polyhydroxybutyrates) can also be produced from algal biomass that is significantly used for production of bioplastics. These polymers can be used successfully in nanotechnology as efficient nanomaterials and scaffolds in tissue engineering (Verma et al. 2019; Bansal et al. 2018).

2.14 Conclusion and Future Prospects

It can be concluded that phycoremediation is one of the safe methods that can be used for treating wastewater. It not only produces the clean water but provides various valuable products as well as better alternative energy sources such as biofuels. Though there are numerous studies on using microalgae in production of these valuable products, still many challenges have been encountered such as land and space requirements, algal contamination with bacterial cells, eutrophication, etc. These problems can be resolved by using photobioreactor which is very efficient novel biotechnological approach. As one of the major problems is availability of clean water for human use, microalgal technology can help humanity in a great way.

Microalgae can be effectively employed for removal of metal ions and can be used as recombinant systems for protein expression for higher plants and animals (Hempel et al. 2011). In some of previous works, *Chlamydomonas reinhardtii* and the diatom *Phaeodactylum tricornutum* have been utilized as model expression systems. They can also be used for preparation of nanoparticles using metal oxides. Microalgae can reduce the pollutant load in environment and avoid the problems that can affect human health care (Fawcett et al. 2017). Moreover, these cells can be efficiently used for alternative energy sources and production of biofuels. Unlike petroleum-based fuels like diesel and petrol, biofuels are rapidly biodegradable. Microalgae can also serve as clean electricity producers for bioenergy production

(Clarens et al. 2011). Along with so many wonderful properties, these cells can be used for production of some of novel compounds such as PHA and PHB which has revolutionized the tissue engineering approaches in human health care.

Acknowledgment Authors would like to thank the Director, Indian Institute of Information Technology Una, for providing the necessary facility to carry out the present work.

References

Abdel-Raouf N, Al-Homaidan AA, Ibraheem IBM (2012) Microalgae and wastewater treatment review. Saudi J Biol Sci 19:257–275

Abdel-Raouf N, Ibraheem IBM, Hammouda O (2003) Eutrophication of river Nile as indicator of pollution. In: Al-Azhar Bull. of Sci Proceeding of 5th Int. Sci. Conf. 25–27 March 2003 pp. 293–306

Abeliovich A (1986) Algae in wastewater oxidation ponds. In: Richmond A (ed) Handbook of microbial mass culture. CRC Press, Boca Raton, pp 331–338

Abraham PJV, Butter RD, Sigene DC (1997) Seasonal changes in whole-cell metal levels in protozoa of activated sludge. Ecotox Environ Safe 38:272–280

Aharon A, Yosef A (1976) Toxicity of ammonia to algae in sewage oxidation ponds. Appl Environ Microbiol 31:801–806

Akpor OB (2011) Wastewater effluent discharge: effects and treatment processes. 3rd international conference on chemical biological and environmental engineering. Biol Environ Eng 20:85–91

Aksu Z, Dönmez G (2000) The use of molasses in copper (II) containing wastewaters: effects on growth and copper (II) bioaccumulation properties of *Kluyveromyces marxianus*. Process Biochem 36:451–458

Aksu Z, Dönmez G (2005) Combined effects of molasses sucrose and reactive dye on the growth and dye bioaccumulation properties of *Candida tropicalis*. Process Biochem 40:2443–2454

Ansa EDO, Lubberding HJ, Ampofo JA, Gijzen HJ (2011) The role of algae in the removal of *Escherichia coli* in a tropical eutrophic lake. Ecol Eng 37(2):317–324

Arvin E (1983) Observations supporting phosphate removal by biologically mediated chemical precipitation: a review. Water Sci Technol 15:43–63

Asulabh KS, Supriya G, Ramachandra TV (2012) Effect of salinity concentrations on growth rate and lipid concentration in *Microcystis* sp., *Chlorococcum* sp. and *Chaetoceros* sp. microalgae for use in tropical aquaculture. Proceedings of the National conference on conservation and management of wetland ecosystem Nov 6–9 lake kottayam Kerala. pp. 27–32

Azov Y (1982) Effect of pH on inorganic carbon uptake in algal cultures. Appl Environ Microbiol 43:1300–1306

Baba M, Snoeck R, Pauwels R, De Clercq E (1998) Sulfated polysaccharides are potent and selective inhibitors of various enveloped viruses, including herpes simplex virus, cytomegalovirus, vesicular stomatitis virus, and human immunodeficiency virus. Antimicrob Agents Chemother 32:1742–1745

Baicha Z, Salar-Garcia MJ, Ortiz-Martinez VM, Hernandez-Fernandez FJ, De los Rios AP, Labjar N, Lotfi E, Elmahi M (2016) A critical review on microalgae as an alternative source for bioenergy production: a promising low cost substrate for microbial fuel cells. Fuel Process Technol 154:104–116

Bansal A, Shinde O, Sarkar S (2018) Industrial wastewater treatment using phycoremediation technologies and co-production of value-added products. J BioremedBiodeg 9(1):1–10

Becker EW (1994) Microalgae, biotechnology and microbiology. Cambridge University Press, Cambridge. 10:1-291. Incomplete

Béress A, Wassermann O, Tahhan S, Bruhn T, Béress L, Kraiselburd N, Gonzales LV, Motta GE, Chavez PI (1993) A new procedure for the isolation of anti-HIV compounds (polysaccharides and polyphenols) from the marine alga *Fucus vesiculosus*. J Nat Prod 56:478–488
Bhaya D, Schwarz R, Grossman AR (2000) Molecular responses to environmental stress, in the ecology of cyanobacteria. Springer, Dordrecht, pp 397–442
Borowitzka MA (1988) Vitamins and fine chemicals from microalgae. In: Borowitzka MA, Borowitzka LJ (eds) Microalgal biotechnology. Cambridge University Press, Cambridge, pp 153–196
Burlew JS (1953) Algal culture from laboratory to pilot plant. Algal Culture 600(1):49–50
Canada Gazzette (2010) Wastewater systems effluent regulations. Regulatory impact analysis statement. Canada Gazzette 144:12–22
Cardozo KHM, Guaratini T, Barros MP, Falcao VR, Tonon AP, Lopes NP, Campos S, Torres MA, Souza AO, Colepicolo C, Pinto E (2007a) Metabolites from algae with economical impact. Comp Biochem Physiol Toxicol Pharmacol 146(2):60–78
Cardozo KHM, Guaratinin T, Barros MP, Falcao VR, Tonon AP, Lopes NP, Campos S, Torres MA, Souza AO, Coepicolo C, Pinto E (2007b) Metabolites from algae with economical impact. Comp Biochem Physiol Part C 146:60–78
Carlsson H, Aspegren H, Lee N, Hilmer A (1997) Calcium phosphate precipitation in biological phphosphorus removal systems. Water Res 31(5):1047–1055
Chaiwong K, Kiatsirioat T, Vorayos N, Thararax C (2013) Study of bio-oil and bio-char production from algae by slow pyrolysis. Biomass Bioenergy 56:600–606
Chevalier P, De la Noüe J (1985a) Wastewater nutrient removal with microalgae immobilized in carrageenan. Enzym Microb Technol 7:621–624
Chevalier P, De la Noüe P (1985b) Efficiency of immobilized hyperconcentrated algae for ammonium and orthophosphate removal from wastewaters. Biotechnol Lett 7:395–400
Chiu SY, Kao CY, Chen TY, Chang YB, Kuo CM, Lin CS (2015) Cultivation of microalgal *Chlorella* for biomass and lipid production using wastewater as nutrient resource. Bioresour Technol 184:179–189
Chojnacka K, Chojnacki A, Gorecka H (2005) Biosorption of cr3þ, cd2þ, and cu2þ ions by blue-green alga *Spirulina* sp.: kinetics, equilibrium and the mechanism of the process. Chemosphere 59:75–84
Christenson L, Sims R (2011) Production and harvesting of microalgae for wastewater treatment, biofuels and byproducts. Biotechnol Adv 29:686–702
Clarens AF, Nassau H, Resurreccion EP, White MA, Colosi LM (2011) Environmental impacts of algae-derived biodiesel and bioelectricity for transportation. Environ Sci Technol 45:7554–7560
Colak O, Kaya Z (1988) A study on the possibilities biological wastewater treatment using algae. Tur J Biol 12(1):18–29
Colley Davies RJ, Donnison AM, Speed DJ (2000) Towards a mechanistic understanding of pond disinfection. Water Sci Technol 42:149–158
Commault AS, Lear G, Novis P (2014) Photosynthetic biocathode enhances the power output of a sediment-type microbial fuel cell. N Z J Bot 52:48–59
Cossich ES, Tavares CRG, Ravagnani TMK (2002) Biosorption of chromium(III) by Sargassum sp. biomass. Electron J Biotechnol 5(2):133–140
Curtis TP, Mara DD, Silva SA (1992) Influence of pH, oxygen and humid substances on ability of sunlight to damage fecal coliforms in waste stabilization pond water. Appl Environ Microbiol 58:1335–1343
Day JG, Gong Y, Hu Q (2017) Microzooplanktonic grazers–a potentially devastating threat to the commercial success of microalgal mass culture. Algal Res 27:356–365
De Bashan LE, Bashan Y (2010) Immobilized microalgae for removing pollutants: review of practical aspects. Bioresour Technol 101(6):1611–1627
De la Noüe J, Laliberete´ G, Proulx D (1992) Algae and wastewater. J Appl Phycol 4:247–254
De la Noüe J, Proulx D (1988) Biological tertiary treatment of urban wastewater by chitosan-immobilized *Phormidium*. Appl Microbiol Biotechnol 29(2):356–365

De la Noüe J, Chevalier P, Proulx D (1990) Effluent treatment with immobilized microalgae and cyanobacteria: a critical assessment. In: Vembuk TRD (ed) Wastewater treatment by immobilized cells. CRC Press, Boca Raton, pp 143–152

De Morias MG, Costa JAV (2007) Carbon dioxide fixation by *Chlorella kessleri, C. vulgaris, Scenedesmus obliquus* and *Spirulina* sp. cultivated in the photobioreactors. Biotechnol Lett 29(9):1349–1352. Incomplete

Deng X, Wilson DB (2001) Bioaccumulation of mercury from wastewater by genetically engineered *Escherichia coli*. Appl Microbiol Biotechnol 56:276–279

DeSiloniz MI, Balsalobre L, Alba C, Valderrama MJ, Peinado JM (2002) Feasibility of copper uptake by the yeast *Pichia guilliermondii* isolated from sewage sludge. Res Microbiol 153:173–180

Doran MD, Boyle WC (1979) Phosphorus removal by activated algae. Water Res 13:805–812

Dubey SK, Dubey JS, Mehra S, Tiwar P (2011) Potential use of cyanobacterial sp. in bioremediation of industrial effluents. Afr J Biotechnol 10(7):1125–1132. Incomplete

El-kassas HY, Mohamed LA (2014) Bioremediation of the textile waste effluent by *Chlorella vulgaris*. Egypt J Aqua Res 40(3):301–308

EPA (2002) Onsite wastewater treatment systems manual. EPA/625/R-00/008/2002. http://www.epa.gov/owmitnet/mtbfact.htm

Fawcett D, Verduin JJ, Shah M, Sharma SB, Poinern GEJ (2017) A review of current research into the biogenic synthesis of metal and metal oxide nanoparticles via marine algae and seagrasses. J Nanosci 8013850:1–16

Fergusson JF, Jenkins D, Eastman J (1973) Calcium phosphate precipitation at slightly alkaline pH values. Water Pollut Cont 45(4):620–631

Fergusson JF, McCarty PL (1971) Effects of carbonate and magnesium on calcium phosphate precipitation. Environ Sci Technol 5(6):534–540

Fernández FGA, Sevilla JMF, Grima EM (2013) Photobioreactors for the production of microalgae. Rev Environ Sci Bio Technol 12(2):131–151

Figueira MMF, Volesky B, Azarian K, Ciminelli VST (1999) Multimetal biosorption in a column using *Sargassum* biomass. In: Amils R, Ballester A (eds) Biohydrometallurgy and the environment toward the mining of the 21st century (part B): international biohydrometallurgy symposium-proceedings. Elsevier Science, Amsterdam/The Netherlands, pp 503–512

Fogg GE (1975) Algal cultures and phytoplankton ecology, 2nd edn. The university of Wisconsin press, Wisconsin

Foladori P, Petrini S, Nesseuzia M, Anderottola G (2018) Enhanced nitrogen removal and energy saving in a microalgal-bacterial consortium treating real municipal wastewater. Water Sci Technol 78:174–182

Fontes AG, Vargas MA, Moreno J, Guerrero MG, Losada M (1987) Factors affecting the production of biomass by a nitrogen-fixing blue-green alga in outdoor culture. Biomass 13:33–43

Garbisu C, Gil JM, Bazin MJ, Hall DO, Serra JL (1991) Removal of nitrate from water by foam-immobilized *Phormidium laminosum* in batch and continuous-flow bioreactors. J Appl Phycol 3:1–14

Garnham GW, Codd GA, Gadd GM (1992) Kinetics of uptake and intracellular location of cobalt, manganese and zinc in the estuarine green alga *Chlorella salina*. Appl Microbiol Biotechnol 37:270–276

Gavrilescu M (2004) Removal of heavy metals from the environment by biosorption. Eng Life Sci 4:219–232

Glazer AN (1994) Phycobiliproteins-a family of valuable, widely used fluorophores. J Appl Phycol 6:105–112

Gomez MA, Gonzalez-Lopez J, Hontoria-Garcia E (2006) Influence of carbon source on nitrate removal of contaminated groundwater in a denitrifying submerged filter. J Hazard Mater B80(1):69–80

Gouveia L, Graça S, Sousa C, Ambrosano L, Ribeiro B, Botrel EP, Neto PC, Ferreira AF, Silva CM (2016) Microalgae biomass production using wastewater: treatment and costs scale-up considerations. Algal Res 16:167–176

Gray FN (2002) Water technology: an introduction for environmental scientists and engineers. Butterworth-Heinemann, Oxford, pp 35–80
Gray NF (1989) Biology of wastewater treatment. Oxford Univ Press, Oxford, pp 1057–1179
Gross M, Zhao X, Mascarenhas V, Wen Z (2016) Effects of the surface physic-chemical properties and the surface textures on the initial colonization and the attached growth in algal biofilm. Biotechnol Biofuels 9:38–52
Gude VG, Kokabian B, Gadhamshetty V (2013) Beneficial bioelectrochemical systems for energy, water, and biomass production. J Microb Biochem Technol S6:005
Guihéneuf F, Khan A, Tran LSP (2016) Genetic engineering: a promising tool to engender physiological, biochemical, and molecular stress resilience in green microalgae. Front Plant Sci 7:400
Gupta PL, Lee SM, Choi HJ (2015) A mini review: photobioreactors for large scale algal cultivation. World J Microbiol Biotechnol 31:1409–1417
Hall DO, Rao KK (1989) Immobilized photosynthetic membranes and cells for the production of fuel and chemicals. Chem Today 3:40–47
Han T, Haifeng L, Shanshan M, Zhang Y, Zhidan L, Na D (2017) Progress in microalgae cultivation photobioreactors and applications in wastewater treatment: a review. Int J Agric Biol Eng 10(1):1–25
NCBI (1995) Handbook N. Simple ncbi directory
Harun R, Singh M, Forde GM, Danquah MK (2010) Bioprocess engineering of microalgae to produce a variety of consumer products. Renew Sust Energ Rev 14:1037–1047
Hashimoto S, Furukawa K (1989) Nutrient removal from secondary effluent by filamentous algae. J Ferment Bioeng 67:62–69
Hempel F, Lau J, Klingl A, Maier UG (2011) Algae as protein factories: expression of a human antibody and the respective antigen in the diatom. *Phaeodactylum tricornutum*. PLoS One 6:e28424
Hlavova M, Turoczy Z, Bisova K (2015) Improving microalgae for biotechnology- from genetics to synthetic biology. Biotechnol Adv 33:1194–1203
Hoffman JP (2002) Wastewater treatment with suspended and nonsuspended algae. J Phycol 34(5):757–763
Horan NJ (1990) Biological wastewater treatment systems. Theory and operation. John Wiley and Sons Ltd, West Sussex
Hurse JT, Connor AM (1999) Nitrogen removal from wastewater treatment lagoons. Water Sci Technol 39(6):191–198
Ibañez E, Herrero M, Mendiola JA, Castro-Puyana M (2012) Extraction and characterization of bioactive compounds with health benefits from marine resources: macro and micro algae, cyanobacteria, and invertebrates. In: Marine bioactive compounds. Springer, Boston, pp 55–98
Ioannou E, Roussis V (2009) Natural products from seaweeds. In: Osbourn AE, Lanzotti V (eds) Plant-derived natural products. Springer, New York, pp 51–81
Ismail H, Azza AM, El-All ABD, Hassanein HAM (2013) Biological influence of some microorganisms on olive oil mill waste water. Egypt J Agric Res 91(1):1–9
Jarvie HP, Neal C, Warwick A, White J, Neal M, Wickham HD, Hill LK, Andrews MC (2002) Phosphorus uptake into algal biofilms in a lowland chalk river. Sci Total Environ 282–283:353–373
Jenkins D, Richard M, Daigger G (2003) Manual on the causes and control of activated sludge bulking foaming and other solids separation problems, 3rd edn. Lewis publishers CRC press, Boca Raton, pp 236–305
Jenkins D, Ferguson JF, Menar AB (1971) Chemical processes for phosphate removal. Water Res 5:369–389
Jinqi L, Houtian L (1992) Degradation of azo dyes by algae. Environ Pollut 75:273–278
Jjemba PK (2004) Interaction of metals and metalloids with microorganisms in the environment (chapter 12). In: Jjemba PK (ed) Environ microbiol—principles and applications. Science Publishers, New Hampshire, pp 257–270
Kim HW, Park S, Rittmann BE (2015) Multicomponent kinetic for the growth of the cyanobacterium *Synechocystis* sp. PCC6803. Environ Eng Res 20(4):347–355

Kocberber N, Donmez G (2007) Chromium (VI) bioaccumulation capacities of adapted mixed cultures isolated from industrial saline wastewaters. BioresourTechnol 98:2178–2183

Kris M (2007) Wastewater pollution in China. http://www.dbc.uci/wsustain/suscoasts/krismin.html

Kujan P, Votruba J, Kamenik V (1995) Substrate-dependent bioaccumulation of cadmium by growing yeast *Candida utilis*. Folia Microbiol 40(3):288–292

Kumar SK, Dahms HU, Won EJ, Lee JS, Shin KH (2015) Microalgae-a promising tool for heavy metal remediation. Ecotoxicol Environ Saf 113:329–352

Laliberte G, Olguin EJ, Noue JD (1997) Mass cultivation and wastewater treatment using Spirulina. In: Vonshak A (ed) *Spirulina platensis*. Physiology, cell biology and biotechnology. Taylor and Francis, London (UK), pp 59–73

Laliberte G, Proulx D, De Pauw N, La Noue J (1994) Algal technology in waste water treatment. Adv Limnol 42:283–302

Larsdotter K (2006) Microalgae for phosphorus removal from wastewater in a Nordic climate. A doctoral thesis from the school of biotechnology royal institute of technology, Stockholm Sweden ISBN: 91-7178-288-5

Laurens LM, Chen-Glasser M, McMillan JD (2017) A perspective on renewable bioenergy from photosynthetic algae as feedstock for biofuels and bioproducts. Algal Res 24:261–264

Lavoie A, De la Noüe J (1983) Harvesting microalgae with chitosan. J World Maricult Assoc 14:685–694

Leadbeater BSC, Callow ME (1992) Formation, composition and physiology of algal biofilms. In: Melo et al (eds) Biofilms science and technology. Kluwer Academic Publishers, Amsterdam Netherlands, pp 149–162

Li YX, Kim SK (2011) Utilization of seaweed derived ingredients as potential antioxidants and functional ingredients in the food industry: an overview. Food Sci Biotechnol 20:1461–1466

Lim S, Chu W, Phang S (2010) Use of *Chlorella vulgaris* for bioremediation of textile wastewater. BioresourTechnol 101:7314–7322

Lloyd JR (2002) Bioremediation of metals: the application of microorganisms that make and break minerals. Microbiol Today 29:67–69

Loukidou MX, Zouboulis AI (2005) Biosorption of toxic metals. Water Encyclopedia 2:68–74

Malik A (2004) Metal bioremediation through growing cells. Environ Int 30(2):261–278

Marbelia L, Bilad HR, Passaris I, Discart V, Vandamme D, Benckels A, Mylaert K, Vankelecom IF (2014) Membrane photobioreactors for integrated microalgae cultivation and nutrient remediation of membrane bioreactors effluent. Bioresour Technol 163:228–235

Markov SA, Bazin MJ, Hall DO (1995) Hydrogen, photoproduction and carbon dioxide uptake by immobilized *Anabaena variabilis* in a hollow-fiber photobioreactor. Enzym Microb Technol 17:306–310

Martin-Gonzalez A, Diaz S, Borniquel S, Gallego A, Gutierrez JC (2006) Cytotoxicity and bioaccumulation of heavy metals by ciliated protozoa isolated from urban wastewater treatment plants. Res Microbiol 157:108–118

Mata TM, Martins AA, Caetano NS (2010) Microalgae for biodiesel production and other applications. Renew Sust Energ Rev 14:217–232

Maynard HE, Ouki SK, Williams SC (1999) Tertiary lagoons: a review of removal mechanisms and performance. Water Res 33:1–13

Mehta SK, Gaur JP (2005) Use of algae for removing heavy metal ions from wastewater: progress and prospects. Crit Rev Biotechnol 25:113–152

Mendis E, Kim SK (2011) Present and future prospects of seaweeds in developing functional foods. Adv Food Nutr Res 64:1–15

Menicucci JA (2010) Algal biofilms, microbial fuel cells, and implementation of state of-the art research into chemical and biological engineering laboratories. (PhD dissertation) Montana State University Bozeman Montana

Mesple FC, Casellas M, Troussellier, Bontoux J (1996) Modelling orthophosphate evolution in a high rate algal pond. Ecol Model 89(1-3):13–21

Miranda AF, Ramkumar N, Andriotis C, Höltkemeier T, Yasmin A, Rochfort S, Wlodkowic D, Morrison P, Roddick F, Spangenberg G, Lal B, Subudhi S, Mouradov A (2017) Applications of microalgal biofilms for bioenergy production and wastewater treatment. Biotechnol Biofuels 10:120–129

Molazadeh M, Ahmadzadeh H, Pourianfar HR, Lyon S, Rampelotto PH (2019) The use of microalgae for coupling wastewater treatment with CO2 biofixation. Front Bioeng Biotechnol 7:42

Moheimani NR (2005) The culture of *Coccolithophorid* algae for carbon dioxide remediation. (PhD dissertation) Murdoch University Murdoch (Australia)

Monteiro CM, Castro PML, Malcata FX (2011) Biosorption of zinc ions from aqueous solutions by the microalga *Scenedesmus obliquus*. Environ Chem Lett 9:169–176

Morales J, De la Noüe J, Picard G (1985) Harvesting marine microalgae species by chitosan flocculation. Aquac Eng 4:257–270

Mostert ES, Grobbelaar JU (1987) The influence of nitrogen and phosphorus on algal growth and quality in outdoor mass algal cultures. Biomass 13:219–233

Mouchet P (1986) Algal reactions to mineral and organic micropollutants, ecological consequences and possibilities for industrial scale application; a review. Water Res 20:399–412

Moutin T, Gal JY, Halouani HE, Picot B, Bontoux (1992) Decrease of phosphate concentration inahighrate pond by precipitation of calcium phosphate:theoretical and experimental results. Water Res 26(11):1445–1450

Ogbonna JC, Yoshizawa H, Tanaka H (2000) Treatment of high strength organic wastewater by a mixed culture of photosynthetic microorganisms. J Appl Phycol 12:277–284

Okoh AT, Odjadjare EE, Igbinosa EO, Osode AN (2007) Wastewater treatment plants as a source of microbial pathogens in receiving water sheds. Afr J Biotechnol 6(25):2932–2944

Oliver RL, Ganf GG (2000) In: Whitton BA, Potts M (eds) Freshwater blooms, in the ecology of cyanobacteria: their diversity in time and space. Kluwer, Dordrecht, pp 149–194

Oswald WJ (1988) Microalgae and wastewater treatment. In: Borowitzka MA, Borowitzka LJ (eds) Microalgal biotechnology. Cambridge University Press, New York, pp 357–394

Otadi M, Poormohamadian S, Zabihi F, Goharrokhi M (2011) Microbial fuel cell production with alga. World Appl Sci 14:91–95

Palmer CM (1974) Algae in American sewage stabilization's ponds. Rev Microbiol (S-Paulo) 5:75–80

Pinto G, Pollio A, Previtera L, Stanzione M, Temussi F (2003) Removal of low molecular weight phenols from olive oil mill wastewater using microalgae. Biotechnol Lett 25(19):1657–1659

Pittman JK, Dean AP, Osundeko O (2011) The potential of sustainable algal biofuel production using wastewater resources. Bioresour Technol 102(1):17–25

Proulx D, LessardP DLNJ (1994) Tertiary treatment of secondarily treated urban wastewater by intensive culture of *Phormidium bohneri*. Environ Technol 15(5):449–458

Pulz O, Gross W (2004) Valuable products from biotechnology of microalgae. Appl Microbiol Biotechnol 65:635–648

Quijano G, Arcila JS, Buitorn G (2017) Microalgal-bacterial aggregates: applications and perspectives for wastewater treatment. Biotechnol Adv 35:772–781

Rai LC, Mallich N (1992) Removal and assessment of toxicity of cu & Fe to *Anabaena doliolum* & *Chlorella vulgaris* using free and immobilized cells. World J Microbial Technol 8:110–114

Rajamani S, Siripornadulsil S, Falcao V, Torres MA, Colepicolo P, Sayre R (2007) Phycoremediation of heavy metals using transgenic microalgae. In: León, R., Galván, Cejudo, a., Fernández, E. (Eds.). Transgenic microalgae as green cell factories. Adv Exp Med Biol 616:99–107

Rajasulochana AP, Dhamotharan R, Krishnamoorthy P, Subbiah M (2009) Antibacterial activity of the extracts of marine red and brown. J Am Sci 5(9):17–22

Ras M, Steyer JP, Bernard O (2013) Temperature effect on microalgae: a crucial factor for outdoor production. Rev Environ Sci Biotechnol 12(2):153–164

Rawat I, Kumar R, Bux F (2013) Phycoremediation by high-rate algal ponds (HRAPs). In: Bux F (ed) Biotechnological applications of microalgae: biodiesel and value-added products. CRC Press, Boca Raton, pp 179–199

Rehnstam Holm AS, Godhe A (2003) Genetic engineering of algal species. Eolss Publishers, Oxford, UK, pp 1–27

Romay C, González R, Ledón N, Remirez D, Rimbau V (2003) C-phycocyanin: a biliprotein with antioxidant, anti-inflammatory and neuroprotective effects. Curr Protein Pept Sci 4:207–216
Romera E, Gonzalez F, Ballester A, Blázquez ML, Muñoz JA (2006) Biosorption with algae: a statistical review. Crit Rev Biotechnol 26:223–235
Rosenberg JN, Oyler GA, Wilkinson L, Betenbaugh MJ (2008) A green light for engineered algae: redirecting metabolism to fuel a biotechnology revolution. Curr Opin Biotechnol 19:430–436
Rybicki S (1997) Advanced wastewater treatment: phosphorus removal from wastewater. Report no. 1. Royal Institute of Technology, Stockholm, Sweden
Saba B, Christy AD, Yu Z, Co AC (2017) Sustainable power generation from bacterio-algal microbial fuel cells (MFCs): an overview. Renew Sust Energ Rev 73:75–84
Sabalowsky AR (1999) An investigation of the feasibility of nitrification and denitrification of a complex industrial wastewater with high seasonal temperatures. Masters thesis from Virginia polytechnic institute and state university Blacksburg
Salama Y, Chennaoui M, Sylla A, Mountadar M, Riha M, Assobhei O (2014) Review of wastewater treatment and reuse in the Morocco: aspects and perspectives. Europ Cent Res Train Develop UK 2(1):9–25
Sawayama S, Minowa T, Dote Y, Yokoyama S (1992) Growth of the hydrocarbon-rich microalga *Botryococcus braunii* in secondarily treated sewage. Appl Microbiol Biotechnol 38:135–138
Sawayama S, Rao KK, Hall DO (1998) Nitrate and phosphate ions removal from water by Phormidium laminosum immobilized on hollow fibres in a photobioreactor. Appl Microbiol Biotechnol 49:463–468
Sebastian S, Nair KVK (1984) Total removal of coliforms and *E. coli* from domestic sewage by high-rate pond mass culture of *Scenedesmus obliquus*. Environ Pollut 34(A):197–206
Sekar S, Chandramohan M (2008) Phycobiliproteins as a commodity: trends in applied research, patents and commercialization. J Appl Phycol 20:113–136
Singh J, Gu S (2010) Commercialization potential of microalgae for biofuels production. Renew Sust Energ Rev 14(9):2596–2610
Singh RN, Sharma S (2012) Development of suitable photobioreactor for algae production - a review. Renew Sust Energ Rev 16(4):2347–2353
Song Y, Hahn HH, Hoffmann E (2002) Effects of solution conditions on the precipitation of phosphate for recovery: a thermodynamic evaluation. Chemosphere 48(10):1029–1034
Soni RA, Sudhakar K, Rana R (2016) Biophotovoltaics and biohydrogen through artificial photosynthesis: an overview. Int J Environ Sust Dev 15:313–325
Su HN, Xie BB, Chen XL, Wang JX, Zhang XY, Cheng Z, Zhang YZ (2010) Efficient separation and purification of allophycocyanin from *Spirulina* (Arthrospira) platensis. J Appl Phycol 22:65–70
Sud D, Mahajan G, Kaur MP (2008) Agricultural waste material as potential adsorbent for sequestering heavy metal ions from aqueous solutions-a review. Bioresour Technol 99:6017–6027
Tam NFY, Wong YS (2000) Effect of immobilized microalgal bead concentrations on wastewater nutrient removal. Environ Pollut 107(1):145–151
Tchobanoglous G, Burton FL, Stensel HD (2003) Wastewater engineering: treatment disposal reuse, 4th edn. Metcalf and Eddy, Inc., McGrwa-Hill Books Company, New York. isbn:0-07-041878-0
Tebbutt THY (1983) Principles of water quality control. Pergammon Press, Oxford USA, p 235
Travieso L, Benitez F, Dupeiron R (1992) Sewage treatment using immobilized microalgae. BioresourTechnol 40:183–187
Ugwu C, Aoyagi H, Uchiyama H (2008) Photobioreactors for mass cultivation of algae. Bioresour Technol 99:4021–4028
Van der Steen P, Brenner A, Shabtai Y, Oron G (2000) The effect of environmental conditions on FC decay in post-treatment of UASB reactor effluent. Water Sci Technol 42:111–118
VanLarsdrecht MC (2005) Role of biological processes in phosphate recovery. Natural History Museum, London
Vasumathi KK, Premalatha M, Subramanian P (2012) Parameters influencing the design of photobioreactor for the growth of microalgae. Renew Sust Energ Rev 16(7):5443–5550

Vela JC, Selles S, Pedreno JN, Bustamante MA, Mataic J, Gomez I (2006) Evaluation of composted sewage sludge as nutritional source for horticultural soils. Waste Manag 26(9):946–952

VenkataMohan S, Rohit MV, Chiranjeevi P, Chandra R, Navneeth B (2015) Heterotrophic microalgae cultivation to synergise biodiesel production with waste remediation: progress and perspectives. Bioresour Technol 184:169–178

Verma ML, Kumar S, Jeslin J, Dubey NK (2019) Microbial production of biopolymers with potential biotechnological applications. Biopolymer-based formulations: biomedical and food applications. Elsevier Publisher, Amsterdam, pp 1–43

Walker JD, Colwell RR, Petrakis L (1975) Degradation of petroleum by an alga, *Prototheca zopfii*. Appl Microbiol 30:79–81

Wang SK, Stiles AR, Guo C, Liu CZ (2014) Microalgae cultivation in photobioreactors: an overview of light characteristics. Eng Life Sci 14(6):550–559

WHO (2004) Guidelines for drinking water quality, vol 1. World health organization press, Geneva Switzerland, pp 1–631

Wijesekara I, Pangestuti R, Kim SK (2011) Biological activities and potential health benefits of sulfated polysaccharides derived from marine algae. Carbohydr Polym 84:14–21

Witvrouw M, De Clercq E (1997) Sulfated polysaccharides extracted from sea algae as potential antiviral drugs. Gen Pharmacol Vasc S 29:497–511

Wu YC, Wanga Z, Zheng Y, Xiao Y, Yang Z, Zhao F (2014) Light intensity affects the performance of photo microbial fuel cells with *Desmodesmus sp.* A8 as cathodic microorganism. Appl Energy 116:86–90

Xu H, Miao X, Wu Q (2006) High quality biodiesel production from a microalga *Chlorella protothecoides* by heterotrophic growth in fermenters. J Biotechnol 126:499–507

Yanna WHH, Hyde KD (2002) Fungal succession on fronds of Phoenix hanceana in Hong Kong. Fungal Divers 10:185–211

Yilmazer P, Saracoglu N (2009) Bioaccumulation and biosorption of copper(II) and chromium(III) from aqueous solutions by *Pichia stipitis* yeast. J ChemTechnol Biotechnol 84:604–610

Yoshida N, Ishii K, Okuno T, Tanaka K (2006) Isolation and characterization of a cyanophage infecting the toxic *Cyanobacterium microcystis aeruginosa*. Curr Microbiol 52(6):460–463

Yu KL, Lau BF, Show PL, Ong HC, Ling TC, Chen WH, Salleh MAM (2017) Recent developments on algal biochar production and characterization. Bioresour Technol 246:2–11

Zhang C, Li X, Kim SK (2012) Application of marine biomaterials for nutraceuticals and functional foods. Food Sci Biotechnol 21:625–631

Zhu W, Ooi VE, Chan PK, Ang POJ (2003) Isolation and characterization of a sulfated polysaccharide from the brown alga *Sargassum patens* and determination of its anti-herpes activity. Biochem Cell Biol 81:25–33

Microbial Remediation for Wastewater Treatment

3

Satyender Singh, Simranjeet Singh, Vijay Kumar, Sanjay Kumar, Daljeet Singh Dhanjal, Romina Romero, Shivika Datta, Pooja Bhadrecha, and Joginder Singh

Abstract

Discharging of industrial effluents directly into water bodies is a global concern for aquatic and terrestrial biota. Various methods like physical and chemical have been implemented so far, but these existing technologies are sometimes restricted of either technical or economic constraints or are expensive and unsustainable approaches. Bioremediation offers a promising means to reclaim such

Satyender Singh, Simranjeet Singh, and Vijay Kumar contributed equally to this work.

S. Singh
Regional Advanced Water Testing Laboratory (RAWTL), Mohali, Punjab, India

S. Singh
Regional Advanced Water Testing Laboratory (RAWTL), Mohali, Punjab, India

Department of Biotechnology, Lovely Professional University, Phagwara, Punjab, India

Punjab Biotechnology Incubators, Mohali, Punjab, India

V. Kumar
Regional Ayurveda Research Institute for Drug Development, Gwalior, Madhya Pradesh, India

S. Kumar
Regional Advanced Water Testing Laboratory (RAWTL), Mohali, Punjab, India

Punjab Biotechnology Incubators, Mohali, Punjab, India

Department of Natural Products, NIPER, Mohali, Punjab, India

D. S. Dhanjal · P. Bhadrecha · J. Singh (✉)
Department of Biotechnology, Lovely Professional University, Phagwara, Punjab, India
e-mail: joginder.15005@lpu.co.in

R. Romero
Technological Development Unit-UDT, Universidad de Concepcion, Coronel, Chile

S. Datta
Department of Zoology, Doaba College, Jalandhar, Punjab, India

P. K. Arora (ed.), *Microbial Technology for Health and Environment*, Microorganisms for Sustainability 22,
https://doi.org/10.1007/978-981-15-2679-4_3

contaminated water bodies in an economical and ecofriendly way. It is an emerging technology and uses living organisms to manage or remediate polluted soils or wastewater. It is defined as the elimination, attenuation, or transformation of polluting or contaminating substances by the use of biological processes. In this book chapter, we will review the potent role of bacterial species to confer remediation of various pollutants in wastewater. Efforts have been made to summarize the new aspects of bioremediation in mitigating the effects of various hazardous contaminants from wastewater and their limitations.

Keywords
Wastewater · Bioremediation · Microbial population · End products

3.1 Introduction

Over the past few decades, water pollution has become a major environmental problem worldwide and has attained considerable attention (Wang and Yang 2016). The rapid growth of population, urbanization, automation, and mining are factors which lead to widespread water pollution (Singh et al. 2017, 2018). Water pollution is increasing day by day because of the discharge of chemical wastes from industries and domestic wastes from home into drains (Florescu et al. 2010). Additionally, regulated and unregulated waste disposal, accidental and process waste spillage, mining and smelting of metalliferous ores, and usage of sewage sludge in agricultural field are few factors which are responsible for discharging pollutants into clean water bodies sites as leachate, which results in the contamination of our ecosystem (Ogbonna et al. 2006; Naruka 2012).

Due to rapid industrialization, the amount of toxic heavy metals in water has elevated and cause physiological, genetic, and ecological problems (Jan et al. 2015). Entering of heavy metals in water has affected the seed mortality and hindered the standard revegetation scheme. As, Cr, Cu, Cd, Pb, and Hg are the heavy metals which are being discharged by agricultural, domestic, and industrial effluents, and as a result, they are contaminating groundwater, freshwater sources, and other water bodies (Kaur et al. 2018; Kumar et al. 2016, 2017; Singh et al. 2016). The ingestion of these toxic metals by drinking contaminated water and eating contaminated food can get start accumulating in human and animal, which can further lead to severe consequences (Wu et al. 2014). Metals like Cd, Cu, Pb and Zn are highly mobile, soluble, and predominantly available in nature, especially in water bodies, and can cause severe damage to the environment as well as human health (Barakat 2011).

Heavy metals behold the major part of inorganic contaminants and present different challenges than organic pollutants (Mishra et al. 2016; Kumar et al. 2015a, b). Even metals are necessarily required in trace amount by both animals and plants, but at high concentration, they become toxic and induce oxidative stress due to free radical formation (Bhat and Khan 2011). Another reason which makes these heavy metals toxic is that they either replace the valuable metal from the pigments or disrupt the functional enzyme (Kumar et al. 2013, 2014a, b). Thus, contamination of heavy metals makes the land barren for plant growth and disturbs the microbial biodiversity (Ayangbenro and Babalola 2017). Therefore, a regulatory mechanism is required to reduce the release of these pollutants in the soil; however reported approaches are still not sufficient to keep the regular checkup of contaminants (Tangahu et al. 2011).

Biological function and chemical properties of heavy metals help in complex group formation, whereas metal toxicity varies according to their concentration (Chibuike and Obiora 2014). Many of them like Cd, Co, Cr, Cu, Hg, Ni, Pb, and Zn are highly hazardous in both soluble and elemental form. Even the trace amount of these heavy metals in the environment can cause a severe problem in living being (Singh et al. 2011). Bioaccumulation of these heavy metals in the food chain has become a significant threat to the health of humans (Agrawal et al. 2007). The route by which these toxic elements enter in the human body is through ingestion of contaminated food or water. As we know, heavy metals can be destroyed but can be converted into less toxic by changing its oxidation state (Jaishankar et al. 2014).

The number of anthropogenic activities in the coastal region has led to the substantial discharge of toxic chemical and heavy metals within industrial effluent into coastal water bodies (Sharifuzzaman et al. 2016). Liberation of these hazardous substances in the environment induces different toxic effects via bioaccumulation or biomagnification (Yu et al. 2013; Genuis and Kelln 2015). The soil has become the conventional place for the disposal of these heavy metals, which treatment has become the major challenge (Chibuike and Obiora 2014). At present, various conventional remediation procedures developed are not ecofriendly but are expensive (Balakrishnan and Velu 2015). Naturally these heavy metals are present in complex or bound form like metal sulfate, metal chloride, and metal oxide in the soil (Wuana and Okieimen 2011). Various human activities like mining, purification of heavy metals, production of steel and other metals, burning of waste material, discharge of industrial effluents, and acute use of chemicals fertilizer, pesticide and sewage water also contaminates farming practices thus affecting ecosystem (Dixit et al. 2015). Table 3.1 enlists the organic pollutants present in domestic, hospital and industrial wastes.

3.2 Wastewater Pollution Sources

Wastewater is an intricate blend of both inorganic and organic materials and is subdivided into sewage and industrial wastewater (Naidoo and Olaniran 2014), and its sources were depicted in Fig. 3.1. Wastewaters comprise of water in which solids exist as settleable particles, scattered as colloids, which are materials that don't settle promptly, or solids immersed in water (Shon et al. 2006). Sewage wastewater originates from households and mainly comprises of water along with small concentrations of dissolved organic and inorganic solids (Praveen et al. 2016). Organic contaminants mostly include lignin, proteins, synthetic detergents, soaps, and other organic chemicals (Oller et al. 2011). Sewage wastewater is also reported for the presence of inorganic substances such as zinc, mercury, lead, chromium, copper, etc., which are likely to affect both aquatic and terrestrial biota (Westerhoff et al. 2015).

Table 3.1 Various organic pollutants in wastewaters

Wastewater source	Organic pollutants	References
Hospital wastewater	Methylene chloride, xylene, octadecanoic acid, butylated hydroxyl toluene, 1-tetradecene	Wyasu and Kure (2012)
Domestic wastewater	Steroid hormones (e.g., 17ß-estradiol, estrone, estriol, progesterone, testosterone, 17α-ethinylestradiol, diethylstilbestrol), detergent residues (e.g., linear alkyl benzene sulfonates, alkylphenols), antimicrobial agents (e.g., triclosan, TCS), musk fragrances, sunscreens, brominated flame retardants, plasticizers, *N*-nitrosodimethylamine, tryptophan-like C2 compound	Shareef et al. (2008)
Textile industry	Volatile organic compounds, halogenated anilines, benzenes, anthraquinones, alkylated phenols, phthalic acid esters	Dsikowitzky and Schwarzbauer (2013)
Tanneries	Naphthalene sulfonates, benzene sulfonates, anthraquinone sulfonates, 6-acylamino-3-amino-naphthalene-2-sulfonic acid, *N*-(bis-hydroxymethyl-phenyl)-acetate, aromatic amines, anilines, nonylphenol, ethoxylates, diethyl phthalate, bis(ethylhexyl) phthalate, carboxylated polyethylene glycols, cyclohexane, aromatic carboxylic acids, phenols, indoles, ethoxylates	Dsikowitzky and Schwarzbauer (2013)
Petrochemical industries	Alkylated benzenes, indane, alkylated derivatives of indane, alkylated naphthalene, quinoline derivatives, xylene, toluene, benzene, indoline, acenaphthene, acenaphthylene, fluorene, styrene, methyl styrene, phenol, nitrobenzene, phthalic acid ester derivatives (Dsikowitzky and Schwarzbauer 2013)	Dsikowitzky and Schwarzbauer (2013)
Paper and pulp industry	Resin acids, lignin, terpene, catechol, hydroxybenzaldehyde, bisphenol A, nonylphenol ethoxy carboxylates, acetyloxy trimethyl bicycloheptanedione, acetylmorpholine	Dsikowitzky and Schwarzbauer (2013)
Rubber and tire production	Benzothiazoles, aniline derivatives, phthalic ester derivatives, diphenylamine, propyldiphenylamine, naphthalene, phenanthrene, pyrene, fluoranthene	Dsikowitzky and Schwarzbauer (2013)
Chemical production sites	Nitrogenous heterocyclic compounds, sulfur compounds, chlorinated compounds, nitroaromatic compounds, alkyl phosphates, trimethyl pentanediol-iso-butyrate, hexathiepene, naphthalene sulfonates, benzene sulfonates	Dsikowitzky and Schwarzbauer (2013)
Pharmaceutical industry	Antihistamine cimetidine, bromazepam, diclofenac, metharbital, cortisol, cortisone, dexamethasone, prednisolone	Dsikowitzky and Schwarzbauer (2013)

Industrial wastewater is produced during cleaning activities or manufacturing process in the industrial sector. Such industries discharge a large amount of chromium, zinc, nickel, cadmium, titanium, iron, and other compounds which are highly toxic and even carcinogenic (Hanchang 2009; Bazrafshan et al. 2015). Wastewater

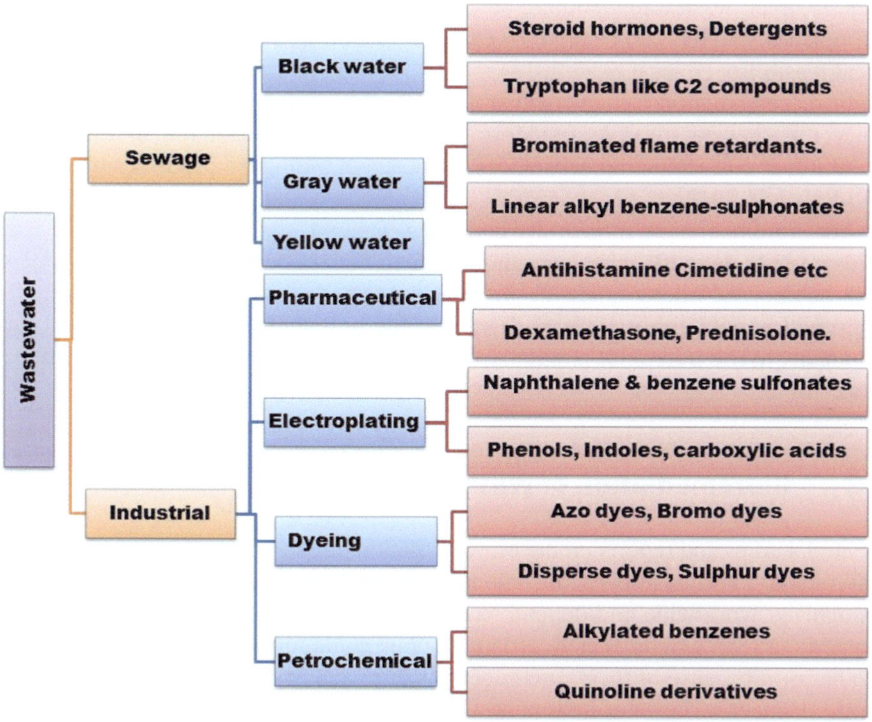

Fig. 3.1 Types of wastewater, their sources, and chemical components present in it

generated from tannery and pharmaceutical industries has a very poor degradability nature because of high biological oxygen demand (BOD) and chemical oxygen demand (COD) (Kavitha et al. 2012).

The lack of wastewater treatment facilities forces large parts of the developing and underdeveloped countries to discharge openly into water bodies such as rivers, streams, and lakes (Cashman et al. 2014). Various diseases and health problems have often been caused by these activities and thus destroy life forms of aquatic ecosystems. The presence of such components (both organic and inorganic) in wastewater poses a worldwide challenge to treat wastewater (Akpor et al. 2014). However, various physical and chemical treatment processes are implemented but are costly, generate wastewater solids (sludge), and are further disposed of in an inappropriate manner (Balakrishnan and Velu 2015). Sewage sludge is sometimes deprived of heavy metals and mostly contains carbon, nitrogen, and nutrients, so it can be used as beneficial manure/fertilizer for plant growth, improving its soil structures, etc. But industrial sludge comprises of high concentration of toxic chemicals or heavy metals which became an increasing problem for treatment plants (Mtshali et al. 2014).

Both of the physical and chemical approaches had certain disadvantages making it a challenging task in terms of ecological and economic terms. Therefore, the first step to treat wastewater should be the evaluation of ecofriendly and possible approaches, and this can be done by checking the desired purification rate and technical feasibility (Bansal et al. 2018).

In bioremediation living organisms are used to manage or remediate polluted soils or wastewater into less toxic forms. It is the deletion, reduction, or alteration of polluting or contaminating substances by means of biological processes (Ojuederie and Babalola 2017).

For bioremediation, a naturally occurring bacterial and fungi or plants are used to degrade or detoxify substances hazardous to human health and the environment. The potency of microorganisms for this process depends on the existence of a microbial population capable of degrading the pollutants, the availability of contaminants to the microbial population, and the environmental factors (Chowdhury et al. 2012).

3.3 Bioremediation of Wastewater

Several studies and reports on the bioremediation of contaminants in wastewater by different bacteria have been documented. A number of microbial populations are recognized worldwide and depend exclusively on the contaminants present in the wastewater (Silva-Bedoya et al. 2016). Organic contaminants present in wastewater are harmful because of their carcinogenic nature and acute toxicity, throughout the world. Hence, wastewater treatment, with low-cost investments, is a must to solve such problems (Zheng et al. 2013). The ability of nature and natural conditions to degrade a substance chemically is termed as biodegradation. Biodegradable pollutants are generally from plants, animals, and various substances from different living organisms (Joutey et al. 2013; Yavari et al. 2015). Activated sludge process helps achieve almost 80% removal of biodegradable matter from the polluted water (Nagwekar 2014). Nature has beforehand gifted these water bodies with microorganisms capable of degrading such organic wastes up to some levels. But, increased organic wastes in water demand for higher oxygen for such microorganisms which diminish the dissolved oxygen in the water, leading to adverse effects on other aquatic organisms and the overall degradation of water quality (Abdel-Raouf et al. 2012). Utilization of microorganisms like bacteria, fungi, algae, protozoa, and rotifers in wastewater treatment plants for degradation of organic pollutants present in wastewater is termed "microbial remediation." Proper maintenance of conditions like pH, dissolved oxygen, and nutrients for the selected microorganisms result in excellent outcomes, as depicted in Table 3.2 (Akpor et al. 2014).

Physical, chemical, and biological processes are the options to degrade organic pollutants. Since the end products of physical and chemical processes are toxic to the environment, bioremediation stands out as an ecofriendly and cost-effective option (Mani and Kumar 2014). Plants, microorganisms, and useful biomolecules produced by them, e.g., enzymes, are used in biodegradation (Ratnakar et al. 2016). Adsorption being a cost-effective method for such problems, and since the requirements are universally available and not expensive, serves as a promising solution for

Table 3.2 Microorganisms used for wastewater treatment (adopted from Akpor et al. 2014)

Microorganism	Use in wastewater treatment
Bacteria	Mainly aerobic bacteria are utilized to degrade organic pollutants in wastewater treatment
Fungi	Enzymes produced by fungi are capable of degrading compounds not easily degraded by bacteria. Also, fungi produce proteins, organic acids, chitins, amylase, glucosamine, antimicrobial compounds, and various other metabolites to adapt in harsh conditions, enhancing their survival rates. They degrade the compounds by adsorption, chemisorption, chelation, and microprecipitation
Protozoa	Ciliated protozoa, capable of growth on water surfaces and feeding on decaying vegetation and microbes, are a great help in purification and matter cycling
Rotifers	These microscopic aquatic animals generally found in freshwater and moist soil are capable of stabilizing organic wastes, stimulating microfloral activities, decomposition, enhanced oxygen, and recycling mineral nutrients. They increase oxygen content in activated sludge
Algae	Capable of assimilating nitrogen as nitrate, nitrite, and ammonium, microalgae also remove organic matter, xenobiotics, and metals in wastewater treatment systems. They can also be easily outcompeted by other microbes for essential nutrients

wastewater treatment. Different physical forces like Van der Waals forces, hydrophobicity, hydrogen bonds, polarity, steric and dipole interactions, etc. controlled the adsorption process. Since the degree of liquid packing in pores determines the extent of adsorption, the adsorbent surface and adsorbate should have comparable pore size (Sabir 2015). Ali et al. (2012) studied the efficiency of household wastes, agricultural products, sea materials, soil and ore materials, metal oxides, and hydroxide waste as adsorbents in wastewater treatment methods. Through this study, they were successful in using these natural waste products as efficient adsorbents for the treatment of wastewater. Pollutants like polycyclic aromatic hydrocarbons, volatile organic compounds, trichloroethylene, trichloroethane cis-DCE (cis-dichloroethylene), cis-DCE vinyl chloride, hexavalent chromium, cadmium, polycyclic aromatic hydrocarbons, pentachlorophenol, benzo(a)pyrene, polychloroethene, and polychlorinated biphenyls are biodegradable (Megharaj et al. 2011). Most of the biodegradable organic matter is readily degraded by biological treatment processes (Fig. 3.2). But non-biodegradable pollutants may require processes like chemical oxidation and ozonation to convert the non-biodegradable matter into biodegradable by-products like aldehydes, ketones, organic acids, and other such small molecules (Van Leeuwen et al. 2009). Applying ozone to activated sludge oxidizes synthetic organic compounds to biodegradable by-products, e.g., formation of biodegradable formic acid while ozonating methylene blue dye (Van Leeuwen et al. 2009; Wang et al. 2018). Bioremediation techniques have become the most popular method of degrading organic pollutants for wastewaters (Azubuike et al. 2016). Few examples are listed in Table 3.3.

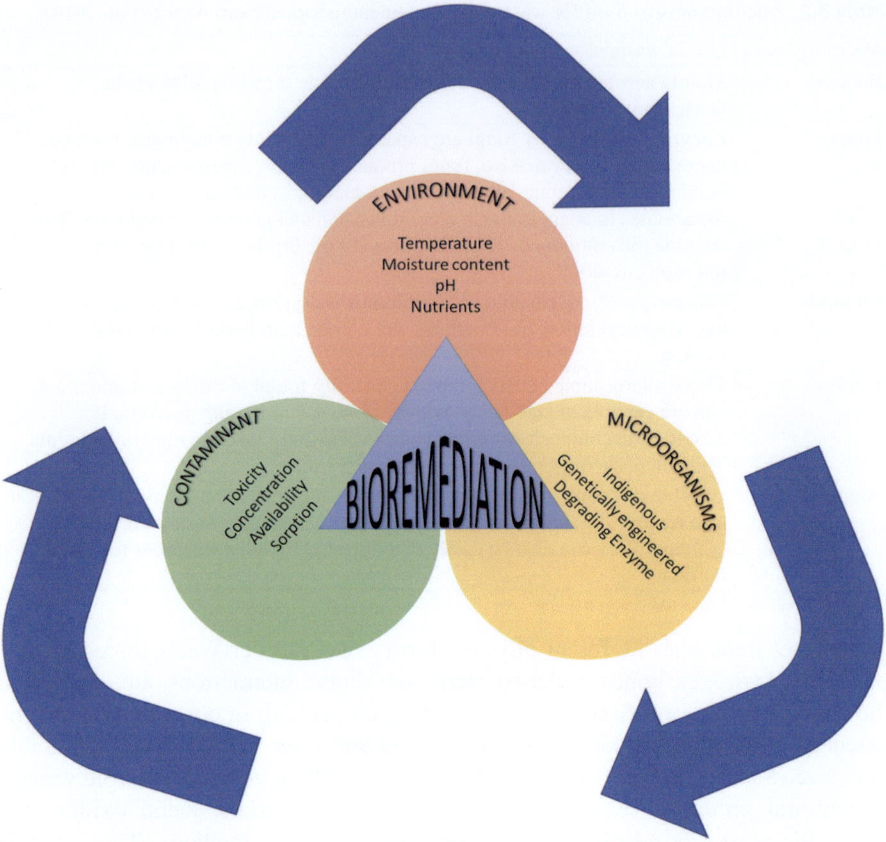

Fig. 3.2 Bioremediation cycle: general overview

3.4 Microbial Enzymes in Biodegradation

Microorganisms' dwelling at extreme environments often produces catabolic enzymes to survive there and utilize the compounds available in the vicinity. A mixed culture offers higher enzyme production and biomass activities. Therefore the use of microorganisms and extracellular enzymes produced by them has gained popularity (Kumar et al. 2011). Enzymes immobilized on the glass and silica beads give even better results and produce compounds which can be easily biodegraded (Dhall et al. 2012; Ratnakar et al. 2016).

Possessing at least one polypeptide moiety, an enzyme can be a protein or a glycoprotein capable of degrading harmful, insoluble compounds into easily degradable molecules with lowered environmental threat (Karigar and Rao 2011). Microbial enzymes capable of degrading organic pollutants present in wastewaters can be broadly classified into two main categories, viz., hydrolases and oxidoreductases (Ratnakar et al. 2016).

Table 3.3 Methods of degrading organic pollutants by use of microorganisms

Method	Pollutant	Microorganisms/compounds	References
Biobarrier system (Bioreactive permeable barrier)	Dieldrin	It utilizes native microorganisms grown over solid particles as a catalyst	Cardona and Suarez (2010)
Microbial consortium treatment of water in a suspension form	Petroleum oil and phenol	*Alcaligene odorans*, *B. subtilis*, *Corynebacterium propinquum*, and *Pseudomonas aeruginosa*	Singh et al. (2013)
Immobilization methodology using the mixed bacterial consortium	Aliphatic and aromatic petroleum hydrocarbons	*Bacillus brevis* and *Pseudomonas aeruginosa* KH6	El-Borai et al. (2016)
Fungal biosorption	Textile effluents	Lyophilized biomass of *Cunninghamella elegans* (MUT2861)	Tigini et al. (2010)
Bioconsortium	Toluene, O-xylene	*Pseudomonas putida*, *Candida membranes*, *Penicillium* sp.	Jecu et al. (2008)
Bioremediation	Lipids	Lipase producing *Bacillus subtilis*, *Bacillus licheniformis*, *Bacillus amyloliquefaciens*, *Serratia marcescens*, *Pseudomonas aeruginosa*, *Staphylococcus aureus*	Prasad and Manjunath (2011)
Enzymatic degradation	Proteinaceous wastes in tannery saline wastewater	Salt tolerant, alkaline protease producing *P. aeruginosa*	Sivaprakasam et al. (2011)
Bioaugmentation by sludge hammer bacteria	TOC (total organic carbon)	*Bacillus subtilis*, *Brevibacillus laterosponus*, *Pseudomonas aeruginosa*	Hesnawi et al. (2014)
Bacterial consortium	COD, BOD, MLSS (mixed liquor suspended solids), TSS	*Bacillus pumilus*, *Brevibacterium* sp., *Pseudomonas aeruginosa*	Dhall et al. (2012)
Decolorization and detoxification using activated sludge	Dyes, additives, salts	*Bjerkandera adusta* MUT3060	Anastasi et al. (2011)
Bioremediation using fungi	Textile azo dyes	White-rot fungi *Marasmius* and *Trametes hirsuta* with lignite xylite and lignite granules	Böhmer et al. (2010)
Sequencing batch reactor	Blue Bezaktiv S-GLD 150 dye	Microbial consortium 'Bx'	Khouni et al. (2012)
Membrane bioreactor, cometabolic transformation	Ketoprofen, bezafibrate, naproxen, ibuprofen	Microbial metabolites: 3-(hydroxy-carboxy-methyl)-hydratropic acid and 3-(keto-carboxymethyl)-hydratropic acid	Quintana et al. (2005)

1. Hydrolase enzymes like cellulase, hemicellulase, and glycosidase act on the chemical linkages in complex toxic pollutants and degrade them to simpler forms (de Lourdes Moreno et al. 2013).
   (a) Lipase: Lipids are present in the plant, animal, and microbial wastes, which are broken down by lipase enzymes that hydrolyze triacylglycerols to glycerol and fatty acids. Bacteria and actinomycetes produce diverge lipases capable of esterification, alcoholysis, and aminolysis (Sharma et al. 2011).
   (b) Cellulase: Microbial cellulases are produced as cell-bound, extracellular, or cell envelope-associated and are capable of breaking down cellulose to simpler reduced sugars. Cellulases are a combination of enzymes like endoglucanase, exoglucanase, and ß-glucosidase, which act together to hydrolyze cellulose (de Lourdes Moreno et al. 2013).
   (c) Protease: Wastes from industries like food, pharmaceuticals, leather, detergent, fishery, and poultry contain proteins, which can be catalyzed by microbial proteases. Bacterial proteases are exclusively used to hydrolyze waste proteins, and the degraded molecules are consumed by bacteria (Beena and Geevarghese 2010).
2. Oxidoreductase enzymes are utilized to degrade xenobiotics, phenolic compounds, and azo dyes. Since a fungus has a greater surface area to contact the pollutants and secretes enzymes, oxidoreductase producing fungus is employed to secrete ligninolytic enzymes like laccases, lignin peroxidase, and manganese peroxidase (Husain 2006).
   (a) Oxygenase: Oxygenase enzymes catalyze oxygenation of reduced substrates by transferring oxygen. They metabolize organic compounds by cleaving aromatic rings and by increasing the compounds' water solubility. They also catalyze dehalogenation reactions of halogenated methanes, ethanes, and ethylenes. Bacterial mono- and dioxygenases are widely explored and utilized (Yagi and Madsen 2009).
   (b) Monooxygenases: Monooxygenases are a versatile group of enzymes which catalyze oxidative reactions of alkanes, steroids, and fatty acids by incorporating single oxygen atoms in the substrate. Dehalogenation, desulfurization, ammonification, denitrification, biotransformation, hydroxylation, and biodegradation of aromatic and aliphatic compounds are done by these enzymes (Chen et al. 2011).
   (c) Dioxygenases: They can catalyze oxygenation of a diverse range of substrates including aromatic compounds and hence are widely utilized in environmental remediation (Karigar and Rao 2011).
   (d) Laccases: Bacteria, fungi, insects, and plants produce laccases which oxidize reduced phenolic and aromatic substrates. They can oxidize ortho- and para-diphenols, aminophenols, polyphenols, polyamines, lignin, aryl diamines, and phenolic and methoxy phenolic acids. These enzymes are also capable of decarboxylation and demethylation (Andualema and Gessesse 2012).
   (e) Peroxidases (hydrogen peroxide oxidoreductases): These heme and non-heme protein-containing enzymes catalyze the oxidation of lignin and phe-

nolic compounds. They are categorized as lignin peroxidases, manganese peroxidases, and versatile peroxidases, based on their activity (Husain 2010).

3.5 Limitations of Bioremediation

Bioremediation utilizes microorganisms for degradation of hazardous compounds and other contaminants like POPs, PCB, PAH, etc. in soil or water (Othman et al. 2011). But biodegradation also has its own limitations and shortcomings, some of which are discussed herein. Biodegradation is a time-consuming process which is applicable only to biodegradable compounds, Biodegradation processes are strictly bound to the maintenance of proper environmental conditions like temperature, pH, and physical interaction beween microbes and compounds. (Azubuike et al. 2016). Sometime the process of biodegradation remains incomplete which results in partially broken compounds of contaminants which get converted to even more toxic by-products (Boopathy 2000).

3.6 Conclusions

Naturally convenient methods of biodegradation have gained popularity nowadays, and these methods are being commercialized because of low investments and better results which are comparatively lesser harmful to the environment as compared to the other available chemical and physical processes. A wide range of bioremediation process has been developed for the treatment of hazardous organic wastes in water using natural and genetically modified microorganisms. Various microorganisms are also genetically modified in the last few decades to be employed for the degradation of toxic organic pollutants present in the environment. Like any other treatment processes, biodegradation also has few limitations like partial degradation of organic wastes, production of toxic by-products, limited availability of chemicals to maintain the required environmental conditions for degradation, etc.

References

Abdel-Raouf N, Al-Homaidan AA, Ibraheem IBM (2012) Microalgae and wastewater treatment. Saudi J Biol Sci 19(3):257–275

Agrawal SB, Singh A, Sharma RK, Agrawal M (2007) Bioaccumulation of heavy metals in vegetables: a threat to human health. Terr Aquat Environ Toxicol 1(2):13–23

Akpor OB, Otohinoyi DA, Olaolu DT, Aderiye BI (2014) Pollutants in wastewater effluents: impacts and remediation processes. Int J Environ Res Earth Sci 3(3):050–059

Ali I, Asim M, Khan TA (2012) Low cost adsorbents for the removal of organic pollutants from wastewater. J Environ Manag 113:170–183

Anastasi A, Parato B, Spina F, Tigini V, Prigione V, Varese GC (2011) Decolourisation and detoxification in the fungal treatment of textile wastewaters from dyeing processes. New Biotechnol 29(1):38–45

Andualema B, Gessesse A (2012) Microbial lipases and their industrial applications. Biotechnology 11(3):100–118
Ayangbenro A, Babalola O (2017) A new strategy for heavy metal polluted environments: a review of microbial biosorbents. Int J Environ Res Public Health 14(1):94
Azubuike CC, Chikere CB, Okpokwasili GC (2016) Bioremediation techniques–classification based on site of application: principles, advantages, limitations and prospects. World J Microbiol Biotechnol 32(11):180
Balakrishnan H, Velu R (2015) Eco-friendly technologies for heavy metal remediation: pragmatic approaches. In: Environmental sustainability. Springer, New Delhi, pp 205–215
Bansal A, Shinde O, Sarkar S (2018) Industrial wastewater treatment using phycoremediation technologies and co-production of value-added products. J Bioremed Biodegr 9(428):2
Barakat MA (2011) New trends in removing heavy metals from industrial wastewater. Arab J Chem 4(4):361–377
Bazrafshan E, Mohammadi L, Ansari-Moghaddam A, Mahvi AH (2015) Heavy metals removal from aqueous environments by electrocoagulation process–a systematic review. J Environ Health Sci Eng 13(1):74
Beena AK, Geevarghese PI (2010) A solvent tolerant thermostable protease from a psychrotrophic isolate obtained from pasteurized milk. Developm Microbiol Molecul Biol 1:113–119
Bhat UN, Khan AB (2011) Heavy metals: an ambiguous category of inorganic contaminants, nutrients and toxins. Res J Environ Sci 5(8):682–690
Böhmer U, Kirsten C, Bley T, Noack M (2010) White-rot fungi combined with lignite granules and lignitic xylite to decolorize textile industry wastewater. Eng Life Sci 10(1):26–34
Boopathy R (2000) Factors limiting bioremediation technologies. Bioresour Technol 74(1):63–67
Cardona S, Suarez E (2010) Biodegradation pathway prediction of POPs (Persistent Organic Pollutants) and biobarrier treatment. Dyna 77(163):115–123
Cashman D, Foster C, McCluskey K, Zhang Y (2014) Identifying Opportunities to Reduce Water Pollution and Encourage Voluntary Compliance in Windhoek, Namibia. Undergraduate Interactive Qualifying Project No. E-project-050814-051919). Retrieved from Worcester Polytechnic Institute Electronic Projects Collection: http://www.wpi.edu/Pubs/E-project/Available/E-project-050814-051919/unrestricted/City_-_Final_IQP_Report.pdf
Chen Y, Patel NA, Crombie A, Scrivens JH, Murrell JC (2011) Bacterial flavin-containing monooxygenase is trimethylamine monooxygenase. Proc Natl Acad Sci 108(43):17791–17796
Chibuike GU, Obiora SC (2014) Heavy metal polluted soils: effect on plants and bioremediation methods. Appl Environ Soil Sci 2014:752708
Chowdhury S, Bala NN, Dhauria P (2012) Bioremediation–a natural way for cleaner environment. Int J Pharmaceut Chem Biol Sci 2(4):600–611
de Lourdes Moreno M, Pérez D, García MT, Mellado E (2013) Halophilic bacteria as a source of novel hydrolytic enzymes. Lifestyles 3(1):38–51
Dhall P, Kumar R, Kumar A (2012) Biodegradation of sewage wastewater using autochthonous bacteria. Sci World J 2012:861903
Dixit R, Malaviya D, Pandiyan K, Singh U, Sahu A, Shukla R, Singh B, Rai J, Sharma P, Lade H, Paul D (2015) Bioremediation of heavy metals from soil and aquatic environment: an overview of principles and criteria of fundamental processes. Sustain For 7(2):2189–2212
Dsikowitzky L, Schwarzbauer J (2013) Organic contaminants from industrial wastewaters: identification, toxicity and fate in the environment. In: Pollutant diseases, remediation and recycling. Springer, Cham, pp 45–101
El-Borai AM, Eltayeb KM, Mostafa AR, El-Assar SA (2016) Biodegradation of industrial oil-polluted wastewater in Egypt by bacterial consortium immobilized in different types of carriers. Pol J Environ Stud 25(5):1901–1909
Florescu D, Ionete RE, Sandru C, Iordache A, Culea M (2010) The influence of pollution monitoring parameters in characterizing the surface water quality from Romania southern area. Rom J Physiol 56:1001–1010
Genuis SJ, Kelln KL (2015) Toxicant exposure and bioaccumulation: a common and potentially reversible cause of cognitive dysfunction and dementia. Behav Neurol 2015:620143

Hanchang SHI (2009) Industrial wastewater-types, amounts and effects. In: Point sources of pollution: local effects and their control, vol 2. EOLSS Publications, Paris, p 191

Hesnawi R, Dahmani K, Al-Swayah A, Mohamed S, Mohammed SA (2014) Biodegradation of municipal wastewater with local and commercial bacteria. Procedia Eng 70:810–814

Husain Q (2006) Potential applications of the oxidoreductive enzymes in the decolorization and detoxification of textile and other synthetic dyes from polluted water: a review. Crit Rev Biotechnol 26(4):201–221

Husain Q (2010) Peroxidase mediated decolorization and remediation of wastewater containing industrial dyes: a review. Rev Environ Sci Biotechnol 9(2):117–140

Jaishankar M, Tseten T, Anbalagan N, Mathew BB, Beeregowda KN (2014) Toxicity, mechanism and health effects of some heavy metals. Interdiscip Toxicol 7(2):60–72

Jan A, Azam M, Siddiqui K, Ali A, Choi I, Haq Q (2015) Heavy metals and human health: mechanistic insight into toxicity and counter defense system of antioxidants. Int J Mol Sci 16(12):29592–29630

Jecu L, Gheorghe A, Popea F, Rosu A, Stoica A, Stroescu M (2008) Potential of microbial species in biodegradation of volatile organic compounds from waters

Joutey NT, Bahafid W, Sayel H, El Ghachtouli N (2013) Biodegradation: involved microorganisms and genetically engineered microorganisms. In: Biodegradation-life of science. InTech, London

Karigar CS, Rao SS (2011) Role of microbial enzymes in the bioremediation of pollutants: a review. Enzym Res 2011:805187

Kaur P, Singh S, Kumar V, Singh N, Singh J (2018) Effect of rhizobacteria on arsenic uptake by macrophyte *Eichhornia crassipes* (Mart.) Solms. Int J Phytoremediation 20(2):114–120

Kavitha RV, Murthy VK, Makam R, Asith KA (2012) Physico-chemical analysis of effluents from pharmaceutical industry and its efficiency study. Int J Eng Res Appl 2(2):103–110

Khouni I, Marrot B, Amar RB (2012) Treatment of reconstituted textile wastewater containing a reactive dye in an aerobic sequencing batch reactor using a novel bacterial consortium. Sep Purif Technol 87:110–119

Kumar L, Awasthi G, Singh B (2011) Extremophiles: a novel source of industrially important enzymes. Biotechnology 10(2):121–135

Kumar V, Upadhyay N, Singh S, Singh J, Kaur P (2013) Thin-layer chromatography: comparative estimation of soil's atrazine. Curr World Environ 8(3):469–472

Kumar V, Upadhyay N, Kumar V, Kaur S, Singh J, Singh S, Datta S (2014a) Environmental exposure and health risks of the insecticide monocrotophos—a review. J Biodivers Environ Sci 5:111–120

Kumar V, Singh S, Manhas A, Singh J, Singla S, Kaur P (2014b) Bioremediation of petroleum hydrocarbon by using Pseudomonas species isolated from petroleum contaminated soil. Orient J Chem 30(4):1771–1776

Kumar V, Singh S, Kashyap N, Singla S, Bhadrecha P, Kaur P (2015a) Bioremediation of heavy metals by employing resistant microbial isolates from agricultural soil irrigated with industrial waste water. Orient J Chem 31(1):357–361

Kumar V, Singh S, Singh J, Upadhyay N (2015b) Potential of plant growth promoting traits by bacteria isolated from heavy metal contaminated soils. Bull Environ Contam Toxicol 94:807–815

Kumar V, Kaur S, Singh S, Upadhyay N (2016) Unexpected formation of N′-phenyl-thiophosphorohydrazidic acid O, S-dimethyl ester from acephate: chemical, biotechnical and computational study. 3 Biotech 6(1):1

Kumar V, Singh S, Singh R, Upadhyay N, Singh J (2017) Design, synthesis, and characterization of 2, 2-bis (2, 4-dinitrophenyl)-2-(phosphonatomethylamino) acetate as a herbicidal and biological active agent. J Chem Biol 10(4):179–190

Mani D, Kumar C (2014) Biotechnological advances in bioremediation of heavy metals contaminated ecosystems: an overview with special reference to phytoremediation. Int J Environ Sci Technol 11(3):843–872

Megharaj M, Ramakrishnan B, Venkateswarlu K, Sethunathan N, Naidu R (2011) Bioremediation approaches for organic pollutants: a critical perspective. Environ Int 37(8):1362–1375

Mishra V, Gupta A, Kaur P, Singh S, Singh N, Gehlot P, Singh J (2016) Synergistic effects of Arbuscular mycorrhizal fungi and plant growth promoting rhizobacteria in bioremediation of iron contaminated soils. Int J Phytoremediation 18(7):697–703

Mtshali JS, Tiruneh AT, Fadiran AO (2014) Characterization of sewage sludge generated from wastewater treatment plants in Swaziland in relation to agricultural uses. Resour Environ 4(4):190–199

Nagwekar PR (2014) Removal of organic matter from wastewater by activated sludge process-review. Int J Sci Eng Technol Res 3(5):1260–1263

Naidoo S, Olaniran A (2014) Treated wastewater effluent as a source of microbial pollution of surface water resources. Int J Environ Res Public Health 11(1):249–270

Naruka K (2012) Impact of municipal solid waste on environment and human health. Bioherald: Int J of Biodiv Environ 2(1):25–30

Ogbonna DN, Igbenijie M, Isirimah NO (2006) Studies on the inorganic chemicals and microbial contaminants of health importance in groundwater resources in Port Harcourt Rivers State Nigeria. J Appl Sci 6:2257–2262

Ojuederie O, Babalola O (2017) Microbial and plant-assisted bioremediation of heavy metal polluted environments: a review. Int J Environ Res Public Health 14(12):1504

Oller I, Malato S, Sánchez-Pérez J (2011) Combination of advanced oxidation processes and biological treatments for wastewater decontamination—a review. Sci Total Environ 409(20):4141–4166

Othman N, Juki MI, Hussain N, Talib SA (2011) Bioremediation a potential approach for soil contaminated with polycyclic aromatic hydrocarbons: an overview. Int J Sust Construct Eng Technol 2:2

Prasad MP, Manjunath K (2011) Comparative study on biodegradation of lipid-rich wastewater using lipase producing bacterial species. Indian J Biotechnol 10:121–124

Praveen PK, Ganguly S, Wakchaure R, Para PA, Mahajan T, Qadri K, Kamble S, Sharma R, Shekhar S, Dalai N (2016) Water-borne diseases and its effect on domestic animals and human health: a review. Int J Emerg Technol Adv Eng 6(1):242–245

Quintana JB, Weiss S, Reemtsma T (2005) Pathways and metabolites of microbial degradation of selected acidic pharmaceutical and their occurrence in municipal wastewater treated by a membrane bioreactor. Water Res 39(12):2654–2664

Ratnakar A, Shankar S, Shikha (2016) An overview of biodegradation of organic pollutants. Int J Scientif Innovat Res 4(1):73–91

Sabir S (2015) Approach of cost-effective adsorbents for oil removal from oily water. Crit Rev Environ Sci Technol 45(17):1916–1945

Shareef A, Kookana R, Kumar A, Tjandraatmadja G (2008) Sources of emerging organic contaminants in domestic wastewater. An assessment based on literature review. *Water for a Healty Country National Research Flagship*

Sharifuzzaman SM, Rahman H, Ashekuzzaman SM, Islam MM, Chowdhury SR, Hossain MS (2016) Heavy metals accumulation in coastal sediments. In: Environmental remediation technologies for metal-contaminated soils. Springer, Tokyo, pp 21–42

Sharma D, Sharma B, Shukla AK (2011) Biotechnological approach of microbial lipase: a review. Biotechnology 10(1):23–40

Shon HK, Vigneswaran S, Snyder SA (2006) Effluent organic matter (EfOM) in wastewater: constituents, effects, and treatment. Crit Rev Environ Sci Technol 36(4):327–374

Silva-Bedoya LM, Sánchez-Pinzón MS, Cadavid-Restrepo GE, Moreno-Herrera CX (2016) Bacterial community analysis of an industrial wastewater treatment plant in Colombia with screening for lipid-degrading microorganisms. Microbiol Res 192:313–325

Singh R, Gautam N, Mishra A, Gupta R (2011) Heavy metals and living systems: an overview. Indian J Pharm 43(3):246

Singh A, Kumar V, Srivastava JN (2013) Assessment of bioremediation of oil and phenol contents in refinery waste water via bacterial consortium. J Petroleum J Environ Biotechnol 4(2):145

Singh S, Singh N, Kumar V, Datta S, Wani AB, Singh D, Singh J (2016) Toxicity, monitoring and biodegradation of the fungicide carbendazim. Environ Chem Lett 14:317–329

Singh S, Kumar V, Upadhyay N, Singh J, Singla S, Datta S (2017) Efficient biodegradation of acephate by Pseudomonas pseudoalcaligenes PS-5 in the presence and absence of heavy metal ions [Cu(II) and Fe(III)], and humic acid. 3 Biotech 7(4):262. https://doi.org/10.1007/s13205-017-0900-9
Singh S, Kumar V, Chauhan A, Datta S, Wani AB, Singh N, Singh J (2018) Toxicity, degradation and analysis of the herbicide atrazine. Environ Chem Lett 16:211–237
Sivaprakasam S, Dhandapani B, Mahadevan S (2011) Optimization studies on production of a salt-tolerant protease from Pseudomonas aeruginosa strain BC1 and its application on tannery saline wastewater treatment. Braz J Microbiol 42(4):1506–1515
Tangahu BV, Abdullah S, Rozaimah S, Basri H, Idris M, Anuar N, Mukhlisin M (2011) A review on heavy metals (As, Pb, and Hg) uptake by plants through phytoremediation. Int J Chem Eng 2011:939161
Tigini V, Prigione V, Giansanti P, Mangiavillano A, Pannocchia A, Varese GC (2010) Fungal biosorption, an innovative treatment for the decolourisation and detoxification of textile effluents. Water 2(3):550–565
Van Leeuwen J, Sridhar A, Harrata AK, Esplugas M, Onuki S, Cai L, Koziel JA (2009) Improving the biodegradation of organic pollutants with ozonation during biological wastewater treatment. Ozone Sci Eng 31(2):63–70
Wang Q, Yang Z (2016) Industrial water pollution, water environment treatment, and health risks in China. Environ Pollut 218:358–365
Wang C, Zhang Q, Jiang L, Hou Z (2018) The organic pollutant characteristics of lurgi coal gasification wastewater before and after ozonation. J Chemother 2018:1461673
Westerhoff P, Lee S, Yang Y, Gordon GW, Hristovski K, Halden RU, Herckes P (2015) Characterization, recovery opportunities, and valuation of metals in municipal sludges from US wastewater treatment plants nationwide. Environ Sci Technol 49(16):9479–9488
Wu B, Wang G, Wu J, Fu Q, Liu C (2014) Sources of heavy metals in surface sediments and an ecological risk assessment from two adjacent plateau reservoirs. PLoS One 9(7):e102101
Wuana RA, Okieimen FE (2011) Heavy metals in contaminated soils: a review of sources, chemistry, risks and best available strategies for remediation. ISRN Ecology 2011:402647
Wyasu G, Kure OA (2012) Determination of organic pollutants in hospital wastewater and food samples within Ahmadu Bello University Teaching Hospital (Abuth), Shika, Zaria-Nigeria. Adv Appl Sci Res 3(3):1691–1701
Yagi JM, Madsen EL (2009) Diversity, abundance, and consistency of microbial oxygenase expression and biodegradation in a shallow contaminated aquifer. Appl Environ Microbiol 75(20):6478–6487
Yavari S, Malakahmad A, Sapari NB (2015) A review on phytoremediation of crude oil spills. Water Air Soil Pollut 226(8):279
Yu Y, Wang YC, Zhou HD, Gao B, Zhao GF (2013) Biomagnification of heavy metals in the aquatic food chain in Daning River of the Three Gorges Reservoir during initial impoundment. Huan jing ke xue 34(10):3847–3853
Zheng C, Zhao L, Zhou X, Fu Z, Li A (2013) Treatment technologies for organic wastewater. In: Water treatment. Intech, London

Quorum Quenching for Sustainable Environment: Biology, Mechanisms, and Applications

4

Naga Raju Maddela, Luz Cecilia García Cruzatty, Daniel Alfredo Leal-Alvarado, Jessenia Castro Olaya, Sagnik Chakraborty, and Anupam Mukherjee

Abstract

Quorum sensing signaling is a hierarchal system in bacteria to communicate with each other and coordinate their activities. Prevention of the QS pathway by disrupting signals is called quorum quenching (QQ), which is essential not just in medicine and healthcare settings but also in membrane bioreactors, aquaculture, and agriculture. QQ could be achieved either by interfering with the QS signaling pathway (e.g., signal generator or receptor) or intercepting the QS molecules. Research on QQ led to the development of strategies that mitigate biofilm-based problems in medicine, agronomy, and water engineering. The QQ-strategy is being given importance in recent times as there is an immediate need to search for an alternative or a complementary approach to phytochemicals and antibiotics. This chapter starts with the historical aspects of QQ; furthermore, it highlights the

N. R. Maddela (✉)
Instituto de Investigación, Universidad Técnica de Manabí, Portoviejo, Ecuador

Faculdad la Ciencias la Salud, Universidad Técnica de Manabí, Portoviejo, Ecuador
e-mail: rmaddela@utm.edu.ec

L. C. García Cruzatty · D. A. Leal-Alvarado
Instituto de Investigación, Universidad Técnica de Manabí, Portoviejo, Ecuador

Faculdad de Agronomía, Universidad Técnica de Manabí, Portoviejo, Ecuador
e-mail: lcgarcia@utm.edu.ec; daleal@utm.edu.ec

J. C. Olaya
Faculdad de Agronomía, Universidad Técnica de Manabí, Portoviejo, Ecuador

S. Chakraborty
School of Energy & Environmental Engineering, Hebei University of Technology, Tianjin, People's Republic of China

A. Mukherjee
Department of Chemical Engineering, Haldia Institute of Technology, Haldia, Purba Medinipur, West Bengal, India

P. K. Arora (ed.), *Microbial Technology for Health and Environment*, Microorganisms for Sustainability 22,
https://doi.org/10.1007/978-981-15-2679-4_4

global research in the area of QQ and the mechanism of quenching pathways. Afterward, applications of QQ-strategies in medicine, agriculture, aquaculture, and water engineering are discussed. Finally, challenges and prospects of QQ technology are delineated.

Keywords

Quorum quenching · Mechanism · Medicine · Agriculture · Aquaculture · Wastewater treatment

Abbreviations

AHLs/HSL	*N*-acyl homoserine lactones
AI-2	Auto Inducer 2
AIPs	Autoinducing peptides
AMR	Anti-microbial resistance
CA	Canada
FO-MBR	Forward osmosis-MBR
GCL	γ-Caprolactone
GMP	Guanosine monophosphate
IDRC	International Development Research Centre
MBBR	Moving bed biofilm reactors
MBfR	Membrane biofilm reactors
MBRs	Membrane bioreactors
MIC	Minimum inhibitory concentration
MRSA	Methicillin-resistant *Staphylococcus aureus*
NIH	National Institute of Health
Ntn	N-terminal nucleophile
ORFs	Open reading frames
PLLs	Phosphotriesterase-like lactonases
PON	Paraoxonase
QQ	Quorum quenching
QQ-AnMBR	QQ-anaerobic MBR
QQ-RO	QQ-reverse osmosis
QS	Quorum sensing
QSIs	Quorum sensing inhibitors
SiNPs	Silver nanoparticles
USA	United States of America
WHO	World Health Organization
WWT	Wastewater treatment

4.1 Introduction

Microorganisms regulate their community behavior by sending and receiving chemical signals through a process called 'quorum sensing' (QS); interruption of such QS-signals is defined as 'quorum quenching' (QQ). QS is a fundamental mechanism behind many functions of bacteria, such as making virulence factors, biofilm formation, exopolysaccharide production, antibiotic production, genetic exchange by conjugation, pigmentation, etc. (Lade et al. 2014). Correspondingly, QQ came to light to control the unwanted functions of QS-related biology. For instance, in the human health-care field, control of bacterial pathogens (so-called biofilms) is still challenging. Pathogenic bacteria are known to utilize QS mechanisms during disease development. The use of antibiotics to control bacterial pathogens has a great threat of drug resistance in bacteria. An attractive and alternative approach is to use QQ-strategy to control pathogens. Several bacteria (e.g., *Acinetobacter baumannii*, *Proteus* spp., *Pseudomonas aeruginosa*, *Serratia* spp. etc.) do produce stable biofilms on medical devices which are difficult to control by traditional approaches that are in practice. Likewise, QS-mediated problems are not uncommon in several other fields such as dental plaque biofilms (Basavaraju et al. 2016), carcinogenesis (Li et al. 2004), agriculture (Wang et al. 2010), aquaculture (Defoirdt et al. 2004), wastewater treatment (WWT) technologies, and marine life (Galloway et al. 2012). Therefore, a high demand exists for QQ-strategy to control QS-mediated problems in fields such as MBRs (membrane bioreactors), medicine, agriculture, aquaculture, etc.

QQ can be achieved by one of the several mechanisms, such as blockage of signal synthesis, interference with signal molecules (called as autoinducers), and inactivation of signal molecules. QQ-strategy has been successfully applied in different environments to control the effects elicited by unwanted biofilms (Bzdrenga et al. 2017; Maddela et al. 2019). Transgenic plants with a gene (e.g., *aiiA*)-encoding QQ enzyme showed tremendous tolerance to a pathogen (e.g., *Pectobacterium carotovorum*) (Dong et al. 2001). Oral administration of purified AHL (*N*-acyl homoserine lactone)-lactonase from *Bacillus* sp. A196 was found to show decrease in *Aeromonas hydrophila* infection in zebrafish (Cao et al. 2012). There are many successful incidences in several wastewater treatment (WWT) facilities for the effective control of biofilm biomass (so-called biofouling) (Paul et al. 2009; Ergön-Can et al. 2017; Lee et al. 2017). Unfortunately, QQ-strategy is far from its commercial perspective which implies that more details of QQ effects on pure-species need to be elucidated in depth. Most studies have assessed the QQ effects in bioreactors exogenously dosed with either QQ-enzymes or QQ-bacteria (Yeon et al. 2009; Oh et al. 2012), and concurrently ignored endogenous QQ activities in bioreactors Also, there is no proper validation of the detrimental effects of the QQ-strategy in bioreactors (Choudhary and Schmidt-Dannert 2010; Jiang et al. 2013). It is also noteworthy that most of the pure culture studies have been conducted for a very few model bacteria (e.g., *Escherichia coli, Pseudomonas* sp., etc.) while uncovering details for other species of Gram-positive bacteria, archaea, fungi, and mixed cultures. Therefore, future research must target QQ effects on many natural isolates at the pure-species level, which would provide more details of QQ mechanisms and would decrease the gap between the lab- and commercial-stage of QQ-strategy.

This chapter draws an overview of bacterial quorum quenching and its applications, covering economically vital fields such as medicine, agriculture, wastewater treatment, etc. This chapter first summarizes the historical aspects of QQ and the current advances in the QQ-technology towards the welfare of society. Consequently, a detailed molecular mechanism of the QQ process is focused on. Special attention is dedicated to applications of QQ-strategy in different sections such as medicine, agriculture, and wastewater treatment. Besides, case studies of QQ-strategy, challenges, and prospects of QQ-technology are discussed considering the recently published literature. Thus, this state-of-the-art chapter is the first compilation of all the critical information and updated knowledge required for understanding the QQ biology and its implications in relevant economically essential fields.

4.2 Biology of Bacterial Quorum Quenching (QQ)

Disruption of the QS communication system is defined as quorum quenching (QQ) (Dong et al. 2001). More than two decades ago, inhibitors and antagonists of the QS system had been reported in a marine macroalga (*Delisea pulchra*) (Givskov et al. 1996). However, in bacteria, the first recorded evidence of QQ has been observed as a part of the biosynthesis of carbapenem antibiotics in a bacterium *Erwinia carotovora* through fluorescence quenching (Welch et al. 2000); importantly the *N*-acyl homoserine lactone (HSL or AHL) signal is known to influence the carbapenem biosynthesis in *E. carotovora*. Since then, different QQ enzymes have been identified in various bacteria, where these enzymes can modify or degrade AHLs (Dong and Zhang 2005; Romero et al. 2011; Torres et al. 2016). Interestingly, QQ-enzymes producing bacteria have been detected both in terrestrial and marine environments (Dong et al. 2001; Romero et al. 2008), implying that these enzymes provide a competitive advantage to the producer in terms of food and space. Notably, two QQ enzymes that have been extensively studied were AHL lactonases and AHL acylases (or AHL amidases) (Christiaen et al. 2011).

It is also important to note that many bacteria genera can quench QS signals. Christiaen and others (Christiaen et al. 2011) isolated 59 different bacterial strains from various types of environmental samples (e.g., rhizosphere, water, and pond water), which belonged to 21 different genera, where the predominant genera were *Pseudomonas*, *Arthrobacter*, *Aeromonas*, *Delftia*, *Stenotrophomonas,* and *Achromobacter*. These results indicate that there is a high competition among the members of the bacterial community in different environments, and the bacteria can overcome such competition by QQ activity. In the same study, it has also been found that the QQ activity in certain bacteria remained resistant to heat and proteinase K, and certain bacteria showed extracellular QQ activity, with different molecular sizes (either $<$ or $>$ 6000 Da) of QQ molecules (Christiaen et al. 2011). Thermal resistance and thermophilicity in the community of microbial lactonases are not uncommon because several heat-stable lactonases have been detected in thermophilic archaea bacteria (Mandrich et al. 2010). On the other side, a broad diversity has been observed in QQ enzymes of bacteria (Fetzner 2015), where 18 AHL lactonases, 17 AHL acylases, 4 oxidoreductases, 1 anti-AQs (2-Alkyl-3-hydroxy4(1*H*)-quinolone 2,4-dioxygenase), and 1 AI-2 kinase enzymes were identified in different bacterial isolates. Also, anaerobic bacteria (e.g., *Microbacterium* sp.) are not exempted from possessing the QQ activity (Liu et al. 2019). All the above results indicate that the widest diversity exists in the QQ molecules of different bacterial species. The variety of QQ molecules needs to be explored in depth for the practical implications of QQ-technology in the nearest future for sustainable development.

According to the 16S rDNA sequence study, it has been found that there is no specific clade associated with the QQ activity in sludge community (Tan et al. 2015). In this study, the bacteria that were identified with QQ activity belonged to four different phyla, such as Actinobacteria, Bacteriodetes, Firmicutes, and Proteobacteria. Besides common bacterial genera (e.g., *Bacillus*, *Ochrobacterium*, *Rhodococcus,* and *Variovorax*), some new genera also came to light as QQ bacteria from this study (e.g., *Acidovorax*, *Flavobacterium*, *Novosphingobium, Rheinheimera,* and *Tsukamurella*). Interestingly, the sludge sample was known to contain both signal-producers and -quenchers in the same community, and some strains showed both activities (e.g., *Rhizobium borbori* N065) too (Tan et al. 2015). These results imply that QQ is known to be associated with a wide range of bacteria; furthermore, there is a co-existence of QS and QQ bacteria in the natural habitats, where some of them have dual properties. At the community level, the QS and QQ activity of each species is essential in eliciting a complex community behavior. Varying the species composition of QS and QQ will have the greatest impact on community behavior. The chronological order of the main breakthrough in QQ biology has been depicted in Fig. 4.1.

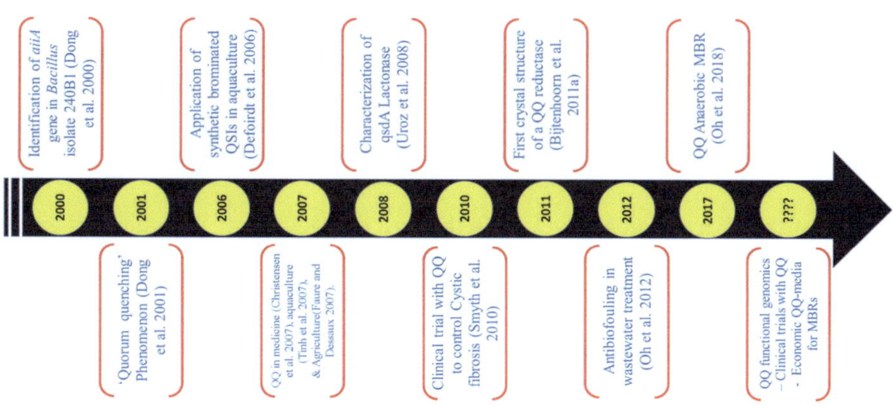

Fig. 4.1 Chronological time scale of QQ biology

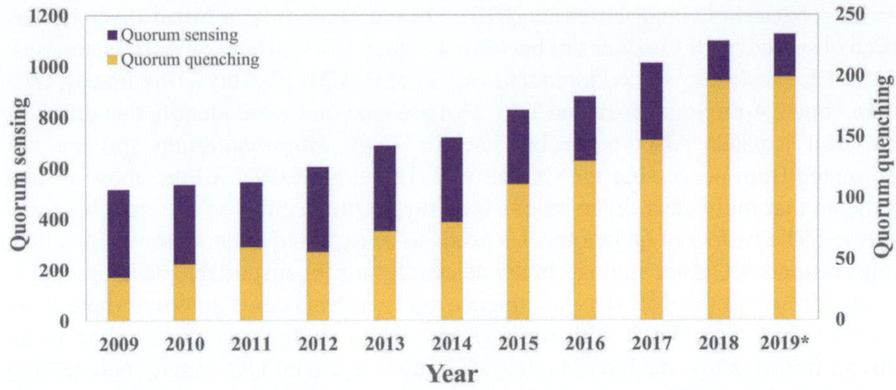

Fig. 4.2 Number of articles published in the last decade (2009–2019) on QS and QQ with an interleaved bar diagram. Values on primary- and secondary-vertical axis are the number of publications (Accessed from ScienceDirect; Keywords 'quorum sensing' vs. 'quorum quenching'). ∗Data as of August 2019

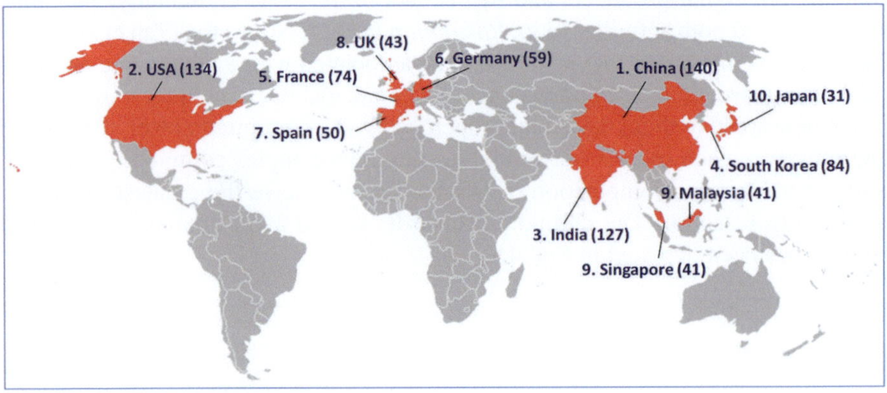

Fig. 4.3 Top 10 countries with the highest research productivity (number of publications) in the field of QQ. Data accessed from Scopus by using the keyword 'quorum quenching', on 13 August 2019. Values in the parenthesis indicate the number of publications

4.3 Global Research in the Area of QQ Technology

There is extensive research going on in the area of 'QQ' to see the implications of QQ-strategies in many economically essential fields such as medicine, agriculture, and industrial MBR facilities. The global research output has increased ~6 times in the past 10 years (Fig. 4.2). Figure 4.3 shows the top ten countries where active research is taking place in the field of QQ.

A program was conducted by the National Institute of Technology (Tiruchi, India) titled, 'Chemical/Phytochemical Mediated Disruption of Bacterial Acyl Homoserine Lactone Mediated, QS Communication Systems' to enlighten the researchers about the threat posed by the use of antimicrobial agents and understand the QS and QQ and the need for developing effective carrier agents (through nanotechnology) as delivery agents (THE HINDU 2019). Novel antibacterial compounds have also emerged from the recent investigations aiming at antibacterial drug design, for example, researchers at the University of Eastern Finland have designed a compound that targets LsrK kinase which is an active protein in bacterial communication (ScienceDaily 2018). Drug design efforts have been made by modeling the LsrK protein structure through computational methods, and this study claimed that this is the first class of LsrK inhibitor reported to date (Medarametla et al. 2018). Such studies will provide useful insights in understanding the behavior of the protein and protein–substrate dynamics, as well as how to interfere with it.

Advances in the QQ research led to the development of modulations in the QQ biofilm to mitigate biofouling on water purification membranes. After knowing the importance of QQ bacteria as a promising approach for membrane biofouling control, several attempts have been made towards the immobilization of QQ bacteria in porous matrices for the prevention of uncontrolled biofilm formation in continuously operated membrane processes. However, these attempts have ignored the QQ biofilm and its interference with membrane performance. Therefore, for the first time, the QQ biofilm has been modulated by dichromatic light, optogenetic c-di-GMP gene circuit, in which QQ bacteria sense near-infrared and blue light to adjust their biofilm formation by regulating the c-di-GMP level (Mukherjee et al. 2018). Such a modulated QQ biofilm could successfully mitigate biofouling on forward osmosis (FO) membranes of water purification facilities, implying that controllable biofilm-enabled applications can be widely implicated in different biofilm-based biocatalysis.

Very recently, an article has been posted on the website of Health Trends in Spanish as '*Posible final de la era de los antibióticos*' which means 'Possible end of era of antibiotics' (Paramá 2019). Researchers from the University of South Carolina published that soon bacterial QS could become a compelling alternative to the conventional treatment of infectious diseases, and they assumed that this could be a more natural way, and they successfully mitigated the bacterial biofilm (of pathogens) by silicon oxide nanoparticles (Si-NPs) coated with β-cyclodextrin. A similar opinion has also been stated by a research group of Biotechnology and Aquaculture headed by Professor Casal at the University of Santiago de Compostela, Spain (Paramá 2019). They opinioned that QQ-strategy could be a future tool to penetrate the biofilms and eliminate infections, which could be challenging to achieve with the current drugs. Researchers from China and America have recently published a paper that they identified a new regulatory structure for the *Rhizobium* QS system, and this could have the greatest agronomic impact. Research is also awaiting generation, regulation, and control of biofilms of cholera producing bacteria, which causes three to five million cases of annual infections according to WHO statistics.

Bacterial biofilms cause significant problems in food packing also. Biofilms adhered to the food products and to the packaging to which it's shipped further cause severe complications in human. According to the latest report of the Centers for Disease Control and Prevention, USA, ~ 48 million illnesses per year are recorded in the USA alone, of which most of them (45%) are caused by bacteria. A research team headed by Professor Doron Steinberg at The Hebrew University of Jerusalem, Israel, found a new synthetic molecule called TZD and incorporated it into biofilm food packaging (PHYSORG 2014). The TZD molecule could mitigate the biofilm formation by bacteria and fungi in corrugated cardboard boxes. This TZD-technology has now been successfully incorporated into industrial acrylic polymers which are used to coat the corrugated cardboard in the fresh food package. Furthermore, TZD has proved to be effective in preventing biofilms in recycled water systems. Thus, the design of TZD like QQ polymers is for their wide applicability in different areas such as frozen food packing, poultry, meat packaging, etc.

In the area of anti-QS research, a research team headed by a Professor Ben Feringa at the University of Groningen, Netherlands, produced a library of 16 compounds with a potential anti-QS activity through a light-operated switch (PHYSORG 2019). These compounds modulated the QS signals in *P. aeruginosa* by light. Furthermore, irradiation of these compounds by light led to enhancement of QS activity by 700-folds, which is attributed to bending of the flexible carbon-based tail in the compound structure. Nonetheless, these studies provide a powerful tool for both clinical and fundamental research to understand the mechanism of QS and QQ. According to the recent reports of the WHO, many pathogenic bacteria (e.g., *Acinetobacter, E. coli, Klebsiella, Proteus, Pseudomonas, Serratia,* etc.) were found to develop resistance to many third-generation antibiotics such as carbapenems and cephalosporins (ID 2019). On the other side, there are no effective antibiotics against specific pathogens such as *Campylobacter* spp., *Enterococcus faecium, Helicobacter pylori,* and *Staphylococcus aureus*. This situation dramatically demands to find new antimicrobial strategies. In this angle, bacterial communication has opened the door to knowledge that was not suspected by anyone in the past. Therefore, track of this type of research is going to dominate in the nearest future in discovering new microbial-mitigation-strategies.

4.4 Inhibition of QS Signals: Biology and Mechanism

Extensive and basic research in the area of 'QQ' has been carried out in the first decade of the twenty-first century, which resulted in the isolation and characterization of many kinds of QQ molecules, and these molecules were found to be significant in the control of several microbial infections. For instance, AHL-lactonases (a type of QQ enzyme) have been found to decrease the disease incidences caused by *P. aeruginosa* (Reimmann et al. 2002), *E. coli* (Lee et al. 2002), *Bacillus thuringiensis* (Dong et al. 2004), *Erwinia amylovora* (Molina et al. 2005) etc. In fact, *P. aeruginosa* was known to be affected by other types of QQ molecules, such as paraoxonase (Ozer et al. 2005), AHL-acylase (Lin et al. 2003), 3-oxo-C12-(2-aminocyclohexanone) (Smith et al. 2003), etc. Nonetheless, several small molecules capable of modulating the AHL-based QS system have been identified over the past 30 years (Galloway et al. 2011); all these insights could help us in understanding

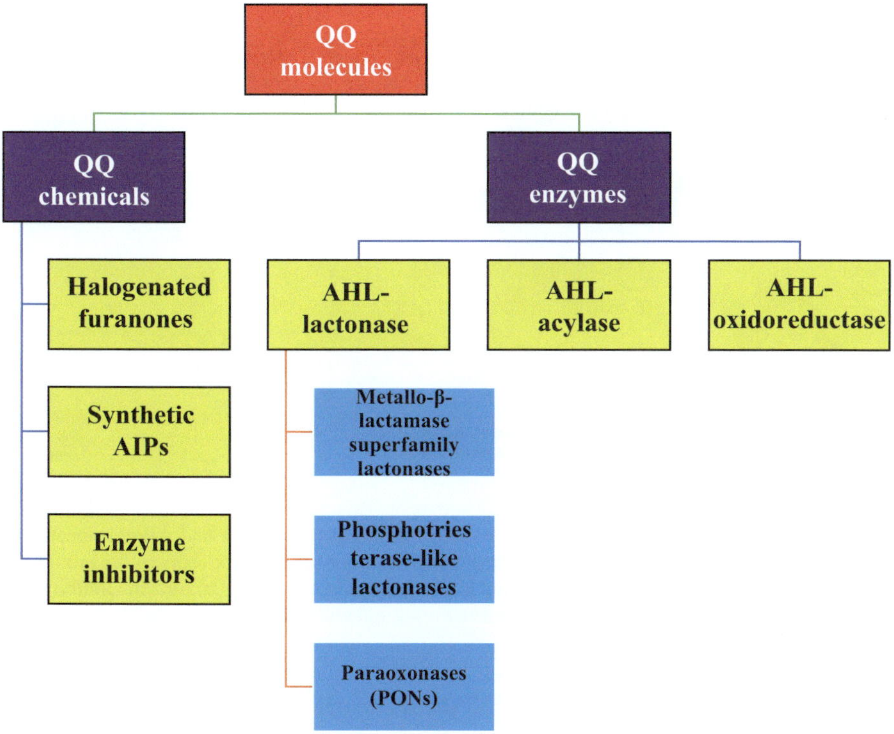

Fig. 4.4 Classification of QQ agents inhibiting QS signal molecules

and implicating the QQ-strategy in the relevant fields. Signal inhibiting molecules are broadly divided into two groups—QQ chemicals and QQ enzymes, and detailed classification of signal inhibitors has been shown in Fig. 4.4. Regarding QQ chemicals, there are three kinds of QQ chemicals as described in this section, two of them are structural analogs, and one is an enzyme inhibitor. Structurally, QQ chemicals are small molecules, and they inhibit the QS signals by one of the several mechanisms. For instance, QQ chemicals those are structurally similar to AHLs and AIPs (autoinducing peptides) either interfere with the corresponding signal binding receptors (Lyon et al. 2000), or they decrease the available receptor concentration (Manefield et al. 2002). Such type of structural mimicry is quite common in the case of halogenated furanones vs. AHLs, and synthetic AIPs vs. AIPs (Fig. 4.5). The third type of QQ chemical is an enzyme inhibitor, which may interrupt the essential events in the biosynthesis of QS molecules. For example, enoyl-ACP reductase is crucial in the formation of an intermediate product during the AHL biosynthesis; however, this reductase is quickly inactivated by triclosan (a small QQ chemical) (Hoang and Schweizer 1999). Like triclosan, closantel interferes with a two-component signal system (is a QS system in gram-positive and gram-negative bacteria) by inhibiting histidine kinase (Stephenson et al. 2000). Thus, the above evidence implies that QQ chemicals elicit QS inhibition either by competing with

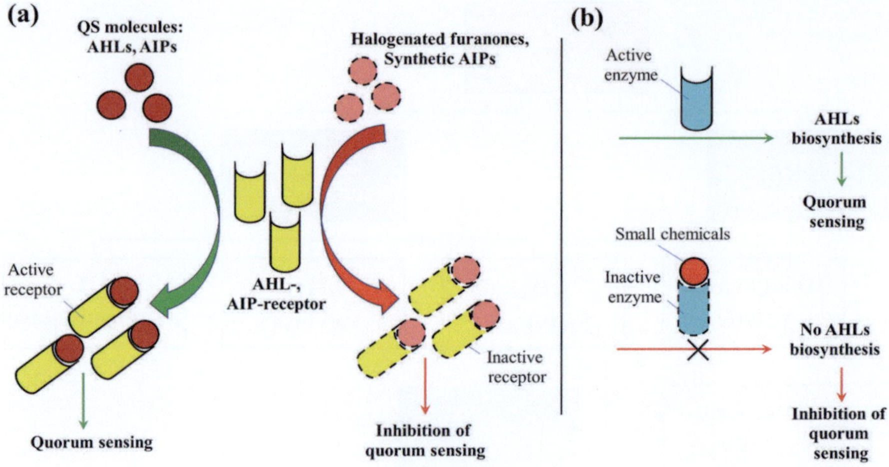

Fig. 4.5 Mechanism of QS inhibition by QQ agents or QSIs. (**a**) Structural Mimicry—QSIs such as halogenated furanones and synthetic AIPs are structurally like AHLs and AIPs, respectively. Binding of QSIs to QS receptors causes either inactivation of receptors or reduces the number of receptors for QS molecules ultimately leading to mitigation of QS signals. (**b**) Enzyme Inhibition—QSIs (small molecules) can bind and inactivate vital enzymes that are responsible for the QS signals

the original QS molecules to bind with the receptors or they may inhibit the vital enzymes of the QS signal pathway.

Unlike QQ chemicals, QQ enzymes achieve QS inhibition by degrading the QS signal molecules. To date, over 20 QQ enzymes have been reported (Dong et al. 2018), and three crucial among them are (Fetzner 2015) namely AHL-lactonase, AHL-acylase, and AHL-oxidoreductase (Fig. 4.4).

4.4.1 AHL-Lactonases

AHL-lactonase (EC 3.1.1.81) is a group of enzymes (QQ enzymes) that can hydrolyze the ester bond of the homoserine lactone ring (C–O bond) of AHL signal molecules. The scaffold of the α/β-hydrolase fold of lactonases was also found to be responsible for the lactonolysis of AHLs. Based on the structure, there are 3 kinds of lactonases, such as metallo-β-lactamase superfamily lactonases, phosphotriesterase type of lactonases (PLLs), and paraoxonase (PON) family. Several bacteria are known to possess AHL-lactonases, e.g., *B. thuringiensis, Burkholderia thailandensis, E. carotovora, E. amylovora, E. coli, P. aeruginosa,* etc.

4.4.2 Metallo-β-Lactamase Superfamily Lactonases

The first AHL-lactonase identified was a product of the *aiiA* gene of *Bacillus* sp. isolate 240B1. Crystallographic study (Kim et al. 2005) has confirmed that

AHL-lactonase is a metalloprotein, and contains two zinc ions in its active site. Since then, AHL-lactonases have been identified in many bacterial species, and refer for a review (Fetzner 2015). More recently, AiiK, an AHL lactonase from a bacterium *Kurthiahuakui*LAM0618^T has been characterized, and its QQ activities have been checked against *P. aeruginosa* PAO1 (Dong et al. 2018). AHL-lactonase is able to hydrolyze both short- and long-chain AHL signal molecules with similar efficiency. More details of different metallo-β-lactamase superfamily lactonases have been presented in Table 4.1. The chemical reaction that is governed by AHL-lactonase is shown below:

N-acyl-L-homoserine lactone + $H_2O \rightleftharpoons$ an *N*-acyl-L-homoserine.

4.4.3 Phosphotriesterase-like Lactonases (PLLs)

PLLs are the members of amidohydrolase superfamily and are characterized by a binuclear metal center within a $(\beta/\alpha)_8$-barrel structural scaffold. Preferably PLLs degrade AHLs of hydrophobic lactones; however, many PLLs are classified as natural lactonases, and they do degrade lactonases other than AHLs (Xiang et al. 2009). Similarly, certain PLLs are active against AHLs containing QsdA (verify) (Afriat et al. 2006), and some of the PLLs are thermostable (Chow et al. 2009), implying that there is existence of wide diversity in the characteristics (particularly substrate specificity, thermostability, etc.) of PLLs. Because of such properties of PLLs, attempts have also been made by structure-based mutagenesis or in vitro evolution to improvise the catalytic efficiency and modify their substrate range (Hiblot et al. 2013). More details about the types and substrate specificities of PLLs have been provided in Table 4.1.

4.4.4 Paraoxonases (PONs)

Paraoxonases (PONs), which are calcium-dependent enzymes, have lactonase-like activity (i.e., hydrolyze the homoserine lactone ring of AHLs). Structurally PONs are distinct from the other lactonases and have a six-bladed β propeller fold (Harel et al. 2004). Three types of PONs have been identified (e.g., PON1, PON2, and PON3) with broad and physiologically relevant hydrolytic activities. There was a first report on the hydrolytic activity of purified PON1 from human serum against 3-oxo-C12 HSL of *P. aeruginosa* (Ozer et al. 2005). PONs were found to be more active against long-chain AHL-signals (e.g., 3-oxo-C12 HSL) than short-chain AHL-signals. In the recent past, bioinformatic approach has been used in the identification of bacterial PONs and have been characterized biochemically. More details of PONs have been shown in Table 4.1.

Table 4.1 QQ enzymes[a]—source and substrate specificity

Enzyme	Source	Substrate	Reference(s)
(A) **AHL-lactonases**			
(A.i.) Metallo-β-lactamase superfamily			
AiiA	*Bacillus* spp. (*Bacillus cereus* group)	C4-, C6-, C8-, C10-HSL; 3OC4-, 3OC6-, 3OC8-, 3OC12-HSL; 3-OH-C4-HSL	Dong et al. (2000), Dong et al. (2002), Lee et al. (2002), Reimmann et al. (2002), Wang et al. (2004), Liu et al. (2005), Morohoshi et al. (2012)
AhIs	*Solibacillus silvestris*StLB046	C10-HSL	Morohoshi et al. (2012)
AhLD	*Arthrobacter* sp. IBN110	C6-, C8-, C10-HSL, 3OC6-, 3OC12-HSL	Park et al. (2003)
AttM (AiiB)	*Agrobacterium tumefaciens* C58, M103	C4-, C6-, C7-, C8-, C10-HSL; 3OC6-, 3OC8-HSL	Zhang et al. (2002), Carlier et al. (2003), Liu et al. (2007)
QsDR1	*Rhizobium* sp. NGR234	3OC8-HSL	Krysciak et al. (2011)
Ah1K	*Klebsiella pneumoniae* KCTC2241	C6-HSL, 3OC6-HSL	Park et al. (2003)
AidC	*Chryseobacterium* sp. StRB126	C6-, C8-, C10-, C12-HSL; 3OC6-, 3OC8-, 3OC10-, 3OC12-HSL	Wang et al. (2012)
Q1cA	Soil metagenome	C6-HSL	Riaz et al. (2008)
(A.ii.) Phosphotriesterase-like lactonases			
MCP	*Mycobacterium avium* ssp. *Paratuberculosis* K-10	C6-, C7-, C8-, C10-, C12-HSL; 3OC8-HSL	Chow et al. (2009)
QsdA	*Rhodococcus erythropolis*W2, SQ1, Mic1, MP50, CECT3008; *Rhodococcus* sp. BH4	C4-, C6- to C14-HSLs, with or without substitution at C3′	Afriat et al. (2006), Uroz and Heinonsalo (2008), Oh et al. (2013)
Ssopox	*Sulfolobus solfataricus*P2	C4-, C6-, C8-, C12-HSL; 3OC6-, 3OC8-, 3OC10-, 3OC12-HSL	Afriat et al. (2006), Hiblot et al. (2013)
SisLac	*Sulfolobus islandicus* M.16.4	C4-, C8-, C12-HSL; 30C8-, 3OC10-, 3OC12-HSL	Hiblot et al. (2012)
(A.iii.) Paraoxonase			
Bacterial PON	*Oceanicaulis alexandrii* HTCC2633	C12-HSL; 3OC10-, 3OC12-HSL	Bar-Rogovsky et al. (2013)

(continued)

Table 4.1 (continued)

Enzyme	Source	Substrate	Reference(s)
(A.iv.) Others			
AiiM	*Microbacterium testaceum* StLB037	C6-, C8-, C10-, C12-HSL; 3OC6-, 3OC8-, 3OC10-, 3OC12-HSL	Wang et al. (2010)
AidH	*Ochrobactrum* sp. T63	C4-, C6-, C10-HSL; 3OC6-, 3OC8-HSL; 3-OH-C6-HSL	Mei et al. (2010)
D1hR	*Rhizobium* sp. NGR234	3OC8-HSL	Krysciak et al. (2011)
BpiB07	Soil metagenome	3OC8-HSL	Schipper et al. (2009)
QsdH	*Pseudoalteromonas byunsanensis* 1A01261	C4-, C6-, C8-, C10-, C12-, C14-HSLs; 3OC6-, 3OC8-HSL	Huang et al. (2012)
BpiB04, BpiB01	Soil metagenome	3OC8-HSL	Schipper et al. (2009)
BpiB05	Soil metagenome	3OC6-, 3OC8-, 3OC12-HSL	Bijtenhoorn et al. (2011b)
(B) **AHL-acylases**			
Ah1M	*Streptomyces* sp. M664	C6-, C8-, C10-HSL; 3OC6-, 3OC8-, 3OC12-HSL	Park et al. (2005)
AibP	*Brucella melitensis* 16 M	C12-HSL; 3OC12-HSL	Terwagne et al. (2013)
AiiD	*Ralstonia* sp. XJ12B	3OC6-, 3OC8-, 3OC10-, 3OC12-HSL	Lin et al. (2003)
Aac	*Ralstonia solanacearum* GMI1000	C7-, C8-, C10-HSL; 3OC8-HSL	Chen et al. (2009)
PvdQ (PA2385)	*P. aeruginosa* PAO1	C7-, C8-, C10-, C11-, C12-, C14-HSL; 3OC10-, 3OC12-, 3OC14-HSL; 3OH-C12-, 3OH-C14-HSL	Huang et al. (2003), Sio et al. (2006), Bokhove et al. (2010), Koch et al. (2010)
QuiP (PA1032)	*P. aeruginosa* PAO1	C7-, C8-, C10-, C12-, C14-HSL; 3OC12-HSL	Huang et al. (2006)
HacB (PA0305)	*P. aeruginosa* PAO1	C6-, C7-, C8-, C10-, C12-, C14-HSL; 3OC10-, 3OC12-, 3OC14-HSL	Wahjudi et al. (2011)
HacB (Psyr.4858)	*P. syringae* B728a	C4-, C6-, C8-, C10-, C12-HSL; 3OC6-, 3OC8-HSL	Shepherd and Lindow (2009)
HacA (Psyr.1971)	*P. syringae* B728a	C8-, C10-, C12-HSL; 3OC8-HSL	Shepherd and Lindow (2009)

(continued)

Table 4.1 (continued)

Enzyme	Source	Substrate	Reference(s)
Aac	*Shewanella* sp. MIB015	C8-, C12-, C12-HSL	Morohoshi et al. (2008)
AiiC	*Anabaena* sp. PCC7120	C4-, C6-, C8-, C10-, C12-, C14-HSL; 3OC4-, 3OC6-, 3OC8-, 3OC10-, 3OC12-, 3OC14-HSL, and corresponding 3OH-C_X-HSLs	Romero et al. (2008)
AiiO	*Ochrobactrum* sp. A44	C4-, C6-, C10-, C12-, C14-HSL; 3OC4-, 3OC6-, 3OC8-, 3OC10-, 3OC12-, 3OC14-HSL and corresponding 3OH-C_X-HSLs	Czajkowski et al. (2011)
QsdB	Soil metagenome	C6HSL; 3OC8-HSL	Tannieres et al. (2013)
(C) **Oxidoreductases**			
CYP102A1	*B. megaterium*	C12- to C16-HSL; 3OC12-HSL (ω-1, ω-2, ω-3-hydroxylation)	Chowdhary et al. (2007), Chowdhary et al. (2008)
BpiB09	Soil metagenome	3OC12-HSL	Bijtenhoorn et al. (2011a)
(D) **Enzymes acting on AQs**			
2-Alkyl-3-OH-4(1*H*)-euinolone-2,4-dioxygenase (Hod)	*Arthrobacter* sp. Rue61a	C1- to C9-*n*-alkyl-3-OH-4(1*H*)-quinolones	Pustelny et al. (2009), Thierbach et al. (2014)
(E) **Enzymes acting on AI-2**			
AI-2 kinase (LsrK)	*E. coli*, other enteric bacteria	Linear form of AI-2 (4,5-di-OH-2,3-pentanedione)	Roy et al. (2010)

[a]Bacterial origin/soil metagenome

4.4.5 AHL-Acylases

Unlike AHL-lactonases, AHL-acylases (EC 3.5.1.97) degrade the AHL molecules by hydrolyzing the amide bond (–NH_2) of AHLs and releasing the corresponding fatty acids and homoserine lactones (Leadbetter and Greenberg 2000). AHL-acylases have been identified in several species of bacteria (Table 4.1), e.g., *P. aeruginosa* PAO1, *Ralstonia* sp., *Streptomyces* sp., *Variovorax paradoxus,* etc. and these AHL-acylases are characteristically similar to Ntn (N-terminal nucleophile) hydrolases but have different substrate specificities. However, AiiO AHL-acylase from *Ochrobactrum* sp. A44 is structurally different and has an α/β-hydrolase fold (Czajkowski et al. 2011). AiiD AHL-acylase (*Ralstonia* sp.) effectively degrades long-chain AHLs, but can also degrade short-chain AHLs with less efficiency (Lin et al. 2003). On the other hand, AHL-acylases PvdQ (*P. aeruginosa* PAO1) and Ah1M (*Streptomyces* sp.) can degrade AHLs longer (Huang et al. 2003) and shorter than eight carbons (Park et al. 2005), respectively. It is important to note that AHL-acylases are more advantageous than AHL-lactonases in the perspective of biotechnological applications. For example, the AHL-lactonase product, i.e. *N*-acyl homoserine, is readily converted back to original AHL by recircularization at acidic pH. Such type of reversal (i.e. reformation of QS molecules from enzyme products) is not possible with AHL-acylase. Furthermore, fatty acids that are formed by the action of AHL-acylases are not long lasting, instead they are metabolized quickly.

4.4.6 AHL-Oxidoreductases

AHL-oxidoreductases are another group of QQ-enzymes, able to catalyze the hydroxylation of native AHLs and products of lactonolysis. Upon the oxidation of AHLs, the product may have still QS properties, but significantly less active than parent AHLs (Chowdhary et al. 2007). One of the examples for the AHL-oxidoreductase is cytochrome P450 monooxygenases CYP102A1 (also known as P450BM-3), isolated from *Bacillus megaterium*. This enzyme interacts with a similar affinity with two substrates such as AHLs and *N*-acyl homoserines. There are also other types of AHL-oxidoreductases isolated from different bacteria and soil metagenome, and the details are shown in Table 4.1.

4.5 QQ Technology in Medicine

Post-antibiotics era is vastly needed in the treatment of bacterial human infections as there is a threat of emergence of multi-drug resistant bacteria. Recent investigations are manifesting that many pathogens are known to use QS to regulate their virulence (Raina et al. 2009; Antunes et al. 2010; Kalia and Purohit 2011), which opened the door to think about an alternative virulence control. The immediately discernible option is 'quorum quenching', which is an attractive alternative approach to attenuate human bacterial infection by the disruption of QS signals (Galloway et al. 2012) (Table 4.2). The first clinical trial has been conducted to evaluate the impact of QQ on the control of human cystic fibrosis, where garlic oil was used as a QS inhibitor. These gave encouraging results in the general improvement in cystic fibrosis symptoms; however, this study hasn't been conducted on a large scale for a better evaluation (Smyth et al. 2010).

Several innovative and ecofriendly approaches have emerged from the QQ-strategies as the awareness on QQ increased. For example, enzymatic QQ technology had shown effective results in the control of several bacterial infections caused by infected catheters or ventilators (*A. baumannii, Proteus* spp., *P. aeruginosa, Serratia* spp. etc.) (Bzdrenga et al. 2017). The virulence of *P. aeruginosa* culture has been decreased by QQ enzymes (e.g., SsoPox or PLLs) immobilized on the nanoalumina membrane.

Table 4.2 Applications of QQ-strategy in medicine

QQ agent (protein superfamily)	Effect	Target pathogen	Reference
AiiA (AHL-lactonase) from *Bacillus* sp.	Inhibition of biofilm formation	*Vibrio cholerae*	Augustine et al. (2010)
Arctic actinomycetes extract (*Streptomyces* sp. and *Nocardiopsis* sp.)	Inhibition of biofilm formation	*V. cholerae*	Augustine et al. (2012)
Resveratrol (AphB)	Inhibition of biofilm formation	*V. cholerae*	Augustine et al. (2014)
BpiB09 (Oxidoreductases)	Inhibition of swarming motility, biofilm formation, and pyocyanin production	*P. aeruginosa*	Brackman et al. (2011a)
Cinnamaldehyde	Inhibition of biofilm formation	*V. cholerae*	Brackman et al. (2011a)
Engineered QQ-lactonase	Biofilm disruption	*A. baumanii* S1	Chow et al. (2014)
Cranberry extract	Block the initial attachment and inhibit cholera toxin (LuxO-HapR)	*V. cholerae*	Dinh et al. (2014)
Transition state analogs of MTANs (5′-Methylth-ioadenosine/*S*-adenosylhomocysteine nucleosidase): MT-, EtT- and BuT-DADMe-ImmA	Decrease the virulence	*V. cholera, E. coli, S. pneumoniae, N. meningitidis, Klebsiella pneumoniae, S. aureus,* and *Helicobacter pylori*	Gutierrez et al. (2009)
Acylase AiiO (acyl-homo-serine lactones 'AHLs')	Inhibition of virulence factors (e.g., elastase and T3SS)	*P. aeruginosa*	Li et al. (2019)
Extract of marine *Bacillus* sp. SS4	Inhibition of AHL-mediated virulence	*P. aeruginosa* PAO1	Musthafa et al. (2011)
Marine bacterial extract (e.g., *Oceanobacillus*, *Halomonas*)	Inhibition of biofilm formation	*P. aeruginosa* PAO1, *Serratia marcescens;*	Nithya et al. (2010)
Phenylbutanoic acid: Bacterial extract of S6-01 (*Bacillus indicus*); S6-15 (*Bacillus pumilus*)	Inhibition of biofilm formation	*E. coli, Shigella flexneri, Proteus mirabilis*	Nithya et al. (2011)
Small molecules quinoline-containing class of mefloquine analogues	Inhibition of biofilm formation	*V. cholerae*	Peach et al. (2011)
Flavonoid compounds (naringenin and quercetin)	Inhibition of biofilm formation	*V. harveyii, V. cholerae*	Vikram et al. (2011)

(continued)

Table 4.2 (continued)

QQ agent (protein superfamily)	Effect	Target pathogen	Reference
QQ proteins (QQ-5, QQ-7) from bacteria	Impaired biofilm formation	*Candida albicans*, *Staphylococcus epidermidis*	Weiland-Bräuer et al. (2019b)
QQ enzymes (lactonases and PON like properties) from mammalian sera	AHLs degradation	*P. aeruginosa*	Yang et al. (2005)
Lactonases and acylases	Reduce biofilm formation	*P. aeruginosa* PAO1	Rehman and Leiknes (2018)
AmiE (*N*-aylhomoserine lactone acylase)	Degradation of AHLs (preferably long chain)	*P. aeruginosa* PAO1	Ochiai et al. (2014)
Homogentisic acid γ-lactone	Inhibit the production of pyocyanin and extracellular matrix	*P. aeruginosa*	Li et al. (2018)
Enzymes acylase and α-amylase	Prevent bacterial biofilm formation on urinary catheters	*P. aeruginosa*, *S. aureus*	Ivanova et al. (2015a)

Coating of catheters with 5-fluorouracil (QSI–QS inhibitor), furanone (synthetic QSI), acylase, acylase + amylase has given effective results in the control of pathogenic bacteria such as *P. aeruginosa* 10145, *S. aureus*, etc. (Choudhary and Schmidt-Dannert 2010; Ivanova et al. 2015a, b; Bzdrenga et al. 2017). A thermostable lactonase has successfully decreased the mortality rate and virulence factors in biofilms 55% and 65%, respectively during the treatment of pulmonary infections such as cystic fibrosis infections caused by *P. aeruginosa* PAO1 (Bzdrenga et al. 2017). Furthermore, QQ-strategies have also been tested at the topical level. Topical application of a lactonase from *Bacillus* sp. ZA12 has prevented the spread of *P. aeruginosa* on burned skin animals and subsequently reduced the mortality of animals. These results were even more effective when lactonase was used in combination with an antibiotic, implying that there is excellent compatibility and synergism between QQ-molecules and antibiotics. It means, no one may doubt that more novel QQ approaches such as enzymatically functionalized dressings and bandages will appear in the future. In the medical field, QQ has crossed the border to dental care. Many dental plaque biofilm-forming bacteria (e.g., *Aggregatibacter actinomycetemcomitans*, *Porphyromonas gingivalis*, and *Streptococcus* sp.) that have been successfully controlled by QQ-strategies resulted in a better oral hygiene (Basavaraju et al. 2016). Unfortunately, QQ-approaches in the field of dentistry are not advancing well which might be due to less knowledge of QSIs.

In recent years, several model experiments have been conducted to decipher the role of QQ-strategies in the control of virulence and disease in animal models (refer for a review Rémy et al. 2018). An aliphatic amidase AmiE (a QQ enzyme) could disrupt the QS followed by reduction of virulence of *P. aeruginosa* in a protozoan model, i.e. *Dictyostelium discoideum* (Clamens et al. 2017). *Caenorhabditis elegans* (a roundworm), is one of the widely used infection models to study the microbial virulence. Decreased virulence in both gram-positive (e.g., *S. aureus*) and gram-negative bacteria (e.g., *Burkholderia cepacian, E. coli, P. aeruginosa* etc.) and enhanced *C. elegans* survival were observed when infected *C. elegans* were treated with three different QQ agents, such as **(i) QSIs** (Baicalin hydrate (Brackman et al. 2011b), broccoli extract (Lee et al. 2011), hamamelitannin (Brackman et al. 2011b), 2,5-piperazinedione (Musthafa et al. 2012a), phenylacetic acid (Musthafa et al. 2012b), clove oil (Husain et al. 2013), menthol (Husain et al. 2015), tea polyphenols (Yin et al. 2015), *Mangifera indica* methanol leaf extract (Husain et al. 2017), etc.), **(ii) QQ enzymes** (AiiA lactonase (Wopperer et al. 2006), PvdQ acylase reductase (Papaioannou et al. 2009), BpiBo9 reductase (Bijtenhoorn et al. 2011a), MomL lactonase (Tang et al. 2015) etc.) and **(iii) QQ bacteria** (*B. cenocepacia* LGM16656, *P. aeruginosa* PAO1 (Christiaen et al. 2014) etc). Furthermore, the mammalian models (e.g., rats or mice) were not exempted from this kind of research. In the beginning, several studies have recognized the significance of site-directed mutagenesis of QS genes in the reduction of severity of several infections such as lungs (Pearson et al. 2000), wound burns (Rumbaugh et al. 1999), peritonitis (Sifri et al. 2002), and prostate gland (Nelson et al. 2009). These results have encouraged us to investigate further level with the direct application of QQ molecules in experimental animals. For example, inhalation of lactonase SsoPox has reduced *P. aeruginosa* colonization or related mortality in lung infection models (Hraiech et al. 2014). AHL degrading enzyme was also found to be effective in controlling virulence caused by *P. aeruginosa* in a burn wound infection model animal (Gupta et al. 2015). The use of tea polyphenols as QSIs was found to be effective in excision injury models (Yin et al. 2015). Similarly, several QSIs have also reduced the *S. aureus* pathogenicity in skin wound model animals (Simonetti et al. 2016; Muhs et al. 2017). Interestingly, co-treatment with a QSI and an antibiotic has drastically reduced colonization of artificial foreign body (such as catheter or implants) by pathogenic bacteria such as *S. aureus* (Simonetti et al. 2016), *P. aeruginosa* (Das et al. 2016), etc.

For identification of successful results with animal models, a couple of human trials have also been conducted with QQ molecules. In the early 2000s, azithromycin was used in clinical trials to treat cystic fibrosis (Wolter et al. 2002) and pulmonary transplanted patients (Gerhardt et al. 2003), but azithromycin usage did not lead to a decrease in bacterial load despite improving patient's quality of life (Saiman et al. 2003). But QSI nature of azithromycin was unveiled during the evaluation of QS activities of *P. aeruginosa* in ventilator-associated pneumonia patients (van Delden et al. 2012). Likewise, a couple of other substances have emerged as QQ molecules in the human clinical trials, e.g., garlic extracts in cystic fibrosis patients (Rasmussen et al. 2005) and 5-FU (a pyrimidine analog) in *P. aeruginosa* (Ueda et al. 2009); overall, very few QQ molecules have reached the stage of human clinical trials, which is attributed to the functional differences between the animal and human system during the interaction of QQ molecules with their targets.

4.6 QQ Technology in Agriculture

There is a mounting demand for crop productivity for two principal reasons, firstly continuous increase in human population and secondly interest of using the crop grain for biofuel production. It is not uncommon that QS plays a vital role in numerous plant pathogens; therefore, inhibition of QS signals could be noteworthy. Transgenic plants with a gene encoding QQ enzyme can protect the plant and crop (Dong et al. 2000, 2001); in this angle, several laboratory-based experiments were conducted to test the efficacy of the QQ gene in transgenic plants. The transfer of QQ enzyme encoding genes from bacterial origin (e.g., *Bacillus*AiiA and *Agrobacterium*AttM lactonases) to different plants has either lowered the symptoms or resulted in the total absence of symptoms induced by a QS-inducing pathogen (e.g., *Pectobacterium*) (Dong et al. 2001; Ban et al. 2009; Vanjildorj et al. 2009; D'Angelo-Picard et al. 2011). Similar results were also observed with the production of transgenic tobacco, and potato plants were designed to release 'lactonase' and one plant-associated bacteria *E. carotovora* was engineered to produce 'Aiia enzyme'(Turan and Engin 2018). Furthermore, it has also been investigated and confirmed that the transgenic process is not going to change the rhizosphere population (D'Angelo-Picard et al. 2011), implying that QQ-strategy is a kind of eco-friendly approach where this strategy can develop disease resistance in the plants without affecting the relationship between plants and the surrounding environment. On the other hand, mutational studies revealed that *aiiA*-defective strain of *B. thuringiensis* did not exhibit any AiiA lactonase activity, as well as the strain failed to protect the plant against potato-tuber caused by *Pectobacterium* (Dong et al. 2004). Furthermore, natural multiplication of a QQ bacterium namely *M. testaceum* on the surface of potato leaf could protect the plant from soft rot disease (Wang et al. 2010). The above results indicate that genes encoding QQ enzymes seem to be stable in their expression, and these genes are quite fit 'transgenics' which encourage researchers to screen a large number of QQ genes to be expressed in a wide range of plants.

Another way to implement QQ strategy in agriculture is 'biostimulation' of QQ bacteria (e.g., *R. erythropolis*) by elective carbon sources (e.g., gamma-caprolactone and gamma-heptanolactone) (Grandclement et al. 2016). Elective carbon sources such as gamma-caprolactone and gamma-heptanolactone have been used to stimulate the growth and root colonization of the introduced and native population of QQ bacteria, i.e. *R. erythropolis*, and this has caused disease resistance in the plants (Cirou et al. 2007, 2012; Barbey et al. 2012). Certain types of QQ enzymes were also found to be effective in controlling plant pathogens, for example, esterase is a specific 3-hydroxypalmitate methyl ester, a kind of QS molecule used by *Ralstonia solanacearum* (Bzdrenga et al. 2017). Also, some QSIs and QQ microorganisms have been used to disrupt the QS-mediated virulence in plant pathogens (soft-rot *Pectobacterium* spp.), but the efficacy of a given QSI was highly dependent on the test pathogen, and this situation dramatically demands that QSIs have to be characterized for each of the targeted plant pathogens. Interestingly, some QQ bacteria namely *Bacillus* spp. and

Rhodococcus spp. found in the soil and rhizosphere regions (of potato) were able to fight against *Proteobacterium* (a plant pathogen) through QQ-strategy without affecting its growth (Uroz et al. 2008), implying that there is a natural existence of QQ activities in the plant environment, but not many species of such bacteria are known. Details about different QQ isolates (if any) will be constructive in the further implications of QQ-strategies in agriculture sector. Also, the activities of QQ enzymes of certain bacteria are not entirely comprehended. For instance, QS-signal inactivation in *R. erythropolis* is governed by several enzyme activities like lactonase, amidase, and reductase (Uroz et al. 2005, 2008); however, only lactonase (encoded by*qsdA* gene) has been characterized so far (Uroz et al. 2008). When *qsdA*-defective mutant of *R. erythropolis* was tested against a pathogen *Pectobacterium*, there was a weak or no effect on QS degradation and potato-tuber protection (Uroz et al. 2008; Barbey et al. 2013), which strongly implies that enzymes other than lactonases (in this case amidase and reductase) cannot be ignored completely. It is noteworthy that QQ-strategies in the plant field raise many doubts, for example, interference between the biocontrol agents that produce and quench QS signal molecules in not completely understood. Because in one study it was found that the QS producers and quenchers had great incompatibility (Molina et al. 2003), suggesting that there is a complex interplay between these two communities in eliciting one final behavior in a given community. Nonetheless, there is intense research in progress to mute the signaling in phytopathogens (Table 4.3), as well as down-regulate the expression of virulence factors.

Table 4.3 Applications of QQ-strategy in agriculture

QQ agent (protein superfamily)	Effect	Target	Reference
AiiA (AHL-lactonase) (e.g., transgenic plants with AHL-lactonase gene)	Enhanced resistance	*E. carotovora*	Dong et al. (2001)
α-D-Galactopyranosyl-(1–2)-glycerol (floridoside); betonicine; isethionic acid (structurally unrelated to AHLs, isolated from red alga, *Ahnfeltiopsis flabelliformis*)	Inhibit the AHL production	Recombinant *A. tumefaciens* liquid culture bioassay	Kim (2007)
QQ-bacteria on the rhizosphere of *Zingiber officinale*(e.g., *Acinetobacter* sp. GG2, *Burkholderia* sp. GG4 and *Klebsiella* sp. Se14)	GG2 and Se14: AHL Degrading activity via lactonolysis GG4: reduced 3-oxo-AHLs to corresponding 3-hydroxy compounds	Attenuate the virulence factors in plant pathogens	Chan et al. (2011)
AttM (AHL-lactonase)	Degradation of acyl-AHLs and reduced virulence	*Erwinia* strain 6276 and other bacterial pathogens of Plants	Carlier et al. (2003)

4.7 QQ-Strategy in Aquaculture

Stringent rules in antibiotic usage in aquaculture and threat of antibiotic resistance development in pathogens have jointly led to the initiation of research to set QQ-strategy as an alternative for the disease control in aquaculture. In the beginning, QQ enzyme (AHLase) of the gut microbial community of fishes (e.g., *Dicentrarchus labrax* L. and *Lates calcarifer*) has been successfully used as a biocontrol agent in prawn (e.g., *Macrobrachium rosenbergii*) larviculture (Nhan et al. 2010), and this approach resulted in the sustainable aquaculture development. *Macrobrachium rosenbergii* larval growth, larval survival, larval quality, and duration of the larval rearing process have dramatically been improved by AHLase, which was attributed to the down regulation of *Vibrio harveyi* infection in larva. Similar results have also been observed with the purified molecules of AiiO-AIO6 (Cao et al. 2012), AHLase of pure-species bacteria (e.g., *Bacillus* sp. QSI-1) (Chu et al. 2014) vs. infected zebrafish. All the above results imply that QQ-strategies are well investigated in the field of aquaculture to combat pathogenic bacteria (Table 4.4), and these strategies will soon replace the antibiotics.

For a sustainable aquaculture industry, novel strategies to control bacterial pathogens are must, and such strategies should replace antibiotic application. In the commercial aquaculture, one of the most critical problems is massive economic loss by

Table 4.4 Applications of QQ-strategy in aquaculture

QQ agent (protein superfamily)	Effect	Target pathogen	Reference
AiiC and Aac (AHL-acylase)	Reduced AHL production and disruption of biofilm formation in fish pathogens	*V. anguillarum*	Morohoshi et al. (2008)
Cinnamaldehyde and cinnamaldehyde derivatives	Reduce virulence (interfered with Al-2 based QS) and inhibit biofilm formation	*V. harveyii*	Brackman et al. (2008)
Hexyl-4,5-dihydroxy-2,3-pentanedione (alkyl-DPD analogues)	Inhibitors of QS	*V. harveyi*, *S. typhimurium*	Lowery et al. (2009)
4-hydroxy cis or trans analogs	Inhibit LuxI/LuxE	*Vibrio fischeri*	Olsen et al. (2002)
N-sulfonyl-HSL	Inhibit the action of 3-oxohexanoyl-L-homoserine lactone	*V. fischeri*	Schaefer et al. (1996)
Furanone C-30	Inhibit the toxicity	*V. anguillarum*	Rasch et al. (2004)
N-acyl homoserine lactone degrader	Increase fish survival (reduce swimming motility and biofilm formation by a fish pathogen)	*Yersinia ruckeri*	Torabi Delshad et al. (2018)
QQ-ORF proteins from metagenomic library from medusa-derived mucus	To maintain a healthy microbiota	Defend specific bacteria	Weiland-Bräuer et al. (2019a)

bacterial diseases; as a result there are high mortality rates in aquacultured organisms such as mollusks, crustaceans, fish, etc. (Defoirdt et al. 2007). Principal infection is vibriosis, transmitted by feeding substances such as algae, rotifer, and *Artemia* (brine shrimps). Two decades ago, the anti-QS activity of microalgae (e.g., *D. pulchra*) had been observed for the first time in the aquaculture system; QQ activity of *D. pulchra* is attributed to the production of furanones (Givskov et al. 1996). After 10 years, the importance of natural and synthetic brominated furanones as QQ-agents came to light in aquaculture, and they were found to protect brine shrimps (*Artemia franciscana*) from the pathogenic vibrios (*V. harveyi, V. campbellii, V. parahaemolyticus,* etc.) (Defoirdt et al. 2006). It is important to note that these compounds were found to be toxic to the higher organisms (e.g., rainbow trout) (Rasch et al. 2004). After knowing the importance of QSIs, attempts have been made to investigate QS-inhibiting algae in both marine and freshwater systems. In one investigation, *Chlorella saccharophila* has been identified as a promising agent for controlling the producers of different QS molecules, such as unsubstituted, oxo- and hydroxyl-substituted AHLs, and AHLs of *V. harveyi* (Natrah et al. 2011).

At a further level, attempts have been made to achieve the QQ through the microorganism or molecules. Gut microflora of the shrimp (*Penaeus vannamei*) had a capacity to degrade QS molecules (e.g., HAI-1), and such microflora (so-called QQ microorganisms) were able to improve the growth rate of rotifers when challenged with a pathogen, i.e. *V. harveyi* (Tinh et al. 2007). Similar results have also been observed in other studies too. Intestinal microflora of certain fish (e.g., *Dicentrarchuslabrax* L. and *Lates calcarifer*) were found to show AHLase activity, and such microflora proved to be an effective biocontrol agent in prawn (*Macrobrachium rosenbergii*) larviculture (Nhan et al. 2010). Furthermore, pure-species of gut mircroflora also showed similar results, e.g., *Bacillus* sp. QSI-1 from a fish *Carassius auratus* showed an increased survival rate of infected zebrafishes (Chu et al. 2014), which is attributed to the AHLase activity of str. QSI-1. Based on these successful results, many patents have also emerged on specific commercial products which contain QQ strains (e.g., *Bacillus* sp.) able to produce AHLase. One such commercial product is AquaStar® from Biomin (https://www.biomin.net/en/products/aquastar/), which improves the larval hatchery production through the strategy of QQ. The mode of application of QQ bacteria in aquaculture is either incorporation of these strains in the rearing water or bioencapsulation in the feedstock. Also, the use of probiotic bacteria alone (D'Alvise et al. 2013) and in combination with QQ-strategy (Prol García et al. 2013) was found to be effective in controlling the diseases in aquaculture, because in certain cases, QS-independent mechanisms are needed to control the pathogens (e.g., *Vibrio anguillarum*) in aquaculture environment. Integrated QQ-strategies seem to play a crucial role in shrimp aquaculture shortly (IDRC 2016). It is important to note that shrimp culture accounted for more than 50% of overall aquaculture production in 2016. This industry is suffering from severe outbreaks of acute hepatopancreatic necrosis disease, caused by *Vibrio parahaemolyticus*, which leads to severe economic losses (IDRC 2016). It is noteworthy that *Vibrio* sp. is found to have multi-resistance towards many known antimicrobials. At the same time vaccination is not effective in shrimps since they lack adaptive immunity. Therefore, QQ-strategies are looking like new alternatives to protect shrimps from infectious diseases.

4.8 QQ Technology in Wastewater Treatment

One of the most eco-friendly and cost-effective approaches to remove contaminants from wastewater is 'biological treatment' (Yang et al. 2017). It is a well-known fact that bacteria regulate their density and community behavior (such as bacterial aggregates in the form of biofilm (Nadell et al. 2016)) in biological wastewater treatment facilities by QS. Many bacterial species in WWT facilities (e.g., sludge) can cause fouling on membranes, which has been studied through single-species cultures by using dead-end filtration settings (Maddela et al. 2018). The role of bacterial biofilms in WWT facilities is either beneficial (i.e. biocatalysis) or detrimental (i.e. biofouling) (Shrout and Nerenberg 2012). Thus, the current challenge is how to manipulate the biofilms without affecting the survival and growth of bacterial community (Huang et al. 2016)? Principal obstacles caused by excess biofilm in membrane biofilm reactors (MBfR) and moving bed biofilm reactors (MBBR) are—limitations in mass transfer, fouling issues, and high liquid headloss, which collectively cause poor performance of biofilm reactors, at the end (Hwang et al. 2010). Therefore, maintaining a constant biofilm thickness during the reactor operation is a challenging issue.

In the area of WWT, the idea of QQ-MBR (membrane bioreactors) was harbored in the year 2002, and then this opened the doors (Table 4.5), and the first practical feasibility of QQ-enzymes for biofilm (so-called biofouling in MBRs) control at the lab-scale level was observed within 5 years of research (i.e. 2007) (Oh and Lee 2018). Since then, research on 'QQ' is being continued through different variations, such as development of immobilized QQ enzymes in 2008, first-time isolation of QQ bacteria from WWT facilities in 2009, and this led to open a new era of bacterial QQ-MBR in the year 2011. In the subsequent years, there was an observation of QQ-beads in 2012, and successful production of QQ-fiber test bead and QQ-sheets in 2015 and 2016, respectively. Then in 2017, AI-2 (autoinducer) QQ (with QQ-RO (reverse osmosis), QQ-AnMBR (anaerobic MBR)) came into practice. Now the QQ-strategy is waiting for its first commercial QQ-MBR. Though plenty of investigation has been done in the fields of QQ-microorganisms, QQ-media, and the size of QQ-MBRs, still commercial QQ-MBR does not have a visible future, which is attributable to the fact that the QQ-media production cost is not economically feasible, information of QS molecules (AHLs, and AIs) is minimal, and there is a difficulty in making a synergism between AHLs-QQ and AI-2-QQ and difficulty in the optimization of design and media for QQ. And beyond QQ-MBR, still there is no idea at all over the commercialization of An-MBR (anaerobic MBR), RO-MBR (reverse osmosis MBR), FO-MBR (forward osmosis MBR), etc. even though all these MBRs have been proven in lab-scale experimentation. Therefore, additional research is obligatory to overcome the hurdles discussed above.

Table 4.5 Applications of QQ-strategy in industry

QQ agent (protein superfamily)	Mechanism	Effect	References
UV photolysis	Inhibit microbial group behaviors	Anti-biofouling in MBRs	Zhang et al. (2019)
AHL-lactonase AiiM	Inhibit quorum sensing	*S. marcescens* AS-1	Okano et al. (2019)
All3924 (homology to the acylase QuiP of *P. aeruginosa* PAO1	Interference of QS signaling	*Anabaena* sp.	Romero et al. (2008)
Complex biostimulating agent (GCL) in core and encapsulated QQ bacteria (*Rhodococcus* sp. BH4)	Inhibition of QS	Anti-biofouling in MBRs	Yu et al. (2019)
QQ-enzymes and QQ-bacteria (e.g., *Afipia* sp., *Acinetobacter* sp., *Pseudomonas* sp., *Micrococcus* sp., *Microbacterium* sp., *Rhodococcus* sp., and *Streptococcus* sp.,)	Inhibition of biofilm formation	Anti-biofouling in MBRs	Kim et al. (2014)
AmiE (*N*-Acylhomoserine lactone acylase)	Degradation of AHLs	*P. aeruginosa* PAO1	Ochiai et al. (2014)
QQ-bacteria (encapsulated in a dumpling-shaped microbial bag)	Degradation of C8-HSL (*N*-Octonoyl-DL-homoserine lactone); C6-HSL (*N*-Hexanoyl-L-homoserine lactone)	Anti-biofouling in MBRs	Gu et al. (2018)
Rhodococcus sp. BH4 (microbial-vessel system)	Hydrolysis oflactone bond of AHL by AHL-lactonase and mitigation of biofilm formation	Anti-biofouling in MBRs	Oh et al. (2012)
The AHL-lactonase (qsdA of *R. erythropolis*)	Degraded AHL molecules intracellularly by hydrolyzing the lactone ring of AHLs	Biofouling inhibition in the continuous MBR.	Oh et al. (2013)
Microbacterium sp. embedded beads (QQ-bacterium)	Degradation of AHLs by the QQB and subsequent suppression of EPS and SMP production	Control membrane biofouling in aerobic membrane bioreactors	Liu et al. (2019)
β-Lactam antibiotic resistant QQ bacteria from penicillin contaminated river sediments	Degradation of a broad range of AHLs including 3-oxo-substitutes	Possible use in biofouling control	Kusada et al. (2019)

In the recent past, several investigations reported that QQ enzymes (Lee et al. 2017) and QQ-bacteria/fungi (Ergön-Can et al. 2017) were effective in disrupting the QS system in different WWT facilities. A QQ enzyme, i.e. acylase I of 5 μg/mL did reduce biofilm formation by 60–73% in *A. hydrophila* and *Pseudomonas putida* (Paul et al. 2009), and quenching effects of these enzymes were even more effective when they were immobilized than in the free form (Lee et al. 2017). However, QQ-microorganisms were found to be more effective than QQ-enzymes, which is attributable to the high cost in the production of enzymes and the less half-life of QQ-enzymes. Attempts have been made to isolate indigenous QQ-bacteria from WWT plants and applied to control the biofilm formation by bacteria isolated from the same environment. By these attempts, many QQ-bacteria came to light (*Acinetobacter* sp., *Afipia* sp., *Microbacterium* sp., *Micrococcus* sp., *Pseudomonas* sp., and *Rhodococcus* sp.) and confirmed that these QQ-bacteria were found to be effective inhibitors of biofilms in both the initial- and established-stage (Kim et al. 2014). Because of such effective results of QQ-strategies, QQ-based methods got wider acceptance in the area of WWT facilities, and are being studied extensively and being applied in the membrane bioreactors (Oh and Lee 2018; Maddela et al. 2019). Yet we have many challenges to be addressed in the area of QQ-based biofouling control in MBRs, and as a result, the current status of QQ-MBR is far from its commercial stage.

It is not surprising to note that WWT settings are the suitable habitats for co-existence of both AHL-producing and AHL-degrading microorganisms. One experimental proof for the co-existence of QS- and QQ-bacteria in activated sludge is that there was no induction of β-galactosidase of a reporter strain *A. tumefaciens* KYC55 when incubated (6 h) with a known AHL (3-oxo-C8-HSL) (Song et al. 2014) which suggest that there was an existence of indigenous QQ activity. Similarly, in gamma-caprolactone (GCL) add-back studies with activated sludge, QS signals were interrupted by several QQ-bacteria namely *Rhodococcus* sp. BH4, *Pseudomonas* sp. IA1 and *Variovorax* paradoxus isolated from sludge, biocake, and nitrifying–denitrifying activated sludge, respectively (Huang et al. 2003; Cheong et al. 2014; Ochiai et al. 2014; Kim et al. 2015). In specific environments, the signal-quencher population is higher than signal-producers' (Tan et al. 2015), implying that there is an excellent level of interplay between QQ- and QS-microorganisms. Now the question is what is the exact role of indigenous QQ in wastewater treatment facilities? The impact of many environmental factors on QQ communities is not fully understood yet. Therefore, to exploit more about their distribution and existence in each ecological niche, additional future studies are necessary, which may reduce the gap between the lab- and commercial-stage of QQ-MBRs.

4.9 Case Studies

Following are some of the case-studies that provide contextual knowledge over QQ in different fields and their progress.

4.9.1 Anti-biofilm and Anti-QS Activities of the Selected Anthocyanidins Against *P. aeruginosa* PAO1

Anti-biofilm activity of three anthocyanidins (e.g., pelargonidin, cyanidin, and delphinidin) was evaluated for the first time under in vitro conditions against biofilm formation potential of *P. aeruginosa* PAO1, for medical reasons (Pejin et al. 2017). The activity of the above listed anthocyanidins was tested at three different concentrations (e.g., 0.5, 0.25 and 0.125 MIC). The percent inhibition of the biofilm of *P. aeruginosa* PAO1 by delphinidin at concentrations (MIC) of 0.5, 0.25, and 0.125 was 85, 77 and 57, respectively. Similarly, the biofilm inhibition (%) values for pelargonidin were continuously 76, 70 and 62, and for the values for cyanidin were 85, 83, and 79, respectively. It means that cyanidin was found to be more effective over the other two anthocyanidins at all the tested concentrations; for instance, the values of percent inhibition of the biofilm (at 0.125 MIC) for cyanidin were 22% and 28% higher than pelargonidin and delphinidin, respectively.

4.9.2 Innovative Veterinary Solutions for Antimicrobial Resistance

Global health agencies (e.g. International Development Research Centre, IDRC) have initiated research to reduce the risk of antimicrobial resistance in livestock, for the sustainable human health and food security (IDRC 2019). A growing problem in animals is the development of antimicrobial resistance (AMR), which threatens our practices that are available to treat bacterial infections in livestock. As a result, livestock and aquaculture industries in low and middle-income countries are affected massively by increased outbreaks of infectious diseases and resultant economic loss in livestock productivity, which ultimately threatens food security and disrupts international trade. Furthermore, contamination of animal products with multi-antimicrobial-resistant bacteria is a risk factor for humans, animals, and the environment. In this direction, very recently, IDRC has announced CA (Canada) $ 21.2 million in research funding for 11 new projects under the name of 'Innovative Veterinary Solutions for Antimicrobial Resistance (InnoVet-AMR) Initiative' (IDRC 2019). The principal objective of this project is to identify innovative veterinary solutions (including vaccines and alternative solutions), and to reduce the utilization of antimicrobials in livestock and aquaculture operations.

4.9.3 These Brazilian Red Berries Could Hold the Key to Fighting Deadly Superbugs

A superbug namely methicillin-resistant *Staphylococcus aureus* (MRSA), which causes mild (e.g., lesions on the skin) and severe life-threatening infections in human, accounts for as many as 11,000 deaths/year in USA. But researchers at Emory University (USA) found that a berry extract of Brazilian peppertrees effectively protected the mice infected with MSRA from developing skin lesions, reported in *ScienceAlert* under the 'Health' section (Dockrill 2017). This is attributed to the disruption of QS in bacteria. Traditional healers in Amazon have used such extract for hundreds of years to treat infections of the skin and soft tissues. A chemical characterization revealed that this berry extract is called as 430D-F5, and is a mixture of 27 chemicals. It has also been found that this compound did not kill the bacteria, rather inhibited the gene expression of cell-to-cell communication, so-called QQ. A single dose of berry extract could show positive effects for up to 2 weeks. Currently, this research team is looking for a safe and effective method for testing this extract in humans, with a possible clinical trial in the nearest future.

4.9.4 Keep Gut Health Challenges at Bay

A commercially available product namely 'Ecobiol', which is composed of *Bacillus amyloliquefaciens* CECT 5940, is found to be effective in preventing the development of 'dysbacteriosis' in birds. Intestinal microbial imbalances or dysbacteriosis is a common disorder in birds (e.g., poultry) which is attributed to changes in feed, poor water quality, heat stress, high stock densities, microbial infections, etc. Therefore, to make the poultry operations profitable, bird performance should be optimized with good gut health. Many in vitro and in vivo tests have confirmed that str. CECT 5940 is potential in the production of secondary metabolites, QQ, modulation of the immune system, and production of lactic acid; all these mechanisms are essential for the maintenance of gut health. QQ activity of *B. amyloliquefaciens* CECT 5940 has practically been proved with a reporter strain i.e. *Chromobacterium violaceum* CV026. Ecobiol® containing str. CECT 5940 successfully prevented the production of violacein by *C. violaceum* CV026, in the presence of C6 HSL (Doranalli and Ortiz 2018). Furthermore, probiotic activities of *B. amyloliquefaciens* CECT 5940 have been confirmed in many case studies, where strain CECT 5940 could inhibit the population of *Clostridium perfringens* and *Salmonella* spp. upon challenging the experiment broiler chickens (PoultryWorld 2018).

4.9.5 Probiotic Good Bacteria Use Fengycins to Eliminate Bad Bacteria

Oral *Bacillus* is found to have QQ activities against the gut colonization of *S. aureus*, which has been confirmed in mice models and human volunteers, revealed in a combined investigation of scientists from NIH (National Institutes of Health, USA) and Thailand (GEN 2018). The pathogenic *S. aureus* leads to tens of thousands of deaths every year; on the other side, there is a great threat of antibiotic resistance development in these strains (e.g., MRSA). An investigation has been carried out with 200 human volunteers in rural Thailand, and it was found that 50% of people carried *Bacillus subtilis* in the gut; furthermore, *S. aureus* was always absent (both in the gut and nose) when *Bacillus* was present implying that *Bacillus* spp. might be an effective alternative to antibiotic treatment. Later,

Agr-based QS was found to be the vital factor for the colonization of *S. aureus*, confirmed in a mouse model (Piewngam et al. 2018). Then, the same investigation reported that an enzyme resistant cyclic lipopeptide namely fengycin B produced by *Bacillus* spp. was responsible for the blocking of Agr-mediated QS in *S. aureus*. These results have been reconfirmed by feeding the mice with *B. subtilis* spores, which resulted in the complete wipe off all *S. aureus* strains in the animal feces (Piewngam et al. 2018). Similar results have also been identified in the Thai human volunteers.

4.10 Challenges and Prospects of QQ Technology

Many challenges remain to be addressed in the area of QQ implications towards a sustainable environment. Some of them are as below:

1. High cost and instability of QQ enzymes limit their applications for controlling QS based problems. More importantly, there is significant uncertainty in the physiological function of QQ-enzymes which is attributed to difficulties in the detection of these enzymes in the environment (Chen et al. 2013). Vast diversity (chemically and physically) in QQ-enzymes is also another limiting factor for their applications. Nonetheless, the current situation of QQ-enzymes demands future research focusing on functional genomics to find out genuine redundancy among the relatively homologous enzymes. Such studies may not only help in the finding of novel QQ- or QS-microorganisms (bacteria) but also encourages the researchers to design QQ or QS like molecules.
2. Because of practical and ethical problems, murine models are not the best for the screening of QQ agents to be used in human medicine. Furthermore, in murine models, the physiological aspects of pathology such as wound healing or inflammation are not fully mimicked with human beings; this greatly discourages the implications of QQ-strategies in human medicine. As of now, very few QQ agents have reached the level of human clinical trials, even though these agents tend to show beneficial effects of QSIs. Plenty of investigation has been done with experimental animals; however, validation of these approaches in clinical phases is obligatory to confirm the therapeutic relevance of QQ agents.
3. Possibility for the development of resistance to QQ agents is a hot topic in most of the reviews published in the recent past. In one of our very recent investigations (Maddela and Meng 2020), biofilm biomass in 6 out of 23 dual-species (bacteria) biofilms has been increased in the presence of *Rhodococcus* sp. BH4, which raises a doubt that the QS-bacteria have developed any resistance to QQ-bacteria. Existence of multiple QS regulatory genes within strains of the same species (Defoirdt et al. 2010) also led to a lot of arguments over the risk of resistance if this diversity induces fitness differences under QQ conditions.
4. How about the self-degradation of AHLs for certain bacteria, such as *Agrobacterium* and *Pseudomonas*?
5. In the plant sector, most QQ approaches have been evaluated under in vitro conditions, and it is hard to find the field-level results in the literature. Results which

are obtained at lab-level may not be close to the reality (Grandclement et al. 2016; Bzdrenga et al. 2017). Though many plants are able to take up and respond to QS signals, only a few plants (e.g., clover and birdfoot deervetch) are known to show QQ activities (Palmer et al. 2014). Still we are far from complete understanding about plant QQ-enzymes and their corresponding genes. Because of this, plant selection and breeding based on QQ-activities are still in their infancy. There is a high-level specificity between QQ agents (e.g., QSIs and QQ-microorganisms) and target plant pathogens; this considerably demands the characterization of QQ agents for each of the targeted plant pathogens (des Essarts et al. 2013). In some instances, QS-mediated functions are beneficial to the plant system. For instance, *Pseudomonas* strains (used as biocontrol agents), do control certain plant pathogens by producing antibiotic and antifungal molecules. It has been found that there was an incompatibility between QS-signal producers and degraders (Molina et al. 2003), which raised many doubts about the interference between the biocontrol agents that can produce and degrade QS signaling molecules. The use of QQ-strategies for pest control increases another problem by affecting beneficial or symbiotic bacteria (e.g., *Pseudomonas* sp.) implying that QQ-strategy seems not to be target-specific. As a result, QQ-strategy is going to alter the positive functions in the plant environment.

6. It is well known that QQ-bacteria (AHL-degraders) are effective biocontrol agents in aquaculture to fight bacterial fish disease. But on the other side, they could have harmful effects on invertebrates. AHLs are good chemoattractants for zoospores of many invertebrates (oysters (Zhao et al. 2003), mussels (Yang 2007), etc.) and allow them to settle on biofilms. It is important to note that rearing or hatching of larva requires an efficient settlement. But application of QQ-enzymes can exert a negative effect on the accumulation of signal molecules and subsequent biofilm formation, which could eventually be a problem to resident invertebrates. Under such circumstances, it is difficult to judge whether QQ-strategy is beneficial or detrimental.
7. Regarding QQ applications in MBRs, currently we have many challenges—the production cost of QQ-media remains high, no details are available about the optimized conditions for QQ-MBR operation, quenching of AI-2 and AIP signal molecules is still at their beginning level, it is not known how to control both AHLs and AI-2 signals simultaneously for biofouling mitigation, and yet optimum design and economical materials for QQ-media are not available. Studies addressing the above challenges will decrease the gap between the lab- and commercial-level of QQ-MBR, and also shape the QQ-technology in a sustainable way for biofouling control in MBRs.

Acknowledgments Dr. Naga Raju Maddela greatly acknowledges the Universidad Téchnica de Manabí, Portoviejo, Manabí, Ecuador, for the facilities and encouragement and his colleagues in the Instituto de Investigación and Facultad la Ciencias la Salud for their help in literature collection. The authors also thank Editor Dr. Pankaj Kumar Arora (Assistant Professor, Babasaheb Bhimrao Ambedkar University, Lucknow, India) for guidance and accepting our request to write this chapter.

References

Afriat L, Roodveldt C, Manco G, Tawfik DS (2006) The latent promiscuity of newly identified microbial lactonases is linked to a recently diverged phosphotriesterase. Biochemistry 45:13677–13686

Antunes LC, Ferreira RB, Buckner MM, Finlay BB (2010) Quorum sensing in bacterial virulence. Microbiology 156:2271–2282

Augustine N, Kumar P, Thomas S (2010) Inhibition of Vibrio cholerae biofilm by AiiA enzyme produced from Bacillus spp. Arch Microbiol 192:1019–1022

Augustine N, Wilson PA, Kerkar S, Thomas S (2012) Arctic actinomycetes as potential inhibitors of Vibrio cholerae biofilm. Curr Microbiol 64:338–342

Augustine N, Goel AK, Sivakumar KC, Kumar RA, Thomas S (2014) Resveratrol—a potential inhibitor of biofilm formation in Vibrio cholerae. Phytomedicine 21:286–289

Ban H, Chai X, Lin Y, Zhou Y, Peng D, Zhou Y, Zou Y, Yu Z, Sun M (2009) Transgenic Amorphophallus konjac expressing synthesized acyl-homoserine lactonase (aiiA) gene exhibit enhanced resistance to soft rot disease. Plant Cell Rep 28:1847–1855

Barbey C, Crepin A, Cirou A, Budin-Verneuil A, Orange N, Feuilloley M, Faure D, Dessaux Y, Burini JF, Latour X (2012) Catabolic pathway of gamma-caprolactone in the biocontrol agent *Rhodococcus erythropolis*. J Proteome Res 11:206–216

Barbey C, Crepin A, Bergeau D, Ouchiha A, Mijouin L, Taupin L, Orange N, Feuilloley M, Dufour A, Burini JF, Latour X (2013) In planta biocontrol of *Pectobacterium atrosepticum* by rhodococcus erythropolis involves silencing of pathogen communication by the rhodococcal gamma-lactone catabolic pathway. PLoS One 8:e66642

Bar-Rogovsky H, Hugenmatter A, Tawfik DS (2013) The evolutionary origins of detoxifying enzymes: the mammalian serum paraoxonases (PONs) relate to bacterial homoserine lactonases. J Biol Chem 288(33):23914–23927

Basavaraju M, Sisnity VS, Palaparthy R, Addanki PK (2016) Quorum quenching: signal jamming in dental plaque biofilms. J Dent Sci 11:349–352

Bijtenhoorn P, Mayerhofer H, Muller-Dieckmann J, Utpatel C, Schipper C, Hornung C, Szesny M, Grond S, Thurmer A, Brzuszkiewicz E, Daniel R, Dierking K, Schulenburg H, Streit WR (2011a) A novel metagenomic short-chain dehydrogenase/reductase attenuates Pseudomonas aeruginosa biofilm formation and virulence on Caenorhabditis elegans. PLoS One 6:e26278

Bijtenhoorn P, Schipper C, Hornung C, Quitschau M, Grond S, Weiland N, Streit WR (2011b) BpiB05, a novel metagenome-derived hydrolase acting on N-acylhomoserine lactones. J Biotechnol 155:86–94

Bokhove M, Nadal Jimenez P, Quax WJ, Dijkstra BW (2010) The quorum-quenching N-acyl homoserine lactone acylase PvdQ is an Ntn-hydrolase with an unusual substrate-binding pocket. Proc Natl Acad Sci U S A 107:686–691

Brackman G, Defoirdt T, Miyamoto C, Bossier P, Van Calenbergh S, Nelis H, Coenye T (2008) Cinnamaldehyde and cinnamaldehyde derivatives reduce virulence in Vibrio spp. by decreasing the DNA-binding activity of the quorum sensing response regulator LuxR. BMC Microbiol 8:149

Brackman G, Celen S, Hillaert U, Van Calenbergh S, Cos P, Maes L, Nelis HJ, Coenye T (2011a) Structure-activity relationship of cinnamaldehyde analogs as inhibitors of AI-2 based quorum sensing and their effect on virulence of Vibrio spp. PLoS One 6:e16084

Brackman G, Cos P, Maes L, Nelis HJ, Coenye T (2011b) Quorum sensing inhibitors increase the susceptibility of bacterial biofilms to antibiotics in vitro and in vivo. Antimicrob Agents Chemother 55:2655–2661

Bzdrenga J, Daude D, Remy B, Jacquet P, Plener L, Elias M, Chabriere E (2017) Biotechnological applications of quorum quenching enzymes. Chem Biol Interact 267:104–115

Cao Y, He S, Zhou Z, Zhang M, Mao W, Zhang H, Yao B (2012) Orally administered thermostable N-acyl homoserine lactonase from Bacillus sp. strain AI96 attenuates Aeromonas hydrophila infection in zebrafish. Appl Environ Microbiol 78:1899–1908

Carlier A, Uroz S, Smadja B, Fray R, Latour X, Dessaux Y, Faure D (2003) The Ti plasmid of Agrobacterium tumefaciens harbors an attM-paralogous gene, aiiB, also encoding N-Acyl homoserine lactonase activity. Appl Environ Microbiol 69:4989–4993

Chan KG, Atkinson S, Mathee K, Sam CK, Chhabra SR, Camara M, Koh CL, Williams P (2011) Characterization of N-acylhomoserine lactone-degrading bacteria associated with the Zingiber officinale (ginger) rhizosphere: co-existence of quorum quenching and quorum sensing in Acinetobacter and Burkholderia. BMC Microbiol 11:51

Chen CN, Chen CJ, Liao CT, Lee CY (2009) A probable aculeacin A acylase from the Ralstonia solanacearum GMI1000 is N-acyl-homoserine lactone acylase with quorum-quenching activity. BMC Microbiol 9:89

Chen F, Gao Y, Chen X, Yu Z, Li X (2013) Quorum quenching enzymes and their application in degrading signal molecules to block quorum sensing-dependent infection. Int J Mol Sci 14:17477–17500

Cheong WS, Kim SR, Oh HS, Lee SH, Yeon KM, Lee CH, Lee JK (2014) Design of quorum quenching microbial vessel to enhance cell viability for biofouling control in membrane bioreactor. J Microbiol Biotechnol 24:97–105

Choudhary S, Schmidt-Dannert C (2010) Applications of quorum sensing in biotechnology. Appl Microbiol Biotechnol 86:1267–1279

Chow JY, Wu L, Yew WS (2009) Directed evolution of a quorum-quenching lactonase from Mycobacterium avium subsp. paratuberculosis K-10 in the amidohydrolase superfamily. Biochemistry 48:4344–4353

Chow JY, Yang Y, Tay SB, Chua KL, Yew WS (2014) Disruption of biofilm formation by the human pathogen Acinetobacter baumannii using engineered quorum-quenching lactonases. Antimicrob Agents Chemother 58:1802–1805

Chowdhary PK, Keshavan N, Nguyen HQ, Peterson JA, Gonzalez JE, Haines DC (2007) Bacillus megaterium CYP102A1 oxidation of acyl homoserine lactones and acyl homoserines. Biochemistry 46:14429–14437

Chowdhary PK, Stewart L, Lopez C, Haines DC (2008) A single mutation in P450BM-3 enhances acyl homoserine lactone: acyl homoserine substrate binding selectivity nearly 250-fold. J Biotechnol 135:374–376

Christiaen SEA, Brackman G, Nelis HJ, Coenye T (2011) Isolation and identification of quorum quenching bacteria from environmental samples. J Microbiol Methods 87:213–219

Christiaen SE, Matthijs N, Zhang XH, Nelis HJ, Bossier P, Coenye T (2014) Bacteria that inhibit quorum sensing decrease biofilm formation and virulence in Pseudomonas aeruginosa PAO1. Pathogen Dis 70:271–279

Chu W, Zhou S, Zhu W, Zhuang X (2014) Quorum quenching bacteria Bacillus sp. QSI-1 protect zebrafish (Danio rerio) from Aeromonas hydrophila infection. Sci Rep 4:5446

Cirou A, Diallo S, Kurt C, Latour X, Faure D (2007) Growth promotion of quorum-quenching bacteria in the rhizosphere of Solanum tuberosum. Environ Microbiol 9:1511–1522

Cirou A, Mondy S, An S, Charrier A, Sarrazin A, Thoison O, DuBow M, Faure D (2012) Efficient biostimulation of native and introduced quorum-quenching Rhodococcus erythropolis populations is revealed by a combination of analytical chemistry, microbiology, and pyrosequencing. Appl Environ Microbiol 78:481–492

Clamens T, Rosay T, Crepin A, Grandjean T, Kentache T, Hardouin J, Bortolotti P, Neidig A, Mooij M, Hillion M, Vieillard J, Cosette P, Overhage J, O'Gara F, Bouffartigues E, Dufour A, Chevalier S, Guery B, Cornelis P, Feuilloley MG, Lesouhaitier O (2017) The aliphatic amidase AmiE is involved in regulation of Pseudomonas aeruginosa virulence. Sci Rep 7:41178

Czajkowski R, Krzyzanowska D, Karczewska J, Atkinson S, Przysowa J, Lojkowska E, Williams P, Jafra S (2011) Inactivation of AHLs by Ochrobactrum sp. A44 depends on the activity of a novel class of AHL acylase. Environ Microbiol Rep 3:59–68

D'Alvise PW, Lillebø S, Wergeland HI, Gram L, Bergh Ø (2013) Protection of cod larvae from vibriosis by Phaeobacter spp.: a comparison of strains and introduction times. Aquaculture 384–387:82–86

D'Angelo-Picard C, Chapelle E, Ratet P, Faure D, Dessaux Y (2011) Transgenic plants expressing the quorum quenching lactonase AttM do not significantly alter root-associated bacterial populations. Res Microbiol 162:951–958

Das MC, Sandhu P, Gupta P, Rudrapaul P, De UC, Tribedi P, Akhter Y, Bhattacharjee S (2016) Attenuation of Pseudomonas aeruginosa biofilm formation by Vitexin: a combinatorial study with azithromycin and gentamicin. Sci Rep 6:23347

Defoirdt T, Boon N, Bossier P, Verstraete W (2004) Disruption of bacterial quorum sensing: an unexplored strategy to fight infections in aquaculture. Aquaculture 240:69–88

Defoirdt T, Crab R, Wood TK, Sorgeloos P, Verstraete W, Bossier P (2006) Quorum sensing-disrupting brominated furanones protect the gnotobiotic brine shrimp Artemia franciscana from pathogenic Vibrio harveyi, Vibrio campbellii, and Vibrio parahaemolyticus isolates. Appl Environ Microbiol 72:6419

Defoirdt T, Boon N, Sorgeloos P, Verstraete W, Bossier P (2007) Alternatives to antibiotics to control bacterial infections: luminescent vibriosis in aquaculture as an example. Trends Biotechnol 25:472–479

Defoirdt T, Boon N, Bossier P (2010) Can bacteria evolve resistance to quorum sensing disruption? PLoS Pathog 6:e1000989

des Essarts YR, Sabbah M, Comte A, Soulere L, Queneau Y, Dessaux Y, Helias V, Faure D (2013) N,N'-alkylated Imidazolium-derivatives act as quorum-sensing inhibitors targeting the Pectobacterium atrosepticum-induced symptoms on potato tubers. Int J Mol Sci 14:19976–19986

Dinh J, Angeloni JT, Pederson DB, Wang X, Cao M, Dong Y (2014) Cranberry extract standardized for proanthocyanidins promotes the immune response of Caenorhabditis elegans to Vibrio cholerae through the p38 MAPK pathway and HSF-1. PLoS One 9:e103290–e103290

Dockrill P (2017) ScienceAlert—Health: These Brazilian Red Berries Could Hold The Key to Fighting Deadly Superbugs

Dong YH, Zhang LH (2005) Quorum sensing and quorum-quenching enzymes. J Microbiol 43:101–109

Dong YH, Xu JL, Li XZ, Zhang LH (2000) AiiA, an enzyme that inactivates the acylhomoserine lactone quorum-sensing signal and attenuates the virulence of Erwinia carotovora. Proc Natl Acad Sci U S A 97:3526–3531

Dong YH, Wang LH, Xu JL, Zhang HB, Zhang XF, Zhang LH (2001) Quenching quorum-sensing-dependent bacterial infection by an N-acyl homoserine lactonase. Nature 411:813–817

Dong YH, Gusti AR, Zhang Q, Xu JL, Zhang LH (2002) Identification of quorum-quenching N-acyl homoserine lactonases from Bacillus species. Appl Environ Microbiol 68:1754–1759

Dong YH, Zhang XF, Xu JL, Zhang LH (2004) Insecticidal Bacillus thuringiensis silences Erwinia carotovora virulence by a new form of microbial antagonism, signal interference. Appl Environ Microbiol 70:954–960

Dong W, Zhu J, Guo X, Kong D, Zhang Q, Zhou Y, Liu X, Zhao S, Ruan Z (2018) Characterization of AiiK, an AHL lactonase, from Kurthia huakui LAM0618T and its application in quorum quenching on Pseudomonas aeruginosa PAO1. Sci Rep 8:6013

Doranalli K, Ortiz A, (2018) Bacillus amyloliquefaciens CECT 5940 (Ecobiol®) expresses quorum quenching activity

Ergön-Can T, Köse-Mutlu B, Koyuncu İ, Lee C-H (2017) Biofouling control based on bacterial quorum quenching with a new application: rotary microbial carrier frame. J Membr Sci 525:116–124

Fetzner S (2015) Quorum quenching enzymes. J Biotechnol 201:2–14

Galloway WRJD, Hodgkinson JT, Bowden SD, Welch M, Spring DR (2011) Quorum sensing in gram-negative bacteria: small-molecule modulation of AHL and AI-2 quorum sensing pathways. Chem Rev 111:28–67

Galloway WR, Hodgkinson JT, Bowden S, Welch M, Spring DR (2012) Applications of small molecule activators and inhibitors of quorum sensing in Gram-negative bacteria. Trends Microbiol 20:449–458

GEN (2018) Genetic Engineering and Biotechnology News—Probiotic Good Bacteria Use Fengycins to Eliminate Bad Bacteria

Gerhardt SG, McDyer JF, Girgis RE, Conte JV, Yang SC, Orens JB (2003) Maintenance azithromycin therapy for bronchiolitis obliterans syndrome: results of a pilot study. Am J Respir Crit Care Med 168:121–125

Givskov M, de Nys R, Manefield M, Gram L, Maximilien R, Eberl L, Molin S, Steinberg PD, Kjelleberg S (1996) Eukaryotic interference with homoserine lactone-mediated prokaryotic signalling. J Bacteriol 178:6618–6622

Grandclement C, Tannieres M, Morera S, Dessaux Y, Faure D (2016) Quorum quenching: role in nature and applied developments. FEMS Microbiol Rev 40:86–116

Gu Y, Huang J, Zeng G, Shi Y, Hu Y, Tang B, Zhou J, Xu W, Shi L (2018) Quorum quenching activity of indigenous quorum quenching bacteria and its potential application in mitigation of membrane biofouling. J Chem Technol Biotechnol 93:1394–1400

Gupta P, Chhibber S, Harjai K (2015) Efficacy of purified lactonase and ciprofloxacin in preventing systemic spread of Pseudomonas aeruginosa in murine burn wound model. Burns 41:153–162

Gutierrez JA, Crowder T, Rinaldo-Matthis A, Ho MC, Almo SC, Schramm VL (2009) Transition state analogs of 5′-methylthioadenosine nucleosidase disrupt quorum sensing. Nat Chem Biol 5:251–257

Harel M, Aharoni A, Gaidukov L, Brumshtein B, Khersonsky O, Meged R, Dvir H, Ravelli RB, McCarthy A, Toker L, Silman I, Sussman JL, Tawfik DS (2004) Structure and evolution of the serum paraoxonase family of detoxifying and anti-atherosclerotic enzymes. Nat Struct Mol Biol 11:412–419

Hiblot J, Gotthard G, Chabriere E, Elias M (2012) Structural and Enzymatic characterization of the lactonase SisLac from Sulfolobus islandicus. PLoS One 7:e47028

Hiblot J, Gotthard G, Elias M, Chabriere E (2013) Differential active site loop conformations mediate promiscuous activities in the lactonase SsoPox. PLoS One 8:e75272

Hoang TT, Schweizer HP (1999) Characterization of Pseudomonas aeruginosa enoyl-acyl carrier protein reductase (FabI): a target for the antimicrobial triclosan and its role in acylated homoserine lactone synthesis. J Bacteriol 181:5489–5497

Hraiech S, Hiblot J, Lafleur J, Lepidi H, Papazian L, Rolain JM, Raoult D, Elias M, Silby MW, Bzdrenga J, Bregeon F, Chabriere E (2014) Inhaled lactonase reduces Pseudomonas aeruginosa quorum sensing and mortality in rat pneumonia. PLoS One 9:e107125

Huang JJ, Han JI, Zhang LH, Leadbetter JR (2003) Utilization of acyl-homoserine lactone quorum signals for growth by a soil pseudomonad and Pseudomonas aeruginosa PAO1. Appl Environ Microbiol 69:5941–5949

Huang JJ, Petersen A, Whiteley M, Leadbetter JR (2006) Identification of QuiP, the Product of Gene PA1032, as the Second Acyl-Homoserine Lactone Acylase of Pseudomonas aeruginosa PAO1

Huang W, Lin Y, Yi S, Liu P, Shen J, Shao Z, Liu Z (2012) QsdH, a novel AHL lactonase in the RND-type inner membrane of marine Pseudoalteromonas byunsanensis strain 1A01261. PLoS One 7:e46587

Huang J, Shi Y, Zeng G, Gu Y, Chen G, Shi L, Hu Y, Tang B, Zhou J (2016) Acyl-homoserine lactone-based quorum sensing and quorum quenching hold promise to determine the performance of biological wastewater treatments: an overview. Chemosphere 157:137–151

Husain FM, Ahmad I, Asif M, Tahseen Q (2013) Influence of clove oil on certain quorum-sensing-regulated functions and biofilm of Pseudomonas aeruginosa and Aeromonas hydrophila. J Biosci 38:835–844

Husain FM, Ahmad I, Khan MS, Ahmad E, Tahseen Q, Khan MS, Alshabib NA (2015) Sub-MICs of Mentha piperita essential oil and menthol inhibits AHL mediated quorum sensing and biofilm of Gram-negative bacteria. Front Microbiol 6:420

Husain FM, Ahmad I, Al-Thubiani AS, Abulreesh HH, AlHazza IM, Aqil F (2017) Leaf extracts of mangifera indica L. inhibit quorum sensing—regulated production of virulence factors and biofilm in test bacteria. Front Microbiol 8:727

Hwang JH, Cicek N, Oleszkiewicz JA (2010) Achieving biofilm control in a membrane biofilm reactor removing total nitrogen. Water Res 44:2283–2291

ID (2019) Enzymes as antibiotics: new tools in the fight against infections

IDRC (2016) International Development Research Centre—Integrated quorum quenching strategies to reduce antimicrobial resistance in shrimp aquaculture (i-QAS)

IDRC (2019) International Development Research Centre—$21.2 million announced for research on Innovative Veterinary Solutions for Antimicrobial Resistance

Ivanova K, Fernandes MM, Francesko A, Mendoza E, Guezguez J, Burnet M, Tzanov T (2015a) Quorum-quenching and matrix-degrading enzymes in multilayer coatings synergistically prevent bacterial biofilm formation on urinary catheters. ACS Appl Mater Interfaces 7:27066–27077

Ivanova K, Fernandes MM, Mendoza E, Tzanov T (2015b) Enzyme multilayer coatings inhibit Pseudomonas aeruginosa biofilm formation on urinary catheters. Appl Microbiol Biotechnol 99:4373–4385

Jiang W, Xia S, Liang J, Zhang Z, Hermanowicz SW (2013) Effect of quorum quenching on the reactor performance, biofouling and biomass characteristics in membrane bioreactors. Water Res 47:187–196

Kalia VC, Purohit HJ (2011) Quenching the quorum sensing system: potential antibacterial drug targets. Crit Rev Microbiol 37:121–140

Kim JS (2007) Quorum sensing inhibitors from the red alga, *Ahnfeltiopsis flabelliformis*. Biotechnol Bioprocess Eng 12:308

Kim MH, Choi WC, Kang HO, Lee JS, Kang BS, Kim KJ, Derewenda ZS, Oh TK, Lee CH, Lee JK (2005) The molecular structure and catalytic mechanism of a quorum-quenching N-acyl-L-homoserine lactone hydrolase. Proc Natl Acad Sci U S A 102:17606–17611

Kim AL, Park SY, Lee CH, Lee CH, Lee JK (2014) Quorum quenching bacteria isolated from the sludge of a wastewater treatment plant and their application for controlling biofilm formation. J Microbiol Biotechnol 24:1574–1582

Kim YW, Sung C, Lee S, Kim KJ, Yang YH, Kim BG, Lee YK, Ryu HW, Kim YG (2015) MALDI-MS-based quantitative analysis for ketone containing homoserine lactones in Pseudomonas aeruginosa. Anal Chem 87:858–863

Koch G, Nadal Jimenez P, Muntendam R, Chen Y, Papaioannou E, Heeb S, Camara M, Williams P, Cool RH, Quax WJ (2010) The acylase PvdQ has a conserved function among fluorescent *Pseudomonas* spp. Environ Microbiol Rep 2:433–439

Krysciak D, Schmeisser C, Preuss S, Riethausen J, Quitschau M, Grond S, Streit WR (2011) Involvement of multiple loci in quorum quenching of autoinducer I molecules in the nitrogen-fixing symbiont Rhizobium (Sinorhizobium) sp. strain NGR234. Appl Environ Microbiol 77:5089–5099

Kusada H, Zhang Y, Tamaki H, Kimura N, Kamagata Y (2019) Novel *N*-acyl homoserine lactone-degrading bacteria isolated from penicillin-contaminated environments and their quorum-quenching activities. Front Microbiol 10:455

Lade H, Paul D, Kweon JH (2014) Quorum quenching mediated approaches for control of membrane biofouling. Int J Biol Sci 10:550–565

Leadbetter JR, Greenberg EP (2000) Metabolism of acyl-homoserine lactone quorum-sensing signals by *Variovorax paradoxus*. J Bacteriol 182:6921

Lee SJ, Park SY, Lee JJ, Yum DY, Koo BT, Lee JK (2002) Genes encoding the N-acyl homoserine lactone-degrading enzyme are widespread in many subspecies of Bacillus thuringiensis. Appl Environ Microbiol 68:3919–3924

Lee KM, Lim J, Nam S, Yoon MY, Kwon YK, Jung BY, Park Y, Park S, Yoon SS (2011) Inhibitory effects of broccoli extract on Escherichia coli O157:H7 quorum sensing and in vivo virulence. FEMS Microbiol Lett 321:67–74

Lee J, Lee I, Nam J, Hwang DS, Yeon KM, Kim J (2017) Immobilization and stabilization of acylase on carboxylated polyaniline nanofibers for highly effective antifouling application via quorum quenching. ACS Appl Mater Interfaces 9:15424–15432

Li L, Hooi D, Chhabra SR, Pritchard D, Shaw PE (2004) Bacterial N-acylhomoserine lactone-induced apoptosis in breast carcinoma cells correlated with down-modulation of STAT3. Oncogene 23:4894–4902

Li S, Leung PHM, Xu X, Wu C (2018) Homogentisic acid γ-lactone suppresses the virulence factors of Pseudomonas aeruginosa by quenching its quorum sensing signal molecules. Chin Chem Lett 29:313–316

Li Y, Zhao J, Quan C, Jin L, Xu Y, Chen M (2019) Production, characterization, and powder preparation of quorum quenching acylase AiiO for pathogen control. Prep Biochem Biotechnol 49:151–157

Lin YH, Xu JL, Hu J, Wang LH, Ong SL, Leadbetter JR, Zhang LH (2003) Acyl-homoserine lactone acylase from Ralstonia strain XJ12B represents a novel and potent class of quorum-quenching enzymes. Mol Microbiol 47:849–860

Liu D, Lepore BW, Petsko GA, Thomas PW, Stone EM, Fast W, Ringe D (2005) Three-dimensional structure of the quorum-quenching N-acyl homoserine lactone hydrolase from Bacillus thuringiensis. Proc Natl Acad Sci U S A 102:11882–11887

Liu D, Thomas PW, Momb J, Hoang QQ, Petsko GA, Ringe D, Fast W (2007) Structure and specificity of a quorum-quenching lactonase (AiiB) from Agrobacterium tumefaciens. Biochemistry 46:11789–11799

Liu J, Eng CY, Ho JS, Chong TH, Wang L, Zhang P, Zhou Y (2019) Quorum quenching in anaerobic membrane bioreactor for fouling control. Water Res 156:159–167

Lowery CA, Abe T, Park J, Eubanks LM, Sawada D, Kaufmann GF, Janda KD (2009) Revisiting AI-2 quorum sensing inhibitors: direct comparison of alkyl-DPD analogues and a natural product fimbrolide. J Am Chem Soc 131:15584–15585

Lyon GJ, Mayville P, Muir TW, Novick RP (2000) Rational design of a global inhibitor of the virulence response in Staphylococcus aureus, based in part on localization of the site of inhibition to the receptor-histidine kinase, AgrC. Proc Natl Acad Sci U S A 97:13330–13335

Maddela NR, Meng F (2020) Discrepant roles of a quorum quenching bacterium (Rhodococcus sp. BH4) in growing dual-species biofilms. Sci Total Environ 713:136402

Maddela NR, Zhou Z, Yu Z, Zhao S, Meng F (2018) Functional determinants of extracellular polymeric substances in membrane biofouling: experimental evidence from pure-cultured sludge bacteria. Appl Environ Microbiol 84:e00756–e00718

Maddela NR, Sheng B, Yuan S, Zhou Z, Villamar-Torres R, Meng F (2019) Roles of quorum sensing in biological wastewater treatment: a critical review. Chemosphere 221:616–629

Mandrich L, Merone L, Manco G (2010) Hyperthermophilic phosphotriesterases/lactonases for the environment and human health. Environ Technol 31:1115–1127

Manefield M, Rasmussen TB, Henzter M, Andersen JB, Steinberg P, Kjelleberg S, Givskov M (2002) Halogenated furanones inhibit quorum sensing through accelerated LuxR turnover. Microbiology 148:1119–1127

Medarametla P, Gatta V, Kajander T, Laitinen T, Tammela P, Poso A (2018) Structure-based virtual screening of LsrK kinase inhibitors to target quorum sensing. ChemMedChem 13:2400–2407

Mei GY, Yan XX, Turak A, Luo ZQ, Zhang LQ (2010) AidH, an alpha/beta-hydrolase fold family member from an Ochrobactrum sp. strain, is a novel N-acylhomoserine lactonase. Appl Environ Microbiol 76:4933–4942

Molina L, Constantinescu F, Michel L, Reimmann C, Duffy B, Defago G (2003) Degradation of pathogen quorum-sensing molecules by soil bacteria: a preventive and curative biological control mechanism. FEMS Microbiol Ecol 45:71–81

Molina L, Rezzonico F, Defago G, Duffy B (2005) Autoinduction in Erwinia amylovora: evidence of an acyl-homoserine lactone signal in the fire blight pathogen. J Bacteriol 187:3206–3213

Morohoshi T, Nakazawa S, Ebata A, Kato N, Ikeda T (2008) Identification and characterization of N-acylhomoserine lactone-acylase from the fish intestinal shewanella sp. strain MIB015. Biosci Biotechnol Biochem 72:1887–1893

Morohoshi T, Tominaga Y, Someya N, Ikeda T (2012) Complete genome sequence and characterization of the N-acylhomoserine lactone-degrading gene of the potato leaf-associated Solibacillus silvestris. J Biosci Bioeng 113:20–25

Muhs A, Lyles JT, Parlet CP, Nelson K, Kavanaugh JS, Horswill AR, Quave CL (2017) Virulence inhibitors from brazilian peppertree block quorum sensing and abate dermonecrosis in skin infection models. Sci Rep 7:42275

Mukherjee M, Hu Y, Tan CH, Rice SA, Cao B (2018) Engineering a light-responsive, quorum quenching biofilm to mitigate biofouling on water purification membranes. Sci Adv 4:eaau1459

Musthafa KS, Saroja V, Pandian SK, Ravi AV (2011) Antipathogenic potential of marine Bacillus sp. SS4 on N-acyl-homoserine-lactone-mediated virulence factors production in Pseudomonas aeruginosa (PAO1). J Biosci 36:55–67

Musthafa KS, Balamurugan K, Pandian SK, Ravi AV (2012a) 2,5-Piperazinedione inhibits quorum sensing-dependent factor production in Pseudomonas aeruginosa PAO1. J Basic Microbiol 52:679–686

Musthafa KS, Sivamaruthi BS, Pandian SK, Ravi AV (2012b) Quorum sensing inhibition in Pseudomonas aeruginosa PAO1 by antagonistic compound phenylacetic acid. Curr Microbiol 65:475–480

Nadell CD, Drescher K, Foster KR (2016) Spatial structure, cooperation and competition in biofilms. Nat Rev Microbiol 14:589–600

Natrah FMI, Kenmegne MM, Wiyoto W, Sorgeloos P, Bossier P, Defoirdt T (2011) Effects of micro-algae commonly used in aquaculture on acyl-homoserine lactone quorum sensing. Aquaculture 317:53–57

Nelson LK, D'Amours GH, Sproule-Willoughby KM, Morck DW, Ceri H (2009) Pseudomonas aeruginosa las and rhl quorum-sensing systems are important for infection and inflammation in a rat prostatitis model. Microbiology 155:2612–2619

Nhan DT, Cam DT, Wille M, Defoirdt T, Bossier P, Sorgeloos P (2010) Quorum quenching bacteria protect Macrobrachium rosenbergii larvae from Vibrio harveyi infection. J Appl Microbiol 109:1007–1016

Nithya C, Aravindraja C, Pandian SK (2010) Bacillus pumilus of Palk Bay origin inhibits quorum-sensing-mediated virulence factors in Gram-negative bacteria. Res Microbiol 161:293–304

Nithya C, Devi MG, Karutha Pandian S (2011) A novel compound from the marine bacterium Bacillus pumilus S6-15 inhibits biofilm formation in gram-positive and gram-negative species. Biofouling 27:519–528

Ochiai S, Yasumoto S, Morohoshi T, Ikeda T (2014) AmiE, a novel N-acylhomoserine lactone acylase belonging to the amidase family, from the activated-sludge isolate Acinetobacter sp. strain Ooi24. Appl Environ Microbiol 80:6919–6925

Oh H-S, Lee C-H (2018) Origin and evolution of quorum quenching technology for biofouling control in MBRs for wastewater treatment. J Membr Sci 554:331–345

Oh H-S, Yeon K-M, Yang C-S, Kim S-R, Lee C-H, Park SY, Han JY, Lee J-K (2012) Control of membrane biofouling in MBR for wastewater treatment by quorum quenching bacteria encapsulated in microporous membrane. Environ Sci Technol 46:4877–4884

Oh HS, Kim SR, Cheong WS, Lee CH, Lee JK (2013) Biofouling inhibition in MBR by Rhodococcus sp. BH4 isolated from real MBR plant. Appl Microbiol Biotechnol 97:10223–10231

Okano C, Murota D, Nasuno E, Iimura KI, Kato N (2019) Effective quorum quenching with a conformation-stable recombinant lactonase possessing a hydrophilic polymeric shell fabricated via electrospinning. Mater Sci Eng C Mater Biol Appl 98:437–444

Olsen JA, Severinsen R, Rasmussen TB, Hentzer M, Givskov M, Nielsen J (2002) Synthesis of new 3- and 4-substituted analogues of acyl homoserine lactone quorum sensing autoinducers. Bioorg Med Chem Lett 12:325–328

Ozer EA, Pezzulo A, Shih DM, Chun C, Furlong C, Lusis AJ, Greenberg EP, Zabner J (2005) Human and murine paraoxonase 1 are host modulators of Pseudomonas aeruginosa quorum-sensing. FEMS Microbiol Lett 253:29–37

Palmer AG, Senechal AC, Mukherjee A, Ane JM, Blackwell HE (2014) Plant responses to bacterial N-acyl L-homoserine lactones are dependent on enzymatic degradation to L-homoserine. ACS Chem Biol 9:1834–1845

Papaioannou E, Wahjudi M, Nadal-Jimenez P, Koch G, Setroikromo R, Quax WJ (2009) Quorum-quenching acylase reduces the virulence of Pseudomonas aeruginosa in a Caenorhabditis elegans infection model. Antimicrob Agents Chemother 53:4891–4897
Paramá A (2019) Tendensias 21: Tendensias de la Salud—Posible final de la era de los antibióticos
Park SY, Lee SJ, Oh TK, Oh JW, Koo BT, Yum DY, Lee JK (2003) AhlD, an N-acylhomoserine lactonase in Arthrobacter sp., and predicted homologues in other bacteria. Microbiology 149:1541–1550
Park SY, Kang HO, Jang HS, Lee JK, Koo BT, Yum DY (2005) Identification of extracellular N-acylhomoserine lactone acylase from a Streptomyces sp. and its application to quorum quenching. Appl Environ Microbiol 71:2632–2641
Paul D, Kim YS, Ponnusamy K, Kweon JH (2009) Applicaiton of quorum quenching to inhibit biofilm formation. Environ Eng Sci 26:1319–1324
Peach KC, Bray WM, Shikuma NJ, Gassner NC, Lokey RS, Yildiz FH, Linington RG (2011) An image-based 384-well high-throughput screening method for the discovery of biofilm inhibitors in Vibrio cholerae. Mol BioSyst 7:1176–1184
Pearson JP, Feldman M, Iglewski BH, Prince A (2000) Pseudomonas aeruginosa cell-to-cell signaling is required for virulence in a model of acute pulmonary infection. Infect Immun 68:4331–4334
Pejin B, Ciric A, Dimitric Markovic J, Glamoclija J, Nikolic M, Sokovic M (2017) An insight into anti-biofilm and anti-quorum sensing activities of the selected anthocyanidins: the case study of Pseudomonas aeruginosa PAO1. Nat Prod Res 31:1177–1180
PHYSORG (2014) Invention prevents contamination of food packaging by bacterial biofilms
PHYSORG (2019) Compound allows bacterial communication to be controlled by light
Piewngam P, Zheng Y, Nguyen TH, Dickey SW, Joo H-S, Villaruz AE, Glose KA, Fisher EL, Hunt RL, Li B, Chiou J, Pharkjaksu S, Khongthong S, Cheung GYC, Kiratisin P, Otto M (2018) Pathogen elimination by probiotic Bacillus via signalling interference. Nature 562:532–537
PoultryWorld (2018) Health Background—Keep gut health challenges at bay
Prol García MJ, Alvise PW, Gram L (2013) Disruption of cell-to-cell signaling does not abolish the antagonism of Phaeobacter gallaeciensis toward the fish pathogen Vibrio anguillarum in algal systems. Appl Environ Microbiol 79:5414
Pustelny C, Albers A, Büldtkarentzopoulos K, Parschat K, Chhabra SR, Cámara M, Williams P, Fetzner S (2009) Dioxygenase-mediated quenching of quinolone-dependent quorum sensing in Pseudomonas aeruginosa. Chem Biol 16:1259–1267
Raina S, De Vizio D, Odell M, Clements M, Vanhulle S, Keshavarz T (2009) Microbial quorum sensing: a tool or a target for antimicrobial therapy? Biotechnol Appl Biochem 54:65–84
Rasch M, Buch C, Austin B, Slierendrecht WJ, Ekmann KS, Larsen JL, Johansen C, Riedel K, Eberl L, Givskov M, Gram L (2004) An inhibitor of bacterial quorum sensing reduces mortalities caused by Vibriosis in rainbow trout (Oncorhynchus mykiss, Walbaum). Syst Appl Microbiol 27:350–359
Rasmussen TB, Bjarnsholt T, Skindersoe ME, Hentzer M, Kristoffersen P, Kote M, Nielsen J, Eberl L, Givskov M (2005) Screening for quorum-sensing inhibitors (QSI) by use of a novel genetic system, the QSI selector. J Bacteriol 187:1799–1814
Rehman ZU, Leiknes T (2018) Quorum-quenching bacteria isolated from red sea sediments reduce biofilm formation by Pseudomonas aeruginosa. Front Microbiol 9:1354–1354
Reimmann C, Ginet N, Michel L, Keel C, Michaux P, Krishnapillai V, Zala M, Heurlier K, Triandafillu K, Harms H, Defago G, Haas D (2002) Genetically programmed autoinducer destruction reduces virulence gene expression and swarming motility in Pseudomonas aeruginosa PAO1. Microbiology 148:923–932
Rémy B, Mion S, Plener L, Elias M, Chabrière E, Daudé D (2018) Interference in bacterial quorum sensing: a biopharmaceutical perspective. Front Pharmacol 9:203–203
Riaz K, Elmerich C, Moreira D, Raffoux A, Dessaux Y, Faure D (2008) A metagenomic analysis of soil bacteria extends the diversity of quorum-quenching lactonases. Environ Microbiol 10:560–570

Romero M, Diggle SP, Heeb S, Camara M, Otero A (2008) Quorum quenching activity in Anabaena sp. PCC 7120: identification of AiiC, a novel AHL-acylase. FEMS Microbiol Lett 280:73–80
Romero M, Martin-Cuadrado AB, Roca-Rivada A, Cabello AM, Otero A (2011) Quorum quenching in cultivable bacteria from dense marine coastal microbial communities. FEMS Microbiol Ecol 75:205–217
Roy V, Fernandes R, Tsao CY, Bentley WE (2010) Cross species quorum quenching using a native AI-2 processing enzyme. ACS Chem Biol 5:223
Rumbaugh KP, Griswold JA, Iglewski BH, Hamood AN (1999) Contribution of quorum sensing to the virulence of Pseudomonas aeruginosa in burn wound infections. Infect Immun 67:5854–5862
Saiman L, Marshall BC, Mayer-Hamblett N, Burns JL, Quittner AL, Cibene DA, Coquillette S, Fieberg AY, Accurso FJ, Campbell PW 3rd (2003) Azithromycin in patients with cystic fibrosis chronically infected with Pseudomonas aeruginosa: a randomized controlled trial. JAMA 290:1749–1756
Schaefer AL, Hanzelka BL, Eberhard A, Greenberg EP (1996) Quorum sensing in Vibrio fischeri: probing autoinducer-LuxR interactions with autoinducer analogs. J Bacteriol 178:2897–2901
Schipper C, Hornung C, Bijtenhoorn P, Quitschau M, Grond S, Streit WR (2009) Metagenome-derived clones encoding two novel lactonase family proteins involved in biofilm inhibition in *Pseudomonas aeruginosa*. Appl Environ Microbiol 75:224
ScienceDaily (2018) Novel antibacterial drugs
Shepherd RW, Lindow SE (2009) Two dissimilar N-acyl-homoserine lactone acylases of Pseudomonas syringae influence colony and biofilm morphology. Appl Environ Microbiol 75:45
Shrout JD, Nerenberg R (2012) Monitoring bacterial twitter: does quorum sensing determine the behavior of water and wastewater treatment biofilms? Environ Sci Technol 46:1995–2005
Sifri CD, Mylonakis E, Singh KV, Qin X, Garsin DA, Murray BE, Ausubel FM, Calderwood SB (2002) Virulence effect of Enterococcus faecalis protease genes and the quorum-sensing locus fsr in Caenorhabditis elegans and mice. Infect Immun 70:5647–5650
Simonetti O, Cirioni O, Cacciatore I, Baldassarre L, Orlando F, Pierpaoli E, Lucarini G, Orsetti E, Provinciali M, Fornasari E, Di Stefano A, Giacometti A, Offidani A (2016) Efficacy of the quorum sensing inhibitor FS10 alone and in combination with tigecycline in an animal model of staphylococcal infected wound. PLoS One 11:e0151956
Sio CF, Otten LG, Cool RH, Diggle SP, Braun PG, Bos R, Daykin M, Cámara M, Williams P, Quax WJ (2006) Quorum quenching by an N-acyl-homoserine lactone acylase from Pseudomonas aeruginosa PAO1. Infect Immun 74:1673–1682
Smith KM, Bu Y, Suga H (2003) Induction and inhibition of Pseudomonas aeruginosa quorum sensing by synthetic autoinducer analogs. Chem Biol 10:81–89
Smyth AR, Cifelli PM, Ortori CA, Righetti K, Lewis S, Erskine P, Holland ED, Givskov M, Williams P, Camara M, Barrett DA, Knox A (2010) Garlic as an inhibitor of Pseudomonas aeruginosa quorum sensing in cystic fibrosis—a pilot randomized controlled trial. Pediatr Pulmonol 45:356–362
Song XN, Cheng YY, Li WW, Li BB, Sheng GP, Fang CY, Wang YK, Li XY, Yu HQ (2014) Quorum quenching is responsible for the underestimated quorum sensing effects in biological wastewater treatment reactors. Bioresour Technol 171:472–476
Stephenson K, Yamaguchi Y, Hoch JA (2000) The mechanism of action of inhibitors of bacterial two-component signal transduction systems. J Biol Chem 275:38900–38904
Tan CH, Koh KS, Xie C, Zhang J, Tan XH, Lee GP, Zhou Y, Ng WJ, Rice SA, Kjelleberg S (2015) Community quorum sensing signalling and quenching: microbial granular biofilm assembly. NPJ Biofilm Microbiom 1:15006
Tang K, Su Y, Brackman G, Cui F, Zhang Y, Shi X, Coenye T, Zhang XH (2015) MomL, a novel marine-derived N-acyl homoserine lactonase from Muricauda olearia. Appl Environ Microbiol 81:774–782

Tannieres M, Beury-Cirou A, Vigouroux A, Mondy S, Pellissier F, Dessaux Y, Faure D (2013) A metagenomic study highlights phylogenetic proximity of quorum-quenching and xenobiotic-degrading amidases of the AS-family. PLoS One 8:e65473

Terwagne M, Mirabella A, Lemaire J, Deschamps C, De Bolle X, Letesson J-J (2013) Quorum sensing and self-quorum quenching in the intracellular pathogen brucellamelitensis. PLoS One 8:e82514

THE HINDU (2019) TIRUCHIRAPALLI—NIT: kick-starts programme under GIAN

Thierbach S, Bui N, Zapp J, Chhabra SR, Kappl R, Fetzner S (2014) Substrate-assisted O2 activation in a cofactor-independent dioxygenase. Chem Biol 21:217–225

Tinh NT, Asanka Gunasekara RA, Boon N, Dierckens K, Sorgeloos P, Bossier P (2007) N-acyl homoserine lactone-degrading microbial enrichment cultures isolated from Penaeus vannamei shrimp gut and their probiotic properties in Brachionus plicatilis cultures. FEMS Microbiol Ecol 62:45–53

Torabi Delshad S, Soltanian S, Sharifiyazdi H, Bossier P (2018) Effect of quorum quenching bacteria on growth, virulence factors and biofilm formation of Yersinia ruckeri in vitro and an in vivo evaluation of their probiotic effect in rainbow trout. J Fish Dis 41:1429–1438

Torres M, Rubio-Portillo E, Anton J, Ramos-Espla AA, Quesada E, Llamas I (2016) Selection of the N-acylhomoserine lactone-degrading bacterium alteromonas stellipolaris PQQ-42 and of its potential for biocontrol in aquaculture. Front Microbiol 7:646

Turan NB, Engin GO (2018) Quorum quenching. Compr Anal Chem 81:117–149

Ueda A, Attila C, Whiteley M, Wood TK (2009) Uracil influences quorum sensing and biofilm formation in Pseudomonas aeruginosa and fluorouracil is an antagonist. Microb Biotechnol 2:62–74

Uroz S, Heinonsalo J (2008) Degradation of N-acyl homoserine lactone quorum sensing signal molecules by forest root-associated fungi. FEMS Microbiol Ecol 65:271–278

Uroz S, Chhabra SR, Camara M, Williams P, Oger P, Dessaux Y (2005) N-Acylhomoserine lactone quorum-sensing molecules are modified and degraded by Rhodococcus erythropolis W2 by both amidolytic and novel oxidoreductase activities. Microbiology 151:3313–3322

Uroz S, Oger PM, Chapelle E, Adeline MT, Faure D, Dessaux Y (2008) A Rhodococcus qsdA-encoded enzyme defines a novel class of large-spectrum quorum-quenching lactonases. Appl Environ Microbiol 74:1357–1366

van Delden C, Kohler T, Brunner-Ferber F, Francois B, Carlet J, Pechere JC (2012) Azithromycin to prevent Pseudomonas aeruginosa ventilator-associated pneumonia by inhibition of quorum sensing: a randomized controlled trial. Intensive Care Med 38:1118–1125

Vanjildorj E, Song SY, Yang ZH, Choi JE, Noh YS, Park S, Lim WJ, Cho KM, Yun HD, Lim YP (2009) Enhancement of tolerance to soft rot disease in the transgenic Chinese cabbage (Brassica rapa L. ssp. pekinensis) inbred line, Kenshin. Plant Cell Rep 28:1581–1591

Vikram A, Jesudhasan PR, Jayaprakasha GK, Pillai SD, Patil BS (2011) Citrus limonoids interfere with Vibrio harveyi cell-cell signalling and biofilm formation by modulating the response regulator LuxO. Microbiology 157:99–110

Wahjudi M, Papaioannou E, Hendrawati O, van Assen AH, van Merkerk R, Cool RH, Poelarends GJ, Quax WJ (2011) PA0305 of Pseudomonas aeruginosa is a quorum quenching acylhomoserine lactone acylase belonging to the Ntn hydrolase superfamily. Microbiology 157:2042–2055

Wang LH, Weng LX, Dong YH, Zhang LH (2004) Specificity and enzyme kinetics of the quorum-quenching N-Acyl homoserine lactone lactonase (AHL-lactonase). J Biol Chem 279:13645–13651

Wang W-Z, Morohoshi T, Ikenoya M, Someya N, Ikeda T (2010) AiiM, a novel class of *N*-acylhomoserine lactonase from the leaf-associated bacterium *Microbacterium testaceum*. Appl Environ Microbiol 76:2524

Wang WZ, Morohoshi T, Someya N, Ikeda T (2012) AidC, a novel N-acylhomoserine lactonase from the potato root-associated cytophaga-flavobacteria-bacteroides (CFB) group bacterium Chryseobacterium sp. strain StRB126. Appl Environ Microbiol 78:7985–7992

Weiland-Bräuer N, Fischer MA, Pinnow N, Schmitz RA (2019a) Potential role of host-derived quorum quenching in modulating bacterial colonization in the moon jellyfish Aurelia aurita. Sci Rep 9:34

Weiland-Bräuer N, Malek I, Schmitz RA (2019b) Metagenomic quorum quenching enzymes affect biofilm formation of Candida albicans and Staphylococcus epidermidis. PLoS One 14:e0211366

Welch M, Todd DE, Whitehead NA, McGowan SJ, Bycroft BW, Salmond GP (2000) N-acyl homoserine lactone binding to the CarR receptor determines quorum-sensing specificity in Erwinia. EMBO J 19:631–641

Wolter J, Seeney S, Bell S, Bowler S, Masel P, McCormack J (2002) Effect of long term treatment with azithromycin on disease parameters in cystic fibrosis: a randomised trial. Thorax 57:212–216

Wopperer J, Cardona ST, Huber B, Jacobi CA, Valvano MA, Eberl L (2006) A quorum-quenching approach to investigate the conservation of quorum-sensing-regulated functions within the Burkholderia cepacia complex. Appl Environ Microbiol 72:1579–1587

Xiang DF, Kolb P, Fedorov AA, Meier MM, Fedorov LV, Nguyen TT, Sterner R, Almo SC, Shoichet BK, Raushel FM (2009) Functional annotation and three-dimensional structure of Dr0930 from deinococcus radiodurans, a close relative of phosphotriesterase in the amidohydrolase superfamily. Biochemistry 48:2237–2247

Yang J-L (2007) Larval settlement and metamorphosis of the mussel Mytilus galloprovincialis on different macroalgae. Mar Biol 152:1121–1132

Yang F, Wang LH, Wang J, Dong YH, Hu JY, Zhang LH (2005) Quorum quenching enzyme activity is widely conserved in the sera of mammalian species. FEBS Lett 579:3713–3717

Yang X, Zhou Z, Raju MN, Cai X, Meng F (2017) Selective elimination of chromophoric and fluorescent dissolved organic matter in a full-scale municipal wastewater treatment plant. J Environ Sci 57:150–161

Yeon KM, Cheong WS, Oh HS, Lee WN, Hwang BK, Lee CH, Beyenal H, Lewandowski Z (2009) Quorum sensing: a new biofouling control paradigm in a membrane bioreactor for advanced wastewater treatment. Environ Sci Technol 43:380–385

Yin H, Deng Y, Wang H, Liu W, Zhuang X, Chu W (2015) Tea polyphenols as an antivirulence compound disrupt quorum-sensing regulated pathogenicity of Pseudomonas aeruginosa. Sci Rep 5:16158

Yu H, Lee K, Zhang X, Choo K-H (2019) Core-shell structured quorum quenching beads for more sustainable anti-biofouling in membrane bioreactors. Water Res 150:321–329

Zhang HB, Wang LH, Zhang LH (2002) Genetic control of quorum-sensing signal turnover in Agrobacterium tumefaciens. Proc Natl Acad Sci U S A 99:4638–4643

Zhang X, Lee K, Yu H, Mameda N, Choo K-H (2019) Photolytic quorum quenching: a new anti-biofouling strategy for membrane bioreactors. Chem Eng J 378:122235

Zhao B, Zhang S, Qian P-Y (2003) Larval settlement of the silver- or goldlip pearl oyster Pinctada maxima (Jameson) in response to natural biofilms and chemical cues. Aquaculture 220:883–901

Antitumor Microbial Products by Actinomycetes Isolated from Different Environments

5

Tábata Cristina Guimarães, Thiara Santana Gomes, Clara Dourado Fernandes, Fernanda Dantas Barros, Kamila Valença Oliveira, Muhammad Bilal, Ram Naresh Bharagava, Luiz Fernando Romanholo Ferreira, and Luciana M. Hollanda

Abstract

Chemotherapy is in many cases the primary treatment against several cancers. However, there has been increasing and uninterrupted resistance to antineoplastic drugs by malignant tumors. Secondary metabolites are bioactive molecules with

T. C. Guimarães · F. D. Barros · K. V. Oliveira
Graduate Program in Industrial Biotechnology, Tiradentes University, Farolândia, Aracaju, Sergipe, Brazil

T. S. Gomes · C. D. Fernandes
Graduate Program in Process Engineering, Tiradentes University, Farolândia, Aracaju, Sergipe, Brazil

M. Bilal
Department of Bioengineering, School of Life Science and Food Engineering, Huaiyin Institute of Technology, Huaian, China

R. N. Bharagava
Laboratory for Bioremediation and Metagenomics Research (LBMR), Department of Microbiology (DM), Babasaheb Bhimrao Ambedkar University (A Central University), Lucknow, Uttar Pradesh, India

L. F. R. Ferreira (✉)
Graduate Program in Process Engineering, Tiradentes University, Farolândia, Aracaju, Sergipe, Brazil

Institute of Technology and Research, Farolândia, Aracaju, Sergipe, Brazil

L. M. Hollanda
Graduate Program in Industrial Biotechnology, Tiradentes University, Farolândia, Aracaju, Sergipe, Brazil

Institute of Technology and Research, Farolândia, Aracaju, Sergipe, Brazil

P. K. Arora (ed.), *Microbial Technology for Health and Environment*, Microorganisms for Sustainability 22,
https://doi.org/10.1007/978-981-15-2679-4_5

diverse functions, used primarily in the pharmaceutical industry serving as the basis for medicines. They can be produced by a variety of plants, animals, and microorganisms. Actinomycetes are ubiquitous bacteria, responsible for producing most antibiotics available on the market and producing essential antitumor agents used in the treatment of cancer. In this review, we show the significant chemicals as well as isolated secondary metabolites that are in the test phase.

Keywords
Chemotherapy resistance · Actinomycetes · Antitumor

5.1 Introduction

Chemotherapy is one of the most used forms of treatment against many types of cancer, due to its low cost when compared to other forms of therapy (Holohan et al. 2013). On chemotherapy, the usage of cytotoxic agents has increased within the last years, reducing the development and progression of tumors. The success of chemotherapy was mainly based on a more in-depth comprehension of different molecular mechanisms that conduct the emergence of neoplasia (Piccolo et al. 2015; Pan et al. 2016).

However, the efficiency of chemotherapy is questioned due to the resistance to the medication and the new forms of cancer, once different organisms with the same type of cancer can react differently, due to genetic mutations and epigenetic interferences (Montazami et al. 2015; Marin et al. 2012; Gerlinger et al. 2012).

Important genetic mechanisms related to chemotherapy resistance are profoundly studied because of their complexity. They mainly are related to metabolic alterations, which promote the inhibition and degradation of drugs, DNA mutations, and epigenetic alterations that increase the expression of the therapeutic targets and activate alternative signalization pathways and the effectors of apoptotic pathways (Ramos and Bentires-Alj 2015; Housman et al. 2014; Hu and Zhang 2016). They are also related to changes on the heterogeneity of tumors and communication between cancerous cells and the microenvironment around them, which can stop therapeutic responses.

The resistance to antineoplastic agents is classified in two major categories: intrinsic and acquired. The intrinsic (primary) resistance occurs when genetic alterations on malignant tumors provide the formation of different resistance factors, which turn the treatment inefficient, even before the administration of the drugs. Acquired (secondary) resistance can occur during the treatment itself. In the beginning, there is an increase in sensitivity to the drug administered, but during the treatment, genetic mutations can occur (cited above) as an adaptive response (Senthebane et al. 2017; Gottesman et al. 2016; Pan et al. 2016; Rebucci and Michiels 2013).

To summarize, resistance to chemotherapy is a result of the organism's own functional dynamic, once the inefficiency of the medicine reaches a threshold too insufficient to cure the patient, stabilize the condition, or eliminate the symptoms (Salehan and Morse 2013). Therefore, the inefficiency of cytotoxic agents results in a treatment flaw that, usually, leads to death. Although the studies focused on the subject are more concentrated on genetics and molecular biology of cancer, the search for new medications is still a major objective on the treatment for this disease (Olano et al. 2009; Alfarouk et al. 2015; Gomathi and Gothandam 2016).

The present work carried out a systematic survey on chemotherapeutic medicine and molecules that presented antitumor/antineoplastic activity and came from actinomycetes on a test phase and are commercially available.

5.2 Search for Bioactive Molecules

Bioactive molecules are secondary metabolites that can be produced by many organisms such as plants, bacteria, fungi, kelp, sponges, and oceanic fauna and flora (Chinembiri et al. 2014; Catalani et al. 2016; Dias et al. 2012). Usually, they are produced in conditions that require adaptive or defensive mechanisms against predators. These compounds have individual characteristics depending on the species producing it; furthermore, they are not vital for growth, development, and/or reproduction of the organism of origin (Dias et al. 2012). Those molecules have anti-inflammatory, antifungal, anticancer, and antibiotic properties (Zhang et al. 2017; Aggarwal et al. 2015; Demain and Vaishnav 2011).

As for the antitumor property, it is worth noticing that bioactive molecules act as regulators of tools for repression of oncogenes (genes with cancer connections) and the cellular cycle (Aggarwal et al. 2015). The bioactive compounds interfere on whatever oncogenic aspects, such as transcription factors, adhesion molecules, growth factors and its receptors, and enzymes that promote inflammatory processes (Shanmugam et al. 2016). They also induce the regular cell death (apoptosis) and the self-destruction of cells (autophagy) and inhibit the topoisomerases I and II and the angiogenesis mechanism (Demain and Vaishnav 2011; Catalani et al. 2016). Last, they are also antioxidant and extinguish free radicals (Kallifatidis et al. 2016).

The most well-known resources are still providing molecules with a larger structural diversity and offer big opportunities for the prospection of new medicine. The microbe diversity represents an important rout for the discovery of new chemical molecules (Shah et al. 2017). The actinomycetes are studied, once they produce a large summary of natural products (NPs) and other bioactive metabolites, including antibiotics, useful enzyme blockers for cancer treatment, enzymes, antifungals, antibiotics and antitumor, immunosuppressants, biosurfactants, and immunomodifiers that increase immune responses (Adegboye and Babalola 2013; Dilip et al. 2013).

5.3 Actinomycetes: An Overview

5.3.1 Morphology and Reproduction

Actinomycetes are Gram-positive bacteria of the *Actinobacteria* phylum (Waksman 1940; Adegboye and Babalola 2013). They are morphologically diverse and vary between coccus (*Micrococcus*) and bacilli (*Mycobacterium*) to a ramified mycelium (*Streptomyces*), a majority of which are also spore formers (Ul-Hassan and Wellington 2009). They differ, in most cases, due to the presence or absence of mycelium (air or substrate), production of melanoid pigments, and their spore structure (Chaudhary et al. 2013).

For a long time, they were mistaken for fungi, because of similar morphologic structures. However, its cell wall has a bigger similarity to other bacteria, because, in its composition, the presence of peptidoglycan and teichoic acid is worth noticing, as well as the absence of cellulose and chitin (which are common in fungi). Facing these facts, new criterions for fungal-bacterial differentiation were created, one that goes beyond cell morphology. These include genetic similarity evaluation between DNA sequences, measurement of the genomic guanine-cytosine content, cell wall analysis, and membrane composition (Aftab et al. 2015; Barka et al. 2016; Behie et al. 2017; Chaudhary et al. 2013; Weber et al. 2015; Polpass and Bhavanath 2016; Ser et al. 2017).

Actinomycetes reproduce via asexual spores. Filamentous species fragment themselves into new cells due to the growth of their hyphae and spore liberation. These spores, once exposed, are called conidia; when wrapped on a sporangium, they are called sporangiospores. Although not resistant to heat, they can endure desiccation, which allows the survival of the species during droughts.

5.3.2 Classification of Actinomycetes

The *Actinobacteria* phylum is one of the biggest taxonomic unities of the *Bacteria* domain. Until 1983, this bacterial group was classified in five subclasses, six orders, and 14 suborders (Goodfellow and Williams 1983). The sequencing and amplification of 16S rRNA gene from microbiotic communities became standard for community comparisons throughout time, space, and environment. The ribossomic DNA sequences of 16S provided actinomycetologists a phylogenetic tree that allowed the investigation of actinomycete evolution and the basis for its identification (Burke and Darling 2016), which lead to the six classes, 15 orders, 43 families, and 130 genera of *Actinobacteria* (Embley and Stackebrandt 1994; Ludwig et al. 2012).

Between actinomycetes, the genus *Streptomyces* is still the focus of systematic research, and it is of great pharmaceutical interest due to its commercial value as a rich source of secondary metabolites (Wei et al. 2017). The *Streptomyces* genus presents itself in nature in the great number of species between all actinomycete genera, with over 500 species. The name *Streptomyces* was introduced in 1943 to actinomycetes of air mycelium production (Dharmaraj 2010).

The genus representatives find themselves mostly on soil. They utilize a large variety of extracellular mechanisms for primary growth and easier associations to other organisms, as well as reaching its attributes regarding development and secondary metabolism. Since Waksman's discovery of streptomycin as the first therapeutically useful *Streptomyces* antibiotic, it was noticed that streptomycetes synthesized an incredible variety of chemically distinct blockers, using many different cellular processes (Chater et al. 2010).

5.3.3 Distribution and Ecological Importance

The *Actinobacteria* species are ubiquitous, which means they are found in different habitats. They are considered the main group of microorganisms that make up the soil but are also present in aquatic environments such as rivers, oceans, lakes, lagoons, swamps, and sewers. Furthermore, they are part of symbiotic associations with plants and animals and are present in the gastrointestinal tract of some mammals (Kumar et al. 2014; Rana and Salam 2014; Barka et al. 2016).

New species were found in atypical environments, including medieval paintings, desert soil, marine sponge, and thermal fonts containing radon. The ability to inhabit the most diverse habitats comes from actinomycete's capacity to produce extracellular hydrolytic enzymes, particularly on soil, where they are responsible for the degradation of organic matter, making them pivotal organisms in the carbon cycle. Some species can break more complex and recalcitrant compounds, which the *Rhodococcus* species are a good example of; they can degrade nitrophenol, dinitrophenol, pyridine, and nitroaromatic compounds (Ul-Hassan and Wellington 2009; Kumar et al. 2014).

They promote the degradation of organic matter, decomposition, and mineralization on sediments and on water, releasing organic and inorganic substances dissolved in it. The mineralization of organic matter, derived from primary producers, results in its cycling, in a way where these substances are returned and, once again, available on soil. This way, *Actinobacteria* not only keep primitive nature of the environment but also act as biologic mediators via their involvement in biogeochemical processes such as nitrogen fixation (Kumar et al. 2014).

5.3.4 Production of Biologically Active Molecule

The *Actinobacteria* produce intra- and extracellular inorganic materials in a nanoscale with a differentiated morphology (Manivasagan et al. 2014). Between the producers of commercially important metabolites, the actinomycetes are the most important prokaryotes in economy and biotechnology (Rambabu et al. 2014).

The filamentous *Actinomycetales* produce more than 10,000 bioactive compounds, 7600 derivatives of *Streptomyces*, and 2500 of the so-called rare species of actinomycetes, which represent the biggest group (45%) of microbial bioactive metabolites (Berdy 2005). The most important fact is that these bacteria have evolved due to agglomerates of biosynthetic genes and, therefore, show an unprecedented potential in the synthesis of natural, biologically active products (Doroghazi and Metcalf 2013; Polpass and Bhavanath 2016).

Now, in the post-genomic era, with thousands of genome sequences for actinomycetes available, we realize this bacterial group possesses the genetic capability to produce an absurd number of natural products (NPs). This realization, allied with the advances in genetic tools, revitalized the interest in exploring actinomycete NPs in innovative and creative ways.

5.3.5 Antitumor Compounds from Actinomycetes

The biotechnological advances improved our comprehension and led to a lot of progress in the development of anticancer medication. Although present in the antibiotic era, major challenges are faced due to the development of resistance in microorganisms and cancer medication. This alarming issue requires immediate action when it comes to discovering and developing new, more potent, and less toxic drugs. Around 60% of antineoplastic agents come from natural sources such as plants, oceanic organisms, and microorganisms (Gomathi and Gothandam 2016).

In Table 5.1, we are able to see 192 bioactive compounds originated from actinomycetes, with antitumor and antineoplastic activity tested against different types of neoplasia. Between the species in this group, the *Streptomyces* genus shows up for the production of the majority of compounds.

In the global market, there are many antitumor agents with major relevance, which were derived from actinomycetes (Table 5.2).

Streptomyces appears in a large summary of the studies as the major chemotherapeutic drug producer. *S. parvulus* and *S. chrysomallus* produce the same compound (dactinomycin) and *S. galilaeus* produces aclacinomycin, which can only be produced in a lab. In the commercial compounds, there are chemotherapeutic drugs, which can be administered by intramuscular, subcutaneous, and intrapleural, not just via intravenous.

5.4 Conclusion

Actinomycetes have a large relevance on cancer treatment, once it has a large production of bioactive molecules. With the increasing progress in science and technology, it is reasonable to predict a higher demand for new bioactive compounds synthesized by *Actinobacteria* from many sources, including earthly and aquatic environments (Azman et al. 2015). There are many other bioactive compounds discovered, but they are still in clinical trials and, therefore, not on the market. The relevance of new drug discovery makes the treatment expectations increase due to chemotherapy resistance.

Table 5.1 Bioactive compounds produced by *Actinobacteria* and its respective biological activity

Compound	Species	Biological activity	Location	Cancer	References
(6R, 10S)-10-Methyl-6-dodecanolide)	*Streptomyces* sp. B6007	Anticancer	Papua New Guinea	Gastric adenocarcinoma, hepatocarcinoma, breast cancer	Stritzke et al. (2004)
(R)-10-Methyl-6-undecanolide	*Streptomyces* sp. B6007	Anticancer	Papua New Guinea	Gastric adenocarcinoma, hepatocarcinoma, breast cancer	Stritzke et al. (2004)
1(10-Aminodecyl) pyridinium	*Amycolatopsis alba* (DVR D4)	Anticancer	Bay of Bengal	Cervix cancer, brain cancer	Dasari et al. (2012)
1,6-Phenazinediol	*Actinomadura* sp. M048	Antitumor, anticancer	Jiaozhou Bay in China	Melanoma (MEXF 462NL), human lung carcinoma (LXFA 629L and LXFL 529L), human breast carcinoma (MAXF 401NL), human kidney tumor (RXF 944L), human uterus cancer (UXF 1138L)	Lombó et al. (2006) and Maskey et al. (2003a, b)
1,8-Dihydroxy-2-ethyl-3-methylanthraquinone	*Streptomyces* sp. FX-58	Antitumor	From a marine plant *Salicornia herbacea* in Qingdao, Shandong Province, China	Cancer cell HL-60, BCTC-823, and MDA-MB-435	Huang et al. (2006)
10′-5″-Methoxycarbonyl) amino-200-methyl-phenylaminocarbonylstaurosporine	*Streptomyces* sp. 172614	Antitumor	Fujian, China	Human colon tumor cell HCT-116	Li et al. (2011a, b)

(continued)

Table 5.1 (continued)

Compound	Species	Biological activity	Location	Cancer	References
1-Ethyl-β-carboline-3-carboxylic acid	*Actinomadura* sp. BCC 24717	Anticancer	Khuean Srinagarindra National Park, Kanchanaburi Province, Thailand	Human epidermoid carcinoma of oral cavity, human small cell lung cancer	Kornsakulkarn et al. (2013)
1-Hydroxy-1-norresistomycin	*Streptomyces chibaensis* AUBN$_1$/7	Antitumor; anticancer	Machilipatnam coast, Bay of Bengal, India	Gastric adenocarcinoma, hepatic carcinoma	Gorajana et al. (2005)
1-Hydroxy-1-norresistomycin	*Streptomyces* sp. B8005	Antitumor	Sediment of the Laguna de Terminos at the Gulf of Mexico	–	Kock et al. (2005)
1-Methyl indole-3-carboxamide	*Actinomadura* sp. BCC 24717	Anticancer	Khuean Srinagarindra National Park, Kanchanaburi Province, Thailand	Human epidermoid carcinoma of oral cavity, human small cell lung cancer	Kornsakulkarn et al. (2013)
1-Vinyl-β-carboline-3-carboxylic acid	*Actinomadura* sp. BCC 24717	Anticancer	Khuean Srinagarindra National Park, Kanchanaburi Province, Thailand	Human epidermoid carcinoma of oral cavity, human small cell lung cancer	Kornsakulkarn et al. (2013)
2,3,4,2’-diO-Succinoyl-di-O-alkanoyl-α-α-trehalose	*Rhodococcus erythropolis* SD-74	Antitumor	–	Basophilic leukemia (KU812)	Isoda et al. 1995

Compound	Species	Biological activity	Location	Cancer	References
3,6-Disubstituted indoles	*Streptomyces* sp. BL-49-58-005	Antitumor	Mexican marine invertebrate	14 different tumor cell lines	Sánchez López et al. (2003)
3-Methyl-β-carboline	*Micromonospora* sp. (M2DG17)	Antitumor	Haikou, China	Human colon tumor cell HCT-116	Huang et al. (2011)
4a,8a-Dimethyl-6-(2-methylpropenyloxy)-3,4,4a,4b,5,6,8a,9-octahydro-1H-phenanthren-2-one	*Actinobacterium* sp. MS1/7	Antitumor	Bay of Bengal	Human leukemia HL-60 cell line	Saha et al. (2006)
4′-Demethylamino-4-hydroxystaurosporine	*Streptomyces* sp. 172614	Antitumor	Fujian, China	Human colon tumor cell HCT-116	Li et al. (2011a, b)
Aclacinomycin A (aclarubicin)	*Streptomyces galilaeus*	Antitumor	–	–	Raty et al. (2002)
Actinofuranones A and B	*Streptomyces* sp. CNQ766	Antitumor	Marine sediment, island of Guam	Mouse splenocyte T-cells	Cho et al. (2006)
Actinomycin C	*Streptomyces chrysomallus*	Antitumor	–	–	Konoshenko et al. (1994)
Albidopyrone	*Streptomyces* sp. NTK 227	Anticancer	Atlantic Ocean sediment	–	Hohmann et al. (2009a)
Alkaloids and quinine	*Actinomycetes* ACT01 and ACT02	Anticancer	Kanyakumari district, Tamilnadu, India	MCF-7 and MDA-MB-231 cell lines	Ravikumar et al. (2012)
Altemicidin	*Streptomyces sioyaensis* SA-1758	Antitumor; Anticancer	Gamo, Japan	Lymphoid leukemia, carcinoma	Takahashi et al. (1989a, b)
Ammosamides	*Streptomyces* sp. CNR-698	Antitumor; anticancer	Bahamas islands	HCT-116 colon carcinoma	Hughes et al. (2009)
Ammosamides A and B	*Streptomyces* sp.	Anticancer	Bahamas	HCT-116 colon carcinoma	Gaudêncio et al. (2008)

(continued)

Table 5.1 (continued)

Compound	Species	Biological activity	Location	Cancer	References
Amorphane sesquiterpenes	*Streptomyces* sp. M491	Antitumor	Jiaozhou Bay in China	Human tumor cell lines	Wu et al. (2007)
Anthracyclines	*Streptomyces galileus*	Antitumor	–	–	Fujii and Ebizuka (1998)
Anthrone and lactones	*Actinomycete* (N2010–37)	Antitumor; cytotoxic	Zhanjiang Mangrove	Human chronic granulocytic leukemia cell line K562	Zhou et al. (2011)
Arcyriaflavin A	*Actinomycete* sp. $Z_2$039–2	Antitumor; anticancer	Qingdao, China	Human chronic myelogenous leukemia K562	Liu et al. (2007)
Arenamide A, B	*Salinispora arenicola* CNT-088	Antitumor, anticancer	Great Astrolabe Reef, in the Kadavu Island chain, Fiji	RAW 264.7 cell line murine	Asolkar et al. (2009)
Arenicolides A	*Salinispora arenicola* CNR-005	Antitumor; anticancer	Sediment sample from Guam	Human colon adenocarcinoma cell line	Williams et al. (2007)
Arenimycin	*Salinispora arenicola*	Anticancer; antibacterial	Mangrove channel at Sweetings Cay, Grand Bahama Island	Human colon adenocarcinoma HCT-116	Asolkar et al. (2010)
Arenimycin	*Streptomyces* CNQ-085	Antibacterial; anticancer	Marine sediment from San Diego, CA	–	Asolkar et al. (2006)
Aureoverticillactam	*Streptomyces aureoverticillatus* NPS001583	Antitumor, anticancer	Mauritius (Indian Ocean)	Colorectal adenocarcinoma, melanoma, leukemia	Mitchell et al. (2004)
Azalomycin F	*Streptomyces* sp. 211726	Antitumor, antimicrobial	Wenchang, China	Human colon tumor cell HCT-116	Yuan et al. (2013)

Compound	Species	Biological activity	Location	Cancer	References
Azalomycin F4a 2-ethylpentyl ester	*Streptomyces* sp. 211726	Antitumor, antimicrobial	Wenchang, China	Human colon tumor cell HCT-116	Yuan et al. (2011)
Azalomycin F5a 2-ethylpentyl ester	*Streptomyces* sp. 211726	Antitumor, antimicrobial	Wenchang, China	Human colon tumor cell HCT-116	Yuan et al. (2011)
Benzastatins	*Streptomyces nitrosporeus* 30643	Anticancer	Sokcho-City, Kangwon-do, Korea	Neuroblastoma	Kim et al. (1996)
Bohemamines	*Streptomyces* sp. CNQ-583	Antitumor, anticancer	Island of Guam	Human colon adenocarcinoma	Bugni et al. (2006)
Butenolides	*Streptoverticillium luteoverticillatum* 11014	Antitumor, anticancer	Taipingjiao, Qingdao, China	Human leukemia and murine lymphoma P388	Li et al. (2006)
Caboxamycin	*Streptomyces* sp. NTK 937	Antitumor; anticancer; antibacterial	Atlantic Ocean, Canary Islands	Human gastric adenocarcinoma, hepatocellular carcinoma, breast carcinoma MCF7 cell lines	Hohmann et al. (2009)
Caerulomycin (F,G, H, I,J,K)	*Actinoalloteichus cyanogriseus* WH1-2216-6	Anticancer	Weihai, China	Human promyelocytic leukemia, human alveolar adenocarcinoma, human chronic myelogenous leukemia, human epidermoid carcinoma	Fu et al. (2011a)

(continued)

Table 5.1 (continued)

Compound	Species	Biological activity	Location	Cancer	References
Caprolactones	*Streptomyces* sp.	Anticancer	Marine actinomycetes at the Alfred-Wegener Institute for Polar and Marine Research in Bremerhaven	HM02 (gastric adenocarcinoma), HepG2 (hepatocellular carcinoma), and MCF7 (breast adenocarcinoma)	Stritzke et al. (2004)
Carboxamycin	*Streptomyces* sp. NTK 937	Antibacterial; anticancer	Atlantic Ocean deep-sea sediment, Canary Basin	Breast carcinoma MCF7 cell lines	Hohmann et al. (2009)
Carminomycin	*Actinomadura carminata*	Antitumor	–	–	Naumova et al. (1986)
Chalcomycin A	*Streptomyces* sp. M491	Antitumor	Qingdao coast, China	HeLa human cervix carcinoma cell line	Wu et al. (2007)
Chalcomycin A	*Streptomycete* B7064	Anticancer	Collection of marine actinomycetes at the Alfred-Wegener Institute for Polar and Marine Research in Bremerhaven, Germany	–	Asolkar et al. (2002)

Compound	Species	Biological activity	Location	Cancer	References
Chalcomycin B	*Streptomyces* sp. B7064	Antitumor	Mangrove sediment near Pohoiki, Hawaii (Pacific Ocean)	–	Asolkar et al. (2002)
Chandrananimycins A, B, C	*Actinomadura* sp. M048	Antitumor; anticancer; antialgal; antibacterial; antifungal	Jiaozhou Bay in China	Human colon carcinoma (CCL HT29), melanoma (MEXF 514L), human lung carcinoma (LXFA 526L and LXFL 529L), human breast carcinoma (CNCL SF268, LCL H460, and MACL MCF-7), human kidney tumor (PRCL PC3M and RXF 631L)	Maskey et al. (2003a, b)
Chartreusin	*Streptomyces* sp. QD518	Antitumor	Jiaozhou Bay of Qingdao, China	Murine (P388 and L1210) leukemia, and melanoma cells (B16)	Wu et al. (2006)
Chartreusin	*Streptomyces* sp. M518	Anticancer	Jiaozhou Bay, China	Melanoma leukemia	Wu et al. (2006) and McGovren et al. (1977)
Chinikomycins A, B	*Streptomyces* sp. M045	Antitumor; anticancer	Jiaozhou Bay in China	Mammary melanoma (MEXF 462NL)	Li et al. (2005)
Chlorinated dihydroquinones	*Actinomycete* CNQ-525	Antitumor; anticancer, antibacterial	La Jolla, California	Human colon carcinoma HCT-116 cells	Soria-Mercado et al. (2005)
Chlorocarcins A, B, C	*Streptomyces lavendulae* No. 314	Antitumor, antibacterial	–	EHRLICH carcinoma, ascites and solid forms, and leukemia (L 1210)	Arai et al. (1976)

(continued)

Table 5.1 (continued)

Compound	Species	Biological activity	Location	Cancer	References
Chloro-dihydroquinones	*Streptomyces* sp.	Anticancer; antibacterial	Ocean sediments collected at a depth of 152 m near La Jolla, California	Human colon carcinoma (HCT-116)	Soria-Mercado et al. (2005)
Chromomycin	*Streptomyces* sp. WBF16	Anticancer	Sea sediments, Bijiatuan, Weihai, China	Human gastric cancer cell line (SGC7901), human liver hepatocellular carcinoma cell line (HepG2), human lung adenocarcinoma epithelial cell line (A549), human colon cancer cell line (HCT116), human ovarian cancer cell line (COC1), and human umbilical vein endothelial cells (HUVEC)	Lu et al. (2012)
Chromomycins B, A2, A3	*Streptomyces coelicolor*	Antitumor	Sea sediments, Bijiatuan, Weihai, China	Human gastric cancer, human liver hepatocellular carcinoma, human colon cancer, human lung adenocarcinoma	Lu et al. (2012)
Chromomycin SA	*Streptomyces sp. KML-2*	Antitumor	Khewra Salt Mines, Pakistan	HeLa and MCF-7	Aftab et al. (2015)
Cyanogrisides A, B, C, D	*Actinoalloteichus cyanogriseus* WH1-2216-6	Anticancer	Weihai, China	Breast cancer, human epidermoid carcinoma, human chronic myelogenous leukemia	Fu et al. (2011b)
Cyanosporasides	*Salinispora pacifica* CNS103	Antitumor	Palau Island	Human colon carcinoma HCT-116	Oh et al. (2006)

Compound	Species	Biological activity	Location	Cancer	References
Cyclomarins	*Streptomycete* CNB-982	Anticancer	Mission Bay, California	Human colon carcinoma	Renner et al. (1999)
Cyclomarins	*Salinispora arenicola* CNS-205	Anticancer	Palau	Human colon carcinoma	Schultz et al. (2008)
Daryamides A, B	*Streptomyces* sp. CNQ-085	Antitumor; anticancer; antifungal	San Diego coast, California	Human colon carcinoma HCT-116	Asolkar et al. (2006)
Daryamides A, B	*Streptomyces* sp. B7064	Anticancer	Pohoiki, Hawaii (Pacific Ocean)	–	Asolkar et al. (2002)
Dermacozines A–G	*Dermacoccus abyssi* sp. MT1.1	Anticancer; antitumor	Mariana Trench sediment	Human chronic myelogenous leukemia	Abdel-Mageed et al. (2010)
Dermacozines A–G	*Dermacoccus abyssi* sp. MR1.2	Anticancer; antitumor	Mariana Trench sediment	Human chronic myelogenous leukemia	Abdel-Mageed et al. (2010)
Diazepinomicin	*Micromonospora* DPJ12	Antibacterial; anticancer; anti-inflammatory	Isolated from *Didemnum proliferum* Kott collected by scuba at Shishijima Island, Japan	–	Charan et al. (2004)
Drimentine G	*Streptomyces* sp. CHQ-64	Anticancer	Guangdong Province	Human colon cancer, human hepatocellular carcinoma, human lung cancer, human ovarian carcinoma	Che et al. (2012), Che et al. (2013)
Duanomycin	*Streptomyces* sp. C5	Antitumor	–	–	Jingsong et al. (1994)

(continued)

Table 5.1 (continued)

Compound	Species	Biological activity	Location	Cancer	References
Echinosporins	*Streptomyces albogriseolus* A2002	Antitumor	Jiaozhou Bay, China	Human myelogenous leukemia (K562)	Cui et al. (2007)
Elaiomycins B and C	*Streptomyces* sp. BK 190	Antitumor	–	–	Helaly et al. (2011)
Elloramycin	*Streptomyces olivaceus*	Antitumor	–	–	Fiedler et al. (1986)
Fredericamycin A	*Streptomyces griseus* FCRC-48	Antitumor	Frederick, Maryland	KB, P388, and L1210 cell lines	Warnick-Pickle et al. (1981)
Fridamycin D	*Streptomyces* sp. B6921	Antitumor	Sediment of a coastal site of Mauritius, Indian Ocean	–	Maskey et al. (2003a, b)
Fridamycins (A, B, D)	*Streptomyces* sp. B6921	Anticancer	Mauritius (Indian Ocean)	–	Maskey et al. (2003a, b)
Furaquinocins C, D, E, F, G, H	*Streptomyces* sp. KO 3988	Antitumor	–	HeLa S3 and B16 melanoma cell	Ishibashi et al. (1991)
Gilvusmycin	*Streptomyces* sp.	Antitumor	Takahashi, Okayama Prefecture, Japan	Murine leukemia (P388), human chronic myelogenous leukemia (K562), human epidermoid carcinoma (A431), and human gastric carcinoma (MKN28)	Tokoro et al. (1999)
Glaciapyrroles	*Streptomyces* sp. NPS008187	Anticancer	Alaska	Colorectal adenocarcinoma, melanoma	Macherla et al. (2005)

Compound	Species	Biological activity	Location	Cancer	References
Grincamycins B–F	*Streptomyces lusitanus*	Anticancer	South China Sea	Human cancer cell lines (HepG2, SW-1990, HeLa, NCI-H460, and MCF-7) and the mouse melanoma cell line (B16)	Huang et al. (2012)
Griseorhodin A	*Streptomyces* sp. JP95	Antitumor	–	–	Tresselt et al. (1978) and Li and Piel (2002)
Gutingimycin	*Streptomyces* sp. B8652	Antitumor	Sediment of the Laguna de Términos at the Gulf of México	Parimycin, trioxacarcins, and gutingimycin showed	Maskey et al. (2004a, b)
Himalomycins A, B	*Streptomyces* sp. B6921	Antitumor	Mauritius (Indian Ocean)	–	Maskey et al. (2003a, b)
Himastatin	*Streptomyces hygroscopicus*	Antitumor	Himachal Pradesh State, India	Melanoma (B16), leukemia (P388)	Lam et al. (1990)
Hydramycin	*Streptomyces violaceus*	Antitumor	Hyderabad, Andhra Pradesh State, India	Melanoma (B16), leukemia (P388)	Hanada et al. (1991)
IB-96212	*Micromonospora* sp. L-25-ES25-008	Antitumor	Sponge collected at the Indian Ocean near the coast of Mozambique	P-388 (ATCC CCL 46), A-549 (ATCC CCL 185), HT-29 (ATCC HTB-38), and MEL-28 (ATCCHTB-72)	Fernández-Chimeno et al. (2000)

(continued)

Table 5.1 (continued)

Compound	Species	Biological activity	Location	Cancer	References
Indimicin B	*Streptomyces* sp. SCSIO 03032	Anticancer	Indian Ocean	Human lymphoblastic leukemia, murine melanoma, human lung cancer, human liver cancer, human breast cancer	Zhang et al. (2014)
Indotertine B	*Streptomyces* sp. CHQ-64	Anticancer	Guangdong Province	Human colon cancer, human hepatocellular carcinoma, human lung cancer, human ovarian carcinoma	Che et al. (2012) and Che et al. (2013)
Iodinin	*Actinomadura* sp. M048	Anticancer; antitumor	Jiaozhou Bay in China	Human colon carcinoma, melanoma, human lung carcinoma, human breast carcinoma, human kidney tumor, human uterus cancer	Maskey et al. (2003a, b)
K252c (Staurosporine aglycon) arcyriaflavin A	*Actinomycete* strain $Z_2$039-2	Anticancer; antitumor	Coast of Qingdao, China	Human chronic myelogenous leukemia (K562)	Liu et al. (2007)
Kazusamycin	*Streptomyces* sp.	Antitumor	–	Female ICR and CDF mice, sarcoma 180 ascites, and leukemia (P388)	Umezawa et al. (1984)
Komodoquinones	*Streptomyces* sp. KS3	Antitumor	Komodo Island, Indonesia	–	Itoh et al. (2003a, b)
Lajollamycin	*Streptomyces nodosus* NPS007994	Anticancer; antitumor	Scripps Canyon, La Jolla, California	Melanoma, murine melanoma cell line B16-F10	Manam et al. (2005)
Landomycin E	*Streptomyces globisporus*	Antitumor	–	–	Zhu et al. (2005)

Compound	Species	Biological activity	Location	Cancer	References
L-Asparagine aminohydrolase	*Streptomycetes parvulus* KUAP106	Antitumor; antineoplastic agent	Southeast coast of India	–	Usha et al. (2011)
Lavendamycin	*Streptomyces lavendulae*	Antitumor	–	Leukemia in mice (P-388)	Balitz et al. (1982)
Leptomycin	*Streptomyces lividans*	Antifungal, antitumor	–	–	Hu et al. (2005)
Levantilides A and B	*Micromonospora* sp.	Anticancer	Isolated from the Eastern Mediterranean deep-sea sediment	Gastric cancer cell line (GXF 251L), lung cancer cell line (LXFL 529L), melanoma cancer cell line (MEXF 462NL), mammary cancer cell line (MAXF 401NL), renal cancer cell line (RXF 486L), and pancreatic cancer cell line (PAXF 1657L)	Gärtner et al. (2011)
Lodopyridone	*Saccharomonospora* sp.	Anticancer	Collected near the mouth of the La Jolla Canyon	Human colon cancer cells (HCT-116)	Maloney et al. (2009)
Lucentamycins A–D	*Nocardiopsis lucentensis* CNR-712	Anticancer; antitumor	Little San Salvador, Bahamas	Colon adenocarcinoma cells (HCT-116)	Cho et al. (2007)

(continued)

Table 5.1 (continued)

Compound	Species	Biological activity	Location	Cancer	References
Luminacin	*Streptomyces* sp.	Antitumor, angiogenesis inhibitor	–	D 6 rat smooth muscle, human fibroblast (WI 38), human colon cancer (WiDr), human lung cancer (H-520), human breast cancer (MDA-MB-435 and MDA-MB-231)	Wakabayashi et al. (2000)
Lynamycins	*Actinomadura* sp. SCSIO 03032	Antitumor	–	–	Gomathi and Gothandam (2016)
Lynamycins	*Marinispora* sp. NPS12745	Antitumor	Mission Bay in San Diego, California	–	McArthur et al. (2008)
Maggiemycin and anhydromaggiemycin	*Streptomyces* sp.	Antitumor	–	Murine cell lines (KB, P388, and LI210)	Pandey et al. (1989)
Mansouramycins A, B, C, D	*Streptomyces* Mei37	Anticancer; antitumor	North Sea coast, Germany	Non-small cell lung cancer, melanoma, breast cancer, prostate cancer	Hawas et al. (2009)
Manumycin	*Streptomyces parvulus*	Antitumor, enzyme inhibitory	–	–	Thiericke and Zeeck (1988)
Manumycin Chinikomycins A,B	*Streptomyces* sp. M045	Anticancer; antitumor	Jiaozhou Bay, China	Renal cancer, mammary cancer, melanoma	Li et al. (2005)
Manumycins	*Streptomyces* sp.	Antitumor, antibacterial, hemolysis trigger activity	–	Human colorectal carcinoma	Sattler et al. (1998); Zhang et al. (2016) and Egler et al. (2016)
Marangucycline B	*Streptomyces* sp. SCSIO 11594	Anticancer	South China Sea	A594, CNE2, HepG2, MCF-7	Song et al. (2015)

Compound	Species	Biological activity	Location	Cancer	References
Marineosin A	*Streptomyces* sp. CNQ-617	Anticancer; antitumor	La Jolla, CA by Alejandra Prieto-Davó	Human colon tumor cell line (HCT-116), melanoma, and leukemia cell lines	Boonlarppradab et al. (2008)
Marineosin B	*Streptomyces* sp. CNQ-617	Antitumor	La Jolla, CA by Alejandra Prieto-Davó	Human colon tumor cell line (HCT-116)	Boonlarppradab et al. (2008)
Marinomycins A, B, C, D	*Marinispora* sp. CNQ-140	Antitumor; anticancer, antibacterial; antifungal; antimicrobial	La Jolla, California	Melanoma SK-MEL-5	Kwon et al. (2006)
Marinones	*Actinomycete* CNH-099	Antitumor	Batiquitos Lagoon, North of San Diego, California	Colon carcinoma (HCT-116)	Pathirana et al. (1992); Hardt et al. (2000); and Kalaitzis et al. (2003)
Marmycins A, B	*Streptomyces* CNH990	Anticancer; antitumor	Sea of Cortez, Baja California Sur, México	Colon adenocarcinoma, breast cancer, prostate cancer, lung cancer, leukemia	Martin et al. (2007)
Mechercharmycins	*Thermoactinomyces* sp. YM3-251	Anticancer; antitumor	Republic of Palau (North Pacific Ocean)	Human leukemia, human lung cancer, human lung adenocarcinoma A549	Kanoh et al. (2005)
Menaquinones	*Micromonospora haikouensi*	Antitumor	Haikou, Hainan, China	–	Xie et al. (2012)

(continued)

Table 5.1 (continued)

Compound	Species	Biological activity	Location	Cancer	References
Metacycloprodigiosin	*Saccharopolyspora* sp. nov.	Antitumor	Coast of Qingdao, China	Mouse lymphoma cell (P388), human peripheral blood promyeloblast cell (HL60), human lung carcinoma cell (A-549 and SPCA4), and human hepatic carcinoma cell (BEL-7402)	Liu et al. (2005)
Methoxyneihumicin	*Nocardiopsis alba* SCSIO 3039	Anticancer	Indian Ocean	SF-268, MCF-7, and NCIH460	Zhang et al. (2013)
Mithramycin and mithramycin SK	*Streptomyces argillaceus*	Antitumor	–	Variety of tumor cell lines, including the 60 human tumor cell lines	Remsinq et al. (2003)
Mitomycin C	*Streptomyces lavendulae* NRRL 2563	Antitumor, binds to double-stranded DNA	–	–	Mao et al. (1999)
N-Acetylquestiomycin	*Actinomadura* sp. M048	Anticancer	Jiaozhou Bay in China	Human colon carcinoma, melanoma, human lung carcinoma, human breast carcinoma, human kidney tumor, human uterus cancer	Maskey et al. (2003a, b)
N-(2-Hydroxyphenyl)-2-phenazinamine (NHP)	*Nocardia dassonvillei* BM-17	Anticancer; antifungal	Arctic Ocean	Human liver hepatocellular carcinoma, human lung adenocarcinoma, human colon adenocarcinoma, human ovarian cancer	Gao et al. (2012)
N-Carboxamido-staurosporines	*Streptomyces* sp. QD518	Antitumor	Qingdao, China	Human tumor cell lines IC50 and human lung adenocarcinoma A549	Wu et al. (2006)

Compound	Species	Biological activity	Location	Cancer	References
Neomarinones	*Actinomycete* CNH-099	Anticancer	Lagoon, North of San Diego, California	Human colon carcinoma	Hardt et al. (2000) and Kalaitzis et al. (2003)
N-Formyl staurosporine	*Streptomyces* sp. QD518	Anticancer	Jiaozhou Bay, China	Bladder, CNS, colon, gastric, head and neck, lung, mammary, ovarian, pancreatic, prostate, renal cancer	Wu et al. (2006)
Nonactin	*Streptomyces* sp. KORDI-3238	Anticancer; antitumor	Sediment at Ayu Trough, in the Pacific Ocean	Human erythroleukemia cell line (K-562)	Jeong et al. (2006)
Okicenone	*Streptomyces* sp.	Antitumor	–	HeLa S3, B16 melanoma, and H69 human lung carcinoma cells	Komiyama et al. (1991)
Oxaprapalines B, D, G	*Streptomyces* sp. G324	Antitumor	Sample collected in Fujieda City, Shizuoka Prefecture, Japan	Murine (P388) and human (A549)	Abe et al. (1993b)
Pacificanones	*Salinispora pacifica* CNS-237	Antitumor	Sediment sample collected on the island of Palau	Human colon cancer (HCT-116)	Oh et al. (2008)

(continued)

Table 5.1 (continued)

Compound	Species	Biological activity	Location	Cancer	References
Parimycin	*Streptomyces* sp. B8652	Antitumor	Sediment of the Laguna de Terminos at the Gulf of Mexico	Stomach cancer (GXF 251L), lung cancer models (H460, LXFA 629L, and LXFL 529L), breast cancer (MCF-7 and MAXF 401NL), melanomas (MEXF 462NL and MEXF 514L)	Maskey et al. (2002)
Piericidins C7 and C8	*Streptomyces* sp. YM14-060	Anticancer; antitumor	Iwayama Bay, Palau Island	RG-E1A-7 cells	Hayakawa et al. (2007a, b)
Piperazimycins	*Streptomyces sp.* CNQ-593	Anticancer; antitumor	Island of Guam.	Human colon carcinoma cell line (HCT-116)	Miller et al. (2007)
Proximicins	*Verrucosispora* sp. MG-37	Anticancer; antitumor	Sea of Japan and Raune Fjord (Norway)	Gastric adenocarcinoma, hepatocellular carcinoma, breast adenocarcinoma	Fiedler et al. (2008) and Schneider et al. (2008)
Proximicins	*Verrucosispora maris* AB-18-032	Anticancer; antitumor	Sea of Japan and Raune Fjord (Norway)	Gastric adenocarcinoma, hepatocellular carcinoma, breast adenocarcinoma	Riedlinger et al. (2004) and Schneider et al. (2008)
Proximicins A, B, C	*Verrucosispora* MG-37	Antitumor	Sea of Japan	Human gastric adenocarcinoma AGS	Schneider et al. (2008)
Proximicins A, B, C	*Verrucosispora maris* AB-18-032	Antitumor	Sea of Japan	Human gastric adenocarcinoma AGS	Schneider et al. (2008)
Deoxynyboquinone	*Pseudonocardia* sp. SCSIO 01299	Anticancer, antibacterial activity	South China Sea	Human breast adenocarcinoma cell line (MCF-7), human non-small cell lung cancer cell line (NCI-H460), and human glioma cell line (SF-268)	Li et al. (2011a, b)

Compound	Species	Biological activity	Location	Cancer	References
Pseudonocardians A–C	*Pseudonocardia* sp. SCSIO 01299	Anticancer, antibacterial activity	South China Sea	Human breast adenocarcinoma cell line (MCF-7), human non-small cell lung cancer cell line (NCI-H460), and human glioma cell line (SF-268)	Li et al. (2011a, b)
Questiomycin A	*Actinomadura* sp. M048	Anticancer; antitumor	Jiaozhou Bay in China	Human colon carcinoma, melanoma, human lung carcinoma, human breast carcinoma, human kidney tumor, human uterus cancer	Maskey et al. (2003a, b)
R-10-Methyl-6-undecanolide (6R,10S)-10-methyl-6-dodeconolide	*Streptomyces* sp. B6007	Phytotoxic, anticancer	Mangrove sediment in Papua New Guinea	Gastric adenocarcinoma (HM02), hepatocellular carcinoma (HepG2), and breast adenocarcinoma (MCF 7)	Stritzke et al. (2004)
Rabelomycin	*Streptomyces* sp. B6921	Antitumor	Mauritius (Indian Ocean)	–	Maskey et al. (2003a, b)
Resitoflavine	*Streptomyces chibaensis* AUBN1/7	Antitumor	Marine sediment, Machilipatnam coast of Bay of Bengal	Gastric adenocarcinoma (HMO2), hepatic carcinoma (HePG2)	Gorajana et al. (2005) and Gorajana et al. (2007)
Resitoflavine	*Streptomyces* sp. B8005	Antitumor	Laguna de Terminos at the Gulf of Mexico	–	Kock et al. (2005)
Resitoflavine	*Streptomyces* sp. B4842	Antitumor	–	–	Kock et al. (2005)
Retamycin	*Streptomyces olindensis*	Antitumor	–	–	Pamoukian and Facciotti (2004)

(continued)

Table 5.1 (continued)

Compound	Species	Biological activity	Location	Cancer	References
Ripromycin	*Streptomyces* sp. Tu 6239	Antibacterial, antitumor	Soil sample, Sao Jose do Rio Preto, Brazil	Human cell lines from gastric adenocarcinoma (HMO2), mamma carcinoma (MCF 7), hepatocellular carcinoma (Hep G2), and hepatoma cells with mutated p53	Bertasso et al. (2003)
Saliniketal	*Salinispora arenicola* and *Streptomyces arenicola*	Cancer chemoprevention	The Bahamas, the Red Sea, Guam, Palau, the US Virgin Islands, and the Sea of Cortez	–	Jensen et al. (2007)
Saliniketals A and B	*Salinispora arenicola* CNR-005	Antitumor; anticancer	Sediment sample from Guam	Human bladder carcinoma T24 cells	Williams et al. (2007) and Jensen et al. (2007)
Salinipyrones A and B	*Salinispora pacifica* CNS-237	Antitumor	Palau Island, western Pacific Ocean	Mouse splenocyte inhibition of interleukin-5	Oh et al. (2008)
Salinosporamides	*Salinispora sp.*	Anticancer	The Bahamas, the Red Sea, Guam, Palau, the US Virgin Islands, and the Sea of Cortez	Myeloma	Jensen et al. (2007) and Prudhomme et al. (2008)
Salinosporamides	*Salinispora* CNB-392	Anticancer	Chub Cay, Bahamas	Human colon carcinoma	Feling et al. (2003)

Compound	Species	Biological activity	Location	Cancer	References
Salinosporamides A	*Salinispora tropica* CNB-440	Antitumor	Bahamas	Human colon carcinoma HCT-116, with murine melanoma cell line B16-F10	Buchanan et al. (2005) Feling et al. (2003) and Williams et al. (2005)
Salinosporamides A	*Salinispora tropica* CNB-476	Antitumor	Bahamas	Human colon carcinoma HCT-116, with murine melanoma cell line B16-F10	Udwary et al. (2007) and Manam et al. (2009)
Saptomycins	*Streptomyces* sp. HP 530	Antitumor, antimicrobial	Isolated from a soil sample collected in Ichikawa City, Chiba Prefecture, Japan	Human (MKN45, MKN74), human (WiDR, SW481), human (PC-312, A54913), and murine (P38814)	Abe et al. (1993a)
Sarkomycin	*Streptomyces* sp.	Antitumor	–	–	Umezawa et al. (1953)
Selina-4(14),7(11)-diene-8,9-diol	*Streptomyces* sp. QD518	Anticancer	Qingdao coast, China	Bladder, CNS, colon, gastric, head and neck, lung, mammary, ovarian, pancreatic, prostate, renal cancer	Wu et al. (2006)

(continued)

Table 5.1 (continued)

Compound	Species	Biological activity	Location	Cancer	References
Sohbumycin	*Streptomyces* sp. 82-85	Antitumor, antibacterial	Kanagawa Prefecture, Japan	HeLa S3 cells	Umezawa et al. (1985)
Spiroindimicins A–D	*Streptomyces* sp. SCSIO 03032	Anticancer	South China Sea	Human lymphoblastic leukemia, murine melanoma, human lung cancer, human liver cancer, human breast cancer	Zhang et al. (2014)
Spiroindimicins A–D	*Streptomyces* sp.	Anticancer	Indian Ocean	Cancer cell lines including MCF-7, HepG2, B16, H460, and CCRF-CEM	Zhang et al. (2012)
Sporolides	*Salinispora tropica* CNB-392	Antitumor	–	–	Buchanan et al. (2005)
SS-228 Y	*Chainia* sp. SS-228	Antitumor	Koajiro inlet in Sagami Bay	Sarcoma 180 solid tumor in mice and EHRLICH ascites tumor	Imamura et al. (1982) and Okazaki et al. (1975)
SS-228 Y	*Streptomyces* sp.	Anticancer	Sagami Bay	Ehrlich breast adenocarcinoma	Okazaki et al. (1975)

Compound	Species	Biological activity	Location	Cancer	References
Staurosporinone	*Streptomyces* sp.	Antitumor; phycotoxicity	Jiaozhou Bay of Qingdao, China	Human tumor cell lines: colorectal (CXF 94L and CXF164L), gastric (GXF 251L), lung (LXFA 289L, LXFA 526L, LXFA 629L, LXFL 529L, LXFL 1072L, LXFL 1121L and LXFS 650L), melanoma (MEXF 276L, MEXF 514L and MEXF 989L), ovarian (OVXF 899L), pleural mesothelioma (PXF 1118L), renal (RXF 486L, RXF 631L, and RXF 944L), renal pelvis (RXF 1218L), soft tissue (SCLS 1442[c]), and uterus (UXF 1138L)	Wu et al. (2006) and Dengler et al. (1995)
Staurosporines	*Streptomyces* sp. KS3	Antitumor	Komodo Island, Indonesia	–	Itoh et al. (2003a, b)
Staurosporines	*Micromonospora* sp. L-31-CLCO-002	Antitumor	Coast of Fuerteventura Island in the Canary Islands archipelago	P388DJ (ATCC CCL-46), A549 (ATCC CCL-185), HT-29 (ATCC HTB-38), and SK-MEL-28 (ATCC HTB-72)	Hernández et al. (2000)

(continued)

Table 5.1 (continued)

Compound	Species	Biological activity	Location	Cancer	References
Staurosporines	*Streptomyces* sp. QD518	Antitumor	–	Human tumor cell lines: colorectal (CXF 94L and CXF164L), gastric (GXF 251L), lung (LXFA 289L, LXFA 526L, LXFA 629L, LXFL 529L, LXFL 1072L, LXFL 1121L, and LXFS 650L), melanoma (MEXF 276L, MEXF 514L, and MEXF 989L), ovarian (OVXF 899L), pleural mesothelioma (PXF 1118L), renal (RXF 486L, RXF 631L, and RXF 944L), renal pelvis (RXF 1218L), soft tissue (SCLS 1442[c]), and uterus (UXF 1138L)	Wu et al. (2006) and Dengler et al. (1995)
Strepsesquitriol	*Streptomyces* sp.	Antitumor	Sediment sample, Bay of Bengal, Indian Ocean	–	Yang et al. (2013)
Streptochlorin	*Streptomyces roseolilacinus*	Antitumor, anti-allergic (dermatitis), antibiotic, anti-inflammatory	Sediment collected at Ayajin, Korea	–	Lee et al. (2013)
Streptochlorin	*Streptomyces* sp. 04DH110	Antitumor; anticancer	Ayajin Bay, Korea	Human leukemia	Shin et al. (2007)

Compound	Species	Biological activity	Location	Cancer	References
Streptokordin	*Streptomyces* sp. KORDI-3238	Antitumor; anticancer	Ayu Trough, Pacific Ocean	Human breast cancer, human renal cancer, human skin cancer, human leukemia, human cancer cell lines	Jeong et al. (2006)
Streptopyrrolidine	*Streptomyces* sp. KORDI-3973	Antitumor; anticancer	Ayu Trough	Human umbilical vein endothelial cells	Shin et al. (2008)
Stretocarbazoles A, B	*Streptomyces* sp. FMA	Anticancer	Sanya, Hainan Province, China	Human promyelocytic leukemia, human alveolar adenocarcinoma	Fu et al. (2012)
Succinoyl trehalose lipids (STLs)	*Rhodococcus* sp. TB-42	Antitumor	–	Human (HL-60) promyelocytic leukemia	Sudo et al. (2000)
Tartrolon D	*Streptomyces* sp. MDG-04-17-069	Anticancer	East coast of Madagascar	Lung carcinoma, colorectal carcinoma, breast adenocarcinoma	Pérez et al. (2009)
Tetracenomycin D	*Streptomyces* sp. B8005	Antitumor	–	–	Kock et al. (2005)
Tetracenomycin D	*Streptomyces corchorusii AUBN(1)/7*	Anticancer, antibacterial	Marine sediment, Bay of Bengal	Against cell lines HMO2 (gastric adenocarcinoma) and HepG2 (hepatic carcinoma)	Adinaryan et al. (2006)
Thiocoraline	*Micromonospora* sp. L-13-ACM2-092	Antitumor	Marine coral, Indian Ocean near the coast of Mozambique	P388 (ATCC CCL-46), A549 (ATCC CCL-185), HT29 (ATCC HTB-38), and MEL28 (ATCC HTB-72)	Romero et al. (1997) and Pérez Baz et al. (1997)
Thiocoraline	*Micromonospora*	Antitumor, antibacterial	Sponge specimens were collected in the Florida Keys	Human pancreatic carcinoid tumor cells (BON) and human bronchopulmonary carcinoid cells (H727)	Wyche et al. (2014)

(continued)

Table 5.1 (continued)

Compound	Species	Biological activity	Location	Cancer	References
T-Muurolol sesquiterpenes	*Streptomyces* M491	Antitumor	Qingdao coast, China	Human tumor cell lines	Ding et al. (2009)
Trioxacarcin	*Streptomyces bottropensis*	Anticancer, antimalarial	Soil sample collected in Sapporo-shi, Hokkaido, Japan	Colon cancer (HT-29), melanoma (MEXF 514L), lung adenocarcinoma (LXFA 526L), large cell lung cancer (LXFL 529L and H-460), central nervous system (SF-268), mammary cancer (MCF-7), prostate cancer (PC3M), and renal cancer (RXF 631L)	Maskey et al. (2004a, b) and Tomita et al. (1981)
Trioxacarcins	*Streptomyces* sp. B8652	Antitumor; antibacterial; anticancer; antimalarial	Soil sample collected in Sapporo-shi, Hokkaido, Japan	Colon cancer (HT-29), melanoma (MEXF 514L), lung adenocarcinoma (LXFA 526L), large cell lung cancer (LXFL 529L and H-460), central nervous system (SF-268), mammary cancer (MCF-7), prostate cancer (PC3M), and renal cancer (RXF 631L)	Maskey et al. (2004a, b) and Tomita et al. (1981)

Compound	Species	Biological activity	Location	Cancer	References
Trioxacarcins	*Streptomyces ochraceus*	Anticancer, antimalarial	Soil sample collected in Sapporo-shi, Hokkaido, Japan	Colon cancer (HT-29), melanoma (MEXF 514L), lung adenocarcinoma (LXFA 526L), large cell lung cancer (LXFL 529L and H-460), central nervous system (SF-268), mammary cancer (MCF-7), prostate cancer (PC3M), and renal cancer (RXF 631L)	Maskey et al. (2004a, b) and Tomita et al. (1981)
Undecylprodigiosin	*Saccharopolyspora* sp. nov.	Antitumor	Coast of Qingdao, China	Mouse lymphoma cell (P388), human peripheral blood promyeloblast cell (HL60), human lung carcinoma cell (A-549 and SPCA4), and human hepatic carcinoma cell (BEL-7402)	Liu et al. (2005)
Urdamycins A to F	*Streptomyces fradiae Tu2717*	Antitumor; anticancer	Soil isolate from Tanzania (Africa)	Stem cells of murine L1210 leukemia	Decker and Haag (1995) and Drautz et al. (1986)

(continued)

Table 5.1 (continued)

Compound	Species	Biological activity	Location	Cancer	References
Urukthapelstatin	*Mechercharimyces asporophorigenes YM11-542*	Anticancer	Sediment sample, marine lake in the northern part of Urukthapel Island in the Republic of Palau	Human cancer cell lines: breast cancer (Br) HBC-4, BSY-1, HBC-5, MCF-7, and MDA-MB-231; brain cancer (CNS) U251, SF-268, SF-295, SF-539, SNB-75, and SNB-78; colon cancer (Co) HCC2998, KM-12, HT-29, HCT-15, and HCT-116; lung cancer (Lu) NCI-H23, NCIH226, NCI-H522, NCI-H460, A549, DMS273, and DMS14; melanoma (Me) LOX-IMVI; ovarian cancer (Ov) OVCAR-3, OVCAR-4, OVCAR-5, OVCAR-8, and SK-OV-3; renal cancer (Re) RXF-631L and ACHN; stomach cancer (St) St-4, MKN1, MKN7, MKN28, MKN45, and MKN74; and prostate cancer (xPg) Du-145 and PC-3	Matsuo et al. (2007)
Usabamycins A–C	*Streptomyces* sp. NPS853	Anticancer	Usa Bay, Kochi Prefecture, Japan	Human epithelial carcinoma	Sato et al. (2011)
Verucopeptin	*Actinomadura verrucosospora*	Antitumor	Root of a tamarind at Mt. Apo, Davao, Mindanao Island, Philippines	Murine melanoma (6-F10), murine leukemia (P388 and LI210), and human colorectal carcinoma (Moser)	Nishiyama et al. (1993)

Compound	Species	Biological activity	Location	Cancer	References
Vicenistatin	*Streptomyces* sp. Tu6239	Antitumor	Sao Jose do Rio Preto, S. P., Brazil	Human cell lines from gastric adenocarcinoma (HMO2), mamma carcinoma (MCF 7), hepatocellular carcinoma (Hep G2), and hepatoma cells with mutated p53 (Huh 7)	Bertasso et al. (2003)
Vicenistatin	*Streptomyces* sp. HC34	Antitumor	Kiryu, Gunma Prefecture, Japan	Human leukemia (HL-60) and human colon carcinoma (COLO205)	Shindo et al. (1993)
Victomycin	*Streptosporangium violaceochromogenes nov., sp.*	Antitumor, antibacterial	–	Solid sarcoma 180 and EHRLICH ascites carcinoma	Takasawa et al. (1975)
Xanthone IB-00208	*Actinomadura* sp.	Anticancer; antitumor	–	P-388 cells	Rodriguez et al. (2003)
YM-216391	*Streptomyces nobilis*	Anticancer	–	–	Sohda et al. (2005)
ZHD-0501 Staurosporine analog	*Actinomadura* sp. 007	Antitumor; anticancer	Jiaozhou Bay, China	Human lung adenocarcinoma (A549), hepatocarcinoma, promyelocytic leukemia	Han et al. (2005)
Zorbamycin	*Streptomyces flavoviridis ATCC21892*	Antitumor	–	–	Wang et al. (2007)

Table 5.2 Antitumors derived from *Streptomyces* sp. available for cancer treatment

Streptomyces sp.	Antitumor agents	Brand name	Laboratory
Streptomyces parvulus *Streptomyces chrysomallus*	Dactinomycin (actinomycin D)	Dacilon ™ Bloicin-S Dacmozen® Dacmozen-Rd Dacticin Dactino Injection Dactocin Dactinoget Lyovac Cosmegen Cosmegen® for injection	Celon Labs Vhb Korea United Pharm CBC Pharma Royal Medical Gls MSD Ovaction Merck Lundbeck
Streptomyces verticillus	Bleomycin	Bleocin ™ Bleocin-S Bleomycin Bleopar Bleomycin Injection Bledmax Blenoxane® Bleocin Bleolem Cbcan Bleostar-S Lyoble Bleoz™	Nippon Kayaku Kalbe Parenteral Drugs India Salius Pharma Max India Limited Klab Lemery Cbc United Biotech Fresenius Kabi Zuvius Life Sciences Ahpl
Streptomyces peucetius	Doxorubicin	Adricin Doxopar Doxolid ™ Anthrasafe® Adriamycin Doxorubicin Doxocin Ribodoxo Doxorrubicina Hikma Adrim Advadox Caelyx Dobicin Doxorubicin Doxilyd Doxolem Cadria Cadria L Doxorubicin Doxulip Doxutec Duxocin®	Korea United Pharm Parenteral Drugs Of India Celon Labs Miracalus Pharma Pfizer Hikma Dbur Sun Pharmaceutical Cbc Teva Cadila Salius Klab United Biotech Biochem Alkem Cipla Meizler Biopharma Zuventus Natco

(continued)

Table 5.2 (continued)

Streptomyces sp.	Antitumor agents	Brand name	Laboratory
		Lipodox® Lyphidox Naprodox® Oncodox-10 Oncodox-50 Oncodox Rubilong Natdox-Lp® Rubex Doxorubicin Hydrochloride Injection Zuvidox 10® Zuvidox 50®	Ahpl Zuvius Lifesciences
Streptomyces peucetius	Daunorubicin	Daunobin Daunocin® Daunomycin Daunorubitec Daunotec Daunoblastina Daunoxome® Daunorubicin Vyxeos™	Oncomed Korea United Pharm Pfizer United biotech Galen Us Cipla Shenzhen Main luck Pharmaceuticals
Streptomyces lavendulae NRRL2564	Mitomycin	Mitonco Lyomit Mitomycin C Mutamycin® Mitosol®	Korea United Pharm United Biotech Biochem Bristol-Myers Squibb Mobius Therapeutics
Streptomyces galilaeus	Aclacinomycin	Aclarubicin®	Shenzhen Main Luck Pharmaceuticals

References

Abdel-Mageed WM, Milne BF, Wagner M, Schumacher M, Sandor P, Pathom-aree W (2010) Dermacozines, a new phenazine family from deep-sea dermacocci isolated from a Mariana Trench sediment. Org Biomol Chem 8(10):2352–2362

Abe N, Nakakita Y, Nakamura T, Enoki N, Uchida H, Munekata M (1993a) Novel antitumor antibiotics, saptomycins. I. Taxonomy of the producing organism, fermentation, HPLC analysis and biological activities. J Antibiot (Tokyo) 46:1530–1535

Abe N, Nakakita Y, Nakamura T, Enoki N, Uchida H, Takeo O, Munekata M (1993b) Novel cytocidal compounds, oxapropalines from Streptomyces sp. G324 producing lavendamycin. I. Taxonomy of the producing organism, fermentation, isolation and biological activities. J Antibiot (Tokyo) 46:1672–1677

Adegboye MF, Babalola OO (2013) Phylogenetic characterization of culturable antibiotic producing Streptomyces from rhizospheric soils. Mol Biol S1:001

Adinaryan G, Venkateshan MR, Bpiraju VV, Sujatha P, Premkumar J, Ellaiah P, Zeeck A (2006) Cytotoxic compounds from the marine actinobacterium. Bio Org Khim 32:328–334

Aftab U, Zechel DL, Sajid I (2015) Antitumor compounds from Streptomyces sp. KML-2, isolated from Khewra salt mines, Pakistan. Biol Res 48(1):58

Aggarwal R, Jha M, Shrivastava A, Jha AK (2015) Natural compounds: role in reversal of epigenetic changes. Biochemistry 80(8):972–989

Alfarouk KO, Stock CM, Taylor S, Walsh M, Muddathir AK, Verduzco D et al (2015) Resistance to cancer chemotherapy: failure in drug response from ADME to P-gp. Cancer Cell Int 15(1):71

Arai T, Yazawa K, Mikami Y, Kubo A, Takahashi K (1976) Isolation and characterization of satellite antibiotics mimosamycin and chlorocarcins from Streptomyces lavendulae, streptothricin source. J Antibiot (Tokyo) 4:398–407

Asolkar RN, Maskey RP, Helmke E, Laatsch H (2002) Chalcomycin B, a new macrolide antibiotic from the marine isolate Streptomyces sp. B7064. J Antibiot (Tokyo) 55:893–898

Asolkar RN, Jensen PR, Kauffman CA, Fenical W (2006) Daryamides A-C, weakly cytotoxic polyketides from a marine-derived actinomycete of the genus Streptomyces strain CNQ-085. J Nat Prod 69:1756–1759

Asolkar RN, Freel KC, Jensen PR, Fenical W, Kondratyuk TP, Park EJ, Pezzuto JM (2009) Arenamides A-C, cytotoxic NF-κB inhibitors from the marine actinomycete Salinispora arenicola. J Nat Prod 72:396–402

Asolkar RN, Kirkland TN, Jensen PR, Fenical W (2010) Arenimycin, an antibiotic effective against rifampin- and methicillin-resistant Staphylococcus aureus from the marine actinomycete Salinispora arenicola. J Antibiot (Tokyo) 63:37–39

Azman AS, Othman I, Velu SS, Chan KG, Lee LH (2015) Mangrove rare actinobacteria: taxonomy, natural compound, and discovery of bioactivity. Front Microbiol 6:856

Balitz DM, Bush JA, Bradner WT, Doyle TW, Herron FAO, Nettleton DE (1982) Isolation of lavendamycin—a new antibiotic from Streptomyces lavendulae. J Antibiot (Tokyo) 3:259–265

Barka EA, Vatsa P, Sanchez L, Gaveau-Vaillant N, Jacquard C, Klenk HP et al (2016) Taxonomy, physiology, and natural products of Actinobacteria. Microbiol Mol Biol Rev 80(1):1–43

Behie SW, Bonet B, Zacharia VM, McClung DJ, Traxler MF (2017) Molecules to ecosystems: actinomycete natural products in situ. Front Microbiol 7:2149

Berdy J (2005) Bioactive microbial metabolites. J Antibiot 58(1):1–26

Bertasso M, Holzenkampfer M, Zeeck A, Stackebrandt E, Beil W, Fiedler HP (2003) Ripromycin and other polycyclic macrolactams from Streptomyces sp. Tu6239. Taxonomy, fermentation, isolation and biological activities. J Antibiot (Tokyo) 56:364–371

Boonlarppradab C, Kauffman CA, Jensen PR, Fenical W (2008) Marineosins A and B, cytotoxic spiroaminals from a marine-derived actinomycete. Org Lett 10:5505–5508

Buchanan GO, Williams PG, Feling RH, Kauffman CA, Jensen PR, Fenical W (2005) Sporolides A and B: structurally unprecedented halogenated macrolides from the marine actinomycete Salinispora tropica. Org Lett 7:2731–2734

Bugni TS, Woolery M, Kauffman CA, Jensen PR, Fenical W (2006) Bohemamines from a marine-derived Streptomyces sp. J Nat Prod 69:1626–1628

Burke CM, Darling AE (2016) A method for high precision sequencing of near full-length 16S rRNA genes on an Illumina MiSeq. PeerJ 4:e2492

Catalani E, Serafini FP, Zecchini S, Picchietti S, Fausto AM, Marcantoni E et al (2016) Natural products from aquatic eukaryotic microorganisms for cancer therapy: perspectives on anti-tumour properties of ciliate bioactive molecules. Pharmacol Res 113:409–420

Charan RD, Schlingmann G, Janso J, Bernan V, Feng X, Carter GT (2004) Diazepinomicin, a new antimicrobial alkaloid from marine Micromonospora sp. J Nat Prod 67:1431–1433

Chater KF, Biró S, Lee KJ, Palmer T, Schrempf H (2010) The complex extracellular biology of Streptomyces. FEMS Microbiol Rev 34(2):171–198

Chaudhary HS, Soni B, Shrivastava AR, Shrivastava S (2013) Diversity and versatility of actinomycetes and its role in antibiotic production

Che Q, Zhu T, Qi X, Mándi A, Kurtán T, Mo X, Li J, Gu Q, Li D (2012) Hybrid isoprenoids from a reeds rhizosphere soil derived actinomycete *Streptomyces* sp. CHQ-64. Org Lett 14:3438–3441

Che Q, Zhu T, Keyzers RA, Liu X, Li J, Gu Q, Li D (2013) Polycyclic hybrid isoprenoids from a reed rhizosphere soil derived *Streptomyces* sp. CHQ-64. J Nat Prod 76:759–763

Chinembiri TN, Du Plessis LH, Gerber M, Hamman JH, Du Plessis J (2014) Review of natural compounds for potential skin cancer treatment. Molecules 19(8):11679–11721

Cho JY, Kwon HC, Williams PG, Kauffman CA, Jensen PR, Fenical W (2006) Actinofuranones A and B, polyketides from a marine-derived bacterium related to the genus Streptomyces (actinomycetales). J Nat Prod 69:425–428

Cho JY, Williams PG, Kwon HC, Jensen PR, Fenical W (2007) Lucentamycins A-D, cytotoxic peptides from the marine-derived actinomycete Nocardiopsis lucentensis. J Nat Prod 70:1321–1328

Cui CB, Liu HB, Gu JY, Gu QQ, Cai B, Zhang DY, Zhu TJ (2007) Echinosporins as new cell cycle inhibitors and apoptosis inducers from marine-derived *Streptomyces albogriseolus*. Fitoterapia 78:238–240

Dasari VRRK, Muthyala MKK, Nikku MY, Donthireddy SRR (2012) Novel Pyridinium compound from marine actinomycete, Amycolatopsis alba var. nov., DVR D4 showing antimicrobial and cytotoxic activities in vitro. Microbiol Res 167:346–351

Decker H, Haag S (1995) Cloning and characterization of a polyketide synthase gene from Streptomyces fradiae Tu2717, which carries the genes for biosynthesis of the anglucycline antibiotic urdamycin A and a gene probably involved in its oxygenation. J Bacteriol 177:6126–6136

Demain AL, Vaishnav P (2011) Natural products for cancer chemotherapy. Microb Biotechnol 4(6):687–699

Dengler WA, Schulte J, Berger DP, Mertelsmann R, Fiebig HH (1995) Development of a propidium iodide fluorescence assay for proliferation and cytotoxicity assays. Anti-Cancer Drugs 6:522–532

Dharmaraj S (2010) Marine Streptomyces as a novel source of bioactive substances. World J Microbiol Biotechnol 26(12):2123–2139

Dias DA, Urban S, Roessner U (2012) A historical overview of natural products in drug discovery. Meta 2(2):303–336

Dilip CV, Mulaje SS, Mohalkar RY (2013) A review on actinomycetes and their biotechnological application. Int J Pharm Sci Res 4(5):1730

Ding L, Pfoh R, Rühl S, Qin S, Laatsch H (2009) T-muurolol sesquiterpenes from the marine Streptomyces sp. M491 and revision of the configuration of previously reported Amorphanes. J Nat Prod 72:99–101

Doroghazi JR, Metcalf WW (2013) Comparative genomics of actinomycetes with a focus on natural product biosynthetic genes. BMC Genomics 14(1):611

Drautz H et al (1986) Metabolic products of microorganisms. 234 urdamycins, new angucycline antibiotics from *Streptomyces Fradiae*. J Antibiot 39(12):1657–1669

Egler J, Zierle J, Lang F (2016) Stimulating effect of manumycin A on suicidal erythrocyte death. Cell Physiol Biochem 38(3):1147–1156

Embley TM, Stackebrandt E (1994) The molecular phylogeny and systematics of the actinomycetes. Annu Rev Microbiol 48(1):257–289

Feling RH, Buchanan GO, Mincer TJ, Kauffman CA, Jensen PR, Fenical W (2003) Salinosporamide A: a highly cytotoxic proteasome inhibitor from a novel microbial source, a marine bacterium of the new genus Salinospora. Angew Chem Int Ed Eng 42:355–357

Fernández-Chimeno RI, Cañedo L, Espliego F, Grávalos D, De La Calle F, Fernández-Puentes JL, Romero F (2000) IB-96212, a novel cytotoxic macrolide produced by a marine Micromonospora. I. Taxonomy, fermentation, isolation and biological activities. J Antibiot (Tokyo) 53:474–478

Fiedler PH, Rohr J, Zeeck A (1986) Elloramycins B, C, D, E and F: minor congeners of the elloramycin producer Streptomyces olivaceus. J Antibiot (Tokyo) 6:856–859
Fiedler HP, Bruntner C, Riedlinger J, Bull AT, Knutsen G, Goodfellow M, Jones A, Maldonado L, Pathom-aree W, Beil W, Schneider K, Keller S, Sussmuth RD (2008) Proximicin A, B and C, novel aminofuran antibiotic and anticancer compounds isolated from marine strains of the actinomycete Verrucosispora. J Antibiot 61:158–163
Fu P, Wang SX, Hong K, Li X, Liu PP, Wang Y (2011a) Cytotoxic bipyridines from the marine-derived actinomycete Actinoalloteichus cyanogriseus WH1-2216-6. J Nat Prod 74:1751–1756
Fu P, Liu PP, Li X, Wang Y, Wang SX, Hong K, Zhu WM (2011b) Cyclic bipyridine glycosides from the marine-derived actinomycete Actinoalloteichus cyanogriseus WH1-2216-6. Org Lett 13:5948–5951
Fu P, Yang C, Wang Y, Liu P, Ma Y, Xu L (2012) Streptocarbazoles A and B, two novel indolocarbazoles from the marine derived actinomycete strain Streptomyces sp. FMA. Org Lett 14:2422–2425
Fujii I, Ebizuka Y (1998) Anthracycline biosynthesis in Streptomyces galilaeus. Chem Inf 29(6)
Gao X, Lu Y, Xing Y, Ma Y, Lu J, Bao W (2012) A novel anticancer and antifungus phenazine derivative from a marine actinomycete BM-17. Microbiol Res 167:616–622
Gärtner A, Ohlendorf B, Schulz D, Zinecker H, Wiese J, Imhoff JF (2011) Levantilides A and B, 20-membered Macrolides from a Micromonospora strain isolated from the Mediterranean deep sea sediment. Mar Drugs 9:98–108
Gaudêncio SP, MacMillan JB, Jensen PR, Fenical W (2008) Ammosamides A and B new cytotoxic alkaloids isolated from a marine Streptomyces sp. Planta Med 74:PB172. https://doi.org/10.1055/s-0028-1084516
Gerlinger M, Rowan AJ, Horswell S, Larkin J, Endesfelder D, Gronroos E et al (2012) Intratumor heterogeneity and branched evolution revealed by multiregion sequencing. New Engl J Med 366(10):883–892
Gomathi A, Gothandam KM (2016) Ocean dwelling actinobacteria as source of antitumor compounds. Braz Arch Biol Technol 59
Goodfellow M, Williams ST (1983) Ecology of actinomycetes. Annu Rev Microbiol 37(1):189–216
Gorajana A, Kurada BV, Peela S, Jangam P, Vinjamuri S, Poluri E, Zeeck A (2005) 1-Hydroxy-1-norresistomycin, a new cytotoxic compound from a marine actinomycete, Streptomyces chibaensis AUBN1/7. J Antibiot 58:526–529
Gorajana A, Venkatesan M, Vinjamuri S, Kurada BV, Peela S, Jangam P, Poluri E, Zeeck A (2007) Resistoflavine, cytotoxic compound from a marine actinomycete, Streptomyces chibaensis AUBN1/7. Microbiol Res 162:322–327
Gottesman MM, Lavi O, Hall MD, Gillet JP (2016) Toward a better understanding of the complexity of cancer drug resistance. Annu Rev Pharmacol Toxicol 56:85–102
Han X, Cui C, Gu Q, Zhu W, Liu H, Gu J, Osada H (2005) ZHD-0501, a novel naturally occurring staurosporine analog from Actinomadura sp. 007. Tetrahedron Lett 46:6137–6140
Hanada M, Kaneta K, Nishiyama Y, Hoshino Y, Konishi M, Oki T (1991) Hydramycin: a new antitumor antibiotic. Taxonomy, isolation, physico-chemical properties, structure and biological activity. J Antibiot (Tokyo) 44:824–831
Hardt IH, Jensen PR, Fenical W (2000) Neomarinone, and new cytotoxic marinone derivatives, produced by a marine filamentous bacterium (actinomycetales). Tetrahedron Lett 41:2073–2076
Hawas UW, Shaaban M, Shaaban KA, Speitling M, Maier A, Kelter G, Mansouramycins A-D (2009) cytotoxic isoquinolinequinones from a marine Streptomycete. J Nat Prod 72(12):2120–2124
Hayakawa Y, Shirasaki S, Kawasaki T, Matsuo Y, Adachi K, Shizuri Y (2007a) Structures of new cytotoxic antibiotics, piericidins C7 and C8. J Antibiot 60(3):201–203
Hayakawa Y, Shirasaki S, Shiba S, Kawasaki T, Matsuo Y, Adachi K (2007b) PiericidinsC7 and C8, new cytotoxic antibiotics produced by a marine Streptomyces sp. J Antibiot 60(3):196–200
Helaly SE, Pesic A, Fiedler HP, Sussmuth RD (2011) Elaiomycins B and C: alkylhydrazide antibiotics from Streptomyces sp. BK 190. Org Lett 13(5):1052–1055

Hernández LM, Blanco JA, Baz JP, Puentes JL, Millán FR, Vázquez FE, Fernández- Chimeno RI, Grávalos DG (2000) 4′-N-methyl-5′-hydroxystaurosporine and 5′-hydroxystaurosporine, new indolocarbazole alkaloids from a marine Micromonospora sp. strain. J Antibiot 53:895–902

Hohmann C, Schneider K, Bruntner C, Irran E, Nicholson G, Bull AT, Jones AL, Brown R, Stach JE, Goodfellow M, Beil W, Krämer M, Imhoff JF, Süssmuth RD, Fiedler HP (2009) Caboxamycin, a new antibiotic of the benzoxazole family produced by the deep-sea strain Streptomyces sp. NTK 937. J Antibiot 62:99–104

Hohmann C, Schneider K, Bruntner C, Brown R, Jones AL, Goodfellow M (2009a) Albidopyrone, a new α-pyrone-containing metabolite from marine-derived Streptomyces sp. NTK 227. J Antibiot 62(2):75–79

Holohan C, Van Schaeybroeck S, Longley DB, Johnston PG (2013) Cancer drug resistance: an evolving paradigm. Nature 13:714–726

Housman G, Byler S, Heerboth S, Lapinska K, Longacre M, Snyder N, Sarkar S (2014) Drug resistance in cancer: an overview. Cancer 6(3):1769–1792

Hu X, Zhang Z (2016) Understanding the genetic mechanisms of cancer drug resistance using genomic approaches. Trends Genet 32(2):127–137

Hu Z, Reid R, Gramajo H (2005) The leptomycin gene cluster and its heterologous expression in Streptomyces lividans. J Antibiot (Tokyo) 58:625–633

Huang YF, Tian L, Fu HW, Hua HM, Pei YH (2006) One new anthraquinone from marine Streptomyces sp. FX-58. Nat Prod Res 20:1207–1210

Huang ZY, Tang JS, Gao H, Li YJ, Hong K, Li J et al (2011) Studies on the cytotoxic constituents from marine actinomycete Micromonospora sp. M2DG17. Chin Mar Drugs 30:29–33. https://doi.org/10.3390/md12052590

Huang HB, Yang TT, Ren XM, Liu J, Song YX, Sun AJ, Ma J, Wang B, Zhang Y, Huang C, Zhang C, Ju J (2012) Cytotoxic Angucycline class glycosides from the deep sea actinomycete Streptomyces lusitanus SCSIO LR32. J Nat Prod 75:202–208

Hughes CC, MacMillan JB, Gaudêncio SP, Jensen PR, Fenical W (2009) The ammosamides: structures of cell cycle modulators from a marine-derived Streptomyces species. Angew Chem Int Ed Eng 48:725–727

Imamura N, Kakinuma K, Ikekawa N, Tanaka H, Omura S (1982) Biosynthesis of vineomycins A1 and B2. J Antibiot 37:602–607

Ishibashi M, Funayama S, Anraku Y, Komiyama K, Omura S (1991) Novel antibiotics furaquinocins C, D, E, F, G and H. J Antibiot (Tokyo) 44:390–395

Isoda H et al (1995) Succinoyl trehalose lipid induced differentiation of human monocytoid leukemic cell line U937 into monocyte–macrophages. Cytotechnology 19:79–88

Itoh T, Kinoshita M, Aoki S, Kobayashi M (2003a) Komodoquinone A, a novel neuritogenic anthracycline, from marine Streptomyces sp. KS3. J Nat Prod 66:1373–1377

Itoh T, Kinoshita M, Wei H, Kobayashi M (2003b) Stereostructure of komodoquinone A, a neuritogenic anthracycline, from marine Streptomyces sp. KS3. Chem Pharm Bull 51:1402–1404

Jensen PR, Williams PG, Oh DC, Zeigler L, Fenical W (2007) Species-specific secondary metabolite production in marine actinomycetes of the genus Salinispora. Appl Environ Microbiol 73:1146–1152

Jeong SY, Shin HJ, Kim TS, Lee HS, Park SK, Kim HM (2006) Streptokordin, a new cytotoxic compound of the methylpyridine class from a marine-derived Streptomyces sp. KORDI-3238. J Antibiot 59:234–240

Jingsong YE, Dickens ML, Plater R, Yun LI, Jessica L, Strohl WR (1994) Isolation and sequence analysis of polyketide synthase genes from the daunomycin producing Streptomyces sp. strain C5. J Bacteriol 176:6270–6280

Kalaitzis JA, Hamano Y, Nilsen G, Moore BS (2003) Biosynthesis and structural revision of neomarinone. Org Lett 5:4449–4452

Kallifatidis G, Hoy JJ, Lokeshwar BL (2016) Bioactive natural products for chemoprevention and treatment of castration-resistant prostate cancer. In: Seminars in cancer biology, vol 40. Academic Press, New York, pp 160–169

Kanoh K, Matsuo Y, Adachi K, Imagawa H, Nishizawa M, Shizuri Y (2005) Mechercharmycins A and B, cytotoxic substances from marine-derived Thermoactinomyces sp. YM3-251. J Antibiot 58:289–292

Kim WG, Kim JP, Kim CJ, Lee KH, Yoo ID (1996) Benzastatins A, B, C, D: new free radical scavengers from Streptomyces nitrosporeus 30643. I. Taxonomy, fermentation, isolation and physico-chemical properties and biological activities. J Antibiot 49:20–25

Kock I, Maskey RP, Biabani MA, Helmke E, Laatsch H (2005) 1-Hydroxy-1-norresistomycin and resistoflavin methyl ether: new antibiotics from marine-derived streptomycetes. J Antibiot 58:530–534

Komiyama K, Funayama S, Anraku Y, Ishibashi M, Takahashi Y, Kawakami T, Omura S (1991) A new antibiotic, Okicenone. I. Taxonomy, fermentation, isolation and biological characteristics. J Antibiot (Tokyo) 44:814–818

Konoshenko GI, Avraleva IV, Anisova LN, Orlova TI (1994) Biologically active substances by a number of strains of the actinomycin C producer Streptomyces chrysomallus. Antibiot Khimioter 39:22–25

Kornsakulkarn J, Saepua S, Boonruangprapa T, Suphothina S, Thongpanchang C (2013) New β-carboline and indole alkaloids from Actinomycete Actinomadura sp. BCC 24717. Phytochem Lett 6:491–494

Kumar V, Naik B, Gusain O, Bisht GS (2014) An actinomycete isolate from solitary wasp mud nest having strong antibacterial activity and kills the Candida cells due to the shrinkage and the cytosolic loss. Front Microbiol 5:446

Kwon HC, Kauffman CA, Jensen PR, Fenical W, Marinomycins A-D (2006) antitumor antibiotics of a new structure class from a marine actinomycete of the recently discovered genus “Marinispora”. J Am Chem Soc 128:1622–1632

Lam KS, Hesler GA, Mattei JM, Mamber SW, Forenza S, Tomita K (1990) Himastatin, a new antitumor antibiotic from Streptomyces hygroscopicus. I. Taxonomy of producing organism, fermentation and biological activity. J Antibiot (Tokyo) 43:956–960

Lee SH, Shin HJ, Kim DY, Shim DW, Kim TJ, Ye SK, Won HS, Koppula S, Kang TB, Lee KH (2013) Streptochlorin suppresses allergic dermatitis and mast cell activation via regulation of lyn/fyn and syk signaling pathways in cellular and mouse models. PLoS One 8(9):e74194

Li A, Piel J (2002) A gene cluster from a marine Streptomyces encoding the biosynthesis of the aromatic spiroketal polyketide griseorhodin A. Chem Biol 9:1017–1026

Li F, Maskey RP, Qin S, Sattler I, Fiebig HH, Maier A, Zeeck A, Laatsch H, Chinikomycins A (2005) B: isolation, structure elucidation, and biological activity of novel antibiotics from a marine Streptomyces sp. isolate M045. J Nat Prod 68:349–353

Li DH, Zhu TJ, Liu HB, Fang YC, Gu QQ, Zhu WM (2006) Four butenolides are novel cytotoxic compounds isolated from the marine-derived bacterium, Streptoverticillium luteoverticillatum 11014. Arch Pharm Res 29:624–626

Li S, Tian X, Niu S, Zhang W, Chen Y, Zhang H, Yang X, Zhang W, Li W, Zhang S, Ju J, Zhang C (2011a) Pseudonocardians AC, new Diazaanthraquinone derivatives from a deep sea actinomycete Pseudonocardia sp. SCSIO 01299. Mar Drugs 9:1428–1439

Li XB, Tang JS, Gao H, Ding R, Li J, Hong K, Yao XS (2011b) A new staurosporine analog from actinomycetes Streptomyces sp. 172614. J Asian Nat Prod Res 13:765–769

Liu R, Cui CB, Duan L, Gu QQ, Zhu WM (2005) Potent in vitro anticancer activity of metacycloprodigiosin and undecylprodigiosin from a sponge-derived actinomycete Saccharopolyspora sp. nov. Arch Pharm Res 28:1341–1344

Liu R, Zhu T, Li D, Gu J, Xia W, Fang Y, Liu H, Zhu W, Gu Q (2007) Two indolocarbazole alkaloids with apoptosis activity from a marine-derived actinomycete Z2039-2. Arch Pharm Res 30:270–274

Lombó F, Velasco A, Castro A, de la Calle F, Braña AF, Sánchez-Puelles JM, Méndez C, Salas JA (2006) Deciphering the biosynthesis pathway of the antitumor thiocoraline from a marine actinomycete and its expression in two Streptomyces species. Chembiochem 7:366–376

Lu J, Ma Y, Liang J, Xing Y, Xi T, Lu Y (2012) Aureolic acids from a marine-derived Streptomyces sp. WBF16. Microbiol Res 167:590–595

Ludwig W, Euzéby J, Schumann P, Busse HJ, Trujillo ME, Kämpfer P, Whitman WB (2012) Road map of the phylum Actinobacteria. In Bergey's manual® of systematic bacteriology. Springer, New York, pp 1–28

Macherla VR, Liu J, Bellows C, Teisan S, Nicholson B, Lam KS (2005) Glaciapyrroles A, B, and C, pyrrolosesquiterpenes from a Streptomyces sp. isolated from an Alaskan marine sediment. J Nat Prod 68(5):780–783

Maloney KN, MacMillan JB, Kauffman CA, Jensen PR, DiPasquale AG, Rheingold AL et al (2009) Lodopyridone, a structurally unprecedented alkaloid from a marine actinomycete. Org Lett 11:5422

Manam RR, Teisan S, White DJ, Nicholson B, Grodberg J, Neuteboom ST, Lam KS, Mosca DA, Lloyd GK, Potts BC (2005) Lajollamycin, a nitro-tetraene spiro-beta-lactone-gamma-lactam antibiotic from the marine actinomycete Streptomyces nodosus. J Nat Prod 68:240–243

Manam RR, Macherla VR, Tsueng G, Dring CW, Weiss J, Neuteboom ST, Lam KS, Potts BC (2009) Antiprotealide is a natural product. J Nat Prod 72:295–297

Manivasagan P, Venkatesan J, Sivakumar K, Kim SK (2014) Pharmaceutically active secondary metabolites of marine actinobacteria. Microbiol Res 169(4):262–278

Mao Y, Varoglu M, Sherman DH (1999) Molecular characterization and analysis of the biosynthetic gene cluster for the antitumor antibiotic Mitomycin C from Streptomyces lavendulae NRRL 2564. Chem Biol 6:251–263

Marin JJ, Briz O, Monte MJ, Blazquez AG, Macias RI (2012) Genetic variants in genes involved in mechanisms of chemoresistance to anticancer drugs. Curr Cancer Drug Targets 12(4):402–438

Martin GD, Tan LT, Jensen PR, Dimayuga RE, Fairchild CR, Raventos-Suarez C, Fenical W (2007) Marmycins A and B, cytotoxic pentacyclic C-glycosides from a marine sediment-derived actinomycete related to the genus Streptomyces. J Nat Prod 70:1406–1409

Maskey RP, Helmke E, Fiebig HH, Laatsch H (2002) Parimycin: isolation and structure elucidation of a novel cytotoxic 2,3-dihydroquinizarin analogue of gamma-indomycinone from a marine streptomycete isolate. J Antibiot 55:1031–1035

Maskey RP, Helmke E, Laatsch H (2003a) Himalomycin A and B: isolation and structure elucidation of new fridamycin type antibiotics from a marine Streptomyces isolate. J Antibiot 56:942–949

Maskey RP, Li F, Qin S, Fiebig HH, Laatsch H (2003b) Chandrananimycins A-C: production of novel anticancer antibiotics from a marine Actinomadura sp. Isolate M048 by variation of medium composition and growth conditions. J Antibiot 56:622–629

Maskey RP, Helmke E, Kayser O, Fiebig HH, Maier A, Busche A et al (2004a) Anti-cancer and antibacterial trioxacarcins with high anti-malaria activity from a marine Streptomycete and their absolute stereochemistry. J Antibiot 57(12):771–779

Maskey RP, Sevvana M, Usón I, Helmke E, Laatsch H (2004b) Gutingimycin: a highly complex metabolite from a marine streptomycete. Angew Chem Int Ed Eng 43:1281–1283

Matsuo Y, Kanoh K, Yamori T, Kasai H, Katsuta A, Adachi K, Shin-Ya K, Shizuri Y (2007) Urukthapelstatin A, a novel cytotoxic substance from a marine derived Mechercharimyces asporophorigenes YM11-542.Fermentation, isolation and biological activities. J Antibiot (Tokyo) 60:251–255

McArthur KA, Mitchell SS, Tsueng G, Rheingold A, White DJ, Grodberg J, Potts BC et al (2008) Lynamicins a–e, chlorinated bisindole pyrrole antibiotics from a novel marine actinomycete. J Nat Prod 71(10):1732–1737

McGovren JP, Neil GL, Crampton SL, Robinson MI, Douros JD (1977) Antitumor activity and preliminary drug disposition studies on chartreusin (NSC 5159). Cancer Res 37:1666–1672

Miller ED, Kauffman CA, Jensen PR, Fenical W (2007) Piperazimycins: cytotoxic hexadepsipeptides from a marine-derived bacterium of the genus Streptomyces. J Organomet Chem 72:323–330

Mitchell SS, Nicholson B, Teisan S, Lam KS, Potts BC (2004) Aureoverticillactam, a novel 22-atom macrocyclic lactam from the marine actinomycete Streptomyces aureoverticillatus. J Nat Prod 67:1400–1402

Montazami N, Aghapour M, Farajnia S, Baradaran B (2015) New insights into the mechanisms of multidrug resistance in cancers. Cell Mol Biol 61(7):70–80

Naumova IB, Diqumbai K, Potekhina NV, Shashkov AS, Terekhova LP (1986) Teichoic acid from the cell wall of Actinomadura carminata a producer of the antibiotic carminomycin. Bio Org Khim 12:670–678

Nishiyama Y, Suqawara K, Tomita K, Yamamoto H, Kamei H, Oki T (1993) Verucopeptin, a new antitumor antibiotic active against B16 melanoma. I. Taxonomy, production, isolation, physico-chemical properties and biological activity. J Antibiot (Tokyo) 46:921–927

Oh DC, Williams PG, Kauffman CA, Jensen PR, Fenical W (2006) Cyanosporasides A and B, chloro- and cyano-cyclopenta[a]indene glycosides from the marine actinomycete "Salinispora pacifica". Org Lett 8:1021–1024

Oh DC, Gontang EA, Kauffman CA, Jensen PR, Fenical W (2008) Salinipyrones and pacificanones, mixed-precursor polyketides from the marine actinomycete Salinispora pacifica. J Nat Prod 71:570–575

Okazaki T, Kitahara T, Okami Y (1975) Studies on marine microorganisms. IV. A new antibiotic SS-228 Y produced by Chainia isolated from shallow sea mud. J Antibiot 28:176–184

Olano C, Méndez C, Salas JA (2009) Antitumor compounds from marine actinomycetes. Marine Drugs 7(2):210–248

Pamoukian CRD, Facciotti MCR (2004) Production of the antitumoral retamycin during continuous fermentations of Streptomyces olindensis. Biochemist 39:2249–2255

Pan ST, Li ZL, He ZX, Qiu JX, Zhou SF (2016) Molecular mechanisms for tumour resistance to chemotherapy. Clin Exp Pharmacol Physiol 43(8):723–737

Pandey RC, Toussaint MW, McGuire JC, Thomas MC (1989) Maggiemycin and anhyromaggiemycin: two novel anthracyclinone antitumor antibiotics: isolation, structures, partial synthesis and biological properties. J Antibiot (Tokyo) 42:1567–1577

Pathirana C, Jensen PR, Fenical W (1992) Marinone and debromomarinone: antibiotic sesquiterpenoid naphthoquinones of a new structure class from a marine bacterium. Tetrahedron Lett 33:7663–7666

Pérez Baz J, Cañedo LM, Fernández Puentes JL, Silva Elipe MV (1997) Thiocoraline, a novel depsipeptide with antitumor activity produced by a marine Micromonospora. II. Physicochemical properties and structure determination. J Antibiot 50:738–741

Pérez M, Crespo C, Schleissner C, Rodríguez P, Zúñiga P, Reyes F (2009) Tartrolon D, a cytotoxic macrodiolide from the marine-derived actinomycete Streptomyces sp. MDG-04-17-069. J Nat Prod 72:2192–2194

Piccolo TM, Menale C, Crispi S (2015) Combined anticancer therapies: an overview of the latest applications. Anti-Cancer Agent Med Chem 15(4):408–422

Polpass AJ, Bhavanath J (2016) New dimensions of research on actinomycetes: quest for next generation antibiotics. Front Microbiol 7:1295

Prudhomme J, McDaniel E, Ponts N, Bertani S, Fenical W, Jensen P et al (2008) Marine actinomycetes: a new source of compounds against the human malaria parasite. PLoS One 3:e2335

Rambabu V, Suba S, Manikandan P, Vijayakumar S (2014) Cytotoxic and apoptotic nature of migrastatin, a secondary metabolite from Streptomyces evaluated on HEP G2 cell line. Int J Pharm Pharm Sci 6:333–338

Ramos P, Bentires-Alj M (2015) Mechanism-based cancer therapy: resistance to therapy, therapy for resistance. Oncogene 34(28):3617–3626

Rana S, Salam MD (2014) Antimicrobial potential of actinomycetes isolated from soil samples of Punjab. India J Microbiol Exp 1(2):00010

Raty K, Hautala A, Torkkell S, Kantola J, Mantsala P, Hakala J, Ylihonko K (2002) Characterization of mutations in aclacinomycin, a non-producing Streptomyces galilaeus strains with altered glycosylation patterns. Microbiology 148:3375–3384

Ravikumar S, Fredimoses M, Gnanadesigan M (2012) Anticancer property of sediment actinomycetes against MCF-7 and MDA-MB-231 cell lines. Asian Pac J Trop Biomed 2(2):92–96

Rebucci M, Michiels C (2013) Molecular aspects of cancer cell resistance to chemotherapy. Biochem Pharmacol 85(9):1219–1226

Remsinq LL, Gonzalez AM, Nur-e-Alam M, Fernandez- Lozano MJ, Brana AF, Rix U, Oliveira MA, Mendez C, Salas JA, Rohr J (2003) Mithramycin SK, a novel antitumor drug with improved therapeutic index, mithramycin SA and demycarosyl-mithramycin SK: three new products generated in the mithramycin producer Streptomyces argillaceus through combinatorial biosynthesis. J Am Chem Soc 125:5745–5753

Renner MK, Shen Y-C, Cheng X-C, Jensen PR, Frankmoelle W, Kauffman CA (1999) New anti-inflammatory cyclic peptides produced by a marine bacterium (Streptomyces sp.). J Am Chem Soc 121(49):11273–11276

Riedlinger J, Reicke A, Zähner H, Krismer B, Bull AT, Maldonado LA, Ward AC, Goodfellow M, Bister B, Bischoff D, Süssmuth RD, Fiedler HP (2004) Abyssomicins, inhibitors of the para-aminobenzoic acid pathway produced by the marine Verrucosispora strain AB-18-032. J Antibiot 57:271–279

Rodriguez JC, Fernandez Puentes JL, Perez Baz J, Canedo LM (2003) IB-00208, a new cytotoxic polycyclic xanthone produced by a marine-derived Actinomadura. II. Isolation, physico-chemical properties and structure determination. J Antibiot (Tokyo) 56:318–321

Romero F, Espliego F, Pérez Baz J, García de Quesada T, Grávalos D, de la Calle F, Fernández-Puentes JL (1997) Thiocoraline, a new depsipeptide with antitumor activity produced by a marine Micromonospora. I. Taxonomy, fermentation, isolation, and biological activities. J Antibiot 50:734–737

Saha M, Jaisankar P, Das S, Sarkar KK, Roy S, Besra SE, Vedasiromani JR, Ghosh D, Sana B, Mukherjee J (2006) Production and purification of a bioactive substance inhibiting multiple drug resistant bacteria and human leukemia cells from a salt-tolerant marine Actinobacterium sp. isolated from the Bay of Bengal. Biotechnol Lett 28:1083–1088

Salehan MR, Morse HR (2013) DNA damage repair and tolerance: a role in chemotherapeutic drug resistance. Br J Biomed Sci 70(1):31–40

Sánchez López JM, Martínez Insua M, Pérez Baz J, Fernández Puentes JL, Cañedo Hernández LM (2003) New cytotoxic indolic metabolites from a marine Streptomyces. J Nat Prod 66:863–864

Sato S, Iwata F, Yamada S, Kawahara H, Katayama M (2011) Usabamycins A–C: new anthramycin-type analogues from a marine-derived actinomycete. Bioorg Med Chem Lett 21(23):7099–7101

Sattler I, Thiericke R, Zeeck A (1998) The manumycin-group metabolites. Nat Prod Rep 15(3):221–240

Schneider K, Keller S, Wolter FE, Röglin L, Beil W, Seitz O, Nicholson G, Bruntner C, Riedlinger J, Fiedler HP, Süssmuth RD, Proximicins A (2008) B, and C-antitumor furan analogues of netropsin from the marine actinomycete Verrucosispora induce upregulation of p53 and the cyclin kinase inhibitor p21. Angew Chem Int Ed Eng 47:3258–3261

Schultz AW, Oh D-C, Carney JR, Williamson RT, Udwary DW, Jensen PR (2008) Biosynthesis and structures of cyclomarins and cyclomarazines, prenylated cyclic peptides of marine actinobacterial origin. J Am Chem Soc 130(13):4507–4516

Senthebane DA, Rowe A, Thomford NE, Shipanga H, Munro D, Al Mazeedi MA et al (2017) The role of tumor microenvironment in chemoresistance: to survive, keep your enemies closer. Int J Mol Sci 18(7):1586

Ser HL, Hern T, Loh T, Law JWF, Chan KGC, Duangjai A et al (2017) Focused review: cytotoxic and antioxidant potentials of mangrove-derived Streptomyces. Front Microbiol 8:2065

Shah AM, Rasool S, Majeed A, Mushtaq S, Khan MH, Hussain A et al (2017) Reappraisal of actinomycetes for novel bioactive metabolites. Ann Phytomed 6(1):13–19

Shanmugam MK, Lee JH, Chai EZP, Kanchi MM, Kar S, Arfuso F et al (2016) Cancer prevention and therapy through the modulation of transcription factors by bioactive natural compounds. In: Seminars in cancer biology, vol 40. Academic Press, New York, pp 35–47

Shin HJ, Jeong HS, Lee HS, Park SK, Kim HM, Kwon HJ (2007) Isolation and structure determination of Streptochlorin, an antiproliferative agent from a marine-derived Streptomyces sp. 04DH110. J Microbiol Biotechnol 17:1403–1406

Shin HJ, Kim TS, Lee HS, Park JY, Choi IK, Kwon HJ (2008) Streptopyrrolidine, an angiogenesis inhibitor from a marine-derived Streptomyces sp. KORDI-3973. Phytochemistry 69:2363–2366

Shindo K, Kamishohara M, Odagawa A, Matsuoka M, Kawai H (1993) Vicenistatin, a novel 20-membered macrocyclic lactam antitumor antibiotic. J Antibiot (Tokyo) 46:1076–1081

Sohda KY, Nagai K, Yamori T, Suzuki K, Tanaka A (2005) YM216391, a novel cytotoxic cyclic peptide from Streptomyces nobilis. I. Fermentation, isolation and biological activities. J Antibiot (Tokyo) 58:27–31

Song Y, Liu G, Li J, Huang H, Zhang X, Zhang H et al (2015) Cytotoxic and antibacterial Angucycline- and Prodigiosin- analogues from the deep-sea derived Streptomyces sp. SCSIO 11594. Mar Drugs 13:1304–1316

Soria-Mercado IE, Prieto-Davo A, Jensen PR, Fenical W (2005) Antibiotic terpenoid chlorodihydroquinones from a new marine actinomycete. J Nat Prod 68:904–910

Stritzke K, Schulz LH, Helmke E, Beil W (2004) Novel caprolactones from a marine streptomycete. J Nat Prod 67:395–401

Sudo T et al (2000) Induction of the differentiation of human HL-60 promyelocytic leukemia cell line by succinoyl trehalose lipids. Cytotechnology 33:259–264

Takahashi A, Ikeda D, Nakamura H, Naganawa H, Kurasawa S, Okami Y, Takeuchi T, Iitaka Y (1989a) Altemicidin, a new acaricidal and antitumor substance. II. Structure determination. J Antibiot 42:1562–1566

Takahashi A, Kurasawa S, Ikeda D, Okami Y, Takeuchi T (1989b) Altemicidin, a new acaricidal and antitumor substance. I. Taxonomy, fermentation, isolation and physico-chemical and biological properties. J Antibiot 42:1556–1561

Takasawa S, Kawamoto I, Okachi R, Kohakura M, Yahashi R, Nara T (1975) A new antibiotic victomycin (XK 49–1-B-2). II. Isolation, purification and physiochemical and biological properties. J Antibiot (Tokyo) XXVIII:366–371

Thiericke R, Zeeck A (1988) Biosynthesis of manumycin: origin of the polyene chains. J Antibiot (Tokyo) 5:694–696

Tokoro Y, Isoe T, Shindo K (1999) Gilvusmycin, a new antitumor antibiotic related to CC-1065. J Antibiot (Tokyo) 52:263–268

Tomita F, Tamaoki T, Morimoto M, Fujimoto K (1981) Trioxacarcins, novel antitumor antibiotics. I. Producing organism, fermentation and biological activities. J Antibiotics 34:1519–1524

Tresselt D, Eckardt K, Ihn W (1978) Antibiotics from Actinomycetes—chemical composition of antibiotic griseorhodin A. Tetrahedron 34:2693–2699

Udwary DW, Zeigler L, Asolkar RN, Singan V, Lapidus A, Fenical W, Jensen PR, Moore BS (2007) Genome sequencing reveals complex secondary metabolome in the marine actinomycete Salinispora tropica. Proc Natl Acad Sci U S A 104:10376–10381

Ul-Hassan A, Wellington EMH (2009) Actinobacteria. Encycl Microbiol 3:25–44

Umezawa H, Takeuchi T, Nitta K, Yamamoto T, Yamaoka S (1953) Sarkomycin, an anti-tumor substance produced by streptomyces. J Antibiot 6:101

Umezawa I, Komiyama K, Oka H, Okada K, Tomisaka S, Miyano T, Takano S (1984) A new antitumor antibiotic, kazusamycin. J Antibiot (Tokyo) 37:706–711

Umezawa I, Tronouet C, Funayama S, Okada K, Komiyama K (1985) A novel antibiotic, sohbumycin. Taxonomy, fermentation, isolation and physico-chemical and biological characteristics. J Antibiot (Tokyo) XXXVIII:967–971

Usha R, Mala KK, Venil CK, Palaniswamy M (2011) Screening of actinomycetes from mangrove ecosystem for L-asparaginase activity and optimization by response surface methodology. Pol J Microbiol 60(3):213–221

Wakabayashi T, Kageyama-Kawase R, Naruse N, Funahashi Y, Yoshimatsu K (2000) Luminacins: a family of capillary tube formation inhibitors from Streptomyces sp. II. Biological activities. J Antibiot (Tokyo) 53(6):591–596

Waksman SA (1940) On the classification of actinomycetes. J Bacteriol 39(5):549

Wang L, Yun BS, Geirge NP, Wendt-Pienkowski E, Galm U, Oh TJ, Couqhlin JM, Zhang G, Tao M, Shen B (2007) Glycopeptide antitumor antibiotic zorbamycin from Streptomyces flavoviridis ATCC 21892: strain improvement and structure elucidation. J Nat Prod 70:402–406

Warnick-Pickle DJ, Byrne KM, Pandey RC, White RJ (1981) Fredericamycin A, a new antitumor antibiotic. II. Biological properties. J Antibiot (Tokyo) 34:1402–1407

Weber T, Charusanti P, Musiol-Kroll EM, Jiang X, Tong Y, Kim HU, Lee SY (2015) Metabolic engineering of antibiotic factories: new tools for antibiotic production in actinomycetes. Trends Biotechnol 33(1):15–26

Wei Z, Xu C, Wang J, Lu F, Bie X, Lu Z (2017) Identification and characterization of Streptomyces flavogriseus NJ-4 as a novel producer of actinomycin D and holomycin. PeerJ 5:e3601

William PG, Asolkar RN, Kondratyuk T, Pezzuto JM, Jensen PR, Fenical W (2007) Saliniketals A and B, bicyclic polyketides from the marine actinomycete Salinispora arenicola. J Nat Prod 70:83–88

Williams PG, Buchanan GO, Feling RH, Kauffman CA, Jensen PR, Fenical W (2005) New cytotoxic salinosporamides from the marine actinomycete Salinispora tropica. J Organomet Chem 70:6196–6203

Williams PG, Miller ED, Asolkar RN, Jensen PR, Fenical W (2007) Arenicolides A-C, 26- membered ring macrolides from the marine actinomycete Salinispora arenicola. J Organomet Chem 72:5025–5034

Wu SJ, Fotso S, Li F, Qin S, Kelter G, Fiebig HH, Laatsch H (2006) 39-N-carboxamidostaurosporine and selina-4(14),7(11)-diene-8,9-diol, new metabolites from a marine Streptomyces sp. J Antibiot 59:331–337

Wu SJ, Fotso S, Li F, Qin S, Laatsch H (2007) Amorphane sesquiterpenes from a marine Streptomyces sp. J Nat Prod 70:304–306

Wyche TP, Dammalapati A, Cho H, Harrison AD, Kwon GS, Chen H, Bugni TS, Jaskula-Sztul R (2014) Thiocoraline activates the notch pathway in carcinoids and reduces tumor progression in vivo. Cancer Gene Ther 21:518–525

Xie QY, Qu Z, Lin HP, Li L, Hong K (2012) Micromonospora haikouensis sp. nov., isolated from mangrove soil. Antonie Van Leeuwenhoek 101(3):649–655

Yang XW, Peng K, Liu Z, Zhang GY, Li J, Wang N, Steinmetz A, Liu Y (2013) Strepsesquitriol: a rearranged zizaane-type sesquiterpenoid from the deep-sea-derived actinomycete Streptomyces sp SCSIO 10355. J Nat Prod 76(12):2360–2363

Yuan G, Lin H, Wang C, Hong K, Liu Y, Li J (2011) 1H and 13C assignments of two new macrocyclic lactones isolated from Streptomyces sp. 211726 and revised assignments of azalomycins F3a, F4a and F5a. Magn Reson Chem 49:30–37

Yuan G, Hong K, Lin H, She Z, Li J (2013) New azalomycin F analogs from mangrove Streptomyces sp. 211726 with activity against microbes and cancer cells. Marine Drugs 11(3):817–829

Zhang WJ, Liu Z, Li SM, Yang TT, Zhang QB, Ma L et al (2012) Spiroindimicins A-D: new Bisindole alkaloids from a deep-sea-derived actinomycete. Org Lett 14:3364–3367

Zhang Q, Li S, Chen Y, Tian X, Zhang X, Zhang G, Zhu Y, Zhang S, Zhang W, Zhang C (2013) New diketopiperazine derivatives from a deep-sea-derived Nocardiopsis alba SCSIO03039. J Antibiot 66:31–36

Zhang W, Ma L, Li S, Liu Z, Chen Y, Zhang H, Zhang G, Zhang Q, Tian X, Yuan C (2014) Indimicins A–E, Bisindole Alkaloids from the Deep-Sea-Derived Streptomyces sp. SCSIO03032. J Nat Prod 77:1887–1892

Zhang J, Jiang H, Xie L, Hu J, Li L, Yang M, Cheng L, Liu B, Qian X (2016) Antitumor effect of manumycin on colorectal cancer cells by increasing the reactive oxygen species production and blocking PI3K-AKT pathway. Onco Targets Ther 9:2885–2895

Zhang MG, Lee JY, Gallo RA, Tao W, Tse D, Doddapaneni R, Pelaez D (2017) Therapeutic targeting of oncogenic transcription factors by natural products in eye cancer. Pharmacol Res 129:365–374

Zhou Z-L, Jin B, Yin W-Q, Fu C-Y, Feng H-F (2011) Components in antineoplastic actinomycete strain (N2010-37) of Bottom Mud in Mangrove. Chin Herb Med 3(3):165–167

Zhu L, Ostah B, Rix U, Nur-E-Alam M, Mayers A, Luzhetskyy A, Mendez C, Salas JA, Bechthold A, Fedorenko V, Rohr J (2005) Identification of the function of gene lndM2 encoding a bifunctional oxygenase-reductase involved in the biosynthesis of the antitumor antibiotic landomycin E by Streptomyces globisporus 1912 supports the originally assigned structure for landomycinone. J Organomet Chem 70:631–638

Microbe-Assisted Phytoremediation in Reinstating Heavy Metal-Contaminated Sites: Concepts, Mechanisms, Challenges, and Future Perspectives

6

Vishal Kumar Deb, Ahmad Rabbani, Shashi Upadhyay, Priyam Bharti, Hitesh Sharma, Devendra Singh Rawat, and Gaurav Saxena

Abstract

Pollution has become a serious matter of environmental and political concerns in the world. Our natural environment has been contaminated by various organic and inorganic contaminants, which are being used in many industrial processes and released along with industrial effluents. Among them, heavy metals are highly toxic pollutants, which cause serious environmental pollution and severe health hazards in living beings, and there is a public outcry to ensure the safest and healthiest environment for living beings. Phytoremediation, a type of bioremediation, has been emerged as an eco-sustainable technology that uses plants and their associated microbes to clean up heavy metal-contaminated soils, water, and wastewaters as compared to various physicochemical remediation technologies currently being applied for environmental restoration. However, in current scenario, phytoremediation assisted by plant-associated microorganisms, i.e., microbe-assisted phytoremediation (use of microbes, i.e., plant growth-promoting rhizobacteria, endophytes, and arbuscular mycorrhizal fungi, in assisted phytoremediation), is highly preferred for the remediation of heavy metal-contaminated sites as they have potential to alleviate the heavy metal toxicity in plants through their own metal resistance system and facilitate and

V. K. Deb · A. Rabbani · S. Upadhyay · P. Bharti · H. Sharma · G. Saxena (✉)
Laboratory of Microbiology, Department of Microbiology, School of Life and Allied Sciences (SLAS), Baba Farid Institute of Technology (BFIT), Dehradun, Uttarakhand, India
e-mail: gaurav10saxena@gmail.com

D. S. Rawat
Laboratory of Biotechnology, Department of Biotechnology, School of Life and Allied Sciences (SLAS), Baba Farid Institute of Technology (BFIT), Dehradun, Uttarakhand, India

P. K. Arora (ed.), *Microbial Technology for Health and Environment*,
Microorganisms for Sustainability 22,
https://doi.org/10.1007/978-981-15-2679-4_6

improve the growth of host plants under heavy metal stress. In this line, this chapter aims to provide an overview on microbe-assisted phytoremediation, illustrate various mechanisms elicited for plant growth promotion and heavy metal phytoremediation (accumulation/detoxification), and discuss drawbacks and future challenges.

Keywords
Heavy metals · Environmental pollution · Toxicity · Microbe-assisted phytoremediation · Contaminated sites

6.1 Introduction

Environmental pollution is of serious ecological concern worldwide with a continually rising public outcry to ensure the safest and healthiest environment. A variety of organic and inorganic pollutants have been reported to cause environmental pollution and severe health hazards in living beings (Maszenan et al. 2011; Saxena and Bharagava 2017). Among them, heavy metals (HMs) are highly notorious pollutants due to their high abundance and nonbiodegradable and persistent nature in the environment. Hence, they cause soil/water pollution and toxic, genotoxic, teratogenic, and mutagenic effects in living beings (Dixit et al. 2015; Sarwar et al. 2017). They also cause endocrine disruption and neurological disorders even at low concentration (Yadav 2010; Maszenan et al. 2011; Dixit et al. 2015; Sarwar et al. 2017). Any naturally occurring metal/metalloid having an atomic number greater than 20 and elemental density greater than 5 g/cm^3 is termed as HM. They include copper (Cu), cadmium (Cd), chromium (Cr), cobalt (Co), zinc (Zn), iron (Fe), nickel (Ni), mercury (Hg), lead (Pb), arsenic (As), silver (Ag), and platinum group elements (Ali et al. 2013; Ali and Khan 2018). Among them, Cd, As, Hg, and Pb do not have any biological function in the body and thus are nonessential elements. They can cause severe health hazards and are listed as priority pollutants by many environmental protection agencies worldwide (Jaishankar et al. 2014; Dixit et al. 2015; Sarwar et al. 2017). Therefore, the removal of HMs from the contaminated matrix is an urgent need to safeguard the environment and human health.

Currently, applied physicochemical approaches are environmentally destructive in nature and are also costly to apply. However, bioremediation is considered as the most eco-friendly approach and employs microbes and plants or their enzymes to degrade/detoxify the organic and inorganic pollutants from contaminated environments. Phytoremediation has been identified as an emerging, low-cost, and eco-sustainable solution for HM pollution prevention and control. It is the most suitable alternative to conventional physicochemical remediation technologies, which are highly expensive and technically more suited to small areas, create secondary pollution and deteriorate soil fertility, and, thus, adversely affect agroecosystem (Ali et al. 2013; Chandra et al. 2015; Mahar et al. 2016; Muthusaravanan et al. 2018).

Phytoremediation is the engineered use of green plants with associated soil beneficial microbes to remove toxic pollutants via degradation and detoxification mechanisms from contaminated soil and water/wastewaters (Bharagava et al. 2017; Mukhopadhyay and Maiti 2010; Ali et al. 2013). It is an eco-friendly, nonintrusive, and aesthetically pleasing remediation technology that removes metal pollutants from the contaminated sites (Lee 2013; Chandra et al. 2015; Chirakkara et al. 2016). It can be commercialized, and income can be generated, if metals removed from contaminated sites could be used to extract usable form of economically viable metals (i.e., phytomining) (Chandra et al. 2015; Mahar et al. 2016). In addition, energy can be generated through the burning of plant biomass, and land restoration could be achieved for sustainable agricultural development or general habitation (Stephenson and Black 2014; Mahar et al. 2016). The rationale, mechanisms, and economic feasibility of phytoremediation have been discussed elsewhere (Ali et al. 2013; Wan et al. 2016; Sarwar et al. 2017). However, a longtime frame required for phytoremediation and physiological damage to remediating plants under toxic metal stress is a major issue. Hence, plant–microbe interactions (PMIs) could be exploited to enhance the plant growth and phytoremediation of HM-contaminated sites. Therefore, this chapter has mainly focused on the microbe-assisted phytoremediation, illustrates various mechanisms elicited for plant growth promotion and heavy metal phytoremediation (accumulation/detoxification), and discusses drawbacks and future challenges with recommendations for further research.

6.2 Heavy Metals: Environmental Pollution and Toxicity Profile

Heavy metals (HMs) can be introduced into the environment either by natural or anthropogenic processes. Natural processes are geological activities, for instance, mineral weathering, erosion, volcanic eruptions, and continental dust. Anthropogenic activities include industrial operations such as mining, smelting, electroplating, and industrial effluent discharge as well as agricultural practices like the use of pesticides and phosphate fertilizers and release of agricultural wastes (Ali et al. 2013; Mahar et al. 2016; Antoniadis et al. 2017). Industrial activities are the major source of HM pollution (water and soil) in the environment. If HMs enter the food chain, they may bioaccumulate and/or biomagnify at higher trophic levels resulting in severe health threats and thus are of serious ecotoxicological concern.

The indiscriminate discharge of toxic metal-rich industrial effluents is one of the major sources of environmental pollution. The effluent discharged from metal-based industries, especially leather industries (Cr used in leather tanning), causes serious soil and water pollution, and hence its treatment and management is a key challenge to pollution control authorities (Sahu et al. 2007; Saxena et al. 2016). A high concentration of HMs has been reported in sediments of Ganga River and its tributaries receiving Cr-loaded tannery effluent (Beg and Ali 2008). In addition, HM beyond the permissible limits also deteriorates water quality and makes it unfit for drinking and irrigation purpose (Nazeer et al. 2014). The effluent released from electroplating and distillery industries also constitutes a highly rich source of HMs and hence is considered as hazardous to living beings (Venkateswaran et al. 2007;

Chandra et al. 2008). Furthermore, effluent released from domestic activities is also responsible for HM pollution and thus is of serious ecotoxicological concerns (Bhardwaj et al. 2017).

In an aquatic ecosystem, HM adversely affects gamete production, sperm quality, and embryonic development; delays hatching; causes physical deformities in fishes; and ultimately leads to the death of newly hatched larvae (Segura et al. 2006; Jezierska et al. 2009; Fatima et al. 2014). HM also causes endocrine disruption, oxidative stress, and genotoxicity in fishes (Jezierska et al. 2009; Luszczek-Trojnar et al. 2014; Javed et al. 2016). Further, HM also causes a reduction in hematological parameters and glycogen reserve and thus makes the fishes weak, anemic, and vulnerable to diseases (Javed and Usmani 2015).

The soil is a nonrenewable resource for sustainable agriculture and acts as a major sink for HMs. The contamination of agricultural soil with toxic metals affects its physicochemical and biological properties and reduces land usability for agricultural farming leading to food insecurity and thus creating land tenure problems (Wuana and Okieimen 2011). Moreover, the coexistence and persistence of HMs in soil is also responsible for the entry of toxic metals into the food chain and thus leads to severe health hazards in living beings (Khan et al. 2008).

HM inhibits several microbial metabolic processes such as respiration, denitrification, and enzymatic activity and, hence, retards the bioremediation processes (Zhuang et al. 2007; Sobolev and Begonia 2008). HM also causes a reduction in the number of specific microbial populations and a shift in the microbial community structure. For instance, Ding et al. (2017) evaluated the effect of Cd and Cr on the microbial community structure in the rhizospheric soil of rice plant during a pot experiment. Results revealed that the relative abundance of a bacterial genus *Longilinea* was significantly higher in the control soil than in Cd- and Cr-treated soils, whereas the relative abundance of the genus *Pseudomonas* was significantly higher in the Cd-treated soils than in the Cr-treated and control soils. However, the relative abundance of a genus *Sulfuricurvum* was also significantly higher in the Cd-treated soil than in the Cr-treated and control soils, whereas the relative abundance of the genus *Bellilinea* was significantly higher in the Cr-treated soil than in the other treated soils. HMs also inhibit the cell division, transcription process, and denaturation of protein and adversely affect the cell membrane distribution in microbes (Jacob et al. 2018). Hexavalent chromium (Cr^{6+}) is also reported to cause DNA damage by exerting oxidative stress in soil bacteria and thus leads to genotoxic effects (Quievryn et al. 2003).

The irrigation of food crops in the agriculture field with water contaminated with toxic metal-rich industrial effluents is a common practice in many developing countries. It may provide a chance for the movement of potentially toxic metals from contaminated soil to edible crops, ultimately reaching into the human/animal body via consumption and, thus, rendering severe toxic effects. HM affects various metal-sensitive enzymes in plants such as alcohol dehydrogenase, nitrogenase, nitrate reductase, and amylase and hydrolytic (phosphatase and ribonuclease) and carboxylating (phosphoenolpyruvate carboxylase and ribulose-1,5-bisphosphate carboxylase) enzymes (Nagajyoti et al. 2010; Yadav 2010). Hence, HM disrupts several biochemical/physiological processes in plants such as seed germination, enzymatic activities, nitrogen metabolism, electron transport system, transpiration, CO_2 assimilation, antioxidant defense system, photosynthesis, photophosphorylation, cellular metabolism, nitrogen fixation, water balance, mineral nutrition, and cellular ionic homeostasis and ultimately leads to plant death (Yadav 2010; Lajayar et al. 2017). Irrigation of agricultural crops with heavy metal-loaded industrial effluents also disrupts several cytological processes in plants such as root growth and elongation, cell membrane permeability, mitotic activity, and the stability of genetic material and also creates chromosomal abnormalities (Nagajyoti et al. 2010; Yadav 2010). For example, the irrigation of agricultural crops with the HM-rich distillery and tannery effluent has been reported to cause a reduction in root/shoot growth and biomass, seed germination, and seedling growth and also induce chlorosis and photosynthetic impairment (Chandra et al. 2009).

HMs may cause oxidative stress by forming reactive oxygen species (ROS), which disrupt the antioxidant defense system and lead to cell damage in humans/animals, and in extreme cases can be fatal (Jaishankar et al. 2014). For instance, hexavalent chromium (Cr^{6+}) has been reported to cause cancer in humans and damage cellular components during its reduction into trivalent chromium (Cr^{3+}), leading to the generation of free radicals that cause DNA damage (Mishra and Bharagava 2016). Therefore, the remediation of HM-contaminated sites is of utmost important for environmental safety.

6.3 Current Remediation Technologies: Status and Drawbacks

Rapid industrialization and urbanization around the world has led to the recognition and understanding of the relationship between environmental contamination and public health. Industries are the key players in the national economies of many developing countries; however, unfortunately, they are also the major polluters of the environment. Among the different sources of environmental pollution, industrial wastewater discharged from different industries is considered the major source of environmental pollution (soil and water). Industries use a variety of chemicals for the processing of raw materials to obtain good-quality products within a short period of time and economically. To obtain good-quality products within a short period of time, industries generally use cheap and poorly or nonbiodegradable chemicals, and their toxicity is usually ignored. However, in the public domain, there are many reports available that confirm the presence of a variety of highly toxic chemicals in industrial wastewaters.

Industrial wastewaters contain a variety of organic and inorganic pollutants that cause serious environmental pollution and health hazards (Maszenan et al. 2011; Megharaj et al. 2011). During production processes, a variety of chemicals with large volumes of water are used to process raw materials in industries. This generates large volumes of high-strength wastewater, which is a major source of environmental pollution (Saxena et al. 2016). The wastewater generated from pollution-causing industries is characterized by high chemical oxygen demand (COD), biological oxygen demand (BOD), total dissolved solids (TDSs), total suspended solids (TSSs), and a variety of recalcitrant organic and inorganic pollutants. Organic pollutants include phenols, chlorinated phenols, endocrine-disrupting chemicals, azo dyes, polyaromatic hydrocarbons, polychlorinated biphenyls, and pesticides, whereas inorganic pollutants include a variety of toxic heavy metals such as cadmium (Cd), chromium (Cr), arsenic (As), lead (Pb), and mercury (Hg). The high concentration and poor biodegradability of recalcitrant organic pollutants and nonbiodegradable nature of inorganic metal pollutants in industrial wastewaters pose a major challenge for environmental safety and human health protection; thus, it is required to adequately treat industrial wastewater before its final disposal in the environment. Although a number of physicochemical methods are applied for the treatment of industrial wastewaters, all of these are costly, use a large amount of chemicals, and generate a large amount of sludge after treatment, which also acts as a secondary pollutant in the environment. Alternatively, biological treatment methods using an array of microorganisms have diverse metabolic pathways and, hence, are regarded as environmentally friendly, cost-effective methods for wastewater treatment with simple structural setup, wider application, operational ease, and less sludge production compared to physicochemical methods (Mendez-Paz et al. 2005; Pandey et al. 2007). Biological methods using microbes are becoming much more popular for the treatment of industrial wastewaters in wastewater treatment plants. Further, most chemical compounds are degraded by acclimated microorganisms during wastewater treatment at wastewater treatment plants; however, some of the chemical compounds are not properly degraded/detoxified due to their recalcitrant nature during wastewater treatment and are discharged along with wastewaters, causing serious environmental pollution (Maszenan et al. 2011). Hence, the application of bioremediation technology using potential microorganisms and their consortia or of phytoremediation technology (use of green plants in constructed wetlands) is required for the degradation and detoxification of such types of recalcitrant industrial wastewaters prior to safe disposal in the environment.

Phytoremediation is considered as the most applicable remediation technology at contaminated sites. Phytoremediation is the engineered use of green plants with associated soil beneficial microbes to remove toxic pollutants via degradation and detoxification mechanisms from contaminated soil and water/wastewaters (Bharagava et al. 2017; Mukhopadhyay and Maiti 2010; Ali et al. 2013). It is an eco-friendly, nonintrusive, and aesthetically pleasing remediation technology that removes metal pollutants from the contaminated sites (Lee 2013; Chandra et al. 2015; Chirakkara et al. 2016). The aim of phytoremediation can be (a) plant-based extraction of metals with financial benefit (phytoextraction), (b) risk minimization (phytostabilization), and (c) sustainable soil management in which phytoremediation steadily increases soil fertility allowing growth of crops with added economic value (Mahar et al. 2016; Vangronsveld et al. 2009). Phytoremediation includes a range of plant-based remediation processes. Phytoremediation reduces the risks of pollutant dispersion, and it is applicable for the decontamination of soils or wastewaters with mixed pollutants (Mahar et al. 2016; Mudhoo et al. 2010). Mechanisms and efficiency of phytoremediation depend on several factors such as the pollutant class, its bioavailability especially in soils, physical and chemical characteristics of the matrix (soil, water, and wastewaters), and plant species (Mahar et al. 2016; Sreelal and Jayanthi 2017). The plants considered more efficient for phytoremediation are the metallophytes. These are able to survive and reproduce on metal-polluted soils (Coninx et al. 2017; Alford et al. 2010). However, a great number of known metallophytes have small biomass and slow growth, characteristics that are not advantageous for phytoremediation technologies (Coninx et al. 2017; Cabral et al. 2015). Further, longtime frame required for phytoremediation and physiological damage to remediating plants under toxic metal stress is a major issue. Therefore, plant–microbe interactions (PMIs) could be exploited to enhance the plant growth and phytoremediation of HM-contaminated sites.

The root-/rhizosphere-colonizing, plant growth-promoting rhizobacteria (PGPR) have been reported to enhance host plant growth in toxic metal-contaminated sites (Yuan et al. 2013; Ma et al. 2015, 2016a). PGPR produces growth hormones such as auxins (IAA, indole-3-acetic acid), cytokinins, gibberellins, and ethylene (Rajkumar et al. 2012; Ma et al. 2015). The mechanisms of plant growth promotion may vary from bacterial strain to strain and depend on various secondary metabolites produced (Ma et al. 2011; Backer et al. 2018). PGPR also produces some other beneficial compounds such as enzymes, osmolytes, biosurfactants, organic acids, metal-chelating siderophores, nitric oxide, and antibiotics (Rajkumar et al. 2012; Ma et al. 2015). These beneficial compounds reduce ethylene production *via* synthesis of ACC (1-aminocyclopropane-1-carboxylate) deaminase that prevents the inhibition of root elongation, lateral root growth, and root hair formation and also improves the mineral (N, P, & K) uptake in acidic soil (Babu et al. 2013; Ma et al. 2015). These compounds also suppress phytopathogens, provide tolerance to abiotic stress, and help in associated nitrogen fixation (Rajkumar et al. 2012; Babu et al. 2013; Ma et al. 2015). Hence, PGPRs are applied in sustainable agriculture development. Besides these, PGPR can lower the metal toxicity to remediating plants through biosorption/bioaccumulation as bacterial cells have an extremely high ratio of surface area to volume (Ma et al. 2016b; Li et al. 2018). PGPR could adsorb high metal concentration by either a metabolism-independent passive or metabolism-dependent active processes. Hence, using PGPR in environmental bioremediation could be a useful strategy for plant survival in the stressed environment. PGPRs reported for the enhanced HM phytoremediation with associated benefits have been reviewed in the past (Ma et al. 2011; Rajkumar et al. 2012; Ullah et al. 2015). Some updated examples from recent studies are summarized in Table 6.1.

Endophytes are the microbes (bacteria/fungi) that reside in the inner tissues of plants without causing harm to host. They also help in plant growth promotion and development under biotic or abiotic stressed environment and exert many beneficial effects than rhizobacteria (Luo et al. 2011; Ma et al. 2011, 2015). They are able to tolerate high metal concentration and hence lower phytotoxicity to remediating plants as well as help in growth promotion enhancing through biocontrol mechanism and induced systemic resistance against phytopathogens (Ma et al. 2011, 2015). They produce phytohormones, organic acids, siderophores, biosurfactants,

Table 6.1 Some studies on microbe-assisted phytoremediation of heavy metal-contaminated soils

Plant growth-promoting rhizobacteria (PGPR)					
Bacterial strain(s)	Host plant	Heavy metal	Medium	Beneficial effects	References
Enterobacter sp. LC1, LC4, & LC6; *Kocuria* sp. LC2 & LC5; and *Kosakonia* sp. LC7	*Solanum nigrum*	As	Soil	IAA and P-solubilization	Mukherjee et al. (2018)
Pseudomonas libanensis and *Pseudomonas reactans*	*Brassica oxyrrhina*	Cu, Zn	Soil	IAA, ACC deaminase, siderophores	Ma et al. (2016a)
Pseudomonas putida, *Rhodopseudomonas* sp.	*Cicuta virosa L.*	Zn	Soil	Metal-chelating compounds	Nagata et al. (2015)
Rhizobium leguminosarum	*Brassica juncea*	Zn	Soil	Metal chelation	Adediran et al. (2015)
Photobacterium spp.	*Phragmites australis*	Hg	Soil	IAA, mercury reductase activity	Mathew et al. (2015)
Bacillus pumilus E2S2 and *Bacillus* sp. E1S2	*Sedum plumbizincicola*	Cd	Soil	IAA, ACC deaminase, siderophores, P-solubilization	Ma et al. (2015)
Pseudomonas sp. LK9	*Solanum nigrum*	Cd	Soil	Biosurfactants, siderophores, organic acids	Chen et al. (2014)
P. aeruginosa	*Triticum aestivum*	Zn	Soil	Antioxidative enzymes (catalase, peroxidase, superoxide dismutase)	Islam et al. (2014)
Mesorhizobium Amorphae	*Robinia pseudoacacia*	Cu, Zn, Cr	Soil	IAA, induced stress Tolerance	Hao et al. (2013)
Acinetobacter sp.	*Cicer arietinum*	As	Soil	IAA production	Srivastava and Singh (2014)
Enterobacter sp. JYX7 and *Klebsiella* sp. JYX10	*Polygonum pubescens*	Cd	Soil	IAA, siderophores, ACC deaminase, P-solubilization	Jing et al. (2014)
Bacillus subtilis, *Bacillus cereus*, *B. megaterium*	*Orychophragmus violaceus*	Cd	Soil	IAA production	Liang et al. (2014)

(continued)

Table 6.1 (continued)

Bacterial strain(s)	Host plant	Heavy metal	Medium	Beneficial effects	References
Plant growth-promoting rhizobacteria (PGPR)					
Phyllobacterium myrsinacearum RC6b	*Sedum plumbizincicola*	Cd, Zn, and Pb	Soil	ACC deaminase, IAA, siderophores, P solubilization	Ma et al. (2013)
Staphylococcus arlettae NBRIEA G-6	*B. juncea*	As	Soil	IAA, siderophores, ACC deaminase	Srivastava et al. (2013)
Rahnella sp.	*Amaranthus hypochondriacu*, *A. mangostanus*, and *S. nigrum*	Cd	Soil	IAA, siderophores, ACC deaminase, P-solubilization	Yuan et al. (2013)
Paenibacillus macerans NBRFT5, *Bacillus endophyticus* NBRFT4, and *Bacillus pumilus* NBRFT9	*Brassica juncea*	Ni	Mix. of fly ash and press mud	Siderophores, organic acids, protons, and other nonspecified enzymes	Tiwari et al. (2012)
Pantoea agglomerans Jp3-3 and *Pseudomonas thivervalensis* Y1-3-9	*Brassica napus*	Cu	Quartz sand	IAA, siderophores, ACC deaminase, P-solubilization	Zhang et al. (2011)
Azotobacter chroococcum and *Rhizobium leguminosarum*	*Zea mays L.*	Pb	Soil	IAA production increased and soil pH decreased	Hadi and Bano (2010)
Bacillus subtilis, *B. cereus*, *Flavobacterium* sp., and *Pseudomonas* sp.	*Orychophragmus violaceus*	Zn	Soil	ACC deaminase, IAA, siderophores	He et al. (2010)
Achromobacter xylosoxidans Ax10	*Brassica juncea*	Cu	Soil	ACC deaminase, IAA, phosphate solubilization	Ma et al. (2009)
Burkholderia sp. J62	*Zea mays and Lycopersicon Esculentum*	Pb, Cd	Soil	IAA, siderophores, ACC deaminase, P solubilization	Jiang et al. (2008)
Burkholderia sp. J62	*B. juncea*	Zn, Pb, Cu	Soil	P, K solubilization	Wu et al. (2006)
Brevibacillus brevis	*Trifolium repens*	Cd, Ni, Pb	Soil	IAA production	Vivas et al. (2006)
Endophytes					

(continued)

Table 6.1 (continued)

Plant growth-promoting rhizobacteria (PGPR)					
Bacterial strain(s)	Host plant	Heavy metal	Medium	Beneficial effects	References
Bacillus thuringiensis GDB-1	*Alnus firma*	As	Mine tailing waste	ACC deaminase, IAA, siderophores, P-solubilization	Babu et al. (2013)
Pseudomonas koreensis AGB-1	*Miscanthus Sinensis*	As, Cd, Cu, Pb, and Zn	Soil	ACC deaminase activity, IAA	Babu et al. (2015)
Staphylococcus, *Curtobacterium*, *Bacillus*, *Pseudomonas*, *Microbacterium*, *Arthrobacter*, *Leifsonia*, *Paenibacillus*	*Alyssum bertolonii*	Ni, Co, Cr, Cu, and Zn	Soil	Production of siderophores	Barzanti et al. (2007)
Serratia nematodiphila LRE07, *Enterobacter aerogenes* LRE17, *Enterobacter* sp. LSE04 *Acinetobacter* sp. LSE06	*Solanum nigrum* L.	Cd	Soil	Production of IAA, siderophores, ACCD, and solubilization of P	Chen et al. (2010)
P. monteilii PsF84, *P. plecoglossicida* PsF610	*Pelargonium graveolens*	Cr	Soil	Production of IAA and siderophores, solubilization of P	Dharni et al. (2014)
Rahnella sp. JN6	*Brassica napus*	Pb	Soil	IAA, ACC deaminase, siderophores, P-solubilization	He et al. (2014)
Actinobacterium	*Salix caprea*	Cd and Zn	Soil	Production of siderophores and ACCD	Kuffner et al. (2010)
Burkholderia cepacia L.S.2.4, *Herbaspirillum seropedicae* LMG2284	*Lupinus luteus* L	Cu, Cd, Co, Ni, Pb, and Zn	Soil	ND	Lodewyckx et al. (2001)

(continued)

Table 6.1 (continued)

Plant growth-promoting rhizobacteria (PGPR)					
Bacterial strain(s)	Host plant	Heavy metal	Medium	Beneficial effects	References
Pseudomonas fluorescens VI8L1, *Bacillus pumilus* VI8L2, *P. fluorescens* II8L4, *P. fluorescens* VI8R2, *Acinetobacter calcoaceticus* II2R3	*Sedum alfredii*	Zn and Cd	Soil	Production of IAA, siderophores, fixation of nitrogen, solubilization of ZnCO3 and Zn3(PO4)2	Long et al. (2011)
Serratia marcescens LKR01, *Arthrobacter* sp. LKS02, *Flavobacterium* sp. LKS03, *Chryseobacterium* sp. LKS04	*Solanum nigrum* L.	Zn, Cd, Pb, and Cu	Soil	Production of IAA, siderophores, ACCD, and solubilization of P	Luo et al. (2011)
Serratia sp. LRE07	*S. nigrum* L	Cd, Cr, Pb, Cu, and Zn	Soil	Production of IAA, siderophores, and solubilization of P	Luo et al. (2011)
Bacillus sp. SLS18	*Sorghum bicolor* L.	Cd and Mn	Soil	Production of IAA, siderophores, and ACCD	Luo et al. (2011)
Pseudomonas sp. A3R3	*Alyssum serpyllifolium*	Ni	Soil	Production of IAA, siderophores, ACCD, and solubilization of P; excreted cellulase and pectinase	Ma et al. (2011)
Methylobacterium oryzae CBMB20, *Burkholderia* sp. CBMB40	*Lycopersicon esculentum*	Ni and Cd	Soil	ND	Madhaiyan et al. (2007)
P. fluorescens G10, *Microbacterium* G16	*Brassica napus*	Pb, Cd, Zn, Cu, and Ni	Soil	Production of IAA, siderophores, ACCD	Sheng et al. (2008)

(continued)

Table 6.1 (continued)

Plant growth-promoting rhizobacteria (PGPR)					
Bacterial strain(s)	Host plant	Heavy metal	Medium	Beneficial effects	References
Bacillus sp. MN3-4	*Alnus firma* and *B. napus*	Pb, Cd, Zn, Ni, and Cu	Soil	Production of IAA and siderophores	Shin et al. (2012)
Endophytes belonged to *Firmicutes*, *Actinobacteria*, Proteobacteria	*Elsholtzia splendens*, *Commelina communis*	Cu	Soil	Production of IAA, siderophores, ACCD, and arginine decarboxylase	Sun et al. (2010)
Microbacterium sp. NCr-8, *Arthrobacter* sp. NCr-1, *Bacillus* sp. NCr-5, *Bacillus* sp. NCr-9, and *Kocuria* sp. NCr-3	*Noccaea caerulescens*, *Thlaspi perfoliatum*	Ni	Soil	Production of IAA, siderophores, and ACCD	Visioli et al. (2014)
Serratia nematodiphila LRE07	*Solanum nigrum* L.	Cd	Soil	ND	Wan et al. (2012)
Rahnella sp. JN27	*Amaranthus hypochondriacus* and *A. mangostanus*	Cd	Soil	Production of IAA, siderophores, ACCD, and solubilization of P	Yuan et al. (2014)
Burkholderia sp. SaZR4, *Burkholderia* sp. SaMR10, *Sphingomonas* sp. SaMR12, and *Variovorax* sp. SaNR1	*Sedum alfredii* Hance	Cd and Zn	Soil	ND	Zhang et al. (2013)
Endophytes belonged to *Firmicutes*, *Proteobacteria*, and *Actinobacteria*	*Pteris vittata* and *P. multifida*	As	Soil	Production of IAA	Zhu et al. (2014)
Arbuscular mycorrhizal fungi					

(continued)

Table 6.1 (continued)

Plant growth-promoting rhizobacteria (PGPR)					
Bacterial strain(s)	Host plant	Heavy metal	Medium	Beneficial effects	References
Glomus mosseae	*Trifolium subterraneum*, *Lolium perenne*	Cd, Zn	Soil	AMF adsorbed up to 0.5 mg Cd per gram of mycelia equivalent to threefold binding capacity of non-tolerant fungi or tenfold higher than reported for *Rhizopus arrhizus* (commonly used as biosorption organism)	Joner et al. (2000)
Glomus intraradices	*Helianthus annuus*	Cr	Soil	AMF increased fivefold root Cr concentration	Davies et al. (2001)
Glomus mosseae, *Glomus caledonium*, and *Glomus claroideum*	*Sorghum vulgare*	Cu	Soil	RM increased Cu-sorption from 2.3 to 13.8 mg Cu g^{-1} dry mycelium	
Glomus mosseae	*Trifolium pratense* L	Zn	Soil	22% of total Zn plant uptake linked to ERM	Chen et al. (2003)
Gigaspora rosea and *Glomus mosseae*	*Zea mays* and *Sorghum vulgare*	Cu	Soil	GRSP produced by *G. rosea* hyphae bound up to 28 mg Cu g^{-1} and *G. mosseae* ranged from 1.0 to 1.6 mg Cu g^{-1}	Gonzalez-Chavez et al. (2004)
Mixed spores of mycorrhizal fungal species isolated from orchard soil	*Kummerowia striata*, *Ixeris denticulata*, *Lolium perenne*, *Trifolium repens*, and *Echinochloa crus-galli*	Pb	Soil	AMF inoculation increased the Pb root concentration from 7.6% to 57.2%	Chen et al. (2005)
Indigenous mycorrhizal populations from polluted soils	*Argemone subfusiformis*, *Baccharis linearis*, *Oenothera affinis*, *Polypogon viridis*	Cu, Zn	Soil	GRSP bound from 1.4% to 28% of total Cu in soil and from 1.4% to 5.8% of total Zn	Cornejo et al. (2008)

(continued)

Table 6.1 (continued)

Plant growth-promoting rhizobacteria (PGPR)					
Bacterial strain(s)	Host plant	Heavy metal	Medium	Beneficial effects	References
Indigenous mycorrhizal populations from polluted soils	Degraded ecosystem with presence of *Sesleria caerulea*	Pb, Zn	Soil	GRSP bound Pb attained until 23.4 mg g^{-1}, which represents about 16% of total soil Pb	Vodnik et al. (2008)

enzymes, and growth regulators that help in water and nutrient (P, N, & K) uptake, osmolyte accumulation, osmotic adjustment, stomatal regulation, and associated nitrogen fixation as additional benefits to host plants (Ma et al. 2011, 2016b). Thus, inoculating plants with endophytes could be an excellent strategy to enhance the phytoremediation of HM-contaminated sites. Endophytes applied to enhance HM phytoremediation with associated benefits have been recently reviewed by several researchers (Afzal et al. 2014; Ma et al. 2016b).

Arbuscular mycorrhizal fungi (AMF: colonize plant roots) have been also reported to protect their host plants against heavy metal toxicity through their mobilization from soil and thus help in phytoremediation (Marques et al. 2009; Meier et al. 2012; Khan et al. 2014). The possible mechanisms by which AMF protect their host plants through metal mobilization from soil include:

(a) Immobilization by chelation;
(b) Binding of metals to biopolymers in the cell wall;
(c) Superficial immobilization in the plasmatic membrane once metals cross the cell wall;
(d) Membrane transportation that mobilizes metals from the soil to the cytosol;
(e) Intracellular chelation through MTs, organic acids, and amino acids;
(f) Export of metals from cytosol by membrane transporters;
(g) Sequestration of metals into vacuoles;
(h) Transportation of metals by means of fungal hyphae;
(i) Storage of metals in fungal spores; and
(j) Exportation by the fungus and access into the plant cells, involving both active and passive transportation into the mycorrhizae (Meier et al. 2012; Cabral et al. 2015).

They confer resistance against drought, high salt, and toxic metal concentration and improve nutrient supply and soil physical properties (Khan et al. 2014). The exact mechanism of plant protection is still not fully understood, and further research is required to explore their role in the phytoremediation.

6.4 Microbe-Assisted Phytoremediation: Concepts and Mechanisms

Most plants growing in polluted environments are often characterized by relatively low growth caused by toxic effects of accumulated substances or their degradation products (Glick 2003). However, the negative effect of the environment can be alleviated by soil microorganisms. The soil is an environment settled by a wide range of genetically diverse microorganisms, which play crucial roles in nutrient cycling and in soil-forming processes (Ahemad and Khan 2013). They include both bacteria, which are the most numerous (9×10^7 in one gram of typical soil), and fungi (2×10^5) (Alexander 1991). Microorganisms inhabiting metalliferous soils often exhibit tolerance to high concentrations of heavy metals (HMs) in the environment. Many studies have confirmed that interactions between plants and metallo-tolerant microorganisms facilitate the recultivation of HM-polluted areas (e.g., Chen et al. 2014; Ma et al. 2015; Złoch et al. 2017). This synergism can accelerate the process of remediation by phytostabilization or phytoextraction of HMs but can also increase plant growth and development under adverse environmental conditions (Khan et al. 2009). The functioning of plant–microorganism associations in HM-contaminated soils depends on both the microorganisms and the plant host (Egamberdieva et al. 2016). The plant roots secrete exudates that are the source of nutrients for microorganisms and also increase the solubility of macro- and microelements affecting the activity of microorganisms associated with plant roots (Iqbal and Ahemad 2015). Plant-associated microorganisms can play significant roles in nutrient cycling, improving soil structure, detoxifying harmful contaminants, modulating plant defense responses to stress factors, and assisting in biological control of phytopathogens and plant growth (Elsgaard et al. 2001; Filip 2002; Giller et al. 1998).

To generalize, the activity of microorganisms inhabiting the roots (endophytes) or rhizosphere can increase the capacity of metalliferous soil phytoremediation as follows:

1. Directly: Plant-associated microorganisms directly increase the uptake and translocation of metals (facilitation of phytoextraction) or reduce the mobility/availability of metals within the rhizosphere (phytostabilization).
2. Indirectly: Microorganisms increase plant tolerance to HMs and/or promote plant biomass production in order to remove/stabilize contaminants. A general outline of plant–microbe–metal interactions for the phytoremediation of heavy metal-contaminated soils is shown in Fig. 6.1.

6.4.1 Direct Mechanisms

In most metalliferous soils, HMs are strongly adsorbed onto soil particles and are therefore hardly available for plant roots during phytoextraction (Gamalero and Glick 2012). Microorganisms can increase their solubility and availability via (a) auto- and heterotrophic leaching (associated with redox reaction), (b) secretion of

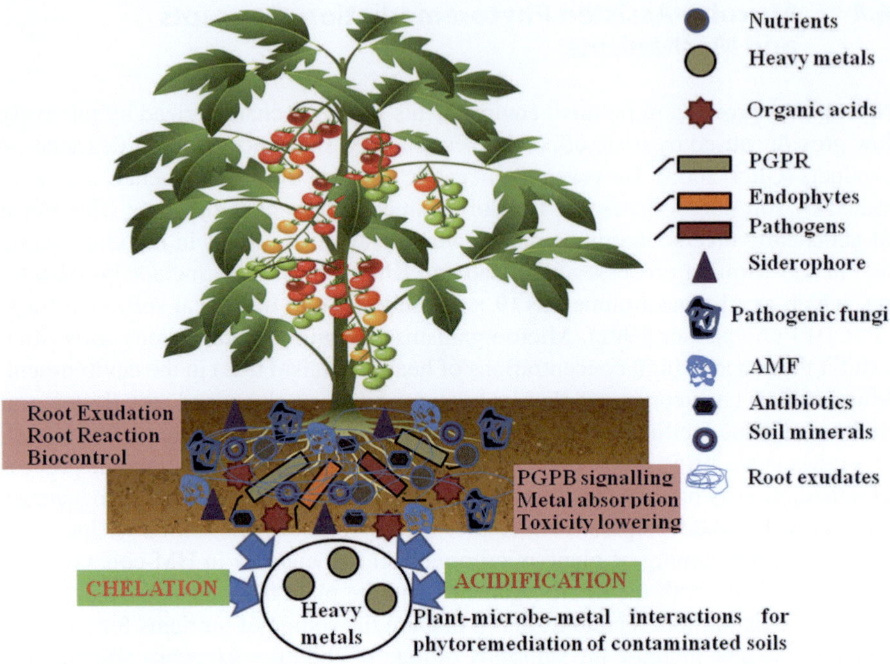

Fig. 6.1 Outline mechanism of plant–microbe–metal interactions for microbe phytoremediation of heavy metal-contaminated sites

organic acids and biosurfactants, and (c) release of siderophores (Gadd 2004; Wenzel 2008; Li et al. 2012). These processes can lead to the dissolution of minimally soluble metal–mineral compounds (including phosphates, sulfates, and more complex ores) as well as metal desorption from the surface of clay minerals or organic matter (Gadd 2004). Microorganisms can acidify the environment by releasing H+ through the transmembrane H^+-ATPase, maintaining the membrane potential or as a result of carbon dioxide accumulation generated during respiratory processes, which leads to the release of free metal cations from their complexes with anions *via* ion exchange occurring between H^+ and metals (Gadd 2004). In most cases, autotrophic leaching of metals is performed by acidophilic bacteria, which assimilate carbon dioxide and produce energy from Fe^{2+} oxidation or sulfur compound reduction (Rawlings 1997; Schippers and Sand 1999). Moreover, many studies have confirmed that rhizosphere bacteria such as *Thiobacillus thiooxidans* are interesting in the context of phytoextraction because they reduce rhizosphere pH through the conversion of reduced sulfur into sulfate, improving the availability of Cu, for example, to plants (Rawlings and Silver 1995; Shi et al. 2011). In recent years, much attention has been paid to the phenomenon of low-molecular-weight organic acids (LMWOAs, compounds with molecular weights ≤300 Da and containing one or more carboxylic groups) being secreted by plant-associated microorganisms and their potential role in the regulation of HM solubility and mobilization of mineral compounds within the rhizosphere (Rajkumar et al. 2012).

Chelators are mainly known to enhance the solubility of HMs and include citric, lactic, malic, oxalic, malonic, 5-ketogluconic, tartaric, succinic, and formic acids (Panhwar et al. 2013). Commonly synthesized oxalates and citrates are known for their ability to form stable complexes with many HMs; furthermore, citrates are highly mobile and highly resistant to degradation (Francis et al. 1992). Saravanan et al. (2007) observed that during secretion of 5-ketogluconic acid by an endophytic bacterium of *Gluconacetobacter diazotrophicus*, various Zn^{2+} sources (e.g., ZnO, $ZnCO_3$, or $Zn_3(PO_4)_2$) are dissolved, which increases the pool of Zn^{2+} readily available for roots. Moreover, Han et al. (2006) revealed stimulatory effects of acetic and malic acid on the Cd^{2+} accumulation in the roots of corn (*Zea mays* L.). Similar observations were noticed in the case of, for example, increased uptake of Cd^{2+} and Zn^{2+} by *Sedum alfredii* due to secretion of formic, acetic, tartaric, succinic, and oxalic acids by rhizosphere bacteria (Li et al. 2010) as well as stimulation of Cd^{2+} uptake by wheat in the presence of citric acid (Panfili et al. 2009). Regarding synthesis of LMWOAs, particularly oxalate, by fungal strains, it has also been suggested that the release of metal ions *via* enhanced mineral weathering plays an important role and leads to the uptake of HMs by plants and microorganisms (Jones 1998; Gadd and Sayer 2000). Such an ability was noted for *Beauveria caledonica*, *Aspergillus niger*, *Penicillium bilaiae*, or *Oidiodendron maius* in the case of cadmium, copper, lead, nickel, or zinc mineral solubilization (Martino et al. 2003; Fomina et al. 2005; Arwidsson et al. 2010). Another important class of metabolites with great potential to increase metal mobility and stimulate the phytoremediation process is the microbial surface-active substances called biosurfactants (Rajkumar et al. 2012). Biosurfactants are amphiphilic molecules consisting of long nonpolar parts (hydrophobic) and polar/ionic (hydrophilic) heads. Their hydrophilic parts consist of mono-, oligo-, or polysaccharides, peptides, and proteins, while their hydrophobic parts usually contain saturated, unsaturated, and hydroxylated fatty acids or fatty alcohols. Siderophores are low-molecular-weight organic compounds (500–1500 Da) with high specificity and affinity for Fe^{3+} chelation (Miethke and Marahiel 2007), which release iron from minerals or organic matter in order to facilitate iron uptake when its availability in the environment is limited (Li et al. 2012). Despite the substantial diversity of chemical structures of siderophores (over 500 diverse siderophores described to date), they can be divided into several groups depending on the presence of metal-binding ligands: (a) hydroxamates, (b) catecholates, (c) phenolates, (d) carboxylates, and (e) mixed (Essen et al. 2006; Saha et al. 2013; Wang et al. 2014; Pluhacek et al. 2016). While the key role of siderophores in iron homeostasis in microorganisms has been well known for over 60 years, there is increasing evidence for the activation of siderophore synthesis by bacteria in the presence of toxic metals, which indicates their potential role in HM homeostasis (Schalk et al. 2011; Złoch et al. 2016). It was suggested that siderophores may form stable complexes with ions such as Ag^{+}, Zn^{2+}, Cu^{2+}, Co^{2+}, Cr^{2+}, Mn^{2+}, Cd^{2+}, Pb^{2+}, Ni^{2+}, Hg^{2+}, Sn^{2+}, Al^{3+}, In^{3+}, Eu^{3+}, Ga^{3+}, Tb^{3+}, and Tl^{+}. Enhanced siderophore synthesis by bacteria (so-called siderophore-producing bacteria, SPB) can protect them from the toxic effects of HMs by, for example, extracellular sequestration, thereby preventing metals from entering into the cells (Saha et al. 2013). Similar observations were noted for fungi; however, the relatively weak ability of fungal siderophores (mainly hexadentate hydroxamate) to chelate HMs other than Fe(III) (Enyedy et al. 2004; Farkas et al. 2008) makes their potential in HM bioremediation rather limited (Pocsi 2011). On the other hand, increased siderophore synthesis can improve the phytoextraction capacity of plants by increasing the mobility of metals and thus their availability for roots (Glick 2003; Rajkumar et al. 2010).

6.4.2 Indirect Mechanisms

The most important mechanisms, and those confirmed so far in the scientific literature, are (a) the synthesis of phytohormones and enzymes (primarily indole-3-acetic acid [IAA], 1-aminocyclopropane-1-carboxylate [ACC] deaminase), (b) increased nutrient uptake (nitrogen fixation, phosphorus, and iron mobilization), and (c) tolerance to biotic (pathogen control) and abiotic (drought, salinity, contamination) stress conditions (Hrynkiewicz and Baum 2012; Ma et al. 2016b). The specific response of nitrogen-fixing legumes in response to Cd, like an overproduction of reactive oxygen species (ROS) in the nodules and its mitigation by PGPB (e.g., by the release of siderophores), was reviewed by Gomez-Sagasti and Marino (2015). IAA is one of the most important phytohormones and regulates many physiological and morphological functions of plants (Glick 2012). In addition to stimulation of root growth, alleviating salt stress, participating in plant–pathogen interactions, and eliciting induced systemic resistance (ISR) against various diseases, IAA is primarily involved in stimulating the proliferation of lateral roots. IAA-synthesizing microorganisms can indirectly increase the extraction of metals and nutrient supplementation of plants by inducing root proliferation and increasing their uptake surface (Glick 2010). Apart from IAA, soil microorganisms demonstrate the ability to synthesize other phytohormones (cytokinins, gibberellins). However, fungi are also known for their ability to secrete compounds similar to phytohormones such as auxins, cytokinins, gibberellic acids, or ethylene (Chanclud and Morel 2016). Ethylene is a crucial phytohormone that regulates plant cell elongation and metabolism (Ping and Boland 2004), and its overproduction induced by stress factors, such as HMs, may inhibit processes involved in plant development (i.e., root elongation, lateral root growth, and formation of root hairs) (Mayak et al. 2004). Microbial ACC deaminase causes the hydrolysis of 1-aminocyclopropane-1-carboxylic acid (an ethylene precursor) to α-ketobutyric acid and ammonia, which can be used as a source of carbon and nitrogen by microorganisms. Thus, inoculation of plants with strains synthesizing ACC deaminase indirectly affects root growth and proliferation and positively influences the plant biomass and efficiency of HM phytoremediation (Gleba et al. 1999; Agostini et al. 2003; Arshad et al. 2007). ACC deaminase-containing bacteria are relatively common in soil (typically free-living pseudomonads) (Glick 2005, 2014), while among fungi, this activity is less frequently observed (although it has been reported in *Penicillium citrinum* and *Trichoderma asperellum* T203) (Jia et al. 2000; Viterbo et al. 2010) and has not been investigated in detail. The presence of elevated amounts of HMs often affects the supplementation of plant roots with Fe, P, Mg, or Ca, leading to plant growth retardation (Ouzounidou et al. 2006; Parida et al. 2003). Under such conditions, plant-associated microorganisms facilitate the uptake of nutrients by increasing their availability for plant roots (Rajkumar et al. 2012). Examples include the bacteria reported by Nautiyal et al. (2000), which demonstrate the ability to increase P availability for plants through phosphate precipitation by acidification of the soil solution, complexation, secretion of organic acids, and ion-exchange reactions or through mineralization of organic phosphorus compounds secreting acid phosphatase (van der Hiejden et al. 2008). Among P-solubilizing microorganisms, fungal strains belonging to *Aspergillus* and *Penicillium* are known for their strong ability to release P from insoluble inorganic compounds, primarily by producing organic acids and preventing the precipitation of P with metals (Jones 1998; Mendes et al. 2014). A similar effect is observed for iron, which is present in the Earth's crust in large quantities; however, iron is found mostly as insoluble hydroxides and oxyhydroxides that are not readily available to plants (Budzikiewicz 2010; Rajkumar et al. 2010). Moreover, plants growing in metalliferous soils are very often exposed to iron deficiency, which produces a decreased photosynthesis rate and consequently a decline in their growth and development (Nagajyoti et al. 2010a, b). In such cases, inoculation of plants with SPB can be a promising method to mitigate iron deficiency (Iqbal and Ahemad 2015). Many studies have confirmed that SPB successfully increased chlorophyll concentration and improved other plant growth parameters in the presence of HM contamination in the soils by facilitating iron uptake (Burd et al. 1998, 2000, Carrillo-Castaneda et al. 2003, Barzanti et al. 2007). It has also been observed that the synthesis of siderophores may stimulate plant growth in metalliferous areas via the following activities: (a) involvement in maintaining an appropriate level of IAA through binding of HMs, thereby reducing the inhibitory effect of metals on the IAA biosynthesis pathways, and through decreased production of reactive oxygen species (ROS), which can degrade IAA molecules; (b) mitigation of oxidative stress by stimulation of peroxidase activity; and (c) phytopathogen control via chelation of iron ions within the rhizosphere and decreasing the availability of iron for pathogens (Dimkpa et al. 2008; Rajkumar et al. 2009).

6.5 Microbe-Assisted Phytoremediation of Heavy Metal-Contaminated Sites

It has been well demonstrated that the inherent ability of endophytic bacteria may help host plants adapt to unfavorable soil conditions and enhance the efficiency of phytoremediation by promoting plant growth, alleviating metal stress, reducing metal phytotoxicity, and altering metal bioavailability in soil and metal translocation in plant (Ma et al. 2011; Ozyigit and Dogan 2015). Overall, the plant-associated microbes promote phytoremediation process in metal-polluted soils by two distinct means, i.e., enhancement of plant metal tolerance and growth and alteration of metal accumulation in plants, as discussed in above sections. Some important studies on the phytoremediation of heavy metal-contaminated soils assisted by plant growth-promoting rhizobacteria, endophytes, and arbuscular mycorrhizal fungi have been summarized in Table 6.1.

6.6 Challenges and Future Perspectives

The success of phytoextraction depends on interactions among soil, metals, and plants. Many plants are not capable of gaining sufficient biomass for noticeable rates of remediation when elevated levels of pollutants are present (Harvey et al. 2002; Chaudhry et al. 2005). The remediation process of contaminated soils is limited and slowed because of their poor nutrient nature. Soil microbes are thought to exert positive effects on plant health via mutualistic relationships between them. However, microbes are sensitive to pollution, and depletion of microbial populations, both in terms of diversity and biomass, often occurs in such contaminated soils (Shi et al. 2002). Biotic or abiotic stress through a small change in the physicochemical–biological properties of rhizosphere soils can cause a dramatic effect on plant–microbe interaction. Further, isolation and characterization of suitable plant-associated beneficial microbes is a time-consuming process. It also requires the analysis of more than thousands of isolates, and thus identification of specific biomarkers may help to select the effective plant–microbe interactions for microbe-assisted phytoremediation (Rajkumar et al. 2012). Further, to ameliorate metal toxicity, plant growth promotion, and metal sequestration, extensive research efforts are also required to explore novel microbial diversity, their distribution, and functions in the autochthonous and allochthonous soil habitats for microbe-assisted phytoremediation of HM-contaminated sites.

6.7 Conclusions and Recommendations

(a) HM pollution in the environment and associated toxicity in living beings is of serious eco-environmental concern.
(b) Inoculation of plants with associated microbes (such as PGPRs, endophytes, and arbuscular mycorrhizal fungi) exhibiting multiple traits could be an excellent strategy to enhance metal detoxification in the rhizosphere. A clear-cut understanding of plant–microbe–metal–soil interactions is crucial for microbe-assisted phytoremediation of HM-contaminated soils.
(c) The effectiveness of co-inoculation of PGPB and AMF in response to multiple biotic and/or abiotic stresses must be assessed for better applicability at field.
(d) Identification of functional genes of beneficial microbes responsible for growth enhancement and metal detoxification should be identified.

(e) Trials for the commercial production of bioinoculants for use in metal decontamination should be performed to make a positive remark toward their field applicability.
(f) Genetic engineering of metal-accumulating plants and associated microbes with required traits could be a very useful strategy for the enhanced phytoremediation, but associated risks should also be considered before field application.
(g) A detailed and accurate characterization of target metal(loid)-contaminated soils is needed before the inoculation of microbes, as well as adequate strategies to enhance inoculant performance by using efficient carrier materials.
(h) The complexity and heterogeneity of soils contaminated with multiple metals and organic compounds requires the design of integrated phytoremediation systems that combine different processes and approaches.
(i) Field trials are required to document time and cost data to provide recommendations and convince regulators, decision-makers, and the general public about the low-cost applicability of microbe-assisted phytoremediation of heavy metal-contaminated sites and for better acceptance in remediation industries.

Conclusively, microbe-assisted phytoremediation technology holds great promise in gaining the sustainable agricultural production in conjunction with phytoremediation of heavy metal-contaminated sites for environmental sustainability.

Acknowledgments Authors are highly thankful to the director of the Baba Farid Institute of Technology (BFIT), Dehradun (UK), India, for providing the infrastructure and facility for the research work.

References

Adediran GA, Ngwenya BT, Frederick J, Mosselmans W, Heal KV, Harvie BA (2015) Mechanisms behind bacteria induced plant growth promotion and Zn accumulation in Brassica juncea. J Hazard Mater 283:490–499
Afzal M, Khan QM, Sessitsch A (2014) Endophytic bacteria: prospects and applications for the phytoremediation of organic pollutants. Chemosphere 117:232–242
Agostini E, Coniglio MS, Milrad SR, Tigier HA, Giulietti AM (2003) Phytoremediation of 2,4-dichlorophenol by *Brassica napus* hairy root cultures. Biotechnol Appl Biochem 37(2):139–144
Ahemad M, Khan MS (2013) Pesticides as antagonists of rhizobia and the legume- Rhizobium symbiosis: a paradigmatic and mechanistic outlook. Biochem Mol Biol Int 1(4):63–75
Alexander M (1991) Introduction to soil microbiology. Krieger Publishing Company, Malabar
Alford ER, Pilon-Smits EAH, Paschke MW (2010) Metallophytes—a view from the rhizosphere. Plant Soil 337:33–50
Ali H, Khan E (2018) What are heavy metals? Long-standing controversy over the scientific use of the term 'heavy metals' – proposal of a comprehensive definition. Toxicol Environ Chem 100(1):6–19
Ali H, Khan E, Sajad MA (2013) Phytoremediation of heavy metals—concepts and applications. Chemosphere 91:869–881

Antoniadis V, Levizou E, Shaheen SM, Sik OY, Sebastian A, Baum C, Prasad MNV, Wenzel WW, Rinklebe J (2017) Trace elements in the soil-plant interface: Phytoavailability, translocation, and phytoremediation–a review. Earth-Sci Rev 171:621–645

Arshad M, Saleem M, Hussain S (2007) Perspectives of bacterial ACC deaminase in phytoremediation. Trends Biotechnol 25(8):356–362

Arwidsson Z, Johansson E, von Kronhelm T, Allard B, van Hees P (2010) Remediation of metal contaminated soil by organic metabolites from fungi—production of organic acids. Water Air Soil Pollut 205:215–226

Babu AG, Kim JD, Oh BT (2013) Enhancement of heavy metal phytoremediation by Alnus firma with endophytic Bacillus thuringiensis GDB-1. J Hazard Mater 250:477–483

Babu GA, Shea PJ, Sudhakar D, Jung I-B, Byung-Taek O (2015) Potential use of Pseudomonas koreensis AGB-1 in association with Miscanthus sinensis to remediate heavy metal(loid)-contaminated mining site soil. J Environ Manag 151:160–166

Backer R, Rokem JS, Ilangumaran G, Lamont J, Praslickova D, Ricci E, Subramanian S, Smith DL (2018) Plant growth-promoting rhizobacteria: context, mechanisms of action, and roadmap to commercialization of biostimulants for sustainable agriculture. Front Plant Sci 9:1473

Barzanti R, Ozino F, Bazzicalupo M, Gabbrielli R, Galardi F, Gonnelli C, Mengoni A (2007) Isolation and characterization of endophytic bacteria from the nickel hyperaccumulator plant *Alyssum bertolonii*. Microb Ecol 53(2):306–316

Bharagava RN, Chowdhary P, Saxena G (2017) Bioremediation: an ecosustainable green technology: its applications and limitations. In: Bharagava RN (ed) Environmental pollutants and their bioremediation approaches, 1st edn. CRC Press, Boca Raton, pp 1–22. https://doi.org/10.1201/9781315173351-2

Budzikiewicz H (2010) Microbial siderophores. In: Kinghorn AD, Falk H, Kobayashi J (eds) Progress in the chemistry of organic natural products. Springer, Vienna, pp 1–75

Burd GI, Dixon DG, Glick BR (1998) A plant growth-promoting bacterium that decreases nickel toxicity in seedlings. Appl Environ Microbiol 64(10):3663–3668

Burd GI, Dixon DG, Glick BR (2000) Plant growth-promoting bacteria that decrease heavy metal toxicity in plants. Can J Microbiol 46(3):237–245

Cabral L, Soares CFRS, Giachini AJ, Siqueira JO (2015) Arbuscular mycorrhizal fungi in phytoremediation of contaminated areas by trace elements: mechanisms and major benefits of their applications. World J Microbiol Biotechnol 31:1655–1664

Carrillo-Castaneda G, Munoz JJ, Peralta-Videa JR, Gomez E, Gardea- Torresdey JL (2003) Plant growth-promoting bacteria promote copper and iron translocation from root to shoot in alfalfa seedlings. J Plant Nutr 26(9):1801–1814

Chanclud E, Morel J-B (2016) Plant hormones: a fungal point of view. Mol Plant Pathol 17(8):1289–1297

Chandra R, Saxena G, Kumar V (2015) Phytoremediation of environmental pollutants: an ecosustainable green technology to environmental management. In: Chandra R (ed) Advances in biodegradation and bioremediation of industrial waste. CRC Press, Boca Raton, pp 1–30. https://doi.org/10.1201/b18218-2

Chaudhry Q, Blom-Zandstra M, Gupta SK, Joner E (2005) Utilising the synergy between plants and rhizosphere microorganisms to enhance breakdown of organic pollutants in the environment (15 pp). Environ Sci Pollut Res Int 12(1):34–48

Chen YX, Lin Q, Luo YM, He YF, Zhen SJ, Yu YL, Tian GM, Wong MH (2003) The role of citric acid on the phytoremediation of heavy metal contaminated soil. Chemosphere 50(6):807–811

Chen X, Wu C, Tang J, Hu S (2005) Arbuscular mycorrhizae enhance metal lead uptake and growth of host plants under a sand culture experiment. Chemosphere 60(5):665–671

Chen L, Luo S, Xiao X, Guo H, Chen J, Wan Y, Li B, Xu T, Xi Q, Rao C, Liu C, Zeng G (2010) Application of plant growth-promoting endophytes (PGPE) isolated from Solanum nigrum L. for phytoextraction of Cd-polluted soils. Appl Soil Ecol 46(3):383–389

Chen L, Luo S, Li X, Wan Y, Chen J, Liu C (2014) Interaction of Cd hyperaccumulator *Solanum nigrum* L. and functional endophyte *Pseudomonas* sp. Lk9 on soil heavy metals uptake. Soil Biol Biochem 68:300–308

Chirakkara RA, Cameselle C, Reddy KR (2016) Assessing the applicability of phytoremediation of soils with mixed organic and heavy metal contaminants. Rev Environ Sci Biotechnol 15(2):299–326

Coninx L, Martinova V, Rineau F (2017) Chapter Four—Mycorrhiza-assisted phytoremediation. Adv Bot Res 83:127–188

Cornejo P, Meier S, Borie G, Rillig MC, Borie F (2008) Glomalin-related soil protein in a Mediterranean ecosystem affected by a copper smelter and its contribution to Cu and Zn sequestration. Sci Total Environ 406(1–2):154–160

Davies FT, Puryear JD, Newton RJ, Egilla JN, Saraiva Grossi JA (2001) Mycorrhizal fungi enhance accumulation and tolerance of chromium in sunflower (Helianthus annuus). J Plant Physiol 158(6):777–786

Dharni S, Srivastava AK, Samad A, Patra DD (2014) Impact of plant growth promoting Pseudomonas monteilii PsF84 and Pseudomonas plecoglossicid a PsF610 on metal uptake and production of secondary metabolite (monoterpenes) by rose-scented geranium (Pelargonium graveolens cv. bourbon) grown on tannery sludge amended soil. Chemosphere 117:433–439

Dimkpa CO, Svatos A, Dąbrowska P, Schmidt A, Boland W, Kothe E (2008) Involvement of siderophores in the reduction of metal-induced inhibition of auxin synthesis in *Streptomyces* spp. Chemosphere 74(1):19–25

Dixit R, Wasiullah MD, Pandiyan K, Singh UB, Sahu A, Shukla R, Singh BP, Rai JP, Sharma PK, Lade H, Paul D (2015) Bioremediation of heavy metals from soil and aquatic environment: an overview of principles and criteria of fundamental processes. Sustainability 7:2189–2212

Egamberdieva D, Abd-Allah EF, Teixeira da Silva JA (2016) Microbially assisted phytoremediation of heavy metal–contaminated soils. In: Ahmad P (ed) Plant metal interaction. Elsevier, Amsterdam, pp 483–498

Elsgaard L, Petersen SO, Debosz K (2001) Effects and risk assessment of linear alkylbenzene sulfonates in agricultural soil. 1. Short-term effects on soil microbiology. Environ Toxicol Chem 20(8):1656–1663

Enyedy EA, Pocsi I, Farkas E (2004) Complexation of desferricoprogen with trivalent Fe, Al, Ga, In and divalent Fe, Ni, Cu, Zn metal ions: effects of the linking chain structure on the metal binding ability of hydroxamate based siderophores. J Inorg Biochem 98(11):1957–1966

Essen SA, Bylund D, Holmstrom SJM, Moberg M, Lundstrom US (2006) Quantification of hydroxamate siderophores in soil solutions of podzolic soil profiles in Sweden. Biometals 19(3):269–282

Farkas E, Batka D, Kremper G, Pocsi I (2008) Structure-based differences between the metal ion selectivity of two siderophores desferrioxamine B (DFB) and desferricoprogen (DFC): why DFC is much better Pb(II) sequestering agent than DFB? J Inorg Biochem 102(8):1654–1659

Filip Z (2002) International approach to assessing soil quality by ecologically-related biological parameters. Agric Ecosyst Environ 88(2):689–712

Fomina MA, Alexander IJ, Colpaert JV, Gadd GM (2005) Solubilization of toxic metal minerals and metal tolerance of mycorrhizal fungi. Soil Biol Biochem 37(5):851–866

Francis AJ, Dodge CJ, Gillow JB (1992) Biodegradation of metal citrate complexes and implications for toxic metal mobility. Nature 356:140–142

Gadd GM (2004) Microorganisms in toxic metal-polluted soils. In: Buscot F, Varma A (eds) Microorganisms in soils: roles in genesis and functions. Springer-Verlag, Berlin, pp 325–358

Gadd GM, Sayer JA (2000) Fungal transformations of metals and metalloids. In: Lovley DR (ed) Environmental microbe-metal interactions. American Society of Microbiology, Washington, pp 237–256

Gamalero E, Glick BR (2012) Plant growth-promoting bacteria and metal phytoremediation. In: Anjum NA, Pereira ME, Ahmad I, Duarte AC, Umar S, Khan NA (eds) Phytotechnologies. Taylor and Francis, Boca Raton, pp 359–374

Giller KE, Witter E, McGrath SP (1998) Toxicity of heavy metals to microorganisms and microbial processes in agricultural soils. Soil Biol Biochem 30(10–11):1389–1414

Gleba D, Borisjuk NV, Borisjuk LG, Kneer R, Poulev A, Skarzhinskaya M, Dushenkov S, Logendra S, Gleba YY, Raskin I (1999) Use of plant roots for phytoremediation and molecular farming. Proc Natl Acad Sci U S A 96(11):5973–5977
Glick BR (2003) Phytoremediation: synergistic use of plants and bacteria to clean up the environment. Biotechnol Adv 21(5):383–393
Glick BR (2005) Modulation of plant ethylene levels by the bacterial enzyme ACC deaminase. FEMS Microbiol Lett 251(1):1–7
Glick BR (2010) Using soil bacteria to facilitate phytoremediation. Biotechnol Adv 28(3):367–374
Glick BR (2012) Plant growth-promoting bacteria: mechanisms and applications [online]. Hindawi Publishing Corporation, Scientifica
Glick BR (2014) Bacteria with ACC deaminase can promote plant growth and help to feed the world. Microbiol Res 169(1):30–39
Gomez-Sagasti MT, Marino D (2015) PGPRs and nitrogen-fixing legumes: a perfect team for efficient Cd phytoremediation? Front Plant Sci 6:81
Gonzalez-Chavez MC, Carrillo-Gonzalez R, Wright SF, Nichols KA (2004) The role of glomalin, a protein produced by arbuscular mycorrhizal fungi, in sequestering potentially toxic elements. Environ Pollut 130(3):317–323
Hadi F, Bano A (2010) Effect of diazotrophs (Rhizobium and Azobactor) on growth of maize (Zea mays L.) and accumulation of Lead (Pb) in different plant parts. Pak J Bot 42:4363–4370
Han F, Shan X, Zhang S, Wen B, Owens G (2006) Enhanced cadmium accumulation in maize roots-the impact of organic acids. Plant Soil 289(1–2):355–368
Hao X, Taghavi S, Xie P, Orbach MJ, Alwathnani HA, Rensing C, Wei G (2013) Phytoremediation of heavy and transition metals aided by Legume-Rhizobia symbiosis. Int J Phytoremediation 16(2):179–202
Harvey PJ, Campanella BF, Castro PML, Harms H, Lichtfouse E, Schäffner AR, Smrcek S, Werck-Reichhart D (2002) Phytoremediation of polyaromatic hydrocarbons, anilines and phenols. Environ Sci Pollut Res 9(1):29–47
He CQ, Tan GE, Liang X, Du W, Chen YL, Zhi GY, Zhu Y (2010) Effect of Zn-tolerant bacterial strains on growth and Zn accumulation in Orychophragmus violaceus. Appl Soil Ecol 44(1):1–5
He L, Yang H, Yu Z, Tang J, Xu L, Chen X (2014) Arbuscular mycorrhizal fungal phylogenetic groups differ in affecting host plants along heavy metal levels. J Environ Sci 26:2034–2040
Hrynkiewicz K, Baum C (2012) The potential of rhizosphere microorganisms to promote the plant growth in disturbed soils. In: Malik A, Grohmann E (eds) Environmental protection strategies for sustainable development. Strategies for Sustainability. Springer, Dordrecht, pp 35–64
Islam F, Yasmeen T, Ali Q, Ali S, Arif MS, Hussain S, Rizvi H (2014) Influence of Pseudomonas aeruginosa as PGPR on oxidative stress tolerance in wheat under Zn stress. Ecotoxicol Environ Saf 104:285–293
Iqbal J, Ahemad M (2015) Recent advances in bacteria-assisted phytoremediation of heavy metals from contaminated soil. In: Chandra R (ed) Advances in biodegradation and bioremediation of industrial waste. CRC Press, Boca Raton, pp 401–423
Jaishankar M, Tseten T, Anbalagan N, Mathew BB, Krishnamurthy N (2014) Beeregowda, toxicity, mechanism and health effects of some heavy metals. Interdiscip Toxicol 7(2):60–72
Jia YJ, Ito H, Matsui H, Honma M (2000) 1-Aminocyclopropane-1-carboxylate (ACC) deaminase induced by ACC synthesized and accumulated in *Penicillium citrinum* intracellular spaces. Biosci Biotechnol Biochem 64(2):299–305
Jiang C-y, Sheng X-f, Qian M, Wang Q-y (2008) Isolation and characterization of a heavy metal-resistant Burkholderia sp. from heavy metal-contaminated paddy field soil and its potential in promoting plant growth and heavy metal accumulation in metal-polluted soil. Chemosphere 72(2):157–164
Jing YX, Yan JL, He HD, Yang DJ, Xiao L, Zhong T et al (2014) Characterization of bacteria in the rhizosphere soils of Polygonum pubescens and their potential in promoting growth and Cd, Pb, Zn uptake by Brassica napus. Int J Phytoremediation 16:321–333

Joner E, Briones R, Leyval C (2000) Metal-binding capacity of arbuscular mycorrhizal mycelium. Plant Soil 226:227–234

Jones DL (1998) Organic acids in the rhizosphere–a critical review. Plant Soil 205(1):25–44

Khan MS, Zaidi A, Wani PA, Oves M (2009) Role of plant growth promoting rhizobacteria in the remediation of metal contaminated soils. Environ Chem Lett 7(1):1–19

Khan A, Sharif M, Ali A, Shah SNM, Mian IA, Wahid F, Jan B, Adnan M, Nawaz S, Ali N (2014) Potential of AM fungi in phytoremediation of heavy metals and effect on yield of wheat crop. Am J Plant Sci 5:1578–1586

Kuffner M, De Maria S, Puschenreiter M, Fallmann K, Wieshammer G, Gorfer M, Strauss J, Rivelli AR, Sessitsch A (2010) Culturable bacteria from Zn- and Cd-accumulating with differential effects on plant growth and heavy metal availability. J Appl Microbiol 108(4):1471–1484

Lee JH (2013) An overview of phytoremediation as a potentially promising technology for environmental pollution control. Biotechnol Bioprocess Eng 18(3):431–439

Li WC, Ye ZH, Wong MH (2010) Metal mobilization and production of short chain organic acids by rhizosphere bacteria associated with a Cd/Zn hyperaccumulating plant *Sedum alfredii*. Plant Soil 326(1–2):453–467

Li HY, Wie DQ, Shen M, Zhou ZP (2012) Endophytes and their role in phytoremediation. Fungal Divers 54(1):11–18

Li X, Li D, Yana Z, Ao Y (2018) Biosorption and bioaccumulation characteristics of cadmium by plant growth-promoting rhizobacteria. RSC Adv 8:30902–30911

Liang X, Chi-Quan HE, Gang NI, Tang G-E, Chen X-P, Lei Y-R (2014) Growth and Cd accumulation of Orychophragmus violaceus as affected by inoculation of Cd-tolerant bacterial Strains. Pedosphere 24(3):322–329

Lodewyckx C, Taghavi S, Mergeay M, Vangronsveld J, Clijsters H, van der Lelie D (2001) The effect of recombinant heavy metal-resistant endophytic bacteria on heavy metal uptake by their host plant. Int J Phytoremediation 3(2):173–187

Long X, Chen X, Chen Y, Wong Jonathan W-C, Wei Z, Wu Q (2011) Isolation and characterization endophytic bacteria from hyperaccumulator Sedum alfredii Hance and their potential to promote phytoextraction of zinc polluted soil. World J Microbiol Biotechnol 27(5):1197–1207

Luo S, Chen L, Chen J, Xiao X, Xu T, Wan Y, Rao C, Liu C, Liu Y, Lai C, Zeng G (2011) Analysis and characterization of cultivable heavy metal-resistant bacterial endophytes isolated from Cd hyperaccumulator *Solanum nigrum* L. and their potential use for phytoremediation. Chemosphere 85:1130–1138

Ma Y, Prasad MNV, Rajkumar M, Freitas H (2011) Plant growth promoting rhizobacteria and endophytes accelerate phytoremediation of metalliferous soils. Biotechnol Adv 29:248–258

Ma Y, Rajkumar M, Freitas H (2009) Inoculation of plant growth promoting bacterium Achromobacter xylosoxidans strain Ax10 for the improvement of copper phytoextraction by Brassica juncea. J Environ Manag 90(2):831–837

Ma Y, Rajkumar M, Luo Y, Freitas H (2013) Phytoextraction of heavy metal polluted soils using Sedum plumbizincicola inoculated with metal mobilizing Phyllobacterium myrsinacearum RC6b. Chemosphere 93(7):1386–1392

Ma Y, Oliviera RS, Nai F, Rajkumar M, Luo Y, Rocha I, Freitas H (2015) The hyperaccumulator Sedum plumbizincicola harbors metal-resistant endophytic bacteria that improve its phytoextraction capacity in multi-metal contaminated soil. J Environ Manag 156:62–69

Ma Y, Rajkumar M, Zhang C, Freitas H (2016a) Inoculation of Brassica oxyrrhina with plant growth promoting bacteria for the improvement of heavy metal phytoremediation under drought conditions. J Hazard Mater 320:36–44

Ma Y, Oliveira RS, Freitas H, Zhang C (2016b) Biochemical and molecular mechanisms of plant microbe-metal interactions: relevance for phytoremediation. Front Plant Sci 7:918

Madhaiyan M, Poonguzhali S, Sa T (2007) Metal tolerating methylotrophic bacteria reduces nickel and cadmium toxicity and promotes plant growth of tomato (Lycopersicon esculentum L.). Chemosphere 69(2):220–228

Mahar A, Wang P, Ali A, Awasthi MK, Lahori AH, Wang Q, Li R, Zhang Z (2016) Challenges and opportunities in the phytoremediation of heavy metals contaminated soils: a review. Ecotoxicol Environ Saf 126:111–121

Marques APGC, Rangel AOSS, Castro PML (2009) Remediation of heavy metal contaminated soils: phytoremediation as a potentially promising clean-up technology. Crit Rev Environ Sci Technol 39:622–654

Martino E, Perotto S, Parsons R, Gadd GM (2003) Solubilization of insoluble inorganic zinc compounds by ericoid mycorrhizal fungi derived from heavy metal polluted sites. Soil Biol Biochem 35(1):133–141

Maszenan AM, Liu Y, Ng WJ (2011) Bioremediation of wastewaters with recalcitrant organic compounds and metals by aerobic granules. Biotechnol Adv 29(1):111–123

Mathew DC, Ho Y-N, Gicana RG, Mathew GM, Chien M-C, Huang C-C, Chauvat F (2015) A rhizosphere-associated symbiont, Photobacterium spp. strain MELD1, and Its targeted synergistic activity for phytoprotection against mercury. PLoS One 10(3):e0121178

Mayak S, Tirosh T, Glick BR (2004) Plant growth-promoting bacteria confer resistance in tomato plants to salt stress. Plant Physiol Biochem 42(6):565–572

Meier S, Borie F, Bolan N, Cornejo P (2012) Phytoremediation of metal-polluted soils by arbuscular mycorrhizal fungi. Crit Rev Environ Sci Technol 42(7):741–775

Mendes G, Freitas ALM, Pereira OL, da Silva IR, Vassilev NB, Costa MD (2014) Mechanisms of phosphate solubilization by fungal isolates when exposed to different P sources. Ann Microbiol 64(1):239–249

Miethke M, Marahiel MA (2007) Siderophore-based iron acquisition and pathogen control. Microbiol Mol Biol R 71(3):413–451

Mudhoo A, Sharma SK, Lin ZQ, Dhankher OP (2010) Phytoremediation of arsenic-contaminated environment an overview. In: Sharma A, Mudhoo SK (eds) Green chemistry for environmental sustainability, vol 127. Taylor and Francis Group, Boca Ratan

Mukherjee G, Saha C, Naskar N, Mukherjee A, Mukherjee A, Lahiri S, Majumder AL, Seal A (2018) An endophytic bacterial consortium modulates multiple strategies to improve arsenic phytoremediation efficacy in Solanum nigrum. Sci Rep 8(1):1–16

Mukhopadhyay S, Maiti SK (2010) Phytoremediation of metal mine waste. Appl Ecol Environ Res 8:207–222

Muthusaravanan S, Sivarajasekar N, Vivek JS, Paramasivan T, Naushad M, Prakashmaran J, Gayathri V, Al-Duaij OK (2018) Phytoremediation of heavy metals: mechanisms, methods and enhancements. Environ Chem Lett 16(4):1339–1359

Nagajyoti PC, Lee KD, Sreekanth TVM (2010a) Heavy metals, occurrence and toxicity for plants: a review. Environ Chem Lett 8:199–216

Nagajyoti PC, Sreekanth TVM, Lee KD (2010b) Heavy metals, occurrence and toxicity for plants: a review. Environ Chem Lett 8(3):199–216

Nagata S, Yamaji K, Nomura N, Ishimoto H (2015) Root endophytes enhance stress-tolerance of L. growing in a mining pond of eastern Japan. Plant Species Biol 30(2):116–125

Nautiyal CS, Bhadauria S, Kumar P, Lal H, Mondal R, Verma D (2000) Stress induced phosphate solubilization in bacteria isolated from alkaline soils. FEMS Microbiol Lett 182(2):291–296

Ouzounidou G, Moustakas M, Symeonidis L, Karataglis S (2006) Response of wheat seedlings to Ni stress: effects of supplemental calcium. Arch Environ Contam Toxicol 50(3):346–352

Ozyigit II, Dogan I (2015) Plant–microbe interactions in phytoremediation. https://doi.org/10.1016/B978-0-12-799937-1.00009-7

Panfili F, Schneider A, Vives A, Perrot F, Hubert P, Pellerin S (2009) Cadmium uptake by durum wheat in presence of citrate. Plant Soil 316(1):299–309

Panhwar QA, Jusop S, Naher UA, Othman R, Razi MI (2013) Application of potential phosphate-solubilizing bacteria and organic acids on phosphate solubilization from phosphate rock in aerobic rice. Sci World J 2013:272409

Parida BK, Chhibba IM, Nayyar VK (2003) Influence of nickel contaminated soils on fenugreek (*Trigonella corniculata* L.) growth and mineral composition. Sci Hortic-Amsterdam 98(2):113–119

Ping L, Boland W (2004) Signals from the underground: bacterial volatiles promote growth in *Arabidopsis*. Trends Plant Sci 9(6):263–266

Pluhacek T, Lemr K, Ghosh D, Milde D, Novak J, Havlicek V (2016) Characterization of microbial siderophores by mass spectrometry. Mass Spectrom Rev 35(1):35–47

Pocsi I (2011) Toxic metal/metalloid tolerance in fungi—a biotechnology-oriented approach. In: Banfalvi G (ed) Cellular effects of heavy metals. Springer, Berlin, pp 31–58

Rajkumar M, Ae N, Freitas H (2009) Endophytic bacteria and their potential to enhance heavy metal phytoextraction. Chemosphere 77(2):153–160

Rajkumar M, Ae N, Prasad MNV, Freitas H (2010) Potential of siderophore producing bacteria for improving heavy metal phytoextraction. Trends Biotechnol 28(3):142–149

Rajkumar M, Sandhya S, Prasad MNV, Freitas H (2012) Perspectives of plant associated microbes in heavy metal phytoremediation. Biotechnol Adv 30(6):1562–1574

Rawlings DE (1997) Mesophilic, autotrophic bioleaching bacteria: description, physiology and role. In: Rawlings DE (ed) Biomining: theory, microbes and industrial processes. Springer, Berlin, Heidelberg, Germany, pp 229–245

Rawlings DE, Silver S (1995) Mining with microbes. Biotech 13:773–738

Saha R, Saha N, Donofrio RS, Bestervelt LL (2013) Microbial siderophores: a mini review. J Basic Microbiol 53(4):303–317

Saravanan VS, Madhaiyan M, Thangaraju M (2007) Solubilization 1447 of zinc compounds by the diazotrophic, plant growth promoting bacterium *Gluconacetobacter diazotrophicus*. Chemosphere 66(9):1794–1798

Sarwar N, Imran M, Shaheen MR, Ishaq W, Kamran A, Matloob A, Rehim A, Hussain S (2017) Phytoremediation strategies for soils contaminated with heavy metals: modifications and future perspectives. Chemosphere 171:710–721

Saxena G, Bharagava RN (2017) Organic and inorganic pollutants in industrial wastes, their ecotoxicological effects, health hazards and bioremediation approaches. In: Bharagava RN (ed) Environmental pollutants and their bioremediation approaches, 1st edn. CRC Press, Boca Raton, pp 23–56

Schalk IJ, Hannauer M, Braud A (2011) New roles for bacterial siderophores in metal transport and tolerance. Environ Microbiol 13(11):2844–2854

Schippers A, Sand W (1999) Bacterial leaching of metal sulfides proceeds by two indirect mechanisms via thiosulfate or via polysulfides and sulfur. Appl Environ Microbiol 65(1):319–321

Sheng X-F, Xia J-J, Jiang C-Y, He L-Y, Qian M (2008) Characterization of heavy metal-resistant endophytic bacteria from rape (Brassica napus) roots and their potential in promoting the growth and lead accumulation of rape. Environ Pollut 156(3):1164–1170

Shi W, Becker J, Bischoff M, Turco RF, Konopka AE (2002) Association of microbial community composition and activity with lead, chromium, and hydrocarbon contamination. Appl Environ Microbiol 68(8):3859–3866

Shi JY, Lin HR, Yuan XF, Chen XC, Shen CF, Chen YX (2011) Enhancement of copper availability and microbial community changes in rice rhizospheres affected by sulfur. Molecules 16:1409–1417

Shin M-N, Shim J, You Y, Myung H, Bang K-S, Cho M, Kamala-Kannan S, Byung-Taek O (2012) Characterization of lead resistant endophytic Bacillus sp. MN3-4 and its potential for promoting lead accumulation in metal hyperaccumulator Alnus firma. J Hazard Mater 199-200:314–320

Sreelal G, Jayanthi R (2017) Review on phytoremediation technology for removal of soil contaminant. Indian J Sci Res 14:127–130

Srivastava S, Singh N (2014) Mitigation approach of arsenic toxicity in chickpea grown in arsenic amended soil with arsenic tolerant plant growth promoting Acinetobacter sp. Ecol Eng 70:146–153

Srivastava S, Verma PC, Chaudhry V, Namrata S, Abhilash PC, Kumar KV, Sharma N, Singh N (2013) Influence of inoculation of arsenic-resistant Staphylococcus arlettae on growth and arsenic uptake in Brassica juncea (L.) Czern. Var. R-46. J Hazard Mater 262:1039–1047

Sun L-N, Zhang Y-F, He L-Y, Chen Z-J, Wang Q-Y, Qian M, Sheng X-F (2010) Genetic diversity and characterization of heavy metal-resistant-endophytic bacteria from two copper-tolerant plant species on copper mine wasteland. Bioresour Technol 101(2):501–509

Tiwari S, Singh SN, Garg SK (2012) Stimulated phytoextraction of metals from fly ash by microbial interventions. Environ Technol 33(21):2405–2413

Ullah A, Heng S, Munis MFH, Fahad S, Yang X (2015) Phytoremediation of heavy metals assisted by plant growth promoting (PGP) bacteria: a review. Environ Exp Bot 117:28–40

van der Hiejden MGA, Bardgett RD, van Straalen NM (2008) The unseen majority: soil microbes as drivers of plant diversity and productivity in terrestrial ecosystems. Ecol Lett 11(3):296–310

Vangronsveld J, Herzig R, Weyens N, Boulet J, Adriaensen K, Ruttens A et al (2009) Phytoremediation of contaminated soils and groundwater: lessons from the field. Environ Sci Pollut Res 16:765–794

Visioli G, D'Egidio S, Vamerali T, Mattarozzi M, Sanangelantoni AM (2014) Culturable endophytic bacteria enhance Ni translocation in the hyperaccumulator Noccaea caerulescens. Chemosphere 117:538–544

Viterbo A, Landau U, Kim S, Chernin L, Chet I (2010) Characterization of ACC deaminase from the biocontrol and plant growth-promoting agent *Trichoderma asperellum* T203. FEMS Microbiol Lett 305(1):42–48

Vivas A, Biro B, Nemeth T, Barea J, Azcon R (2006) Nickel-tolerant Brevibacillus brevis and arbuscular mycorrhizal fungus can reduce metal acquisition and nickel toxicity effects in plant growing in nickel supplemented soil. Soil Biol Biochem 38(9):2694–2704

Vodnik D, Grčman H, Maček I, van Elteren JT, Kovăcevič M (2008) The contribution of glomalin related soil protein to Pb and Zn sequestration in polluted soil. Sci Total Environ 392:130–136

Wan Y, Luo S, Chen J, Xiao X, Liang C, Zeng G, Liu C, He Y (2012) Effect of endophyte-infection on growth parameters and Cd-induced phytotoxicity of Cd-hyperaccumulator Solanum nigrum L. Chemosphere 89(6):743–750

Wang W, Qiu Z, Tan H, Cao L (2014) Siderophore production by actinobacteria. Biometals 27(4):623–631

Wenzel WW (2008) Rhizosphere processes and management in plant-assisted bioremediation (phytoremediation) of soils. Plant Soil 321(1–2):385–408

Wu CH, Wood TK, Mulchandani A, Chen W (2006) Engineering Plant-Microbe symbiosis for rhizoremediation of heavy metals. Appl Environ Microbiol 72(2):1129–1134

Yadav SK (2010) Heavy metals toxicity in plants: an overview on the role of glutathione and phytochelatins in heavy metal stress tolerance of plants. S Afr J Bot 76(2):167–117

Yuan M, He H, Xiao L, Zhong T, Liu H, Li S, Deng P, Ye Z, Jing Y (2013) Enhancement of Cd phytoextraction by two Amaranthus species with endophytic Rahnella sp. JN27. Chemosphere 103:99–104

Yuan M, He H, Xiao L, Zhong T, Liu H, Li S, Deng P, Ye Z, Jing Y (2014) Enhancement of Cd phytoextraction by two Amaranthus species with endophytic Rahnella sp. JN27. Chemosphere 103:99–104

Zhang Y-f, He L-y, Chen Z-j, Wang Q-y, Qian M, Sheng X-f (2011) Characterization of ACC deaminase-producing endophytic bacteria isolated from copper-tolerant plants and their potential in promoting the growth and copper accumulation of Brassica napus. Chemosphere 83(1):57–62

Zhang X, Lin L, Zhu Z, Yang X, Wang Y, An Q (2013) Colonization and modulation of host growth and metal uptake by endophytic bacteria of Sedum alfredii. Int J Phytoremediation 15(1):51–64

Zhu L-J, Guan D-X, Luo J, Rathinasabapathi B, Ma LQ (2014) Characterization of arsenic-resistant endophytic bacteria from hyperaccumulators Pteris vittata and Pteris multifida. Chemosphere 113:9–16

Złoch M, Thiem D, Gadzała-Kopciuch R, Hrynkiewicz K (2016) Synthesis of siderophores by plant-associated metallotolerant bacteria under exposure to Cd^{2+}. Chemosphere 156:312–325
Złoch M, Kowalkowski T, Tyburski J, Hrynkiewicz K (2017) Modeling of phytoextraction efficiency of microbially stimulated *Salix dasyclados* L. Int J Phytoremediation 19(12):1150–1164

Bioprospecting and Biotechnological Applications of Microbial Endophytes

7

Sneh Sharma, Varsha Rani, Raj Saini, and Madan L. Verma

Abstract

Endophytes are a family of microbes which grow inter/intracellularly in the tissues of higher plants without causing any kind of harm to the host plant in which they reside. Endophytic microbes are representing a potential source of natural bioactive compounds which are highly useful in agriculture, medicine and industries, such as antioxidants, anticancerous agents, antidiabetic, antibiotics, biological control agents and others. A broad variety of bioactive secondary metabolites are being provided by the endophytes with unique structural properties, including steroids, phenolic acids, alkaloids, flavonoids, benzopyranones, terpenoids, quinines, xanthones, tetralones, etc. These bioactive secondary metabolites find a wider range of applications as immunosuppressants, antibiotics, agrochemicals, antioxidants, anticancerous agents and antiparasitic. Novel antimicrobial metabolite discovery from the endophytes is an alternative way to overcome the problem of drug resistance in human and plant pathogens. Novel compound production via the process of biotransformation by endophytes is an

S. Sharma
Department of Biotechnology, Dr. Y.S. Parmar University of Horticulture and Forestry, Neri, Hamirpur, Himachal Pradesh, India

V. Rani
Department of Biotechnology, Shoolini University, Solan, Himachal Pradesh, India

R. Saini
Department of Basic Sciences, Dr. Y.S. Parmar University of Horticulture and Forestry, Neri, Hamirpur, Himachal Pradesh, India

M. L. Verma (✉)
Department of Biotechnology, School of Basic Sciences, Indian Institute of Information Technology, Una, Himachal Pradesh, India
e-mail: madanverma@iiitu.ac.in

P. K. Arora (ed.), *Microbial Technology for Health and Environment*, Microorganisms for Sustainability 22,
https://doi.org/10.1007/978-981-15-2679-4_7

interesting phenomenon, providing a number of advantages over the chemical synthesis as well as enhancing the productivity of the desired products.

Endophytes have the ability to produce similar secondary bioactive metabolites as produced by their host plants, thus promoting good yield and growth, and enable host plant to tolerate the abiotic as well as biotic stress conditions and disease resistance. This field is attracting a lot of interest, and therefore it can be utilised for novel natural products in medicinal, food and agricultural industries. This chapter is dealing with the endophytic microorganisms, their applications and phytochemicals produced via endophytes.

Keywords
Microorganisms · Plants · Interactions · Biotransformation · Metabolites · Bioactive

7.1 Introduction

Plants are the major source of medicinal bioactive compounds against various forms of ailments. These products have been exploited for human use for centuries (Subbulakshmi et al. 2012). The term endophytes includes a suite of microorganisms or an endosymbiotic group of microorganisms that grow in inter- and/or intracellular locations of plants without causing symptoms of disease on the plants (Pimentel et al. 2011; Singh and Dubey 2015). They exhibit complex interactions like mutualism, antagonism and parasitism with their host plants (Nair and Padmavathy 2014). De Bary (1866) provided the first definition of an endophyte as "any organism that lives within plant tissues or these are plant-colonizing microorganisms". Endophytes provide protection and survival conditions to enhance host growth and also improve the tolerance ability against biotic and abiotic stresses and enhance the resistance to insects and pests by stimulating immune responses (Ek-Ramos et al. 2019). They produce novel bioactive compounds such as alkaloids, phenolic acids, quinines and tannins for use in industry, agriculture and medicine (Strobel 2003; Strobel and Daisy 2003).

Endophytic microorganisms are important prolific producers of natural bioactive compounds and play a significant role in the drug discovery and various developmental processes (Zhang et al. 2018). The rich diversity of endophytic microbes, their metabolite production and their adaptation to various environmental stresses is continuously explored for isolation of novel bioactive compounds to reduce agrochemical usage in food and drug production (Singh et al. 2017). The symbiotic nature of endophytic microbes indicates that bioactive compounds are less toxic to the cell and these compounds do not harm the host system and may not adversely affect human cells (Chutulo and Chalanaavar 2018).

Endophytic fungi have been reported first time in grasses (De Bary 1866) and in leaves, bark, roots and xylem of almost all plant species (Petrini 1991). Endophytes are associated with plants in various forms, including bacteria and fungi which live asymptomatically inside the plant tissues (Golinska et al. 2015). Hawksworth and Rossman in 1987 reported approximately 1 million fungal species, and only 100,000 have been described yet. More than 200 genera of bacterial species have been reported such as *Agrobacterium, Bacillus, Brevibacterium, Pseudomonas* and *Xanthomonas* (Sun et al. 2013). The presence of actinomycetes that belong to the phylum *Actinobacteria* and possess mycelium like fungus and form spores was reported by Chaudhary et al. (2013) and Barka et al. (2016). *Streptomyces* is most commonly isolated as endophytic actinomycetes and known to produce bioactive metabolites that act as antimicrobial and anticancer compounds (Berdy 2012; Golinska et al. 2015). Many important microbial metabolites such as paclitaxel extracted from *Kitasatospora* sp. associated with *Taxus baccata* and tyrosol from *Emblica officinalis* are reported to inhibit foodborne microbes (Zhao et al. 2011; Gangwar et al. 2014). Fungal endophytes produced large numbers of anticancer agents against cell lines, and it was recorded that more than 60% of the anticancer compounds are natural products or their derivatives (Rajamanikyam et al. 2017).

Endophytic mycoplasma species are also considered as plant endophytes which show endobiotic bacterial–algal interactions with *Bryopsis pennata* and *B. hypnoides*, and this suggests close association of endophytic bacterial communities with the algal host (Hollants et al. 2011). Thus, endophytic microorganisms play an important role for the production of natural bioactive compounds, with promising potential in human health and drug discovery (Lam 2007).

Endophytes represent a subset of microorganisms that are found in specific/particular nitches and proved as a reservoir of bioactive compounds, if explored properly (Kapoor et al. 2019). Now it is the time to evaluate and elucidate the significance of these microorganisms applied on biotechnological processes for the production of bioactive compounds to combat various pathogens associated with health sector and other possible medicinal uses. Their diversity and adaptability to extreme stress conditions make them excellent novel metabolite source for environment-friendly and sustainable food and drug production (Ek-Ramos et al. 2019).

This chapter discusses the interaction between the microorganism and plant system with special attention to biodiversity and phytochemistry of the endophytes. Screening, isolation and applications of endophytes at commercial level are discussed in detail.

7.2 Biodiversity and Phytochemistry of Endophytes

Microbial endophytes provide protection and adaptation to the host plant in which they are residing, under the stressed conditions by producing plethora of substances (Adhikari and Pandey 2018; Rani 2016). Medicinal plants have been one of the best, cost-effective, easily available and unique sources of drugs since ancient times. Nature is a hub of various kinds of organisms, blessing us with favourable climatic conditions and a very rich flora. There are various plant species and varieties of plants growing in the different climatic conditions in different regions of the world, producing several bioactive compounds (Alsheikh et al. 2009). Medicinal plants are highly useful for curing several health-related ailments. Bioactive molecules which are produced by medicinal plants are being used in Ayurvedic system of medicines to treat various life-threatening diseases (Alsheikh et al. 2009). But continuous use of such medicinal plants leads to reduction in the population of that particular plant and name of such plant species included in the red book (Ahmedullah and Nayar 1999). Extensive use of a particular plant can lead to its extinction, so this problem of extensive use of plants can be solved via using endophytes instead of plants. Endophytes include a suit of microorganisms growing inter−/intracellularly inside the tissues of higher plants producing bioactive natural compounds. Substances produced by the endophytes residing inside the medicinal plants are highly useful in agriculture, industry and medicine (Rani 2016). It has been proved that endophytes are rich source of natural bioactive compounds with various biological and pharmacological activities (Kogel et al. 2006). Every plant growing on Earth acts as a host for one or many endophytic microorganisms (Strobel et al. 2002). Endophytes act as a selection system for microorganisms in producing the bioactive compounds

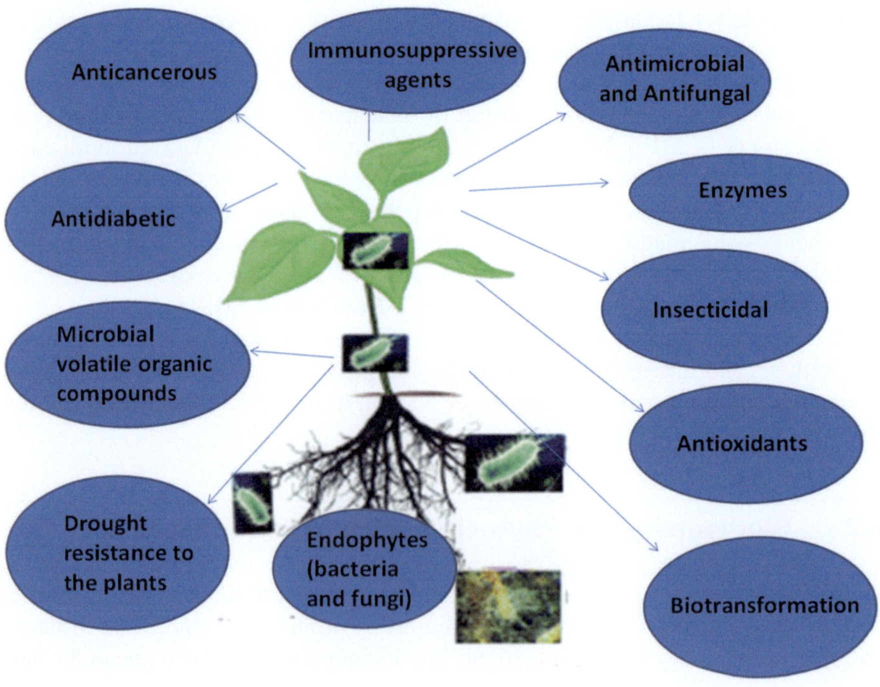

Fig. 7.1 Applications of endophytes

having lower toxicity towards higher organisms (Rani 2016). Bioactive compounds produced from endophytes are being utilised by the plants protecting themselves from human pathogens, and some of these compounds can be utilised for novel drug discovery. Natural products such as alkaloids, terpenoids, steroids, flavonoids, etc., have been reported from the endophytes. Natural bioactive compounds produced from endophytes are antimicrobial, antifungal, antibiotics, insecticidal, antioxidants, antidiabetic, anticancerous, immunosuppressive agents, etc. (Ezra et al. 2004; Yu et al. 2010; Huang et al. 2007; Berkodia et al. 2018). Enzymes, produced by the endophytes, provide resistance to the plant against the pathogens, and endophytes trigger the plant growth hormones, thus playing an important role in plant growth and development. Endophytes enable the plants to tolerate the salt stress and drought-like conditions. The eco-friendly nature of the endophytes makes them suitable to be used in pharmaceutical industries as well as for the sustainable agriculture production, as shown in Fig. 7.1 (Berkodia et al. 2018).

Crude extracts of *Alternaria alternata* have been reported to contain several bioactive compounds, responsible for the antimicrobial activity, and can be utilised for controlling the antimicrobial infections (Jagadevi and Vidyasagar 2018). The medicinal plant *Achillea millefolium* found in the Western Himalayas had been explored for the presence of endophytic fungi *Aspergillus niger*, *Aspergillus terreus* and *Aspergillus flavus*, respectively. These strains of genus *Aspergillus* showed potential antioxidant potential and some phytochemical constituents (Satari et al. 2018). *Pinus roxburghii* which is commonly a timber plant having medicinal properties has been explored for the endophytes, and endophytes present in *Pinus roxburghii* showed the presence of various bioactive secondary metabolites (Sharma and Baunthiyal 2018). Several novel antibiotics have been reported from the endophytes (Zou et al. 2000; Shiono et al. 2005; Gu et al. 2007; Losgen et al. 2008). Anticancer chemicals, such as Hsp90 inhibitors (Turbyville et al. 2006), camptothecin (Amna et al. 2006), paclitaxel (Stierle et al. 1993) and sequoiatones A and B (Stierle et al. 1999), have been reported from the endophytes.

Endophytes residing in the plant tissues influence the volatile compounds production by the host plant (Mucciarelli et al. 2007). *Acremonium strictum*, a root endophyte, has been found to modify the volatile profile of the host plant and thus affects the selection of host as well as the oviposition behaviour of polygamous moth (Jallow et al. 2008). Endophyte, *Xylaria* sp., has been found to produce lactones which have the potential to function as an antimicrobial drug (Jimenez-Romero et al. 2008). Endophytes can also perform stereoselective biotransformation of certain chemicals in spite of producing different bioactive compounds, thus aiding in the process of drug modification (Borges et al. 2007). Current agricultural practices are using a wide range of bactericides, fungicides and pesticides globally which is affecting our health adversely. So, there is need to reduce the use of these harmful chemicals. Volatile organic compounds from microorganisms, like fungi and bacteria, respectively, are very dynamic and complex and are able to modulate the physiology of microbes and plants by regulating the genomic, metabolomic and proteomic status. So microbial volatile compounds can be utilised as a cost-effective, eco-friendly and sustainable strategy in the agricultural practices. Microbial volatile organic compounds can play a very important role in the signalling process. Important signalling roles for the ecological interactions between bacteria and plants, fungi and plants, plant and plants and arthropods and plants, respectively, can be performed by microbial volatile compounds (Kanchiswam et al. 2015). Metabolism of the host plant can be affected by endophytes, as they are expected to improve the quality of the crop especially those crops which have organoleptic products such as wine grape. Fruit flesh cells of grape were treated with endophytic fungal strains, and metabolites were analysed by using high-pressure liquid chromatography. Quantities of detected metabolites after treating with the endophytic fungal strains were found to be increased from 6 to 17, and 1 to 11 novel metabolites have been introduced into the grape wine metabolome. Thus, it can be concluded from this study that introduction of fungal endophytes into grape wine can improve the grape quality and its characteristics and novel metabolites can also be introduced in the grape wine (Huang et al. 2018).

7.3 Screening and Isolation Techniques of Endophytes

There is rich microbial diversity within the intracellular and intercellular spaces of plant tissues which not only enhances the ability of plants to cope with various types of abiotic and biotic stresses but also constitutes an important source of compounds of biotechnological relevance (Gouda et al. 2016). Isolation of endophytic microorganisms from plants is an important step to explore their prospects as sources of enzymes, secondary metabolites and bioactive molecules of industrial and medical importance and for applications as bio-inoculants in agriculture sector (Abdalla and McGaw 2018). Various culture-based techniques are generally used for isolation of microorganisms from endophytic plants, whereas culture-independent techniques allow studying the diversity of both culturable and non-culturable endophytic communities and the variations that occur due to various environmental fluctuations (Chen et al. 2019b). The various factors regulating the isolation of endophytes depend on the plant itself, environmental conditions, agricultural practices, the isolation techniques and also the microbes being isolated (Fuchs et al. 2017; Golinska et al. 2015; Verma et al. 2014).

7.4 Selection of Plant and Collection of Plant Materials

Plant selection for isolation of endophytes should be based on their economic importance. Healthy living plants are selected taking into consideration their location and age. As endophytes show some degree of tissue preference (Manjunatha et al. 2019), different plant tissues should be screened for more number of endophytic strains. Old and clean leaves from branches reported to harbour more endophytic diversity (Praptiwi et al. 2018), including branches and side roots, may be cut and collected for woody plants, and whole plant may be uprooted for herbaceous plants.

Plant tissues such as nodules, leaves, twigs, stem, roots, rhizomes, fruits and seeds are home to different microbial endophytes (Gupta et al. 2019; Kandel et al. 2017) and may be used for their isolation. The nodules of *Mimosa pudica* were used to isolate non-rhizobial endophytes with plant growth-promoting traits (Sánchez-Cruz et al. 2019). Investigators prefer unique locations and strategies for selection of plants for isolation of endophyte (Abdalla and McGaw 2018) like plants occurring in areas with a great natural biodiversity, plants growing in saline and acidic soils (Szymańska et al. 2018; Postma et al. 2007), plants growing in unique environments such as hot and cold areas (Li et al. 2012; Massimo et al. 2015; Hassan et al. 2018), traditional medicinal plants (Jia et al. 2016) and plants existing in particular geographic region, i.e. endemic plants (Ferreira et al. 2017).

7.5 Pretreatment of Plant Tissue Sample for Isolation of Endophytes

The freshly collected plant samples are processed without delay or can be stored at 4 °C for isolation within 24–48 h of collection or preserved at −80 °C if the samples are to be processed after 1 or 2 weeks to 1 month. The collected plant samples need to be washed thoroughly with tap water to remove adhering soil particles and debris (Hassan et al. 2018). After washing the freshly collected samples with the running tap, Potshangbam et al. (2017) also soaked the samples for 10 min in distilled water containing a few drops of Tween 80.

7.5.1 Surface Sterilisation of Plant Tissues

Besides being a time-consuming process, the surface sterilisation of host plant tissues is essential to study endophytes. Prior to the isolation of endophytes, this crucial step ensures the removal of epiphytic microbes from plant samples. In general, optimisation of the sterilisation procedure should be carried out for different plant tissues, especially time given for sterilisation, since the sensitivity to the sterilants may vary with the age, species and part of the host plant (Qin et al. 2011). Solutions of various sterilising agents including mercuric chloride, ethanol, sodium hypochlorite, hydrogen peroxide, sodium thiosulphate, hydrochloric acid, etc., in different concentrations, have been used for surface sterilisation, in a sequential manner for the purpose of surface sterilisation of plant samples; the time period given for treatment may vary depending on the tissue being treated (Ramalakshmi et al. 2018; Mufti et al. 2015). The rinsing with sterile distilled water is given in between the sequence and at the end of the treatment for complete removal of the sterilants. Variations in the selection of sterilants can be found in different studies related to the isolation of endophytes from different plants (Waheeda and Shyam 2017).

Leaves were reported to be treated for surface sterilisation by the following solution sequence—distilled water for 1 min, ethanol 70% for 1 min, sodium hypochlorite 2.5–3% for 3–4 min and ethanol 70% for 30 s—and finally washed in sterile distilled water for 5–10 times (Hassan et al. 2018; da Silva Ribeiro et al. 2018).

For surface sterilisation, root tissues were treated with 70% ethanol (1–3 min) and 5% aqueous solution of sodium hypochlorite (1 min), followed by 70% ethanol (2 min) and then with 0.1% mercury chloride (1 min), and rinsed with sterile distilled water (Strobel et al. 2002). *T. wallichiana* roots were immersed in 99% ethanol for 1 min, followed by 5% sodium hypochlorite for 5 min, and then washed 4–5 times in sterilised water (Adhikari and Pandey 2019). The treatment procedure for peanut roots includes using 75% ethanol for 3 min, sodium hypochlorite 3% for 6 min and ethanol 75% for 30 s and then rinsing six times with water (Chen et al. 2019a; Gao et al. 2017).

The leaves, stems and roots of *Oryza sativa* L. and *Zea mays*, after cutting into small segments and washing twice with sterile distilled water, were immersed in ethanol 80% for different time periods of 1 min, 2 min and 3 min, respectively. This was followed by treatment with NaOCl 4% and alcohol 70% for 1 min giving rinsing in between and at the end with sterile distilled water (Potshangbam et al. 2017). Ullah et al. (2018) conducted surface disinfection of plant tissues of *Withania coagulans* and *Olea ferruginea* with 70% ethanol for 5 min and 0.1% $HgCl_2$ for 1–2 min, whereas Sánchez-Cruz et al. (2019) used 70% ethanol and 2% sodium hypochlorite for 10 and 20 min, respectively, for surface sterilisation of nodules of *Mimosa pudica*. The surface sterilisation of stems of *Bixa orellana* L. plants was conducted by immersing stem segments of 1 cm firstly in 75% ethanol for 1 min, followed by 4% sodium hypochlorite for 2 min and then again in 75% ethanol for 30 s.

7.5.2 Checking the Effectiveness of Surface Sterilisation

A sterility check measure of the plant samples ensures the effectiveness of the surface sterilisation procedure. For this purpose, an aliquot from the final rinse water is plated onto the media specific for the growth of bacteria, actinomycetes and fungi and incubated; absence of any microbial growth indicates the effectiveness of the protocol and confirms that the subsequent isolates will be true endophytes (Hassan et al. 2018; da Silva Ribeiro et al. 2018).

The validation of surface sterilisation procedure can also be done by imprint method where a surface-sterilised plant sample imprint is made by gently pressing the tissue on the media and incubating to check for removal of microbial epiphytes (Greenfield et al. 2015).

7.6 Isolation of Endophytes

The growth of endophytes under laboratory conditions is highly dependent on the media composition and incubation conditions. Endophytic microorganisms include prokaryotes like bacteria, archaea, actinomycetes and eukaryotic microorganisms like fungi (Compant et al. 2019). The right selection of culture media makes it possible to isolate the maximum diversity of culturable endophytes. Different culture media, often selective, are initially prepared for isolation of a particular group of endophytes. Addition of a certain amount of plant extracts into the isolation medium has also been found effective as reported by Qin et al. (2011).

7.6.1 Media for Isolation of Endophytic Bacteria

The commonly used media for isolation of bacterial endophytes comprise of the following complex media: Luria–Bertani agar, nutrient agar, 869 medium, trypticase soy agar (TSA), casein–starch medium, peptone yeast extract agar and minimal media: 284+ C medium and M1 to M10 minimal media (Sánchez-Cruz et al. 2019; Liu et al. 2017; Eevers et al. 2015).

7.6.2 Media for Isolation of Endophytic Actinobacteria

Different media have been assessed by researchers for selective isolation of actinomycetes such as Gause's synthetic agar, tap water-yeast extract (TWYE) agar, starch–casein agar, sodium propionate agar, International Streptomyces Project (ISP) medium 5 agar, trehalose–proline agar, MOPS–amino acid agar, CMC agar, humic acid vitamin agar, glycerol–asparagine agar and chitin medium (Hassan et al. 2018; Chen et al. 2019b; Singh et al. 2018). The starch–casein and sodium propionate media cultures have been reported to yield maximum number and highest diversity of endophytic actinobacteria, followed by ISP 5 medium (Chen et al. 2019b). Various inhibitors including cycloheximide, nalidixic acid, nystatin and $K_2Cr_2O_7$ can be used for selective isolation of endophytic actinomycete at the rate of 50 mg/L, 25–50 mg/L, 50 mg/L and 25 mg/L, singly or in combination in the media (Chen et al. 2019b).

7.6.3 Media for Isolation of Endophytic Fungi

Investigators have tested various media for isolation of endophytic fungi from plants including potato dextrose agar, water agar, Sabouraud dextrose agar with chloramphenicol, corn meal agar, malt extract agar, Czapek–Dox agar and oatmeal agar with or without inhibitors such as penicillin G, streptomycin sulphate, tetracycline and ampicillin for suppressing the growth of bacteria (Adhikari and Pandey 2019; Potshangbam et al. 2017; Garcia et al. 2012).

After preparation of the media, the surface-sterilised plant samples are dissected or ground into small pieces or pulverised aseptically. Chen et al. (2019a) used phosphate-buffered saline (PBS) for mashing surface-sterilised roots. In another study (da Silva Ribeiro et al. 2018), surface-sterilised leaf fragments of *Pachystachys lutea* were crushed in 1 mL of an aqueous solution of 0.01% Tween 80. The isolation can be carried in different ways (Zhao et al. 2011; Potshangbam et al. 2017). Small pieces (0.1 g) of sample are homogeneously dispersed on the freshly prepared isolation medium in Petri dishes which are then incubated at 28 °C (for bacteria) and 25 °C (for fungi). In another method, tenfold serial dilutions of the ground sample (1 g) in 9 mL sterile water are prepared up to 10^{-3} thorough stirring. Aliquots of 0.1 or 1.0 mL are plated into Petri plates containing the isolation media and then incubated. The growth of bacterial and fungal colonies is monitored during the incubation period up to 1–2 weeks. Isolation of endophytic actinomycetes may require incubation at 28 °C for 2–8 weeks (Chen et al. 2019b). The next way is to prepare thin slices of plant samples after removing the outer cover carefully and thereafter make direct impression of these sterilised tissues on the plates containing the fungal isolation media and incubate until the growth is observed (Hassan et al. 2018) plated surface-sterilised leaves of *Oxalis corniculata* L., cut into small pieces (0.5–1.0 cm) onto starch nitrate agar media supplemented with fungal inhibitor (nystatin, 25 μg/mL) for isolation of actinomycetes.

7.7 Screening of Endophytes for Different Biological Activities of Biotechnological Importance

Screening of endophytes is of major significance as it enables selection of microbes with desired trait(s) which can ultimately be used at the production level in industries. Table 7.1 gives an overview of the various biological activities of biotechnological relevance screened in the endophytic microorganisms by different investigators. Here follows some plate-based/qualitative screening of microbial activities of agricultural and industrial importance.

7.7.1 Antagonistic Activity

Antagonistic activity of the endophytes against plant pathogenic bacteria and fungi highlights their potential for biological control and indicates their contribution to their ecological adaptation within the host plant tissues.

7.7.1.1 Antibacterial Activity

Antibacterial activity of a bacterial strain can be tested by cup-plate assay/cylinder-plate assay. Chen et al. (2019a, b) tested endophytic bacterial strain isolated from peanut roots against peanut pathogenic bacteria by this assay. In this method, 50 mL of nutrient broth inoculated with pathogenic bacterium after 24 h incubation at 30 °C was added to cooled, melted nutrient agar, mixed rapidly and poured into

Table 7.1 Some common surface sterilant and media used for isolation of bacterial and fungal endophytes including the activities screened

Plant tissue sterilised/ plant used	Surface sterilants used (in given sequence)	Media for isolation of endophytic bacteria, actinomycetes and fungi	Bacteria and fungi isolated	Activities screened	References
Nodules (*Mimosa pudica*)	70% ethanol (10 min), 2% sodium hypochlorite (20 min)	Peptone yeast extract agar	*Enterobacter* sp. *Serratia* sp.	P-solubilisation, siderophore production, auxin, cellulose, chitinase degradation, antifungal activity	Sánchez-Cruz et al. (2019)
Leaves (*Sapindus saponaria* L.)	NaOCl at 5% (3 min)	Potato dextrose agar (PDA) with 50 μg/ mL of tetracycline	*Cochliobolus, Alternaria, Curvularia, Phomopsis, Diaporthe, Phoma*	Antagonistic activity, amylase, cellulose, pectinase	Santos et al. (2019); Garcia et al. (2012)
Roots, stem and leaves *(Oxalis corniculata)*	3% (v/v) H_2O_2 for 4 min, 0.1% $HgCl_2$ for 5 min	Dobereiner nitrogen (DN)-free semisolid media	*Agrobacterium tumefaciens, Pantoea agglomerans, Kocuria rhizophila and Bacillus pumilus* *Kocuria rhizophila*	P solubilisation, IAA, ammonia, HCN, heavy metal tolerance Salt tolerance assay, IAA, P solubilisation, catalase, siderophore, HCN, ACC deaminase	Mufti et al. (2015) Afridi et al. (2019)
Leaves, stems and roots (*Dysphania ambrosioides*)	0.5% sodium hypochlorite (2 min), 70% ethanol (2 min)	PDA with 0.5 g/L streptomycin sulphate	237 isolates; dominant genera: *Plectosphaerella, Cladosporium, Verticillium* *Phoma, Peyronellaea, Alternaria, Penicillium* *Phomopsis columnaris*	Heavy metal (HM) tolerance, determination of minimum inhibitory concentration (MIC) of HM	Li et al. (2016)

(continued)

Table 7.1 (continued)

Plant tissue sterilised/plant used	Surface sterilants used (in given sequence)	Media for isolation of endophytic bacteria, actinomycetes and fungi	Bacteria and fungi isolated	Activities screened	References
Leaves, stem and root (*Ocimum basilicum, Withania somnifera and Rauvolfia tetraphylla*)	Method I: 70% ethanol for 5 min. Method II: solution of cycloheximide, 50 μg/mL (4 h), 3.15% of sodium hypochlorite (15 min) Method III: 70% ethanol (5 min), 0.9% NaOCl (20 min), 10% sodium hydrogen carbonate (10 min) Method IV: 3.15% calcium hypochlorite (10 min), 10% sodium hydrogen carbonate (15 min), 1% sodium azide (2 min)	Starch casein agar, yeast extract-malt extract agar Actinomycete isolation agar, humic acid vitamin B agar with cycloheximide 50 μg/mL and nystatin 50 μg/mL; ISP-4; modified ISP-4 with ZnSO4, FeSO4 and MnCl2	32 isolates *Streptomyces* sp. *Streptomyces flavoviridis*	Antibacterial activity	Waheeda and Shyam (2017)
Roots, stems, leaves and seeds (*Panax notoginseng*)	75% ethanol solution for 2–3 min, 5.5% sodium hypochlorite solution for 1–2 min	Potato dextrose agar with 100 mg/L ampicillin	*Cladosporium, Phialophora, Phomopsis, Penicillium, Aspergillus, Colletotrichum, Botryotinia, Acremonium, Fusarium, Ilyonectria, Myrothecium, Plectosphaerella, Trichoderma, Periconia, Alternaria*	In vitro antagonistic activity	Zheng et al. (2017)

Plant tissue sterilised/ plant used	Surface sterilants used (in given sequence)	Media for isolation of endophytic bacteria, actinomycetes and fungi	Bacteria and fungi isolated	Activities screened	References
Leaves (*Oxalis corniculata* L.)	Distilled water (1 min), ethanol 70% (1 min) sodium hypochlorite 2.5% (3 min), ethanol 70% for 30 s	Starch nitrate agar with fungal inhibitor (nystatin, 25 μg/mL)	*Streptomyces zaomyceticus, S. pseudogriseolus*	Biosynthesis of copper oxide nanoparticles, antimicrobial activity of biosynthesised CuO-NPs (MIC), antifungal assay of CuO-NPs against phytopathogenic fungi; antioxidant activity for biosynthesised CuO-NPs; larvicidal activity	Hassan et al. (2018)
Roots (*Arachis hypogaea L.*)	75% ethanol for 3 min, 3% sodium hypochlorite for 6 min and 75% ethanol for 30 s	Nutrient Agar	*Bacillus velezensis*	Antifungal activity Antibacterial activity	Chen et al. (2019a)
Root (*Taxus wallichiana* Zucc.)	99% ethanol (1 min), 5% solution of sodium hypochlorite (5 min)	Potato dextrose agar	*Penicillium Aspergillus*	P-solubilisation, phytases, organic acid production, alkaline and acid phosphatases	Adhikari and Pandey (2019)
Leaves (*Pachystachys lutea*)	70% ethanol for 1 min, in 3% sodium hypochlorite for 4 min and in 70% ethanol again for 30 s	PDA supplemented with tetracycline	*Diaporthe, Alternaria* sp., *Phyllosticta capitalensis, Xylaria berteroi, Nemania* sp. and *Colletotrichum fructicola*	In vitro antagonist activity, cellulase, endoglucanase activity	da Silva Ribeiro et al. (2018)

(continued)

Table 7.1 (continued)

Plant tissue sterilised/ plant used	Surface sterilants used (in given sequence)	Media for isolation of endophytic bacteria, actinomycetes and fungi	Bacteria and fungi isolated	Activities screened	References
Roots, stems and leaves of soybean	75% ethanol for 1 min, sodium hypochlorite (2.5% HCl) for 4 min, ethanol for 30 s	Nutrient agar (NA) amended with 20% glycerol and trypticase soy agar	223 isolates *Enterobacter, Burkholderia, Kosakonia, Agrobacterium, Rhizobium, Pantoea, Variovorax, Serratia, Bacillus*	In vitro antagonistic activity Antifungal activity Antibacterial activity	
Leaf and stem (*Pistacia atlantica* subsp. *kurdica*)	5% sodium hypochlorite for 5 min	Nutrient agar, nutrient agar plus 5% sucrose, King B agar, LB agar media	*Pseudomonas*, *Stenotrophomonas*, *Serratia*, *Pantoea* and *Bacillus*	Biocontrol activity Solubilisation of phosphate siderophore production ability to fix nitrogen on nitrogen-free NFb medium and HCN production Protease production IAA production Production of gibberellic acid	
Root, leaf and seed (*Paullinia cupana*)	70% ethanol for 30 s and 2.5% NaOCl for 8 min (roots), 5 min (leaves) or 20 min (seeds)	YS, MYGP, TSA and ACA culture media; MYGP and ACA supplemented with chloramphenicol (10 mg/L) and nistatin (100 mg/L), respectively	*Microbacterium* *Burkholderia* *Bacillus* *Pseudomonas* *Pantoea* *Burkholderia nodosa and Leifsonia* sp.	IAA, phosphate solubilisation, enzyme cellulase, protease, esterase and amylase secretion of siderophore production of hydrocyanic acid and ammonia, antibiosis against the phytopathogens Biological nitrogen fixation ability	Liotti et al. (2018)

Plant tissue sterilised/plant used	Surface sterilants used (in given sequence)	Media for isolation of endophytic bacteria, actinomycetes and fungi	Bacteria and fungi isolated	Activities screened	References
Leaves, stems and roots (wheat and maize)	80% ethanol for 1 min (leaf), 2 min (stem) and 3 min (root); 4% sodium hypochlorite, 70% alcohol for 1 min	Potato dextrose agar, Sabouraud dextrose agar with chloramphenicol, corn meal agar, malt extract agar, Czapek–Dox agar, yeast extract mannitol agar, oatmeal agar	*Fusarium, Aspergillus, Penicillium, Acremonium, Gibberella*	Stress tolerance activity, in vitro antagonism assay, IAA production, phosphate solubilisation and protease production assay, chitinase, b-1, 3-glucanase, siderophore, cellulase and amylase, HCN, root colonisation assay	Potshangbam et al. (2017)
Roots (*Simmondsia chinensis*)	Sodium hypochlorite (30%) for 30 min	Nutrient agar for total bacteria Czapek medium for actinomycetes	Total 101 isolates (24 actinomycetes) *Bacillus* sp., *Methylobacterium aminovorans*, *Rhodococcus pyridinivorans*, *Streptomyces* sp. and *Oceanobacillus kimchii*	Phosphorus-solubilising activity, nitrogen fixation, indole acetic acid, ACC deaminase, siderophore, antifungal activity	

(continued)

Table 7.1 (continued)

Plant tissue sterilised/plant used	Surface sterilants used (in given sequence)	Media for isolation of endophytic bacteria, actinomycetes and fungi	Bacteria and fungi isolated	Activities screened	References
Stems and roots *(Ferula sinkiangensis)*	1-min wash in 75% ethanol, 8-min wash in 5% NaOCl	M1 to M8 and M10	125 isolates *Bacillus*, *Kocuria*, *Nocardiopsis*, *Streptomyces*, *Sphingomonas* and *Williamsia, Acinetobacter*, *Arthrobacter*, *Bacillus*, *Brevibacterium*, *Brevundimonas*, *Microbacterium*, *Micrococcus*, *Nocardiopsis, Pseudonocardia, Porphyrobacter*, *Paracoccus* and *Amycolatopsis*	Phosphate solubilisation, nitrogen fixation activity, IAA synthesis, siderophore production, antagonism assay	Liu et al. (2017)
Leaves (*Teucrium polium*)	Ethanol 70% for 1 min, sodium hypochlorite 2.5% for 5 min, ethanol 70% for 30 s	Luria–Bertani and nutrient agar media; PDA and CD media	12 isolates (7 bacterial and 5 fungal) *Bacillus cereus, Bacillus subtilis, Penicillium chrysogenum, Penicillium crustosum*	P-solubilisation, ammonia production, amylase, cellulose, protealase, pectinase and xylanase, antimicrobial activity, IAA production	Hassan et al. (2018)
Roots and shoots (*Lavandula dentata*)	75% (*v*/*v*) ethanol for 2 min and in 25% (*v*/*v*) commercial bleach for 2 min	TSA and plate count agar	*Pseudomonas, Variovorax, Rhizobium, Caulobacter, Bacillus and Paenibacillus, Pseudomonas and Xanthomonas*	Ammonia, siderophores, indol-3-acetic acid and hydrogen cyanide and phosphate solubilisation, plant cell wall-degrading enzymes (lipases, cellulases, proteases, pectinases)	Pereira et al. (2016)

Plant tissue sterilised/ plant used	Surface sterilants used (in given sequence)	Media for isolation of endophytic bacteria, actinomycetes and fungi	Bacteria and fungi isolated	Activities screened	References
Young leaves, the main stems and the taproots (*Brassica napus*)	70% ethanol (v/v) for 1 min, then in 5% sodium hypochlorite (v/v) for 5 min, again in a 70% ethanol for 30 s	Potato dextrose agar	97 endophytic fungal isolates *Alternaria alternata* *Rhizopus oryzae, C. globosum, Chaetomium* sp., *Clonostachys rosea, F. oxysporum, Fusarium proliferatum, Periconia* sp., *A. mali, B. cinerea, Leptosphaeria biglobosa, Guignardia vaccinii, Simplicillium lamellicola, Rhizoctonia solani and Sporobolomyces* sp.	Antifungal activity Antifungal volatile organic compounds	Zhang et al. (2014)

Petri dishes. After the agar cools down, a sterilised cylinder is placed at the centre of Petri dish, and then 200 μL broth (10^8 cfu/mL) of the test strain is added into the cylinder. Sterile water served as control. The plates were then incubated and the presence of inhibition zone around the cylinder indicated antibacterial activity, and the diameter of inhibition zone was measured. Antimicrobial activity of extracts of the endophytic actinomycetes grown in broth was evaluated by disc diffusion method by Waheeda and Shyam (2017) where filter paper discs impregnated with an ethyl acetate extracts were used.

7.7.1.2 Antifungal Activity

Dual culture plate assay can be used to check the antifungal activity of the endophytes (Sánchez-Cruz et al. 2019). The plate coculture assay was carried out to evaluate the antifungal activity of a bacterial strain against various peanut pathogenic fungi including *Aspergillus tenuissima, A. flavus, A. niger, Fusarium oxysporum, F. moniliforme, Rhizoctonia solani, Rhizopus* sp. *and Ralstonia solanacearum.* In this method, 6 mm mycelial bits of 4–6-day-old culture of each pathogenic fungus were placed at the centre of potato dextrose agar plates, and the endophytic bacterium was inoculated on the same plate about 25–30 mm apart from the centre and kept for incubation until the control plate containing only fungus had grown to the whole plate. A zone of inhibition between the endophytic bacterium and the test fungus indicated antifungal activity, the width of which was measured from the edges of both the fungal and bacterial colonies. The ratio of inhibition was determined according to Fokkema which is equal to C−T/C where C refers to diameter of the pathogenic fungus in the control plate and T is the diameter of fungus on the plate, where both the endophyte and the pathogen were inoculated. The inhibition ratio can be used to calculate the inhibition percentage (Chen et al. 2019a, b; Potshangbam et al. 2017; Hassan et al. 2018).

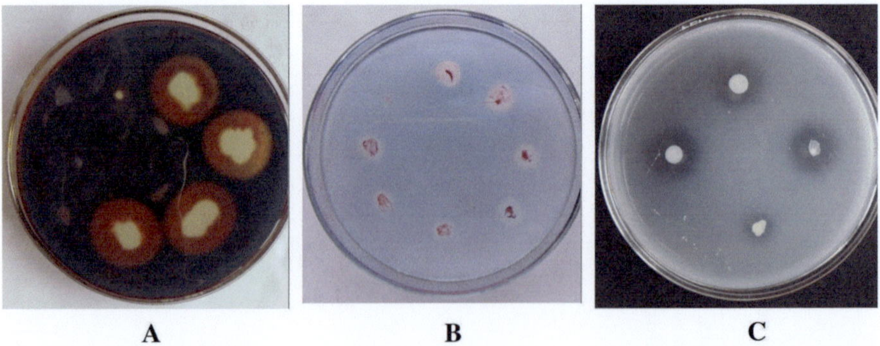

Fig. 7.2 Qualitative screening of bacterial isolates for amylase, siderophore production and phosphate solubilisation. (**a**) Amylase activity on starch agar. (**b**) Siderophore production on CAS medium. (**c**) Phosphate solubilisation on Pikovskaya's agar

7.7.1.3 Chitinase Production

Since chitin is a major component of fungal cell walls and biocontrol of phytopathogens by fungal cell wall-degrading enzymes depends on chitinase production by the endophytic isolates (Golinska et al. 2015). Basal chitinase detection medium comprising of colloidal chitin, $MgSO_4{\bullet}7H_2O$, $(NH_4)_2SO_4$, KH_2PO_4, citric acid monohydrate, Tween 80 and agar along with pH indicator bromocresol purple is used for this assay; pH of the medium is adjusted to 4.7 (Agrawal and Kotasthane 2012). Plates are inoculated with fresh cultures of fungal/bacterial endophytes and incubated at 28 °C for 5 days. The presence of colour change from yellow to purple colour around the colony indicates positive chitinase activity.

7.7.1.4 Siderophore Production (Qualitative Detection)

Endophytic microorganisms synthesise siderophores which have high affinity for ferric iron (Sánchez-Cruz et al. 2019; Liaqat and Eltem 2016). The different types of siderophores can enhance plant growth and yield by increasing the uptake of iron. They act as a potential biocontrol agent against phytopathogens and thus can act as substitute for hazardous pesticides (Saha et al. 2016). The fresh cultures of endophytic bacterial isolates are spot inoculated on chrome azurol sulphonate (CAS) agar medium, and in case of fungi, mycelia disc of one isolate per plate is placed and incubated at 28 °C for 7–10 days for detection of siderophore production as described by Schwyn and Neilands (1987). The orange halos around the bacterial colonies or fungal colonies indicating the production of siderophores are measured (Fig. 7.2b). Zone of siderophore production is determined by subtracting the diameter of bacterial/fungal colony from the diameter of total zone inclusive of the colony.

7.7.1.5 Tolerance of Endophytes to Heavy Metals

Endophytes also occur widely in heavy metal-contaminated environments. Some endophytes reduce phytotoxicity, improve plant growth and enhance the phytoremediation effectiveness of host plants which can be source of bioremediation of heavy metals (Shen et al. 2013).

Li et al. (2016) screened fresh cultures of 62 fungal endophytes by placing three fungal discs, diameter 4.4 mm, of each isolate on fresh PDA plate containing different heavy metals, such as lead, zinc and cadmium as $Pb(NO_3)_2$, $ZnSO_4{\cdot}7H_2O$ or $CdSO_4{\cdot}8H_2O$. The concentration of heavy metals in the medium was 9.66 mmol/L (Pb), 46.15 mmol/L (Zn) and Cd 1.00 mmol/L (Cd). PDA without heavy metal served as control. The plates were incubated at 25 °C, and diameter of the fungal colony was measured every other day up to 6 days to assess the effect of heavy metals on microbial growth.

7.7.1.6 Indole Acetic Acid Production (Qualitative Assay)

Endophytes can produce phytohormone indole acetic acid (IAA) to promote plant growth (Ali et al. 2019). Luria–Bertani broth incorporated with L-tryptophan (5.0 mM), sodium dodecyl sulphate (0.05%) and glycerol (1.0%) can be used for screening potential bacterial/fungal isolates for qualitative IAA production

(Potshangbam et al. 2017). They overlaid sterile nitrocellulose membrane discs of 8.2 cm on the above-mentioned medium that was inoculated with the endophyte and incubated inversely at 28 °C until the desired growth was observed; the membrane disc was taken out and treated with Salkowski reagent. There was a characteristic red halo zone within the nitrocellulose membrane immediately surrounding the colony which indicated IAA production by endophytic cultures. Production of IAA can be determined quantitatively by colorimetric estimation at 530 nm using Salkowski reagent (Liaqat and Eltem 2016).

7.7.1.7 Phosphate Solubilisation and Phytase Activity (Plate Assay)

Inorganic phosphate solubilisation through microorganisms is one of the main mechanisms involved in the availability of soluble phosphorus to plants and hence promotes their growth. The endophytic isolates are screened for phosphate solubilisation on Pikovskaya's agar with tricalcium phosphate as the inorganic form of phosphate which has a milky white opaque appearance in solidified state (Adhikari and Pandey 2019; Ali et al. 2018). The spot inoculation of each culture on Pikovskaya's agar plates is carried out and incubation is done at 28 °C for 5 days. The appearance of a clear zone around the colonies is an indicator of phosphate solubilisation by the endophytes (Fig. 7.2c). The solubilisation zone is determined by subtracting the diameter of bacterial colony from the diameter of total zone.

To release inorganic phosphate from various phosphate-containing organic compounds, the enzyme phytase plays a key role as it hydrolyses phytate. Production of phytases by the endophytic isolates can be determined by the method of Richardson and Hadobas (1997) where the isolates are inoculated on phytase-screening medium in Petri plates and phytase activity is determined by subtracting the diameter of bacterial colony from the diameter of total zone after incubation at 28 °C for 3 days. Adhikari and Pandey (2019) screened endophytic fungal isolates from *Taxus wallichiana* Zucc. roots for their ability to hydrolyse phytate. They used phytase-screening medium containing two different substrates – calcium phytate and sodium phytate – and incubated at different temperatures 5, 15, 25 and 35 °C, for an incubation period of 7 days. The isolates positive for phytase enzyme produced a clear zone, due to the hydrolysis of calcium and sodium phytate, around the colony.

7.7.1.8 Growth on Nitrogen-Free Media

Bacteria are the only microorganisms having capacity of biological nitrogen fixation (BNF) which serves as the primary source of fixed nitrogen in land-based agriculture systems. The endophytic bacterial isolates have been reported to play an important role in BNF (Oses et al. 2018) and can be screened for this activity by their growth on nitrogen-free media such as Burk's nitrogen-free media, Norris glucose nitrogen-free media, Dobereiner's semisolid N-free bromothymol blue media and Jensen's media after 48 h of incubation at 28 °C which is a qualitative indicator of nitrogen fixation (Shabanamol et al. 2018; Potshangbam et al. 2017).

7.7.1.9 Hydrogen Cyanide (HCN) Production

HCN production by the endophytic bacteria is an important trait that is involved in inhibiting fungal growth (Verma et al. 2018). The method of Castric is used for screening of endophytic bacterial isolates for HCN production (Vyas and Kaur 2019). In this method, one bacterial culture is streaked per plate on trypticase soy agar or nutrient agar amended with 4.4 g/L glycine. The discs of Whatman filter paper No. 1 are soaked in a solution of 2% Na_2CO_3 prepared in 0.5% picric acid solution. A single treated disc is then placed inside the lid of the Petri dishes streaked with the cultures which are further sealed with parafilm and incubated at 28 °C for 4 days in an inverted position. Development of orange to brown colour indicates HCN production. Potshangbam et al. (2017) followed a modified protocol of Miller and Higgins for HCN production by endophytic fungal isolates where treated strips of Whatman filter paper were placed on PDA slants inoculated with the fungal cultures, closing the lids tightly with parafilm and incubating for 7–14 days. The change in the colour of filter paper strips from the yellow colour to brown or reddish brown was scored as HCN production.

7.7.1.10 Ammonia Production

Production of ammonia is another important plant growth-promoting trait of the endophytes (Rohini et al. 2018; Ullah et al. 2018). Bacterial endophyte with broad-spectrum antifungal activity against phytopathogens of grapevine and fungal endophytes associated with medicinal plant *Asclepias sinaica* were screened for the production of ammonia by Andreolli et al. (2019) and Fouda et al. (2015), respectively. For this purpose, they inoculated respective fresh cultures in peptone water (10 mL) taken in tubes and incubated at 28 °C for 48–72 h in shaking conditions. The tubes positive for ammonia production lead to the development of faint yellow to dark brown colour after the addition of 0.5 mL Nessler's reagent.

7.7.1.11 Screening for Herbicidal Activity

Microorganisms have been reported to produce metabolites that aid in the biological control of weeds and can prove to be environmentally friendly alternates to the chemical herbicides (Saad 2019; van Lenteren et al. 2018; Mallik 2001). To screen the herbicidal activity of endophytic actinomycetes isolated from leaves, stem and roots of six different plants, Singh et al. (2018) used surface-sterilised seeds (treated with 2–3 min in 3% solution of hydrogen peroxide and rinsed with sterilised distilled water) of *Parthenium hysterophorus, Bidens biternata* and *Ageratum conyzoides* (6–8 of each per plate). These seeds were then placed on either sides of the culture of endophytic actinomycete streaked in the form of a strip on media plate, which had incubated earlier at 27 °C for 1–2 weeks to allow the growth of actinomycetes and diffusion of metabolites in the medium, and also on uninoculated control plates. Plates were incubated in the dark at 28 °C for 4 days. Seed germination was monitored. Saad (2019) also assessed the herbicidal activities of endophytic fungi and its secondary metabolites.

7.7.1.12 ACC Deaminase

The endophytic bacteria can improve plant tolerance against various biotic and abiotic stresses by the production of 1-aminocyclopropane-1-carboxylate (ACC) deaminase (Afridi et al. 2019; Lumactud and Fulthorpe 2018). The qualitative assay used to screen bacterial strains checks for the utilisation of ACC as a nitrogen source (Glick et al. 1995). For this purpose, the bacterial cells (endophytic bacterial isolates grown in trypticase soy broth, 5 mL with shaking at 120 rpm for 24 h at 28 °C) obtained by centrifugation at 3000 g for 5 min were washed twice and resuspended in Tris–HCl (0.1 M; pH 7.5) by Afridi et al. (2019). Then plates containing Dworkin and Foster media supplemented with and without ACC were spot inoculated with the cultures and plates with ammonium sulphate as the N source served as a positive control. After 3 days of incubation at 28 °C, the growth on ACC-supplemented plates was examined and compared to positive and negative controls.

7.8 Screening for Enzyme Production

Endophytes produce numerous extracellular enzymes such as amylases, cellulases, lipases, pectinases, laccases and proteases (Khan et al. 2017). These enzymes may play the key role in providing protection against phytopathogens and meet nutritional needs from the host plants (Fouda et al. 2015; Sunitha et al. 2013). These enzymes have applications in food processing, the leather industry, manufacturing of detergents, pharmaceuticals, medical therapy and the field of molecular biology (Santos et al. 2019; Corrêa et al. 2014).

7.8.1 Proteases

Proteases have extensive applications and are used in laundry, food, leather, brewing, photography and other industries (Delgado-García et al. 2019). Endophytes can be a potential source of this enzyme (Dorra et al. 2018). Pure fungal or bacterial endophytic isolates can be screened for protease enzyme production by inoculating on skim milk agar and incubating at 28 °C until the colony is observed. Visible clearance around the colony qualitatively indicates protease activity (López et al. 2018).

7.8.2 Cellulases

To select endophytes with cellulolytic activity in the extracellular medium, all isolates are inoculated in minimal medium and supplemented with 5 g/L yeast extract and 1% carboxymethylcellulose (CMC). CMC plates are incubated for 7 days at 28 °C and then visualised with 0.1% Congo red dye. Enzyme production is indicated by halo formation (da Silva Ribeiro et al. 2018; Sánchez-Cruz et al. 2019).

7.8.3 Pectinases

Pectinases have massive industrial applications, and endophytes can serve a good source of these enzymes (Singh et al. 2019; Khan et al. 2018). Pectinolytic activity of the endophytic fungi was determined by growing them in pectin agar medium containing 1% pectin (Uzma et al. 2016; Fouda et al. 2015). A clear zone around the fungal colony on flooding the plates with 1% hexadecyltrimethyl ammonium bromide aqueous solution after incubation confirms the pectinase activity of the isolate.

7.8.4 Lipases

The microbial lipases find numerous applications in processes of biotechnological and industrial importance including food, pharmaceutical, paper and oleochemical industries (Naik et al. 2019). For screening lipase activity of the endophytic fungi, Sunitha et al. (2013) and Bezerra et al. (2015) used peptone agar medium and PDA, respectively, supplemented with Tween 20 that was sterilised separately and added at the rate of 1% in the medium. After the incubation of 7 days at 28 °C, the presence of precipitates around the colony indicated positive lipase activity; the visible precipitates are a result of formation of Ca^{2+}salts of the lauric acid released by the enzyme.

7.8.5 Amylases

Microbes are a preferred source of amylases, and screening of bacterial and fungal endophytes for these extracellular enzymes is a crucial step in their commercial production and applications (Khan et al. 2017; Sindhu et al. 2017). The protocol for screening involves the spot inoculation of endophytic bacterial or fungal cultures on agar medium supplemented with soluble starch (Santos et al. 2019; Potshangbam et al. 2017). Endophytes hydrolyse starch showing zone of hydrolysis around the colonies when flooded with iodine solution (Fig. 7.2a).

7.9 Natural Products from Endophytic Microbes

Endophytes have been acknowledged as a source of materials and products with high potential (Table 7.2) as compared to the plants in pharmaceutical and agrochemical areas and also capable of producing various bioactive compounds of biotechnological application (Sato and Kumagai 2013).

Table 7.2 Bioactive compounds produced by endophytic microbes

S. No.	Source of endophytes	Bioactive compounds	Applications	References
1.	Fusarium species	Xularosides, munumbicins	Antifungal activity	Jalgaonwala et al. (2011)
2.	*Hypericum perforatum*	Hypericin	Antibacterial activity	Joseph and Priya (2011)
3.	*Chloridum* sp.	Javanicin	Antibacterial activity	Jalgaonwala et al. (2011)
4.	*Cladosporium* sp.	Cardiac glycosides, phenolic compounds	Antiviral activity	Selvi and Balagengatharathilagam (2014)
5.	*Boesenbergia rotunda*	Munumbicins	Antibacterial activity	Golinska et al. (2015)
6.	*Streptomyces* sp.	Munumbicins	Antibacterial activity	Kumar et al. (2014)
7.	*Xylaria* sp.	Phenolic compounds	Antibacterial activity	Selvi and Balagengatharathilagam (2014)
8.	*Cytonaema* sp.	Cytonic acids A and B	Against hepatitis virus	Bhardwaj and Agrawal (2014)
9.	*Pseudomonas viridiflava*	Ecomycin	Antifungal activity	Miller et al. (1998)
10.	*Streptomyces caespitosus*	Mitomycin C	Chemotherapeutic agent	Danshiitsoodol et al. (2006)
11.	*Streptomyces* sp.	Coronamycin	Antimalarial activity	Ezra et al. (2004)
12.	*Streptomyces* NRRL 30562	Munumbicin	Antibacterial activity	Castillo et al. (2002)
13.	*Bacillus thuringiensis*	β-exotoxin	Insecticidal activity	Espinasse et al. (2002)
14.	*Streptomyces* sp.	Xiamycin	Anti-HIV activity	Ding et al. (2010)
15.	*Bacillus subtilis*	Subtilin	Antibacterial activity	Stein (2005)
16.	*Staphylococcus aureus*	EtOAc (ethyl acetate)	Antibiotic and anticancer activities	Handayani et al. (2018)
17.	*Streptomyces* sp. *strain BO-07*	Biphenyls	Anticancer activity	Taechowisan et al. (2017)
18.	*Bacillus* sp.	Lipopeptides, polysaccharides	Antimicrobial and antitumour activity	Villarreal-Delgado et al. (2018)
19.	*Bacillus subtilis* and *B. amyloliquefaciens*	Lipopeptides	Antifungal activity	Gond et al. (2015)
20.	*Actinomycetes*	Anthracyclines, quinoxalines	Antitumour agents	Cardoso-Filho (2018)

(continued)

Table 7.2 (continued)

S. No.	Source of endophytes	Bioactive compounds	Applications	References
21.	*Penicillium* sp.	Octadecanoic acid	Antidiabetic activity	Murugan et al. (2017)
22.	*Alternaria* sp.	Cetene, 1,2-benzenedicarb-oxylic acid	Antioxidant activity and antimicrobial activity	Elgorban and Abdel-Wahab (2018)

7.9.1 Endophytic Microbial Products as Antibiotics

Antibiotics are natural products made by microorganisms as secondary metabolites that are active against other microorganisms (Demain 1981). The majority of endophytic bacteria are a source of different kinds of antibiotics. Ecomycin and pseudomycin are some of the antibiotics produced by *Pseudomonas viridiflava* (Christina et al. 2013). Ecomycin represents a family of lipopeptides and is used for the treatment of respiratory, urinary tract and gut infections. It incorporates common amino acids like alanine, serine, glycine and threonine (Miller et al. 1998) and some unusual amino acids such as homoserine and β-hydroxyaspartic acid.

The pseudomycins are active against plant and human pathogenic fungi as *Cryptococcus albicans, C. neoformans* and *Ceratocystis ulmi* (Ballio et al. 1994; Harrison et al. 1991). *Streptomyces* sp. strain NRRL 30562, an endophyte of snake vine plant, produced munumbicins with a broad-spectrum biological activity against several human diseases. Kakadumycin, a novel antibiotic produced by endophytic *Streptomyces* sp. strain NRRL 30566, is isolated from a fern-leaved *Grevillea pteridifolia* tree. Kakadumycin A and echinomycin are chemically related but differ due to their spectral qualities and biological activities (Castillo et al. 2003).

Three endophytic bacteria CER5, CER6 and CER11 were isolated from the roots of *Capsicum annum* and showed biological activity against pathogenic bacteria (Syed et al. 2017).

7.9.2 As Antimicrobial Compounds

The discovery of antimicrobial metabolites from endophytes is an alternative option to overcome the increasing levels of drug resistance (Ferlay et al. 2010; Taechowisan et al. 2012). *Streptomyces* sp. is involved in the production of bioactive metabolite xiamycin against anti-HIV activity (Ding et al. 2010). These compounds can be used as drugs by humankind and also represent broad-spectrum activity against foodborne and food-spoilage microorganisms such as *E. coli*, *S. aureus*, *Aspergillus niger*, etc.

Cryptocin produced by endophytic fungus *Cryptosporiopsis quercina* is isolated from the inner bark of *Tripterygium wilfordii* that inhibit the growth of *Pyricularia oryzae* and other plant pathogens (Li et al. 2000). Antimicrobial activity was also shown by *Citrus nobilis* fruit against *Colletotrichum truncatum, Fusarium oxysporum* and *Geotrichum candidum* (Hong-Thao et al. 2016). Endophytic bacteria synthesise various nanoparticles which play a significant role for the treatment of various types of cancer which emerge as novel antimicrobial compounds in the research area of pharmaceutical engineering (Sunkar and Nachiyar 2012). Silver nanoparticles have antibacterial properties against HIV-1, hepatitis B virus and herpes virus (Sun et al. 2005; Taylor et al. 2005; Baram-Pinto et al. 2009).

Endophytic fungi isolated from leaf, bark and roots of mangrove *Sonneratia alba* Sm showed antimicrobial and cytotoxic activities against tumour cell lines T47D (Handayani et al. 2018). An endophytic fungus *Cinnamomum zeylanicum* isolated from *Muscodor albus* produced volatile organic compounds used for preserving fruits and vegetables during storage (Kapoor et al. 2019).

7.9.3 Endophytes as Anticancer Compounds

Endophytes produce many bioactive compounds which have been identified as anticancer agents (Firakova et al. 2007). Paclitaxel, a bioactive compound, has been isolated from *Taxus* species, and this molecule is the world's first anticancer drug. Endophytic bacteria from ginseng produced ginsenosides that possess anticancerous property (Gao et al. 2015). Bacillus species serves as a source of antitumoural EPS for cancer treatment (Chen et al. 2013).

Another bioactive compound torreyanic acid was isolated from *Pestalotiopsis microspora* strain as an anticancer agent (Lee et al. 1996). A large group of substances known as cytochalasin have been found in endophytic fungal genera such as *Xylaria*, *Phoma*, *Hypoxylon*, *Chalara*, etc. These compounds have antitumour and antibiotic activities other than cytotoxicity (Wagenaar et al. 2000).

L-asparaginase has also been introduced to the multidrug chemotherapy in acute lymphoblastic leukaemia for the treatment of patients. Endophytic actinomycetes are involved in the production of potent antitumour agents like maytansinoids, lupinacidin, 6-alkyl salicylic acids and salaceyins A and B (Snipes et al. 2007; Powell and Smith 1980). Another compound naphthomycin A was extracted from *Streptomyces* sp. and was found cytotoxic against human tumour cells P388 and A549 (Qin et al. 2011; Li et al. 2010). Similarly, seven new macrolides were reported from endophytic fungus *P. microspora* isolated from fresh fruits of *Drepanocarpus lunatus*. They showed cytotoxicity against human ovarian cancer cell lines (Liu et al. 2016).

A *Streptomyces* sp. strain BO-07 produced endophytic biphenyls against human HepG2 and Huh7 liver cell lines (Taechowisan et al. 2017). Another endophytic *Streptomyces cavourensis* strain YBQ59 was isolated from *Cinnamomum cassia* against human lung adenocarcinoma and resistant cell A549 and H1299 growth (Vu et al. 2018). Ek-Ramos et al. in 2019 reported that endophytic genes are associated with specific metabolite production that are involved in plant growth promotion and have antimicrobial, anticancer and insecticidal activities.

7.9.4 Endophytes as Antioxidants

Natural antioxidants are found in vegetables, fruits and other plants to provide protection from harmful free radicals. Compounds like pestacin and isopestacin have been isolated from *Pestalotiopsis microspora.* These compounds displayed antioxidant and antimicrobial activities (Strobel et al. 2002). The antioxidant activity of pestacin was found greater than trolox, a vitamin E derivative (Harper et al. 2003). Another endophytic *Streptomyces* species isolated from the plant *Alpinia oxyphylla* produced a compound 2,6-dimethoxy-terephthalic acid, which possesses antioxidant activity (Jasmine and Agastian 2013).

Chutulo and Chalanaavar in 2018 reported antioxidant activity of endophytic fungi isolated from *Azadirachta indica* and suggested that these bioactive compounds have great potential to combat various diseases, to prevent cell damage and insects-pests and to target pathogenic microbes.

7.9.5 Endophytes as Antidiabetic Agents

Endophytes have been explored for their antidiabetic activity. A nonpeptidal fungal metabolite (L-783, 281) was extracted from an endophytic fungus (*Pseudomassaria* sp.). This compound acts as an insulin to lower blood glucose level (Zhang et al. 1999).

Immunosuppressive compounds subglutinols A and B have been isolated from endophytic fungus *Fusarium subglutinam* from *T. wilfordii* (Lee et al. 1995). Another compound, cyclosporine, has also been extracted from endophytic fungus *Tolypocladium inflatum,* which exhibits immunosuppressant activity (Borel and Kis 1991).

Endophytic *Actinomycetes* sp. were collected from the roots of *Caesalpinia sappan*, which produced α-glucosidase inhibitor, a target for antidiabetic treatment (Savi et al. 2015; Irawan 2009). *Streptomyces* species have also displayed a significant antidiabetic potential (Pujiyanto et al. 2012; Christhudas et al. 2013).

Murugan et al. (2017) reported the antidiabetic activity of endophytic fungi, *Penicillium* sp., isolated from *Tabebuia argentea* and suggested that octadecanoic acid is responsible for inducing antidiabetic activity and the compound has the ability to inhibit all diabetic protein activity.

Elgorban and Abdel-Wahab in 2018 evaluate the potential of endophytic fungi, *Alternaria* sp., isolated from *Salvadora persica* for the production of bioactive compounds such as cetene and 1,2-benzenedicarboxylic acid which shows antidiabetic activity.

7.9.6 Endophytes as Bioinsecticides

Endophytes are found to be used as bioinsecticides and have great potential in anti-insecticidal activity. They provide an eco-friendly approach and help in reducing the load of synthetic pesticides. An endophyte, *Nodulisporium* sp., from the plant *Bontia daphnoides* produced nodulisporic acid that exhibits insecticidal properties against the larvae of blowfly, by activating insect glutamate-gated chloride channel (Smith et al. 2000). Two bioactive compounds, 5-hydroxy-2-benzofuron (1′-hydroxy-5′-methyl-4′-hexenyl) and 5-hydroxy-2-benzofuron(1′-oxo-5′-methyl-4′-hexenyl), have also been isolated from endophytic fungus from *Gaultheria procumbens* that display insecticidal properties against spruce budworm (Findlay et al. 1997). An endophyte *Phomopsis oblonga* produced insect toxins against the beetle *Physocnemum brevilinenu* from elm tree (Webber 1981). Another endophytic fungus, *Muscodor vitigenus*, isolated from a liana (*Paullina paullinioides*) produced naphthalene, an active ingredient in common mothballs, which exhibit insect repellent property against wheat stem sawfly (Daisy et al. 2002).

Actinomycetes including *Actinomyces*, *Micromonospora* and *Streptomyces* are of great importance as biocontrol agents (Ek-Ramos et al. 2019).

7.10 Conclusion and Future Directions

It has been proved from the studies that endophytes are a rich source of so many natural bioactive compounds with higher level of structural diversity as well as a wide spectrum of biological activities. Biotransformation of natural products by the biocatalysts present inside the endophytes is a very interesting phenomenon. Medicinal plants for curing various diseases and health-related problems have been in use since ancient times. But continuous use of medicinal plants is leading to reduction in their population. So the use of microbial endophytes is the best way to cure several ailments in cost-effective, eco-friendly and sustainable way. But there is a need to understand the endophytes' physiology, defensive role, their biochemical pathways and secondary metabolite production. So, attention should be drawn towards this emerging field of research and possible exploitation of the sources which are available for their use in various fields like pharmaceutical industry, food industry, cosmetics and medical industry.

References

Abdalla MA, McGaw LJ (2018) Bioprospecting of South African plants as a unique resource for bioactive endophytic microbes. Front Pharmacol 9:456. https://doi.org/10.3389/fphar.2018.00456

Adhikari P, Pandey A (2018) Diversity of endophytic fungi associated with Himalayan Yew roots. Proc Himalayan Res Consort 1:1–9. https://www.researchgate.net/publication/332170697_Diversity_of_Endophytic_Fungi_Associated_with_Himalayan_Yew_Taxus_wallichiana_Zucc_Roots/citations

Adhikari P, Pandey A (2019) Phosphate solubilization potential of endophytic fungi isolated from *Taxus wallichiana* Zucc. roots. Rhizosphere 9:2–9. https://doi.org/10.1016/j.rhisph.2018.11.002

Afridi MS, Sumaira A, Mahmood T, Salam A, Mukhtar T, Mehmood S, Ali J, Khatoon Z, Bibi M, Javed Sultan MT, Chaudhary HJ (2019) Induction of tolerance to salinity in wheat genotypes by plant growth promoting endophytes: Involvement of ACC deaminase and antioxidant enzymes. Plant Physiol Biochem 139:569–577. https://doi.org/10.1016/j.plaphy.2019.03.041

Agrawal T, Kotasthane AS (2012) Chitinolytic assay of indigenous *Trichoderma* isolates collected from different geographical locations of Chhattisgarh in Central India. Springer Plus 1:73. https://doi.org/10.1186/2193-1801-1-73

Ahmedullah M and Nayar MP (1999) Red data book of Indian plants (Peninsular India). Botanical Survey of India, Calcutta. Vol 4. http://agris.fao.org/agris-search/search.do?recordID=US201300602313

Ali S, Isaacson J, Kroner Y, Saldias S, Kandasamy S, Lazarovits G (2018) Corn sap bacterial endophytes and their potential in plant growth-promotion. Environ Sustain 1:341–355. https://doi.org/10.1007/s42398-018-00030-4

Ali S, Khan SA, Hamayun M, Iqbal A, Khan A, Hussain A, Shah M (2019) Endophytic fungi from *Caralluma acutangula* can secrete plant growth promoting enzymes. Fresenius Environ Bull 28:2688–2696. https://www.researchgate.net/publication/332291226

Alsheikh M, Saleh AI, Mohammad AI, Dosari S, Abdul M, Maged Abdel-Kader S (2009) Evaluation of the hepatoprotective effect of fumaria parviflora and *Momordica balsamina* from Saudi folk medicine against experimentally induced liver injury in rats. Res J Med Plant 3:9–15. https://doi.org/10.3923/rjmp.2009.9.15

Amna T, Puri SC, Verma V, Sharma JP, Khajuria RK, Musarrat J, Spiteller M, Qazi GN (2006) Bioreactor studies on the endophytic fungus Entrophospora infrequens for the production of an anticancer alkaloid camptothecin. Can J Microbiol 52:189–196. https://doi.org/10.1139/w05-122

Andreolli M, Zapparoli G, Angelini E, Lucchetta G, Lampis S, Vallini G (2019) Pseudomonas protegens MP12: a plant growth-promoting endophytic bacterium with broad-spectrum antifungal activity against grapevine phytopathogens. Microbiol Res 219:123–131. https://doi.org/10.1016/j.micres.2018.11.003

Ballio A, Bossa F, DiGiogio P, Ferranti P, Paci M, Pucci P, Scaloni A, Segre A, Strobel GA (1994) Structure of the pseudomycins, new lipodepsipeptides produced by Pseudomonas syringae MSU 16H. FEBS Lett 355:96–100. https://doi.org/10.1016/0014-5793(94)01179-6.

Baram-Pinto D, Shukla S, Perkas N, Gedanken A, Sarid R (2009) Inhibition of herpes simplex virus type 1 infection by silver nanoparticles capped with mercaptoethane sulfonate. Bioconjug Chem 20:1497–1502. https://doi.org/10.1021/bc900215b

Barka EA, Vatsa P, Sanchez L, Gaveau-Vaillant N, Jacquard C, Klenk HP (2016) Taxonomy, physiology, and natural products of Actinobacteria. Microbiol Mol Biol Rev 80:1–43. https://doi.org/10.1128/MMBR.00019-15

Berdy J (2012) Thoughts and facts about antibiotics: where were now and where were heading. J Antibiot 65:385–395. https://doi.org/10.1038/ja.2012.27

Berkodia M, Joshi U, Rami NV, Wati L (2018) Endophytes: a hidden treasure inside plant. Int J Chem Stud 6:1660–1665

Bezerra JDP, Nascimento CCF, Barbosa R do N, da Silva DCV, Svedese VM, Silva-Nogueira EB, Gomes BS, Paiva LM, Souza-Motta CM (2015) Endophytic fungi from medicinal plant *Bauhinia forficata*: Diversity and biotechnological potential. Braz J Microbiol 46(1):49–57. https://doi.org/10.1590/S1517-838246120130657

Bhardwaj A, Agrawal P (2014) A review fungal endophytes: as a storehouse of bioactive compound. World. J Pharm Pharm Sci 3:228–237. www.wjpps.com

Borel JF, Kis ZL (1991) The discovery and development of cyclosporine. Transplant Proc 23:1867–1874. https://doi.org/10.1007/978-1-4615-9846-6_2

Borges KB, Borges WDS, Pupo MT, Bonato PS (2007) Endophytic fungi as models for the stereoselective biotransformation of thioridazine. Appl Microbiol Biotechnol 77:669–674. https://doi.org/10.1007/s00253-007-1171-x

Cardoso-Filho JA (2018) Endophytic microbes as a novel source for producing anticancer compounds as multidrug resistance modulators. In: Akhtar M, Swamy M (eds) Anticancer plants: natural products and biotechnological implements. Springer, Singapore, pp 343–381. https://doi.org/10.1007/978-981-10-8064-7_15

Castillo UF, Strobel GA, Ford EJ, Hess WM, Porter H, Jensen JB, Albert H, Robinson R, Condron MA, Teplow DB et al (2002) Munumbicins, wide-spectrum antibiotics produced by

Streptomyces NRRL 30562, endophytic on Kennedia nigricans. Microbiologica 148:2675–2685. https://doi.org/10.1099/00221287-148-9-2675
Castillo U, Harper JK, Strobel GA, Sears J, Alesi K, Ford E, Lin J, Hunter M, Maranta M, Ge H, Yaver D, Jensen JB, Porter H, Robison R, Millar D, Hess WM, Condron M, Teplow D (2003) Kakadumycins, novel antibiotics from Streptomyces sp. NRRL 30566, an endophyte of Grevillea pteridifolia. FEMS Lett 224:183–190. https://doi.org/10.1016/S0378-1097(03)00426-9
Chaudhary HS, Soni B, Shrivastava AR, Shrivastava S (2013) Diversity and versatility of actinomycetes and its role in antibiotic production. Int J Pharm Sci 3:S83–S94. https://doi.org/10.7324/JAPS.2013.38.S14
Chen YT, Yuan Q, Shan LT, Lin MA, Cheng DQ, Li CY (2013) Antitumor activity of bacterial exopolysaccharides from the endophyte Bacillus amyloliquefaciens sp. isolated from Ophiopogon japonicus. Oncol Lett 5:1787–1792. https://doi.org/10.3892/ol.2013.1284
Chen L, Shi H, Heng J, Wang D, Bian K (2019a) Antimicrobial, plant growth-promoting and genomic properties of the peanut endophyte *Bacillus velezensis* LDO2. Microbiol Res 218:41–48. https://doi.org/10.1016/j.micres.2018.10.002
Chen P, Zhang C, Ju X, Xiong Y, Xing K, Qin S (2019b) Community composition and metabolic potential of endophytic actinobacteria from coastal salt marsh plants in Jiangsu, China. Front Microbiol 10:1063. https://doi.org/10.3389/fmicb.2019.01063
Christhudas IN, Kumar PP, Agastian P (2013) In vitro α-glucosidase inhibition and antioxidative potential of an endophyte species (Streptomyces sp. Loyola UGC) isolated from Datura stramonium L. Curr Microbiol 67(1):69–76. https://doi.org/10.1007/s00284-013-0329-2
Christina A, Christapher V, Bhore SJ (2013) Endophytic bacteria as a source of novel antibiotics: an overview. Pharmacogn Rev 7:11–16. https://doi.org/10.4103/0973-7847.112833
Chutulo EC, Chalanaavar RK (2018) Endophytic microflora and their bioactive compounds from Azadirachta indica: a comprehensive review. J Fungi 4:42. https://doi.org/10.3390/jof4020042
Compant S, Samad A, Faist H, Sessitsch A (2019) A review on the plant microbiome: ecology, functions, and emerging trends in microbial application. J Adv Res 19:29–37. https://doi.org/10.1016/j.jare.2019.03.004
Corrêa RCG, Rhoden SA, Mota TR, Azevedo JL, Pamphile JA (2014) Endophytic fungi: expanding the arsenal of industrial enzyme producers. J Ind Microbiol Biotechnol 41(10):1467–1478. https://doi.org/10.1007/s10295-014-1496-2
da Silva Ribeiro A, Polonio JC, Costa AT, dos Santos CM, Rhoden SA, Azevedo JL, Pamphile JA (2018) Bioprospection of culturable endophytic fungi associated with the ornamental plant *Pachystachys lutea*. Curr Microbiol 75:588–596. https://doi.org/10.1007/s00284-017-1421-9
Daisy BH, Strobel GA, Castillo U, Ezra D, Sears J, Weaver DK, Runyon JB (2002) Naphthalene, an insect repellent, is produced by Muscodor vitigenus, a novel endophytic fungus. Microbiology 148:3737–3741. https://doi.org/10.1099/00221287-148-11-3737
Danshiitsoodol N, Pinho CA, Matoba Y, Kumagai T, Sugiyama M (2006) The mitomycin C (MMC)-binding protein from MMC producing microorganisms protects from the lethal effect of bleomycin: crystallographic analysis to elucidate the binding mode of the antibiotic to the protein. J Mol Biol 360:398–408. https://doi.org/10.1016/j.jmb.2006.05.017
De Bary A (1866) Morphologie und physiologie der pilze, flechten und myxomyceten. Leipzig: hofmeister's handbook of physiological. Botany 2:20. https://archive.org/details/bub_gb_-I0NAAAAYAAJ/page/n5
Delgado-García M, Flores-Gallegos AC, Kirchmayr M, Rodríguez JA, Mateos-Díaz JC, Aguilar CN, Muller M, Camacho-Ruíz RM (2019) Bioprospection of proteases from *Halobacillus andaensis* for bioactive peptide production from fish muscle protein. Electron J Biotechnol 39:52–60. https://doi.org/10.1016/j.ejbt.2019.03.001
Demain AL (1981) Industrial microbiology. Science 214:987–994. https://doi.org/10.1126/science.6946560
Ding L, Münch J, Goerls H, Maier A, Fiebig H-H, Lin W-H, Hertweck C (2010) Xiamycin, a pentacyclic indolosesquiterpene with selective anti-HIV activity from a bacterial mangrove endophyte. Bioorg Med Chem Lett 20(22):6685–6687. https://doi.org/10.1016/j.bmcl.2010.09.010

Dorra G, Ines K, Imen BS, Laurent C, Sana A, Olfa T, Pascal C, Thierry J, Ferid L (2018) Purification and characterization of a novel high molecular weight alkaline protease produced by an endophytic *Bacillus halotolerans* strain CT2. Int J Biol Macromol 111:342–351. https://doi.org/10.1016/j.ijbiomac.2018.01.024

Eevers N, Gielen M, Sánchez-López A, Jaspers S, White JC, Vangronsveld J, Weyens N (2015) Optimization of isolation and cultivation of bacterial endophytes through addition of plant extract to nutrient media. Microb Biotechnol 8(4):707–715. https://doi.org/10.1111/1751-7915.12291

Ek-Ramos MJ, Gomez-Flores R, Orozco-Flores AA, Rodríguez-Padilla C, González-Ochoa G, Tamez-Guerra P (2019) Bioactive products from plant-endophytic Gram-positive bacteria. Front Microbiol 10:463. https://doi.org/10.3389/fmicb.2019.00463

Elgorban AM, Abdel-Wahab MA (2018) Natural products of Alternaria sp. an endophytic fungus isolated from *Salvadora persica* from Saudi Arabia. Saudi J Biol Sci 26:1068–1077. https://doi.org/10.1016/j.sjbs.2018.04.010

Espinasse S, Gohar M, Lereclus D, Sanchis V (2002) An ABC transporter from Bacillus thuringiensis is essential for beta-exotoxin I production. J Bacteriol 184:5848–5854. https://doi.org/10.1128/jb.184.21.5848-5854.2002

Ezra D, Castillo UF, Strobel GA, Hess WM, Porter H (2004) Coronamycins, peptide antibiotics produced by a verticillate *Streptomyces* sp. MSU-2110 endophytic on Monstera sp. Microbiologica 150:785–793. https://doi.org/10.1099/mic.0.26645-0

Ferlay J, Shin HR, Bray F, Forman D, Mathers C, Parkin DM (2010) Cancer incidence and mortality worldwide: IARC cancer base No. 10 Lyon, France: International Agency for Research on Cancer. Globocan 46:765–781. https://doi.org/10.1016/j.ejca.2009.12.014

Ferreira MC, Cantrell CL, Wedge DE, Gonçalves VN, Jacob MR, Khan S, Rosa CA, Rosa LH (2017) Diversity of the endophytic fungi associated with the ancient and narrowly endemic neotropical plant *Vellozia gigantea* from the endangered Brazilian rupestrian grasslands. Biochem Syst Ecol 71:163–169. https://doi.org/10.1016/j.bse.2017.02.006

Findlay J, Bethelezi AS, Li G, Sevek M (1997) Insect toxins from an endophyte fungus from wintergreen. J Nat Prod 60:1214–1215. https://doi.org/10.1021/np970222j

Firakova S, Sturdikova M, Muckova M (2007) Bioactive secondary metabolites produced by microorganisms associated with plants. Biologia 62:251–257. https://doi.org/10.2478/s11756-007-0044-1

Fouda AH, Hassan SED, Eid AM, Ewais EED (2015) Biotechnological applications of fungal endophytes associated with medicinal plant *Asclepias sinaica* (Bioss.). Ann Agric Sci 60(1):95–104. https://doi.org/10.1016/j.aoas.2015.04.001

Fuchs B, Krischke M, Mueller MJ, Krauss J (2017) Plant age and seasonal timing determine endophyte growth and alkaloid biosynthesis. Fungal Ecol 29):52–58. https://doi.org/10.1016/j.funeco.2017.06.003

Gangwar M, Dogra S, Gupta UP, Kharwar RN (2014) Diversity and biopotential of endophytic actinomycetes from three medicinal plants in India. Afr J Microbiol Res 8:184–191. https://doi.org/10.5897/AJMR2012.2452

Gao Y, Lu Q, Pu Zang P, Li X, Ji Q, He Z et al (2015) An endophytic bacterium isolated from Panax ginseng CA Meyer enhances growth, reduces morbidity, and stimulates ginsenoside biosynthesis. Phytochem Lett 11:132–138. https://doi.org/10.1016/j.phytol.2014.12.007

Gao Z, Zhang B, Liu H, Han J, Zhang Y (2017) Identification of endophytic *Bacillus velezensis* ZSY-1 strain and antifungal activity of its volatile compounds against *Alternaria solani* and *Botrytis cinerea*. Biol Control 105:27–39. https://doi.org/10.1016/j.biocontrol.2016.11.007

Garcia A, Rhoden SA, Rubin Filho CJ, Nakamura CV, Pamphile JA (2012) Diversity of foliar endophytic fungi from the medicinal plant *Sapindus saponaria* L. and their localization by scanning electron microscopy. Biol Res 45(2):139–148. https://doi.org/10.4067/S0716-97602012000200006

Glick BR, Karaturovíc DM, Newell PC (1995) A novel procedure for rapid isolation of plant growth promoting pseudomonads. Can J Microbiol 41:533–536. https://doi.org/10.1007/s10482-015-0502-7.

Golinska P, Wypij M, Agarkar G, Rathod D, Dahm H, Rai M (2015) Endophytic actinobacteria of medicinal plants: diversity and bioactivity. Antonie Van Leeuwenhoek 108:267–289. https://doi.org/10.1007/s10482-015-0502-7

Gond SK, Bergen MS, Torres MS (2015) Endophytic bacillus spp. Produce antifungal lipopeptides and induce host defence gene expression in maize. Microbiol Res 172:79–87. https://doi.org/10.1016/j.micres.2014.11.004

Gouda S, Das G, Sen SK, Shin H-S, Patra JK (2016) Endophytes: a treasure house of bioactive compounds of medicinal importance. Front Microbiol 7:1538. https://doi.org/10.3389/fmicb.2016.01538

Greenfield M, Pareja R, Ortiz V, Gómez-Jiménez MI, Vega FE, Parsa S (2015) A novel method to scale up fungal endophyte isolations. Biocont Sci Technol 25:1208–1212. https://doi.org/10.1080/09583157.2015.1033382

Gu W, Ge HM, Song YC, Ding H, Zhu HL, Zhao XA, Tan RX (2007) Cytotoxic Benzo fluoranthene metabolites from Hypoxylon truncatum IFB-18, an endophyte of Artemisia annua. J Nat Prod 70:114–117. https://doi.org/10.1021/np0604127

Gupta A, Verma H, Singh PP, Singh P, Singh M, Mishra V, Kumar A (2019) Rhizome endophytes: roles and applications in sustainable agriculture. In: Verma S, White J (eds) Seed endophytes. Springer, Cham, pp 405–421. https://doi.org/10.1007/978-3-030-10504-4_19

Handayani D, Rivai H, Mulyana R, Suharti N, Rasyid R, Hertiani T (2018) Antimicrobial and cytotoxic activities of endophytic fungi isolated from mangrove plant Sonneratia alba Sm. J Appl Pharmaceut Sci 8:049–053. https://doi.org/10.7324/JAPS.2018.8207

Harper JK, Ford EJ, Strobel GA, Arif A, Grant DM, Porco J, Tomer DP, Oneill K (2003) Pestacin: a 1,3-dihydro isobenzofuran from *Pestalotiopsis microspora* possessing antioxidant and antimycotic activities. Tetrahedron 59:2471–2476. https://doi.org/10.1021/ja010997l.

Harrison L, Teplow D, Rinaldi M, Strobel GA (1991) Pseudomycins, a family of novel peptides from Pseudomonas syringae, possessing broad spectrum antifungal activity. J Gen Microbiol 137:2857–2865. https://doi.org/10.1099/00221287-137-12-2857

Hassan A, Radwan U, El-Zyat S, El-Sayed M (2018) Desert plant-fungal endophytic association: the beneficial aspects to their hosts. Biolog Forum An Int J 10:138–145. https://www.researchgate.net/publication/325945526

Hollants J, Leroux O, Leliaert F, Decleyre H, De Clerck O, Willems A (2011) Exploration of endophytic bacteria within the siphonous green seaweed Bryopsis (*Bryopsidales, Chlorophyta*). PLoS One 6:e26458. https://doi.org/10.1371/journal.pone.0026458

Hong-Thao PH, Mai-Linh NV, Hong-Lien NT, Hieu NV (2016) Biological characteristics and antimicrobial activity of endophytic Streptomyces sp. TQR12-4 isolated from elite Citrus nobilis cultivar Ham Yen of Vietnam. Int J Microbiol 2016:7207818. https://doi.org/10.1155/2016/7207818

Huang WY, Cai YZ, Xing J, Corke H, Sun M (2007) A potential antioxidant resource: Endophytic fungi from medicinal plants. Econ Bot 61:14–30. https://doi.org/10.1663/0013-0001(2007)61[14:APAREF]2.0.CO;2

Huang L, Yuan M, Ao X, Ren A, Zhang H, Yang M (2018) Endophytic fungi specifically introduce novel metabolites into grape flesh cells *in vitro*. PLoS One 13(5):e019699. https://doi.org/10.1371/journal.pone.0196996

Irawan D (2009) Isolation of endophytic actinomycetes in medicinal plants and their potency as an antidiabetes based on α-glucosidase activity, IPB Scient Repos—Bogor Agricultural University 69. http://repository.ipb.ac.id/handle/123456789/12450

Jagadevi S, Vidyasagar GM (2018) Phytochemical and antimicrobial evaluation of endophytic alternaria alternata isolated from Terminalia arjuna (Roxb.) Wight & Arn. Asian. J Pharm Pharmacol 4:456–461. https://doi.org/10.31024/ajpp.2018.4.4.13.

Jalgaonwala RE, Mohite BV, Mahajan RT (2011) Natural products from plant associated endophytic fungi. J Microbiol Biotechnol Res 1:21–32. http://ajpp.in/uploaded/p163.pdf

Jallow MFA, Dugassa-Gobena D, Vidal S (2008) Influence of an endophytic fungus on host plant selection by a polyphagous moth via volatile spectrum changes. Arthropod Plant Interact 2:53–62. https://doi.org/10.1007/s11829-008-9033-8

Jasmine DJ, Agastian P (2013) In vitro antioxidant activity and in vivo alpha glucosidase activity of endophytic actinomycetes isolated from *Catharanthus roseus* (l.) G. Don. J Pharm Res 6:674–678. https://doi.org/10.1016/j.jopr.2013.06.007.

Jia M, Chen L, Xin H-L, Zheng C-J, Rahman K, Han T, Qin L-P (2016) A friendly relationship between endophytic fungi and medicinal plants: a systematic review. Front Microbiol 7:906. https://doi.org/10.3389/fmicb.2016.00906

Jimenez-Romero C, Ortega-Barria E, Arnold AE, Cubilla-Rios L (2008) Activity against Plasmodium falciparum of lactones isolated from the endophytic fungus Xylaria sp. Pharm Biol 46:1–4. https://doi.org/10.1080/13880200802215859

Joseph B, Priya RM (2011) Bioactive compounds from endophytes and their potential in pharmaceutical effect: a review. Am J Biochem Mol Biol 1:291–309. https://doi.org/10.3923/ajbmb.2011.291.309

Kanchiswam CN, Malnoy M, Maffei M (2015) Bioprospecting bacterial and fungal volatiles for sustainable agriculture. Trends Plant Sci 20:206–211. https://doi.org/10.1016/j.tplants.2015.01.004

Kandel SL, Joubert PM, Doty SL (2017) Bacterial endophyte colonization and distribution within plants. Microorganisms 5:77–102. https://doi.org/10.3390/microorganisms5040077

Kapoor N, Jamwal VL, Gandhi SG (2019) Endophytes as a source of high-value, bioactive metabolites. Phytochemistry:427–458. https://doi.org/10.1007/978-3-319-90484-9_9

Khan AL, Shahzad R, Al-Harrasi A, Lee IJ (2017) Endophytic microbes: a resource for producing extracellular enzymes. In: Maheshwari D, Annapurna K (eds) Endophytes: crop productivity and protection. Sustainable development and biodiversity, vol 16. Springer, Cham, pp 95–110. https://doi.org/10.1007/978-3-319-66544-3_5

Khan N, Martínez-Hidalgo P, Ice TA, Maymon M, Humm EA, Nejat N, Sanders ER, Kaplan D, Hirsch AM (2018) Antifungal activity of *Bacillus* species against *Fusarium* and analysis of the potential mechanisms used in biocontrol. Front Microbiol 9:2363. https://doi.org/10.3389/fmicb.2018.02363

Kogel KH, Franken P, Huckelhoven R (2006) Endophyte or parasite what decides? Curr Opin Plant Biol 9:358–363. https://doi.org/10.1016/j.pbi.2006.05.001

Kumar S, Agarwal RP, Shukla H, Rajak RC, Sandhu SS (2014) Endophytic fungi: as a source of antimicrobials bioactive compounds. World J Pharm Pharm Sci 3:1179–1197

Lam KS (2007) New aspects of natural products in drug discovery. Trends Microbiol 15:279–289. https://doi.org/10.1016/j.tim.2007.04.001

Lee J, Lobkovsky E, Pliam NB, Strobel GA, Clardy J (1995) Subglutinols A and B: immunosuppressive compounds from the endophytic fungus *Fusarium subglutinans*. J Organomet Chem 60:7076–7077. https://doi.org/10.1021/acs.accounts

Lee JC, Strobel GA, Lobkovsky E, Clardy JC (1996) Torreyanic acid: a selectively cytotoxic quinine dimer from the endophytic fungus *Pestalotiopsis microspora*. J Organomet Chem 61:3232–3233

Li J, Strobel G, Harper J, Lobkovsky E, Clardy J (2000) Cryptocin, a potent tetramic acid antimycotic from the endophytic fungus Cryptosporiopsis cf. quercina. Org Lett 2(6):767–770. https://doi.org/10.1021/ol000008d

Li J, Lu C, Shen Y (2010) Macrolides of the bafilomycin family produced by Streptomyces sp. CS. J Antibiot 63:595–599. https://doi.org/10.1038/ja.2010.95

Li H-Y, Shen M, Zhou Z-P, Li T, Wei Y-L, Lin L-B (2012) Diversity and cold adaptation of endophytic fungi from five dominant plant species collected from the Baima Snow Mountain, Southwest China. Fungal Divers 54:79–86. https://doi.org/10.1007/s13225-012-0153-1

Li X, Li W, Chu L, White JF Jr, Xiong Z, Li H (2016) Diversity and heavy metal tolerance of endophytic fungi from *Dysphania ambrosioides*, a hyperaccumulator from Pb–Zn contaminated soils. J Plant Interact 11(1):186–192. https://doi.org/10.1080/17429145.2016.1266043

Liaqat F, Eltem R (2016) Identification and characterization of endophytic bacteria isolated from in vitro cultures of peach and pear rootstocks. 3 Biotech 6:120. https://doi.org/10.1007/s13205-016-0442-6

Liotti RG, da Silva Figueiredo MI, da Silva GF, de Mendonça EAF, Soares MA (2018) Diversity of cultivable bacterial endophytes in *Paullinia cupana* and their potential for plant growth promotion and phytopathogen control. Microbiol Res 207:8–18. https://doi.org/10.1016/j.micres.2017.10.011

Liu S, Dai H, Makhloufi G, Heering C, Janiak C, Hartmann R, Mándi A, Kurtán T, Müller WE, Kassack MU (2016) Cytotoxic 14-membered macrolides from a mangrove-derived endophytic fungus, *Pestalotiopsis microspora*. J Nat Prod 79(9):2332–2340. https://doi.org/10.1021/acs.jnatprod.6b00473

Liu Y, Guo J, Li L, Mipeshwaree DA, Zhang Y, Osama AM, Nimaichand S, Li W (2017) Endophytic bacteria associated with endangered plant *Ferula sinkiangensis* K. M. Shen in an arid land: diversity and plant growth-promoting traits. J Arid Land 9(3):432–445. https://doi.org/10.1007/s40333-017-0015-5

López JL, Alvarez F, Príncipe A, Salas ME, Lozano MJ, Draghi WO, Jofré E, Lagares A (2018) Isolation, taxonomic analysis, and phenotypic characterization of bacterial endophytes present in alfalfa (*Medicago sativa*) seeds. J Biotechnol 267:55–62. https://doi.org/10.1016/j.jbiotec.2017.12.020

Losgen S, Magull J, Schulz B, Draeger S, Zeeck A (2008) Isofusidienols: novel chromone-3-oxepines produced by the endophytic fungus Chalara sp. Eur J Org Chem 4:698–703. https://doi.org/10.1002/ejoc.200700839

Lumactud R, Fulthorpe RR (2018) Endophytic bacterial community structure and function of herbaceous plants from petroleum hydrocarbon contaminated and non-contaminated sites. Front Microbiol 9:1926. https://doi.org/10.3389/fmicb.2018.01926

Mallik MAB (2001) Selective isolation and screening of soil microorganisms for metabolites with herbicidal potential. Allelopathy Agrosystems 4:219–236. https://doi.org/10.1300/J144v04n02_07

Manjunatha BS, Paul S, Aggarwal C, Bandeppa S, Govindasamy V, Dukare AS, Rathi MS, Satyavathi CT, Annapurna K (2019) Diversity and tissue preference of osmotolerant bacterial endophytes associated with pearl millet genotypes having differential drought susceptibilities. Microb Ecol 77(3):676–688. https://doi.org/10.1007/s00248-018-1257-2

Massimo NC, Nandi Devan MM, Arendt KR, Wilch MH, Riddle JM, Furr SH, Steen C, U'Ren JM, Sandberg DC, Arnold AE (2015) Fungal endophytes in aboveground tissues of desert plants: infrequent in culture, but highly diverse and distinctive symbionts. Microb Ecol 70(1):61–76. https://doi.org/10.1007/s00248-014-0563-6

Miller RV, Miller CM, Kinney DG, Redgrave B, Sears J, Condron M et al (1998) Ecomycins, unique antimycotics from Pseudomonas viridiflava. J Appl Microbiol 84:937–944. https://doi.org/10.1128/MMBR.67.4.491-502.2003

Mucciarelli M, Camusso W, Maffei M, Panicco P, Bicchi C (2007) Volatile terpenoids of endophyte-free and infected peppermint (Mentha piperita L.): chemical partitioning of a symbiosis. Microb Ecol 54:685–696. https://doi.org/10.1007/s00248-007-9227-0

Mufti R, Amna Rafiqu M, Haq F, Hussain M, Munis Masood S, Mumtaz AS, Chaudhary HJ (2015) Genetic diversity and metal resistance assessment of endophytes isolated from *Oxalis corniculata*. Soil Environ 34(1):89–99. https://www.researchgate.net/publication/277233421

Murugan KK, Poojari CC, Ryavalad C, Melappa G (2017) Antidiabetic activity of endophytic fungi, Penicillium species of *Tabebuia argentea*; *in silico* and experimental analysis. Res J Phytochem 11:90–110. https://doi.org/10.3923/rjphyto.2017.90.110

Naik BS, Abrar S, Krishnappa M (2019) Industrially important enzymes from fungal endophytes. In: Yadav A, Mishra S, Singh S, Gupta A (eds) Recent advancement in white biotechnology through Fungi. Fungal biology. Springer, Cham, pp 263–280. https://doi.org/10.1007/978-3-030-10480-1_7

Nair DN, Padmavathy S (2014) Impact of endophytic microorganisms on plants, environment and humans. Sci World J 2014:250693. https://doi.org/10.1155/2014/250693

Oses R, Frank AC, Valenzuela S, Rodríguez J (2018) Nitrogen fixing endophytes in forest trees. In: Pirttilä A, Frank A (eds) Endophytes of forest trees. Forestry sciences, vol 86. Springer, Cham, pp 191–204. https://doi.org/10.1007/978-3-319-89833-9_9

Pereira SIA, Monteiro C, Vega AL, Castro PML (2016) Endophytic culturable bacteria colonizing *Lavandula dentata* L. plants: Isolation, characterization and evaluation of their plant growth-promoting activities. Ecol Eng 87:91–97. https://doi.org/10.1016/j.ecoleng.2015.11.033

Petrini O (1991) Fungal endophytes of tree leaves. In: Andrews JH, Hirano SS (eds) Microbial ecology of leaves. Springer-Verlag, New York, NY, pp 179–197. https://doi.org/10.1007/978-1-4612-3168-4_9

Pimentel MR, Molina G, Dionisio AP, Maróstica MR, Pastore GM (2011) Use of endophytes to obtain bioactive compounds and their application in biotransformation process. Biotechnol Res Int. https://doi.org/10.4061/2011/576286

Postma JWM, Olsson PA, Falkengren-Grerup U (2007) Root colonisation by arbuscular mycorrhizal, fine endophytic and dark septate fungi across a pH gradient in acid beech forests. Soil Biol Biochem 39(2):400–408. https://doi.org/10.1016/j.soilbio.2006.08.007

Potshangbam M, Devi SI, Sahoo D, Strobel GA (2017) Functional characterization of endophytic fungal community associated with *Oryza sativa* L. and *Zea mays* L. Front Microbiol 8:325–330. https://doi.org/10.3389/fmicb.2017.00325

Powell RG, Smith CR (1980) Antitumor agents from higher plants. In: The resource potential in phytochemistry, vol 14. Springer, New York, pp 23–51. https://doi.org/10.1007/978-1-4684-8309-3_2

Praptiwi MR, Wulansari D, Fathoni A, Agusta A (2018) Antibacterial and antioxidant activities of endophytic fungi extracts of medicinal plants from Central Sulawesi. J Appl Pharmaceut Sci 8(08):069–074. https://doi.org/10.7324/JAPS.2018.8811

Pujiyanto S, Lestari Y, Suwanto A, Budiarti S, Darusman LK (2012) Alpha-glucosidase inhibitor activity and characterization of endophytic actinomycetes isolated from some Indonesian diabetic medicinal plants. Int J Pharm Pharm Sci 4(1):327–333. https://www.researchgate.net/publication/262728745

Qin S, Xing K, Jiang J-H, Xu L-H, Li W-J (2011) Biodiversity, bioactive natural products and biotechnological potential of plant-associated endophytic actinobacteria. Appl Microbiol Biotechnol 89(3):457–473. https://doi.org/10.1007/s00253-010-2923-6

Rajamanikyam M, Vadlapudi V, Amanchy R, Upadhyayula SM (2017) Endophytic fungi as novel resources of natural therapeutics. Braz Arch Biol Technol 60:e17160542. http://www.plantsjournal.com/archives/?year=2018&vol=6&issue=1&part=C&ArticleId=764

Ramalakshmi K, Prasanna VK, Magesh K, Sanjana R, Siril JS, Ravibalan KA (2018) potential surface sterilization technique and culture media for the isolation of endophytic bacteria from *Acalypha indica* and its antibacterial activity. J Med Plant Stud 6:181–184. http://www.plantsjournal.com/archives/?year=2018&vol=6&issue=1&part=C&ArticleId=764

Rani V (2016) Endophytes isolated from the plants of Himachal Pradesh and their importance. Int J Curr Microbiol App Sci 5:105–110. https://doi.org/10.20546/ijcmas.2016.509.012

Richardson AE, Hadobas PA (1997) Soil isolates of *Pseudomonas* spp. that utilize inositol phosphates. Can J Microbiol 43:509–516. https://doi.org/10.1139/m97-073

Rohini S, Aswani R, Kannan M, Sylas VP, Radhakrishnan EK (2018) Culturable endophytic bacteria of ginger rhizome and their remarkable multi-trait plant growth-promoting features. Curr Microbiol 75:505–511. https://doi.org/10.1007/s00284-017-1410-z

Saad MMG (2019) Herbicidal activities of endophytic fungi and its secondary metabolites. Allelopath J 46:51–60. https://doi.org/10.26651/allelo.j/2019-46-2-1208

Saha M, Sarkar S, Sarkar B, Sharma BK, Bhattacharjee S, Tribedi P (2016) Microbial siderophores and their potential applications: a review. Environ Sci Pollut Res 23:3984–3999. https://doi.org/10.1007/s11356-015-4294-0

Sánchez-Cruz R, Vázquez IT, Batista-García RA, Méndez-Santiago EW, del Rayo Sánchez-Carbente M, Leija A, Lira-Ruan V, Hernández G, Wong-Villarreal A, Folch-Mallol JL (2019) Isolation and characterization of endophytes from nodules of *Mimosa pudica* with biotechnological potential. Microbiol Res 218:76–86. https://doi.org/10.1016/j.micres.2018.09.008

Santos CM, Ribeiro AS, Garcia A, Polli AD, Polonio JC, Azevedo JL, Pamphile JA (2019) Enzymatic and antagonist activity of endophytic fungi from *Sapindus saponaria* L. (*Sapindaceae*). Acta Biol Colomb 24(2):322–330. https://doi.org/10.15446/abc.v24n2.74717

Satari AH, Zargar MI, Shah WA, Bansal R, Bhat MF (2018) Isolation, molecular identification, phytochemical screening and in vitro antioxidant activity of endophytic fungi from *Achillea millefolium* Linn. J Pharmacogn Phytochem 7:87–92. http://www.phytojournal.com/archives/?year=2018&vol=7&issue=4&ArticleId=4878

Sato F, Kumagai H (2013) Microbial production of isoquinoline alkaloids as plant secondary metabolites based on metabolic engineering research. Proc Jpn Acad Ser B 89:165–181. https://doi.org/10.2183/pjab.89.165

Savi DC, Shaaban KA, Vargas N, Ponomareva LV, Possiede YM, Thorson JS, Glienke C, Rohr J (2015) Microbispora sp. LGMB259 endophytic actinomycete isolated from Vochysia divergens (Pantanal, Brazil) producing β-carbolines and indoles with biological activity. Curr Microbiol 70:345–354. https://doi.org/10.1007/s00284-014-0724-3

Schwyn B, Neilands JB (1987) Universal chemical assay for the detection and determination of siderophores. Anal Biochem 160:45–56. https://doi.org/10.1016/0003-2697(87)90612-9

Selvi BK, Balagengatharathilagam P (2014) Isolation and screening of endophytic fungi from medicinal plants of Virudhunagar district for antimicrobial activity. Int J Sci Nat 5:147–155. https://pdfs.semanticscholar.org/9ff9/9206cbf3bc73d3f129e102a5968a26ec7ed7.pdf

Shabanamol S, Divya K, George TK, Rishad KS, Sreekumar TS, Jisha MS (2018) Characterization and in planta nitrogen fixation of plant growth promoting endophytic diazotrophic *Lysinibacillus sphaericus* isolated from rice (*Oryza sativa*). Physiol Mol Plant Pathol 102:45–54. https://doi.org/10.1016/j.pmpp.2017.11.003

Sharma P, Baunthiyal M (2018) Endophytic *Actinobacteria* from *Pinus roxburghii*: isolation, diversity and antimicrobial potential against human pathogens. J Pharmacogn Phytochem 7:3021–3027. http://www.phytojournal.com/archives/2018/vol7issue5/PartAY/7-5-175-234.pdf

Shen M, Liu L, Li DW, Zhou WN, Zhou ZP, Zhang CF, Luo YY, Wang HB, Li HY (2013) The effect of endophytic *Peyronellaea* from heavy metal-contaminated and uncontaminated sites on maize growth, heavy metal absorption and accumulation. Fungal Ecol 6:539–545. https://doi.org/10.1016/j.funeco.2013.08.001

Shiono Y, Murayama T, Takahashi K, Okada K, Katohda S, Ikeda M (2005) Three oxygenated Cyclohexenone derivatives produced by an endophytic fungus. Biosci Biotechnol Biochem 69:287–292. https://doi.org/10.1271/bbb.69.287

Sindhu R, Binod P, Pandey A (2017) 1—α-Amylases. In: Pandey A, Negi S, Soccol CR (eds) Current developments in biotechnology and bioengineering. Elsevier, Amsterdam, pp 3–24. https://doi.org/10.1016/B978-0-444-63662-1.00001-4

Singh R, Dubey AK (2015) Endophytic actinomycetes as emerging source for therapeutic compounds. Indo Global J Pharm Sci 5:106–116. https://www.researchgate.net/publication/298423646

Singh M, Kumar A, Singh R, Pandey KD (2017) Endophytic bacteria: a new source of bioactive compounds. 3 Biotech 7:315. https://doi.org/10.1007/s13205-017-0942-z

Singh H, Naik B, Kumar V, Bisht GS (2018) Screening of endophytic actinomycetes for their herbicidal activity. Annal Agrarian Sci 16:101–107. https://doi.org/10.1016/j.aasci.2017.11.001

Singh J, Kundu D, Das M, Banerjee R (2019) Enzymatic processing of juice from fruits/vegetables: an emerging trend and cutting edge research in food biotechnology. In: Kuddus M (ed) Enzymes in food biotechnology. Academic Press, New York, pp 419–432. https://doi.org/10.1016/B978-0-12-813280-7.00024-4

Smith MM, Warren VA, Thomas BS, Brochu RM, Ertel EA, Rohrer S, Schaeffer J, Schmatz D, Petuch BR, Tang YS (2000) Nodulisporic acid opens insect glutamate-gated chloride channels: identification of a new high affinity modulator. Biochemistry 39:5543–5554. https://doi.org/10.1021/bi992943i

Snipes CE, Duebelbeis DO, Olson M, Hahn DR, Dent Iii WH, Gilbert JR, Werk TL, Davis GE, Lee-Lu R, Graupner PR (2007) The ansacarbamitocins: polar ansamitocin derivatives. J Nat Prod 70(10):1578–1581

Stein T (2005) *Bacillus subtilis* antibiotics: structures, syntheses and specific functions. Mol Microbiol 56:845–857. https://doi.org/10.1111/j.1365-2958.2005.04587.x

Stierle A, Strobel G, Stierle D (1993) Taxol and taxane production by Taxomyces andreanae, an endophytic fungus of Pacific yew. Science 260:214–216. https://doi.org/10.1126/science.8097061

Stierle AA, Stierle DB, Bugni T (1999) Sequoiatones A and B: novel antitumour metabolites isolated from a redwood endophyte. J Organomet Chem 64:5479–5484. https://doi.org/10.1021/jo990277l

Strobel GA (2003) Endophytes as sources of bioactive products. Microbes Infect 5:535–544. https://www.ncbi.nlm.nih.gov/pubmed/12758283

Strobel GA, Daisy B (2003) Bioprospecting for microbial endophytes and their natural products. Microbiol Mol Biol Rev 67:491–502. https://doi.org/10.1128/mmbr.67.4.491-502.2003

Strobel G, Ford E, Worapong J, Harper JK, Arif AM, Grant DM, Fung PC, Chau RMW (2002) Isopestacin, an isobenzofuranone from *Pestalotiopsis microspora*, possessing antifungal and antioxidant activities. Phytochemistry 60(2):179–183. https://doi.org/10.1016/s0031-9422(02)00062-6

Subbulakshmi GK, Thalavaipandian A, Bagyalakshmi RV, Rajendran A (2012) Bioactive endophytic fungal isolates of Biota orientalis (L) Endl., *Pinus excelsa* Wall. and *Thujaoccidentalis* L. Int J Adv Life Sci 4:9–15. http://www.unitedlifejournals.com/ijals/view-pdf.php?id=34

Sun RW, Chen R, Chung NP, Ho CM, Lin CL, Che CM (2005) Silver nanoparticles fabricated in Hepes buffer exhibit cytoprotective activities toward HIV-1 infected cells. Chem Commun (Camb) 40:5059–5061. https://doi.org/10.1039/b510984a

Sun H, He Y, Xiao Q, Ye R, Tian Y (2013) Isolation, characterization, and antimicrobial activity of endophytic bacteria from *Polygonum cuspidatum*. Afr J Microbiol Res 7:1496–1504. https://doi.org/10.5897/AJMR12.899.

Sunitha VH, Devi DN, Srinivas C (2013) Extracellular enzymatic activity of endophytic fungal strains isolated from medicinal plants. World J Agricult Sci 9(1):01–09. https://doi.org/10.5829/idosi.wjas.2013.9.1.72148.

Sunkar S, Nachiyar CV (2012) Biogenesis of antibacterial silver nanoparticles using the endophytic bacterium Bacillus cereus isolated from Garcinia xanthochymus. Asian Pac J Trop Biomed 2:953–959. https://doi.org/10.1016/S2221-1691(13)60006-4

Syed CS, Mounika PP, Mounika Y, Mounika S, Kumar S, Bai VT, Audipudi AV (2017) Evaluation of antimicrobial and antibiotic sensitivity of Chilli root endophytic bacteria for eco-friendly biofertilizer. Int J Curr Microbiol App Sci 5:45–53. https://www.ijcmas.com/special/5/Chandini%20S.%20Syed2,%20et%20al.pdf

Szymańska S, Borruso L, Brusetti L, Hulisz P, Furtado B, Hrynkiewicz K (2018) Bacterial microbiome of root-associated endophytes of *Salicornia europaea* in correspondence to different levels of salinity. Environ Sci Pollut Res Int 25:25420–25431. https://doi.org/10.1007/s11356-018-2530-0

Taechowisan T, Chanaphat S, Ruensamran W, Phutdhawong WS (2012) Anti-inflammatory effect of 3-methylcarbazoles on RAW 2647 cells stimulated with LPS, polyinosinic-polycytidylic acid and Pam3CSK. Adv Microbiol 2:98–103. https://doi.org/10.3390/molecules16087132.

Taechowisan T, Chaisaeng S, Phutdhawong WS (2017) Antibacterial, antioxidant and anticancer activities of biphenyls from Streptomyces sp. BO-07: an endophyte in Boesenbergia rotunda (L.) Mansf A. Food Agric Immunol 28:1330–1346. https://doi.org/10.1080/09540105.2017.1339669

Taylor PL, Omotoso O, Wiskel JB, Mitlin D, Burrell RE (2005) Impact of heat on nanocrystalline silver dressings. Part II: physical properties. Biomaterials 26:7230–7240. https://doi.org/10.1016/j.biomaterials.2005.05.041.

Turbyville TJ, Wijeratne EMK, Liu MX, Burns AM, Seliga CJ, Luevano LA, David CL, Faeth SH, Whitesell L, Gunatilaka AAL (2006) Search for HSP 90 inhibitors with potential anticancer activity: isolation and SAR studies of radicicol and monocillin I from two plant-associated fungi of the Sonoran desert. J Nat Prod 69:178–184. https://doi.org/10.1021/np058095b.

Ullah A, Mushtaq H, Ali U, Hakim AE, Mubeen S (2018) Screening, isolation, biochemical and plant growth promoting characterization of endophytic bacteria. Microbiol Curr Res 2(3):62–68. https://doi.org/10.4066/2591-8036.18-368.

Uzma F, Konappa NM, Chowdappa S (2016) Diversity and extracellular enzyme activities of fungal endophytes isolated from medicinal plants of Western Ghats, Karnataka. Egypt J Basic Appl Sci 3(4):335–342. https://doi.org/10.1016/j.ejbas.2016.08.007

van Lenteren JC, Bolckmans K, Köhl J, Ravensberg WJ, Urbaneja A (2018) Biological control using invertebrates and microorganisms: plenty of new opportunities. Biol Control 63:39–59. https://doi.org/10.1007/s10526-017-9801-4

Verma SK, Gond SK, Mishra A, Sharma VK, Kumar J, Singh DK, Kumar A, Goutam J, Kharwar RN (2014) Impact of environmental variables on the isolation, diversity and antibacterial activity of endophytic fungal communities from *Madhuca indica* Gmel. at different locations in India. Ann Microbiol 64(2):721–734. https://doi.org/10.1007/s13213-013-0707-9

Verma SK, Kingsley KL, Bergen MS, Kowalski KP, White JF (2018) Fungal disease prevention in seedlings of rice (*Oryza sativa*) and other grasses by growth-promoting seed-associated endophytic bacteria from invasive *Phragmites australis*. Microorganisms 6(1):E21. https://doi.org/10.3390/microorganisms6010021

Villarreal-Delgado MF, Villa-Rodríguez ED, Cira-Chávez LA, Estrada Alvarado MI, Parra-Cota FI, de los Santos-Villalobos S (2018) The genus bacillus as a biological control agent and its implications in the agricultural biosecurity. Mex J Phytopathol 36:95–130. https://doi.org/10.18781/R.MEX.FIT.1706-5

Vu HT, Nguyen DT, Nguyen HQ, Chu HH, Chu SK, Chau MV, Phi QT (2018) Antimicrobial and cytotoxic properties of bioactive metabolites produced by Streptomyces cavourensis YBQ59 isolated from *Cinnamomum cassia* prels in yen bai province of vietnam. Curr Microbiol 75:1247–1255. https://doi.org/10.1007/s00284-018-1517-x

Vyas P, Kaur R (2019) Culturable stress-tolerant plant growth-promoting bacterial endophytes associated with *Adhatoda vasica*. J Soil Sci Plant Nutr 19:290–298. https://doi.org/10.1007/s42729-019-00028-9

Wagenaar M, Corwin J, Strobel GA, Clardy J (2000) Three new cytochalasins produced by an endophytic fungus in the genus Rhinocladiella. J Nat Prod 63:1692–1695. https://doi.org/10.1021/np0002942

Waheeda K, Shyam KV (2017) Formulation of novel surface sterilization method and culture media for the isolation of endophytic actinomycetes from medicinal plants and its antibacterial activity. J Plant Pathol Microbiol 8:399. https://doi.org/10.4172/2157-7471.1000399

Webber J (1981) A natural biological control of Dutch elm disease. Nature 292(5822):449–451. http://agris.fao.org/agris-search/search.do?recordID=GB19820813296

Yu H, Zhang L, Li L, Sun P, Qin L (2010) Recent developments and future prospects of antimicrobial metabolites produced by endophytes. Microbiol Res 165:437–449. https://doi.org/10.1016/j.micres.2009.11.009

Zhang B, Salituro G, Szalkowski D, Li Z, Zhang Y, Royo I, Vilella D, Dez M, Pelaez F, Ruby C, Kendall RL, Mao X, Griffin P, Calaycay J, Zierath JR, Heck JV, Smith RG, Moller DE (1999) Discovery of small molecule insulin mimetic with antidiabetic activity in mice. Science 284:974–981. https://doi.org/10.1126/science.284.5416.974.

Zhang Q, Zhang J, Yang L, Zhang L, Jiang D, Chen W, Li G (2014) Diversity and biocontrol potential of endophytic fungi in *Brassica napus*. Biol Control 72:98–108. https://doi.org/10.1016/j.biocontrol.2014.02.018

Zhang H, Bai X, Zhang M, hen J, Wang H (2018) Bioactive natural products from endophytic microbes. Nat Product J 8:86–108. https://doi.org/10.2174/2210315508666180103160508

Zhao K, Penttinen P, Guan T, Xiao J, Chen Q, Xu J (2011) The diversity and anti-microbial activity of endophytic actinomycetes isolated from medicinal plants in Panxi Plateau China. Curr Microbiol 62:182–190. https://doi.org/10.1007/s00284-010-9685-3

Zheng YK, Miao CP, Chen HH, Huang FF, Xia YM, Chen YW, Zhao LX (2017) Endophytic fungi harbored in *Panax notoginseng*: diversity and potential as biological control agents against host plant pathogens of root-rot disease. J Ginseng Res 41:353–360. https://doi.org/10.1016/j.jgr.2016.07.005

Zou WX, Meng JC, Lu H, Chen GX, Shi GX, Zhang TY, Tan RX (2000) Metabolites of Colletotrichum gloeosporioides, an endophytic fungus in Artemisia mongolica. J Nat Prod 63:1529–1530. https://doi.org/10.1021/np000204t

8 Applications of Microorganisms in Agriculture

Khirood Doley, Ajinkya Terkar, and Mahesh Borde

Abstract

At present, the major global challenge is to accomplish future food security without interfering with the present environment or ecosystem. The global crop production suffers largely due to several pests, insects, or diseases which are being controlled widely by the use of detrimental agrochemicals which are now being considered to damage our health and ecosystem. Therefore, various other alternatives of biological origin are being looked upon for their application as bio-fertilizer or biological control agents such as arbuscular mycorrhizal fungi (AMF), *Trichoderma* spp., plant growth-promoting rhizobacteria (PGPR), and endophytes. Moreover, many other probable microorganisms are still being discovered, and their ecological roles are being studied as well. Therefore, appropriate selection and investigation for applying them effectively with the use of novel technologies have huge potential to safeguard our future food and environment. In addition, many underlying mechanisms which are previously unknown during interaction for crop health improvement can be unveiled by the use of modern technologies such as clustered regularly interspaced short palindromic repeats (CRISPR/Cas), transcriptomics, proteomics, genomics, etc. Even plant growth-promoting traits are being tailored by the use of modern gene engineering techniques which will definitely improve overall plant health, thereby leading to food security. Thus, this chapter presents a brief overview of recent trends in application of various microbial interactions with the twenty-first century technology for crop productivity and overall sustainability of our agricultural ecosystem for our future generation.

Keywords

Agrochemicals · Biological · Ecosystem · Environment · Novel technology

K. Doley (✉) · A. Terkar · M. Borde
Department of Botany, Savitribai Phule Pune University, Pune, Maharashtra, India
e-mail: khirood.doley@unipune.in

P. K. Arora (ed.), *Microbial Technology for Health and Environment*,
Microorganisms for Sustainability 22,
https://doi.org/10.1007/978-981-15-2679-4_8

8.1 Introduction

In the near future, the population of billions will need ample amount of food grains as food. In addition, the major challenges ahead include safeguarding of food from any pathogen infection and harmful chemical pesticides. Therefore, the crop productivity needs to be increased with rise in food demand along with maintaining the overall health of the plant as a whole. On the contrary, currently, the conventional practices that are being adopted by growers especially in developing and less developed countries to optimize food production have been the excessive or less concentration of chemicals probably due to ignorance (Gahukar 2014). The indiscriminate applications of chemicals that are produced in industries are already reported to cause several varieties of pollution to our ecosystem (Youssef and Eissa 2014; Singh et al. 2017). In this respect, the feeding of our ever-growing population sustainably taking into account the limited resources, reducing the use of inorganic fertilizers, maintaining agricultural productivity, biosafety, and novel innovations in agricultural practices has been the priority at present. As per the Food and Agriculture Organization (FAO 2019), India has recently achieved food grain self-sufficiency but with serious concerns in terms of sustainability. Therefore, there is an increased need for innovative farm production in a sustainable manner where suitable alternatives to chemical fertilizers that have an ability to enhance crop production by improving plant nutrition, stress tolerance, and protection from plant pathogens attract the growing demand (Olanrewaju et al. 2017). Even though the government agencies worldwide derive various schemes for growers at socioeconomic levels, there is an ultimate need at the ground level, which includes fertility of soil, plant pathogen, and crop production. So, for that to happen, a suitable scientific awareness and novel technologies are needed for agricultural practices where various other disciplines of biotechnology, nano-biotechnology, nanotechnology, microbial technology, etc., may be incorporated (Srilatha 2011). And, historically, the use of these biotechnological techniques has been evidenced since 7000 BC in various fields of food, agricultural, fuel, medical, or environment technologies (Vitorino and Bessa 2017). Since the microbial application has several implications in a wide variety of fields, hence, it seems to be the most suitable application without further distressing our natural resources or ecosystems for future generations to come (Sengupta and Gunri 2015; Singh et al. 2017; Woo and Pepe 2018). However, here mostly the perspective of beneficial microbes and their roles in alleviation in overall health of plant either by direct or indirect manner and related novel studies are being emphasized.

8.2 Microbe–Plant Interaction

8.2.1 Rhizosphere

The plant rhizosphere consists of numerous microorganisms that inhabit above- or underground and possess functional diversity which proves to be beneficial for the plants. The root of a plant in rhizosphere is accompanied by various other microorganisms such as plant growth-promoting bacteria (PGPR), rhizobia, *Trichoderma* sp., endophytes, and arbuscular mycorrhizal fungi (AMF). The equilibrium between these microorganisms and plants promotes better disease-free growth and development. Therefore, microbial communities associated with plant play a significant role in plant nutrition, plant growth, and carbon or nitrogen cycling and during biotic or abiotic stress conditions (Van Der Heijden et al. 2008; Vacheron et al. 2015). In ecosystem, crop productivity is largely manipulated by interactions between plant and microbes (Emmett et al. 2017). The microbial community compositions are distinct when compared to the bulk soil and are supported by the physiochemical environment of the rhizosphere (Mendes et al. 2014). So far, various investigations have undertaken on the association of plant with microbes that signifies the very importance of these microbes in the determination of plant health (Gill et al. 2016). But understanding the biological root system and its associated microbiota is still at its initial stage largely due to astounding numbers of interactions, enormous species diversity, and complex community structure that exists within the rhizosphere. Apart from that, there are reports that among microbial communities social behaviors are exhibited, viz., cooperation, colonization of ecological niches, host infection, resistance to invaders, and invasion by parasites and pathogens (Besset-Manzoni et al. 2018). In addition, remarkable cooperation among the microbial communities has been observed which may be mutualistic or altruistic in nature where donor or recipient may be positively or negatively affected (West et al. 2007). The overall global microbial diversity is largely unexplored which is considered to be one of the largest reservoirs of biodiversity on Earth. Thus, it may prove to be an important frontier in biology which requires intensive further investigations. Hence, for sustainable function of agrosystems, the interactions among several organisms are extremely essential factors.

8.2.1.1 Role of Bacteria

Microbial antagonists are finding its way into biological control of several plant pathogens which will serve as an alternative to agrochemicals. The rhizosphere around plant is being seen as resource for nutritional and water requirements which will be helpful in future agricultural practices. The microbes present in the rhizosphere have the potential in providing fertility to soil, thereby elevating the crop productivity. One of them is the bacteria which are known as plant growth-promoting rhizobacteria (PGPR), and it is one of the most abundant groups of microorganisms that coexist with algae, actinomycetes, fungi, and protozoa in the rhizosphere. The prominent presence of bacterial population in the rhizosphere has been expected to be influential on various physiological processes of plants as they are present nearest to the plant roots (Barriuso et al. 2008). These beneficial bacteria belong to the several genera of *Acetobacter, Acinetobacter, Alcaligenes, Arthrobacter, Achromobacter, Azoarcus, Azospirillum, Azotobacter, Bacillus, Beijerinckia, Burkholderia, Derxia, Enterobacter, Gluconacetobacter, Herbaspirillum, Klebsiella, Microbacterium, Ochrobactrum, Pantoea, Paenibacillus, Pseudomonas, Rhodococcus, Serratia, Stenotrophomonas, Zoogloea,* etc., and have been subject of extensive research and marked as biological control agents throughout the years (Babalola 2010; Spaepen and Vanderleyden 2011; El-Borollosy and Oraby 2012; Oku et al. 2014; Lee et al. 2015; Kumar and Verma 2018). The abundance of PGPR in the vicinity of plant has a vast role in enhancing potential plant growth in various important crops of agricultural importance (Santoyo et al. 2016). These PGPR assist in the development of plant by various mechanisms such as triggering of plant growth hormones, antioxidant system, siderophore production, and augmentation of nutritional capacity of the plants (Kloepper 2003; Kumar and Verma 2018). The ability of siderophore production in bacteria has been correlated with its biocontrol against phytopathogenic fungi (Buysens et al. 1996). The siderophore-producing bacterium, *Bacillus subtilis* CAS15, has been reported to promote biocontrol of pathogen *F. oxysporum* that causes *Fusarium* wilt in pepper plant (Yu et al. 2011). Therefore, considering the benefits of PGPR mediation in agricultural system, it may soon become an alternative strategy against synthetic chemicals that are doing more harm to our ecosystem than any good.

8.2.1.2 Role of Fungi

Fungi are a well-known group that are often being associated in agriculture for alleviation in crop production, and partially they have been introduced in the present agricultural systems. However, the full-fledged application still needs further breakthrough due to its low acceptance among actual growers despite it being more sustainable for the future agrosystems. One such fungi *Trichoderma* spp. (teleomorph *Hypocrea*) is asexual fungal genus which is ubiquitous and often predominant component of the mycoflora in soil, organic matter, litter, and rhizospheric ecosystem of almost all climatic zones as saprophytes. The effects of genus *Trichoderma* have been attributed to its biological control activities and during various biotic or abiotic stress conditions which belong to species *T. asperellum, T. atroviride, T. harzianum, T. resseyi, T. virense, and T. longibrachiatum* (Alizadeh et al. 2013; Contreras-Cornejo et al. 2014; Devi et al. 2017; Mona et al. 2017; Sabbagh et al. 2017; Saravanakumar et al. 2017; Téllez-Vargas et al. 2017). Different genus of *Trichoderma* sp. has long been known to be contributing to crop productivity by mediating with several mechanisms which include antibiotic activities, mycoparasitism, cell wall lytic enzymes, improving health of plant, enhancing plant growth, and alleviation of defense mechanisms against several pathogens. *Trichoderma* sp. contributes to plant growth by increasing soil fertility (Harman 2000; Shoresh et al. 2010); they contribute to the release of plant growth regulators, thereby promoting root growth and making nutrient available for uptake (Benitez et al. 2004). Moreover, recent investigations revealed that *Trichoderma* sp. are opportunistic, avirulent plant symbionts apart from being parasitic to other fungi as well. *Trichoderma* sp. establishes long-lasting colonizations which are of vigorous nature on the root surfaces which leads penetration into the epidermis. The fungi also induce resistance against various other pathogens and nematodes and bioremediation of heavy metals and environmental pollutants. But the main reason for *Trichoderma* sp. applications as biopesticides has been the ability of these fungi to sense, invade, and obliterate other fungi.

In addition, another fungus such as arbuscular mycorrhizal fungi (AMF) that belongs to subphylum *Glomeromycotina* engages significantly with several plants in their development mainly to overcome phosphorous (P) deficiency (Smith and Smith 2011). Phosphorous is an important nutrient for growth and overall well-being of plant as it is the fundamental component of nucleic acids and phospholipids. Plants require it in the form of inorganic phosphate (Pi) in much higher amount. Due to this higher assimilation in soil, an area of Pi depletion zone is observed. So, to overcome the situation of Pi limitation, plants involve itself in widespread branching by increasing their root lengths and making use of organic acid and phosphatase secretions for the solubilization of Pi present in soil. In this mutuality, the host plant provides carbohydrate source to their partner which is essential for fungal growth, and nutrition is improved in host plant (Luginbuehl and Oldroyd 2017). The AM fungi have been distinguished for its ability to incur biological protection against several soilborne fungal plant pathogens or abiotic stresses (Pereira et al. 2015). However, it has been observed that non-host AM plant can inhibit AMF colonization and recognition of signals on the surfaces of roots that induces hyphal growth and branching are found to be in off condition (Giovannetti and Sbrana 1998). Nevertheless, the mycorrhizal colonized plants have conferred resistance against a variety of pathogens (Pozo and Azcón-Aguilar 2007; Jung et al. 2012; Cangelosi et al. 2017). The mechanisms that are employed by AM fungi for biocontrol of pathogen involve AMF's direct effect on pathogen or indirect effect via plant and competition for space and nutrients and morphological alterations in the host root tissues (Dalpe 2005; Schouteden et al. 2015). But the actual mediation by mycorrhiza and its molecular mechanisms is still not very clear. However, despite the several investigations that suggest employment of AMF as biological control agents, it is still not viewed practically applicable in actual field among growers (Salvioli and Bonfante 2013). Thus, a clear insight into their modes of action will lead to definite employment in the near future.

8.3 Biopesticides

Due to population growth, there is ever-increasing demand for more food (Keinan and Clark 2012). Therefore, several nations worldwide had undertaken "Green Revolution" programs to double up the productions. Despite increase in huge production of food crops, there is a constant threat of various insects, pests, and diseases which contributes to significant loss in yields. To overcome these threats, growers invariably have applied chemical-based fertilizers, insecticides, pesticides, fungicides, etc. As a result, we were able to achieve enormous production of food crops but at the cost of huge ecological impact on our current and future resources. This implied search for the innovations in current practices which is causing huge impact on sustainability of environment and life for the future. Thus, the phenomenon of integrated pest management (IPM) came into existence as an alternative to exploitations that was undertaken during Green Revolution, where application of microbial-based agrochemical was considered as environment-friendly and less hazardous to life on Earth as a whole. Even though the appropriate products under "biopesticides" are themselves contentious, several strategies from various disciplines are contributing to overall development of biopesticides for crop yield and protection from diseases. The preferential implication of biopesticide in utilizing them has been associated with its biological origin which necessitates contemporary and future demands. The biopesticides derived may include bacteria, fungi, entomopathogenic viruses, plant secondary metabolites, and nematodes. For instance, biopesticide such as nematophin from *Xenorhabdus nematophila* YL001 has been used in effective inhibition of mycelia growth of pathogenic *Rhizoctonia solani* (Zhang et al. 2019). Likewise, *Bacillus thuringiensis* (Bt)-derived crystalline proteins have been known to inhibit insect pest species like lepidopteran. In that, the determination of target insect gut receptor occurs by the binding of *Bt* crystalline proteins (Kumar 2012). Similarly, fungi, virus, and nematodes have shown its ability to act against several pests (Mnyone et al. 2010; Prater et al. 2006; Loya and Hower Jr 2002). Therefore, the worldwide use of the biopesticide-related products (1400 approx.) has been commercially available that accounts for 2.5% of the total pesticide market (Balog et al. 2017). And they are undergoing further research consequently to incorporate them in our agricultural systems owing to its several benefits.

8.4 Host–Microbe Interaction in Pathogenesis

Since the first evolution of plants 700 million years ago (Heckman et al. 2001), plants have been interacting with epiphytic, symbiotic, and pathogenic microbes. In those interactions, pathogenic establishment occurs firstly by accessing the interior side of plant cell by direct penetration or attack or by natural openings such as stomata present underside of leaves which functions as gaseous exchange or by wounds. Afterward, when it gains entry inside the plant cell, it encounters obstacle of cellulose wall on plant cell. The microbe that reaches host plasma by crossing cellulose is received with extracellular surface receptors that recognize pathogen-associated molecular patterns (PAMPs) by pattern recognition receptor (PRR) which triggers pattern or PAMP-triggered immunity (PTI) that inhibits their further spread which leads to MAPK cascade in transcription of defense-related genes. But the pathogens have developed systems to suppress PTI responses by secreting effector proteins. In view of the ability evolved in pathogen to contain primary defense, the plant then deals with pathogen by effector-triggered immunity (ETI) which is more advanced system to detect microbial proteins that were used to sabotage the PTI by plant resistance (R) proteins (Chisholm et al. 2006). However, the pathogens have so far been able to evolve to get away from PAMP/MAMP-like defenses by injecting effector molecules in host cells. Therefore, studying these effectors will help understand how one avoids or decreases defense systems in the host.

8.4.1 Types of Plant Disease Management

8.4.1.1 Chemical Control

The current disease management has been largely dependent on chemical insecticides, pesticides, or fungicides as the main components in integrated pest management (IPM). Until now, they are used indiscriminately irrespective of whether it has been controlling the disease or not which are posing potential threat to our ecosystem. The most affected are the microbial communities besides soil and groundwater pollution which is a serious concern to organisms as a whole. Nowadays, environment-friendly management is preferred instead of synthetic or chemical based due to its detrimental effects (Shafique et al. 2016). Moreover, the pathogens are providing increased resistance to these chemical control strategies (Ma and Michailides 2005). Therefore, it is been recommended to phase out the use of any chemical-based agrochemicals (EU 2009). Thus, the need for the development of novel sustainable alternative technologies is then currently required to control plant disease so as to address the future of food security.

8.4.1.2 Biological Control

The major alternative strategy that is to avail at present is to employ organisms that possess characteristics of low developmental cost-effectiveness, environment friendliness, and effective inhibition of major plant disease-causing pathogens for a sustainable agriculture. In that direction, the use of microbes is perceived to be an inevitable tool for controlling of plant diseases in the future because it provides very first defense-related activities against several pathogens. In that direction, various plant-associated microbial populations have been studied extensively. The emphasis has been made toward employing biocontrol agents such as rhizospheric or endophytic bacteria, *Trichoderma* sp., AM fungi, etc. (Kiely et al. 2006; Jung et al. 2012; Guzmán-Guzmán et al. 2017). The advantage of utilizing these microbial populations not only provides disease suppression but also helps in overall development of plants. Besides, it provides pollutant-free sustainable agricultural system to our ecosystem. The typical characteristics of mechanisms exhibited by these microbes are antibiotic production, well-organized root colonization, competition, mycoparasitism, secondary metabolite production, nutrient acquisitions, etc. (Bonfante and Genre 2010; Sharif and Claassen 2011; Gveroska and Ziberoski 2012; Harman 2000; Khalili et al. 2012). However, despite many advantages in incorporating these microbes into our agricultural systems, still further extension in research is needed to fully exploit it to level of growers that remains a constant challenge.

8.4.1.3 Resistant Varieties

There has been increase in procreation for resistant varieties as they are considered to provide dependable protection against several plant pathogens (Kamthan et al. 2016). This has been achieved mainly by genetic manipulation of plants by means of chromosomal or extrachromosomal DNA modification directly or indirectly, resulting into the formation of genetically modified organisms (GMOs) or genetically modified plants or crops (Zhang et al. 2016). The approach has gained significant success among growers as they are perfect for providing good-quality plant yields and sustenance ability during unfavorable environment (Lopez-Arredondo et al. 2015). Often the resistant variety is preferred as a means to control several diseases of plant because it is more economical and environment-friendly when compared with industrially developed agrochemicals (Gupta et al. 2014). However, there has been reluctance in employing it entirely as some plants that come under the term of "genetically modified" have to come under the scrutiny of various regulatory approvals, ethics, and consumer recognition (Garcia-Ruiz et al. 2018). For instance, in third world countries like India where technical and other infrastructural constraints contribute to significant yield losses to growers due to several pests. Therefore, genetically modified pest-resistant crops alleviate the damage incurred upon by the pests, and the introduction of *Bacillus thuringiensis* (Bt) cotton in India has been found to increase not only total output but also in reduction of pest-related losses. Furthermore, the yield has been found to be elevated when compared to other countries where introduction of genetically modified crops was undertaken to substitute chemical pest control (Qaim and Zilberman 2003). Despite all the recent achievements, the rise in unscrupulous experimentation has led to the foundation of new phenomenon of genetically resistant among organisms due to numerous causes, viz., mutations in pathogens toward virulence and sexual and asexual recombination events causing unique diseases in both plants and animals (Maghari and Ardekani 2011).

8.5 Emerging Biocontrol Strategies

The plant–microbiome interaction where plant has selective approaches toward certain microbes in rhizosphere or phyllosphere proves to be beneficial for plant growth and development (Vorholt 2012). The rhizospheric microbiomes assist directly or indirectly in the development of plant by nutrient acquisition or plant pathogen inhibition. Thus, it provides novel developmental opportunities in the field of valuable interaction between plant and microbiome for modern-day biocontrol methodologies (Mueller and Sachs 2015). The low-molecular-weight volatile organic compounds (VOCs) are carbon-containing compounds that play a significant role in conferring plant with basal immune system known as induced systemic resistance (ISR) as opposed to several pathogens (Bailly and Weisskopf 2017). There have been suggestions that these VOC-interceded responses are present in plants, bacteria, and fungi (Bailly and Weisskopf 2017; Tyc et al. 2015; Werner et al. 2016). However, the exact molecular mechanism lays steps for further investigations.

Bacteriophage-based biocontrol against bacterial wilt is another strategy in agricultural systems which is environment-friendly. Bacteriophages (phages) are viruses which inhibit bacterial infections by infecting them. The life cycle of bacteriophage has been described as lytic or lysogenic. In lytic phage, there is direct destruction of host bacterial cell, while the lysogenic phage works by integration of bacterial genome into the host without destruction of host bacterial cell (Alvarez and Biosca 2017). A study by Bhunchoth et al. (2015) showed that the lytic life cycle of bacteriophage was able to inhibit *Ralstonia solanacearum* causing bacterial wilt in tomato plants. Therefore, it is one of the most promising biological control strategies in an organic way.

The phyllosphere (aerial part of plants)-associated biocontrol is also an important aspect because it has been considered to be the habitat for diverse community of microbes (Vorholt 2012). The microorganisms of phyllosphere contribute to the promotion of plant growth by production of growth hormones, increase in photosynthetic activity, and elongation of root structure (Mwajita et al. 2013). Even though the demonstration by phyllosphere communities has resulted into significant applications such as biocontrol agents (Michavila et al. 2017), yet it has not received much attention as in the case of vegetables and fruit plants (Leff and Fierer (2013). Moreover, the cooperation of phyllosphere communities has been complicated because of its complete exposure to atmosphere which makes it susceptible to external factors, viz., air particulates, light, UV radiation, and biological inoculants (Williams et al. 2013; Carvalho and Castillo 2018). There have been reports that various isolates from the phyllosphere and rhizosphere of *Drosera spatulata* Lab. were found to be producing siderophores which are known as an alternative for the biocontrol of many pathogens (Fu et al. 2016). In addition, the phyllosphere-associated *Pseudomonas syringae* pv. *syringae* have demonstrated the ability of biocontrol by resisting *P. syringae* pv. *glycinea* via indirect siderophore arbitration (Wensing et al. 2010). Hence, the phyllospheric microorganism opens up novel frontier for their contribution in overall fitness and development of plant.

8.6 Microbiological Technology Application in Agriculture

The contemporary necessity for microbiological involvement in agriculture is significant because it is supportive for attainment of higher productivity with sustainability in agriculture in a number ways. Therefore, the agricultural microbiology has its importance concerning the improvement in yield and disease management in crops (Pelczar et al. 1988). Microbiological technology may assist in increasing nutrient availability to plants, inhibition of soilborne plant pathogens, solubilization of nutrients, the use of microorganism in industries, food and pharmaceutical industry, bioremediation, organic waste recycling, production of antibiotics, nanotechnology, probiotics, nitrogen fixation and other bioactive compounds, and many more.

The advent of this microbiological technology helps in unraveling exact complex signaling mechanisms that were difficult to understand in the past for phenomenon such as of mycoparasitism in case of *Trichoderma* sp. which has been considered to be an effective biocontrol tool against many of plant pathogens. During mycoparasitism, morphological changes occur after recognition of pathogen, and killing of pathogen occurs via release of hydrolytic enzymes and structure known as appressorium by penetration (Benitez et al. 2004). Initially, receptors present on the cell surface lead to internal cascade of signal transduction by transcription of genes related to mycoparasitism upon ligand attachment. G-protein-coupled receptors (GPCRs) that are heterodimeric in structures send signals to cyclic adenosine monophosphate (cAMP) and mitogen-activated protein kinase (MAPK) which mediates in modifiable infections (Omann and Zeilinger 2010). The G-protein alpha-subunit gene *tga 1* has been found to be involved in an important mechanism of mycoparasitism where direct killing by *Trichoderma* sp. occurs and it is now helping in understanding the complicated signaling pathways (Rocha-Ramirez et al. 2002). In the study of Mukherjee et al. (2003), the MAPK homologue which was the product of *tmkA* gene belonging to YERK1 class was isolated from *Trichoderma virens*, and construction of loss-of-function mutants was performed and sporulated in the dark. This resulted into attenuated mycoparasitism on the sclerotia of *Rhizoctonia solani* and *Sclerotium rolfsii*. Also, in mycoparasitic interactions, novel effector proteins are being identified by using computational methods for gene expression because effectors are molecules which have the ability to alter cellular function of plants in allowance of pathogen to infect the plant (Hogenhout et al. 2009). So far, approximately 233 putative effector proteins have been studied from *Trichoderma* sp., and the pattern in which these genes are expressed has undergone analysis in *Trichoderma–Arabidopsis* interaction, and observation of upregulating genes was grouped into LysM (lysin motif) proteins, CFEM, cerato-platanin, hydrophobins, serine proteases, and thioredoxin families (Guzmán-Guzmán et al. 2017) (Table 8.1). However, the characterization of LysM encoding genes present in *Trichoderma* genomes still needs further investigations (Mendoza-Mendoza et al. 2018). The induction of effector proteins occurs when microbes sense the host plant (Lanver et al. 2017). The ability to identify effector proteins in fungi has been carried out by latest software like EffectorP where classification is based on protein net charge, molecular weight, and sequence length. Also, the content of cysteine,

Table 8.1 *Trichoderma* strains and their protein effectors when interacting with host plants

Strain	Effector proteins	Mechanism of interaction with plants	References
T. atroviride	Epl1	Induces of defense-related genes	Salas-Marina et al. (2015)
T. virens	Sm1	Induces of defense-related genes	Salas-Marina et al. (2015)
T. virens	Sm2	Role in plant root colonization and protection	Crutcher et al. (2015)
T. asperellum	*TasHyd1*	Role in plant root colonization	Viterbo and Chet (2010)
T. asperellum	*HBF2-6*	Role in plant root colonization, JA and SA pathway induction	Huang et al. (2015)
T. longibrachiatum	HYTLO1	Growth promotion and defense-related responses	Moscatiello et al. (2018)
T. harzianum	Thph1 and Thph2	Activation of defense-related gene	Saravanakumar et al. (2016)
T. virens	*TVHYDII1*	Role in plant root colonization	Guzmán-Guzmán et al. (2017)
T. harzianum	ThPG1	Role in plant root colonization and induction of ISR	Moran-Diez et al. (2009)

serine, and tryptophan is considered for identification purposes (Sperschneider et al. 2016). Therefore, search for novel effector proteins will pave way for better insight into the functioning of plant–beneficial fungi interactions.

In case of mycorrhiza fungi, the major contribution of this association toward host plant has been the nutrient acquisition. Since most land plants are associated with AMF, the advancement in understanding the regulation of nutrient exchange serves better purpose of implementing AMF in actual field. For instance, the recent understanding has highlighted signal pathways involved in controlling plant infection by AMF and transfer of lipids from the plant host to AMF as a major carbon source (Wang et al. 2017a, b). The molecular perspective has shown that the AM marker gene *LePT4* has a preference in its expression which is a mycorrhiza-specific phosphate transporter in arbuscule-containing cells of mycorrhiza-colonized roots in tomato plant (Fiorilli et al. 2009). The DELLA proteins are considered to be promising in regulation of nutrient signaling during AM associations (Jin et al. 2016). In addition, the viral disease is threatening the crops worldwide, and AM fungi have been looked upon as a potential biocontrol agent for their eradication which is a complex plant–fungus–virus interaction (Hao et al. 2019). The continuous improvement in technologies at avail will let us have better insights of various aspects that are involved in AM associations which will definitely become a milestone in setting up sustainable agricultural system for our next generation.

At present, the primary focus is being given to exploit plant microbiome which is considered to be the next-generation global agricultural production system to meet growing demand in ecological way. In this regard, transgenic approaches and the recent discovery of molecular tools such as clustered regularly interspaced short palindromic repeats (CRISPR)/Cas-mediated genome editing (GE), metabolomics, transcriptomics, proteomics, genomics, etc., are of great significance to investigate molecular interactions in plant–microbe associations and crop improvement programs (Belhaj et al. 2015; Jansson and Hofmockel 2018; Jaganathan et al. 2018). The initial discovery of CRISPR/Cas9 gene editing system in prokaryotes has transformed research in plant and animal biology with its genome editing technique (Jinek et al. 2012). Now the mandatory design for cloning is being on replacement. Generally, the sophisticated CRISPR-Cas9 system is an adaptive immune response in microbial organism that utilizes short RNA-guided nucleases for degradation of foreign DNA or RNA by either insertion or deletion (Sorek et al. 2011). The CRISPRs are locus that consists of a series of short repeating sequences (20–50 bp) that are separated by unique spacers. These short conserved sequences which are proto-spacer adjacent motif (PAM) are required for recognizing the target that provides self- and non-self-identification of sequences. The CRISPR repeat sequences are mostly diverse in their nature, and it consists of GAAA (C/G) motif toward 3’ end for binding of Cas proteins (Godde and Bickerton 2006). The Cas proteins are specific sets that are associated for their interaction with CRIPR loci (Grissa et al. 2007). In short, the functioning of CRISPR-Cas9 system in bacteria was firstly observed where bacteria incorporate some DNA segments (CRISPR arrays) from the invader.

When bacteria is encountered in the future with similar type of invader or that particular invader only, then Cas9 nuclease or related enzymes are guided to the target sites of invader which complements to 20-nucleotide sequence. The DNA endonuclease Cas9 may consist of two nuclease domains, viz., HNH and RuvC, for introduction of double-stranded breaks (DSBs) in the target DNA leading to inhibition of the invader (Soda et al. 2018). Thus, using these kinds of modern techniques such as CRISPR/Cas9 and other intragenic technologies, several applications are now possible that involve gene modification system that may be considered in various plant-breeding approaches. For instance, Xie et al. (2014) demonstrated crop improvement by exact editing of plant genome with the help of CRISPR–Cas9 system where they suggested that specifically three guide RNAs (gRNAs) could be designed to target more than 90% of rice genes. In biotic stress, the citrus canker disease caused by *Xanthomonas citri* had been controlled by using CRISPR/Cas9-targeted editing by altering host disease susceptible gene *CsLOB1* promoter (Peng et al. 2017). In cucumber, for the first time, resistance was provided against *Cucumber vein yellowing virus* (*Ipomovirus*) and potyviruses *Zucchini yellow mosaic virus* and *Papaya ringspot mosaic virus-W* infection where Cas9/subgenomic RNA (sgRNA) technology was used for disrupting the functional aspect of recessive *eIF4E* (*eukaryotic translation initiation factor 4E*) gene (Chandrasekaran et al. 2016). In tomato plant, *SlMAPK3* gene has been target by nonhomologous end joining (NHEJ) method which revealed its role in protection during abiotic stress condition such as drought (Wang et al. 2017a, b). In addition, many significant improvements in trait of major crops have been achieved for quality, yield, and biological pathogen control (Gupta et al. 2012; Baltes et al. 2015; Kim et al. 2017; Li et al. 2017, 2018; Lu et al. 2018; Macovei et al. 2018) (Table 8.2). As of late, there is substantial significant increase in employment of microorganisms in agriculture sector, and it has been mainly considered for its compounds that are anti-pesticidal, insecticidal, herbicidal, and nematicidal. Moreover, the filamentous ascomycete *Trichoderma reesei* that has been utilized for producing several cellulolytic or hemicellulolytic enzymes or recombinant proteins in various biotechnological industries (Singh et al. 2015) has undergone genome sequencing in an Illumina-based whole-genome sequencing approach (Yang et al. 2015). Also, until now, the genome editing system of CRISPR-Cas9 has been reported in more than 40 different species of filamentous fungi or oomycetes (Schuster and Kahmann 2019). Thus, the rapid development in molecular biology tools has been noteworthy due to its high effectiveness and precision which can contribute a lot beyond our imagination.

Table 8.2 CRISPR/Cas9 technology applications for resistance of various bacterial, fungal, and viral diseases in plants

Host plant	Pathogen	Disease	Gene of interest	References
Citrus	*Xanthomonas citri* subsp. *Citri*	Citrus canker	*CsLOB1*	Jia et al. (2017)
Banana	Endogenous *banana streak virus* (eBSV)	Banana streak disease	Target sites in microbial viral genome	Tripathi et al. (2019)
Tomato	*Oidium neolycopersici*	Powdery mildew	*SlMlo1*	Nekrasov et al. (2017)
Lychee	*Peronophythora litchi*	Downy blight	Pectin acetylesterase, *PAE4* and *PAE5*	Kong et al. (2019)
Tomato	*Phytophthora capsici*, *Pseudomonas syringae*, *Xanthomonas* spp.	Bacterial speck, blight, and spot	*SlDMR6-1* gene deletion	Thomazella et al. (2016)
Cucumber	*Zucchini yellow mosaic virus* and *Papaya ringspot mosaic virus-W*	*Cucumber vein yellowing virus*(CVYV)	*eIF4E* mutation	Chandrasekaran et al. (2016)
Cotton, legumes, tomato	*Fusarium oxysporum*	Fusarium wilt	*FoSso1* and *FoSso2*	Wang and Coleman (2019)
Cassava	CBSV	Brown streak	*nCBP-1* and *nCBP-2*	
Grape	*Botrytis cinerea*	Gray mold	*WRKY52*	Wang et al. (2019)
Wheat	*Blumeria graminis* f. sp. tritici	Powdery mildew	*MLO-A1, B1,* and *D1*	Wang and Coleman (2019)
Rice	*Xanthomonas oryzae* pv. oryzae	Bacterial blight	Susceptible genes *OsSWEET14* and *OsSWEET11*	Jiang et al. (2013)

8.7 Conclusion and Future Perspective

The present agricultural systems suggest that there are many forms of constraints present in the form of abiotic or biotic stresses which affect plant growth and development, thereby causing significant losses in global food production. The major challenges that are encountered by plants are alkalinity, salinity, drought, ion toxicity, nutrient immobilization, and several direct/indirect injuries or diseases caused by pests or pathogens. The added burdens of using agro-based chemicals are harming our ecosystem than any good in longer perspective. Therefore, the focus on various other strategies has been developed that suggests application of microbial population for the future global food production which will be helpful in maintaining health of environment as well. Along with novel strategies in the integrated management that takes microbial consortia into account, the modern-day genome engineering also carries a significant role which was not possible before for establishing various beneficial traits to our crops that promises to attain a minimum threshold in crop productivity in sustainable manner in the near future.

Acknowledgments We wish to thank the University Grants Commission (UGC), New Delhi, for financial support.

References

Alizadeh H, Behboudi K, Ahmadzadeh M, Javan-Nikkhah M, Zamioudis C, Pieterse CMJ, Bakker PAHM (2013) Induced systemic resistance in cucumber and Arabidopsis thaliana by the combination of *Trichoderma harzianum* Tr6 and Pseudomonas sp. Ps14. Biol Control 65:14–23

Alvarez B, Biosca EG (2017) Bacteriophage-based bacterial wilt biocontrol for an environmentally sustainable agriculture. Front Plant Sci 8:1218. https://doi.org/10.3389/fpls.2017.01218

Babalola O (2010) Beneficial bacteria of agricultural importance. Biotechnol Lett 32:1559–1570. https://doi.org/10.1007/s10529-010-0347-0

Bailly A, Weisskopf L (2017) Mining the volatilomes of plant-associated microbiota for new biocontrol solutions. Front Microbiol 8:1638. https://doi.org/10.3389/fmicb.2017.01638

Balog A, Hartel T, Loxdale HD, Wilson K (2017) Differences in the progress of the biopesticide revolution between the EU and other major crop-growing regions. Pest Manag Sci 73:2203–2208. https://doi.org/10.1002/ps.4596

Baltes NJ, Hummel AW, Konecna E, Cegan R, Bruns AN, Bisaro DM, Voytas DF (2015) Conferring resistance to geminiviruses with the CRISPR–Cas prokaryotic immune system. Nat Plants 1:15145

Barriuso J, Solano BR, Mañero FJG (2008) Protection against pathogen and salt stress by four plant growth-promoting rhizobacteria isolated from *Pinus* sp. on *Arabidopsis thaliana*. Phytopathology 98:666–672

Belhaj K, Chaparro-Garcia A, Kamoun S, Patron NJ, Nekrasov V (2015) Editing plant genomes with CRISPR/Cas9. Curr Opin Biotechnol 32:76–84

Benitez T, Rincon AM, Limon MC, Codon AC (2004) Biocontrol mechanisms of *Trichoderma* strains. Int Microbiol 7:249–260

Besset-Manzoni Y, Rieusset L, Joly P, Comte G, Prigent-Combaret C (2018) Exploiting rhizosphere microbial cooperation for developing sustainable agriculture strategies. Environ Sci Pollut Res 25:29953–29970. https://doi.org/10.1007/s11356-017-1152-2

Bhunchoth A, Phironrit N, Leksomboon C, Chatchawankanphanich O, Kotera S, Narulita E, Kawasaki T, FujieM YT (2015) Isolation of *Ralstonia solanacearum* infecting bacteriophages from tomato fields in Chiang Mai, Thailand, and their experimental use as biocontrol agents. J Appl Microbiol 118:1023–1033. https://doi.org/10.1111/jam.12763

Bonfante P, Genre A (2010) Mechanisms underlying beneficial plant—fungus interactions in mycorrhizal symbiosis. Nat Commun 1:48. https://doi.org/10.1038/ncomms1046

Buysens S, Heungens K, Poppe J, Hofte M (1996) Involvement of Pyochelin and pioverdin in suppression of *Pseudomonas aeruginosa* 7NSK2. Appl Environ Microbiol 62:865–871

Cangelosi B, Curir P, Beruto M, Monroy F, Borriello R (2017) Protective effects of arbuscular mycorrhizae against *Fusarium oxysporum* F. Sp. *Ranunculi* in *Ranunculus Asiaticus* cultivations for flower crop. J Hortic Sci Res 1:36–41

Carvalho SD, Castillo JA (2018) Influence of light on plant–phyllosphere interaction. Front Plant Sci 9:1482. https://doi.org/10.3389/fpls.2018.01482

Chandrasekaran J, Brumin M, Wolf D, Leibman D, Klap C, Pearlsman M, Sherman A, Arazi T, Gal-On A (2016) Development of broad virus resistance in non-transgenic cucumber using CRISPR/Cas9 technology. Mol Plant Pathol 17:1140–1153. https://doi.org/10.1111/mpp.12375

Chisholm ST, Coaker G, Day B, Staskawicz BJ (2006) Host-microbe interactions: shaping the evolution of the plant immune response. Cell 124:803–814. https://doi.org/10.1016/j.cell.2006.02.008

Contreras-Cornejo HA, Macias-Rodriguez L, Alfaro-Cuevas R, Lopez-Bucio J (2014) *Trichoderma* spp. improve growth of arabidopsis seedlings under salt stress through enhanced root development, osmolite production, and Na+ elimination through root exudates. Mol Plant-Microbe Interact 27:503–514. https://doi.org/10.1094/MPMI-09-13-0265-R

Crutcher FK, Moran-Diez ME, Ding SL, Liu J, Horwitz B, Mukherjee PK, Kenerley CM (2015) A paralog of the proteinaceous elicitor SM1 is involved in colonization of maize roots by *Trichoderma virens*. Fungal Biol 119:476–486. https://doi.org/10.1016/j.funbio.2015.01.004

Dalpe Y (2005) Mycorrhizae: a potential tool for plant protection but not a panacea. Phytoprotection 86:53–59

Devi S, Sreenivasulu Y, Rao K (2017) Protective role of *Trichoderma logibrachiatum* (WT2) on Lead induced oxidative stress in *Helianthus annus* L. Indian J Exp Biol 55:235–241

El-Borollosy AM, Oraby MM (2012) Induced systemic resistance against *Cucumber mosaic cucumovirus* and promotion of cucumber growth by some plant growth-promoting rhizobacteria. Ann Agric Sci 57:91–97

Emmett BD, Youngblut ND, Buckley DH, Drinkwater LE (2017) Plant phylogeny and life history shape rhizosphere bacterial microbiome of summer annuals in an agricultural field. Front Microbiol 8:2414. https://doi.org/10.3389/fmicb.2017.02414

European Union (2009) Directive 2009/128/EC of the European Parliament and of the Council establishing a framework for Community action to achieve the sustainable use of pesticides. Official J EU http://eur-lex.europa.eu/search.html?qid=1456841738609&text=2009/128&scope=EURLEX&type=quick&lang=en. Accessed August 2018

FAO (2019) The state of food and agriculture 2019. Moving forward on food loss and waste reduction. Rome. Licence: CC BY-NC-SA 3.0 IGO

Fiorilli V, Catoni M, Miozzi L, Novero M, Accotto GP, Lanfranco L (2009) Global and cell-type gene expression profiles in tomato plants colonized by an arbuscular mycorrhizal fungus. New Phytol 184:975–987

Fu SF, Sun PF, Lu HY, Wei JY, Xiao HS, Fang WT, Cheng BY, Chou JY (2016) Plant growth-promoting traits of yeasts isolated from the phyllosphere and rhizosphere of *Drosera spatulata* Lab. Fungal Biol 120:433–448. https://doi.org/10.1016/j.funbio.2015.12.006

Gahukar RT (2014) Potential and utilization of plant products in pest control. In: Abrol DP (ed) Insect pest management: current concepts and ecological perspectives. Elsevier Inc., New York, NY, pp 125–139

Garcia-Ruiz MT, Knapp AN, Garcia-Ruiz H (2018) Profile of genetically modified plants authorized in Mexico. GM Crops and Food 9:152–168. https://doi.org/10.1080/21645698.2018.1507601

Gill SS, Gill R, Trivedi DK, Anjum NA, Sharma KK (2016) Piriformospora indica: potential and significance in plant stress tolerance. Front Microbiol 7:332. https://doi.org/10.3389/fmicb.2016.00332

Giovannetti M, Sbrana C (1998) Meeting a non-host: the behavior of AM fungi. Mycorrhiza 8:123–130

Godde JS, Bickerton A (2006) The repetitive DNA elements called CRIPRs and their associated genes: evidence of horizontal transfer among prokaryotes. J Mol Evol 62:718–729

Grissa I, Vergnaud G, Pourcel C (2007) CRISPRFinder: a web tool to identify clustered regularly interspaced repeats. Nucleic Acids Res 35:W52–W57

Gupta M, DeKelver RC, Palta A, Clifford C, Gopalan S, Miller JC, Novak S, Desloover D, Gachotte D, Connell J, Flook J, Patterson T, Robbins K, Rebar EJ, Gregory PD, Urnov FD, Petolino JF (2012) Transcriptional activation of Brassica napus beta-ketoacyl-ACP synthase II with an engineered zinc finger protein transcription factor. Plant Biotechnol J 10:783–791

Gupta S, Gupta R, Sharma S (2014) Impact of pesticides on plant growth promotion of *Vigna radiata* and non-target microbes: comparison between chemical- and bio-pesticides. Ecotoxicology 23:1015–1021. https://doi.org/10.1007/s10646-014-1245-3

Guzmán-Guzmán P, Alemán-Duarte MI, Delaye L, Herrera-Estrella A, Olmedo-Monfil V (2017) Identification of effector-like proteins in *Trichoderma* spp. and role of hydrophobin in the plant-fungus interaction and mycoparasitism. BMC Genet 18:16. https://doi.org/10.1186/s12863-017-0481-y

Gveroska B, Ziberoski J (2012) *Trichoderma harzianum* as a biocontrol agent against *Alternaria alternata* on tobacco. Appl Technol Innovat 7:67–76. https://doi.org/10.15208/ati.2012.9

Hao Z, Xie W, Chen B (2019) Arbuscular mycorrhizal symbiosis affects plant immunity to viral infection and accumulation. Viruses 11:1–12. https://doi.org/10.3390/v11060534

Harman GE (2000) Myths and dogmas of biocontrol-changes in perceptions derived from research on *Trichoderma harzianum* T-22. Plant Dis 84:377–393

Heckman DS, Geiser DM, Eidell BR, Stauffer RL, Kardos NL, Hedges SB (2001) Molecular evidence for the early colonization of land by fungi and plants. Science 293:1129–1133

Hogenhout SA, Van der Hoorn RAL, Terauchi R, Kamoun S (2009) Emerging concepts in effector biology of plant-associated organisms. Mol Plant-Microbe Interact 22:115–122

Huang Y, Mijiti G, Wang Z, Yu W, Fan H, Zhang R, Liu Z (2015) Functional analysis of the class II hydrophobins gene HFB2-6 from the biocontrol agent *Trichoderma asperellum* ACCC30536. Microbiol Res 171:8–20. https://doi.org/10.1016/j.micres.2014.12.004

Jaganathan D, Ramasamy K, Sellamuthu G, Jayabalan S, Venkataraman G (2018) CRISPR for crop improvement: an update review. Front Plant Sci 9:985. https://doi.org/10.3389/fpls.2018.00985

Jansson JK, Hofmockel KS (2018) The soil microbiome—from metagenomics to metaphenomics. Curr Opin Microbiol 43:162–168

Jia H, Zhang Y, Orbovic V, Xu J, White FF, Jones JB, Wang N (2017) Genome editing of the disease susceptibility gene *CsLOB1* in citrus confers resistance to citrus canker. Plant Biotechnol J 15:817–823

Jiang W, Zhou H, Bi H, Fromm M, Yang B, Weeks DP (2013) Demonstration of CRISPR/Cas9/sgRNA-mediated targeted gene modification in *Arabidopsis*, tobacco, sorghum and rice. Nucleic Acids Res 41:e188

Jin Y, Liu H, Luo D, Yu N, Dong W, Wang C, Zhang X, Dai H, Yang J, Wang E (2016) DELLA proteins are common components of symbiotic rhizobial and mycorrhizal signalling pathways. Nat Commun 7:12433

Jinek M, Chylinski K, Fonfara I, Hauer M, Doudna JA, Charpentier E (2012) A programmable dual-RNA–guided DNA endonuclease in adaptive bacterial immunity. Science 337:816–821. https://doi.org/10.1126/science.1225829

Jung SC, Martinez-Medina A, Lopez-Raez JA, Pozo MJ (2012) Mycorrhiza-induced resistance and priming of plant defenses. J Chem Ecol 38:651–664

Kamthan A, Chaudhuri A, Kamthan M, Datta A (2016) Genetically modified (GM) crops: milestones and new advances in crop improvement. Theor Appl Genet 129:1639. https://doi.org/10.1007/s00122-016-2747-6

Keinan A, Clark AG (2012) Recent explosive human population growth has resulted in an excess of rare genetic variants. Science 336:740–743

Khalili E, Sadravi M, Naeimi S, Khosravi V (2012) Biological control of rice brown spot with native isolates of three trichoderma species. Braz J Microbiol 43:297–305. https://doi.org/10.1590/S1517-83822012000100035

Kiely P, Haynes J, Higgins C, Franks A, Mark G, Morrissey J, O'gara F (2006) Exploiting new systems-based strategies to elucidate plant-bacterial interactions in the rhizosphere. Microb Ecol 51:257–266

Kim H, Kim ST, Ryu J, Kang BC, Kim JS, Kim SG (2017) CRISPR/Cpf1-mediated DNA-free plant genome editing. Nat Commun 8:14406

Kloepper JW (2003) A review of mechanisms for plant promotion by PGPR. In: Reddy MS, Anandaraj M (eds) Abstracts and short papers. 6th International PGPR workshop, 5-10 October 2003. Indian Institute of Spices Research, Calicut, India, pp 81–92

Kong G, Wan L, Deng YZ, Yang W, Li W, Jiang L, Situ J, Xi P, Li M, Jiang Z (2019) Pectin acetylesterase PAE5 is associated with the virulence of plant pathogenic oomycete *Peronophythora litchii*. Physiol Mol Plant Pathol 106:16–22

Kumar S (2012) Biopesticides: a need for food and environmental safety. J Biofertil Biopestic 3:4
Kumar A, Verma JP (2018) Does plant—microbe interaction confer stress tolerance in plants: a review? Microbiol Res 207:41–52
Lanver D, Tollot M, Schweizer G, Lo Presti L, Reissmann S, Ma LS, Schuster M, Tanaka S, Liang L, Ludwig N, Kahmann R (2017) *Ustilago maydis* effectors and their impact on virulence. Nat Rev Microbiol 15:409–421
Lee BD, Dutta S, Ryu H, Yoo SJ, Suh DS, Park K (2015) Induction of systemic resistance in *Panax ginseng* against *Phytophthora cactorum* by native *Bacillus amyloliquefaciens* HK34. J Ginseng Res 39:213–220. https://doi.org/10.1016/j.jgr.2014.12.002
Leff JW, Fierer N (2013) Bacterial communities associated with the surfaces of fresh fruits and vegetables. PLoS One 8(3):e59310. https://doi.org/10.1371/journal.pone.0059310
Li X, Zhou W, Ren Y, Tian X, Lv T, Wang Z, Fang J, Chu C, Yang J, Bu Q (2017) High-efficiency breeding of early-maturing rice cultivars via CRISPR/Cas9-mediated genome editing. J Genet Genomics 44:175–178
Li J, Zhang Y, Zhang Y, Yu PL, Pan H, Rollins JA (2018) Introduction of large sequence inserts by CRISPR-Cas9 to create pathogenicity mutants in the multinucleate filamentous pathogen *Sclerotinia sclerotiorum*. MBio 9:1–19
Lopez-Arredondo D, Gonzalez-Morales SI, Bello-Bello E, Alejo-Jacuinde G, Herrera L (2015) Engineering food crops to grow in harsh environment [v1; ref status: indexed, http://f1000r.es/5f1]. F1000Res 4:651. https://doi.org/10.12688/f1000research.6538.1
Loya LJ, Hower AA Jr (2002) Population dynamics, persistence, and efficacy of the entomopathogenic nematode *Heterorhabditis bacteriophora* (Oswego strain) in association with the clover root curculio (Coleoptera: Curculionidae) in Pennsylvania. Environ Entomol 31:1240–1250
Lu K, Wu B, Wang J, Zhu W, Nie H, Qian J, Huang W, Fang Z (2018) Blocking amino acid transporter OsAAP3 improves grain yield by promoting outgrowth buds and increasing tiller number in rice. Plant Biotechnol J 16:1710–1722
Luginbuehl LH, Oldroyd GED (2017) Understanding the arbuscule at the heart of endomycorrhizal symbioses in plants. Curr Biol 27:R952–R963
Ma Z, Michailides TJ (2005) Advances in understanding molecular mechanisms of fungicide resistance and molecular detection of resistant genotypes in phytopathogenic fungi. Crop Prot 24:853–863
Macovei A, Sevilla NR, Cantos C, Jonson GB, Slamet-Loedin I, Cermák T, Voytas DF, Choi IR, Chadha-Mohanty P (2018) Novel alleles of rice *eIF4G* generated by CRISPR/Cas9-targeted mutagenesis confer resistance to *Rice tungro spherical virus*. Plant Biotechnol J 16:1918–1927
Maghari BM, Ardekani AM (2011) Genetically modified foods and social concerns. Avicenna J Med Biotech 3:109–117
Mendes LW, Kuramae EE, Navarrete AA, van Veen JA, Tsai SM (2014) Taxonomical and functional microbial community selection in soybean rhizosphere. ISME J 8:1577–1587. https://doi.org/10.1038/ismej.2014.17
Mendoza-Mendoza A, Zaid R, Lawry R, Hermosa R, Monte E, Horwitz BA, Mukherjee PK (2018) Molecular dialogues between *Trichoderma* and roots: role of the fungal secretome. Fungal Biol Rev 32:62–85. https://doi.org/10.1016/j.fbr.2017.12.001
Michavila G, Adler C, De Gregorio PR, Lami MJ, Di Santo MCC, Zenoff AM, de Cristobal RE, Vincent PA (2017) *Pseudomonas protegens* CS1 from the lemon phyllosphere as a candidate for citrus canker biocontrol agent. Plant Biol 19:608–617
Mnyone LL, Koenraadt CJM, Lyimo IN, Mpingwa MW, Takken W, Russell TL (2010) Anopheline and culicine mosquitoes are not repelled by surfaces treated with the entomopathogenic fungi *Metarhizium anisopliae* and *Beauveria bassiana*. Parasit Vectors 3:80
Mona SA, Hashem A, Abdulaziz EF, Alqarawi C, Wafi D, Soliman K, Wirth S, Egamberdieva D (2017) Increased resistance of drought by *Trichoderma harzianum* fungal treatment correlates with increased secondary metabolites and proline content. J Integr Agric 16:1751–1757
Moran-Diez E, Hermosa R, Ambrosino P, Cardoza RE, Gutierrez S, Lorito M, Monte E (2009) The ThPG1 endopolygalacturonase is required for the *Trichoderma harzianum*-plant beneficial interaction. Mol Plant Microbe In 22:1021–1031

Moscatiello R, Sello S, Ruocco M, Barbulova A, Cortese E, Nigris S, Baldan B, Chiurazzi M, Mariani P, Lorito M, Navazio L (2018) The hydrophobin HYTLO1 secreted by the biocontrol fungus *Trichoderma longibrachiatum* triggers a NAADP-mediated calcium signalling pathway in *Lotus japonicus*. Int J Mol Sci 19:E2596. https://doi.org/10.3390/ijms19092596

Mueller UG, Sachs JL (2015) Engineering microbiomes to improve plant and animal health. Trends Microbiol 23:606–617. https://doi.org/10.1016/j.tim.2015.07.009

Mukherjee PK, Latha J, Hadar R, Horwitz BA (2003) TmkA, a mitogen-activated protein kinase of *Trichoderma virens*, is involved in biocontrol properties and repression of conidiation in the dark. Eukaryot Cell 2:446–455

Mwajita M, Murage H, Tani A, Kahangi EM (2013) Evaluation of rhizosphere, rhizoplane and phyllosphere bacteria and fungi isolated from rice in Kenya for plant growth promoters. Springerplus 2:606. https://doi.org/10.1186/2193-1801-2-606

Nekrasov V, Wang C, Win J, Lanz C, Weigel D, Kamoun S (2017) Rapid generation of a transgene-free powdery mildew resistant tomato by genome deletion. Sci Rep 7:482

Oku S, Komatsu A, Nakashimada Y, Tajima T, Kato J (2014) Identification of *Pseudomonas fluorescens* chemotaxis sensory proteins for malate, succinate, and fumarate and their involvement in root colonization. Microbes Environ 29:413–419

Olanrewaju OS, Glick BR, Babalola OO (2017) Mechanisms of action of plant growth promoting bacteria. World J Microbiol Biotechnol 33:197. https://doi.org/10.1007/s11274-017-2364-9

Omann M, Zeilinger S (2010) How a mycoparasite employs g-protein signaling: using the example of *Trichoderma*. J Signal Transduct 2010:123126. https://doi.org/10.1155/2010/123126

Pelczar MJ, Chan ECS, Creig NR (1988) Microbiology. Tata Mcgraw-Hill, Publishing Company Limited, New Delhi, pp 560–580

Peng A, Chen S, Lei T, Xu L, He Y, Wu L, Yao L, Zou X (2017) Engineering canker-resistant plants through CRISPR/Cas9-targeted editing of the susceptibility gene CsLOB1 promoter in citrus. Plant Biotechnol J 15:1509–1519. https://doi.org/10.1111/pbi.12733

Pereira JAP, Vieira IJC, Freitas MSM, Prins CL, Martins MA, Rodrigues R (2015) Effects of arbuscular mycorrhizal fungi on *Capsicum* spp. J Agric Sci 154:828–849

Pozo MJ, Azcón-Aguilar C (2007) Unraveling mycorrhiza-induced resistance. Curr Opin Plant Biol 10:393–398. https://doi.org/10.1016/j.pbi.2007.05.004

Prater CA, Redmond C, Barney W, Bonning BC, Potter DA (2006) Microbial control of black cutworm (Lepidoptera: Noctuidae) in turfgrass using *Agrotis ipsilon* multiple nucleopolyhedrovirus. J Econ Entomol 99:1129–1137

Qaim M, Zilberman D (2003) Yield effects of genetically modified crops in developing countries. Science 299:900–902. https://doi.org/10.1126/science.1080609

Rocha-Ramirez V, Omero C, Chet I, Horwitz BA, Herrera-Estrella A (2002) *Trichoderma atroviride* G-protein α-subunit gene *tga 1* is involved in mycoparasitic coiling and conidiation. Eukaryot Cell 1:594–605. https://doi.org/10.1128/EC.1.4.594-605.2002

Sabbagh SK, Roudini M, Panjehkeh N (2017) Systemic resistance induced by *Trichoderma harzianum* and *Glomus mosseae* on cucumber damping-off disease caused by *Phytophthora melonis*. Arch Phytopathol Plant Protect 50:375–388

Salas-Marina MA, Isordia-Jasso MI, Islas-Osuna MA, Delgado-Sánchez P, Jiménez-Bremont JF, Rodríguez-Kessler M, Rosales-Saavedra MT, Herrera-Estrella A, Casas-Flores S (2015) The Epl1 and Sm1 proteins from *Trichoderma atroviride* and *Trichoderma virens* differentially modulate systemic disease resistance against different life style pathogens in *Solanum lycopersicum*. Front Plant Sci 23:77. https://doi.org/10.3389/fpls.2015.00077

Salvioli A, Bonfante P (2013) Systems biology and "omics" tools: a cooperation for next-generation mycorrhizal studies. Plant Sci 203–204:107–114. https://doi.org/10.1016/j.plantsci.2013.01.001

Santoyo G, Moreno-Hagelsieb G, del Carmen Orozco-Mosqueda M, Glick BR (2016) Plant growth-promoting bacterial endophytes. Microbiol Res 183:92–99

Saravanakumar K, Fan L, Fu K, Yu C, Wang M, Xia H, Sun J, Li Y, Chen J (2016) Cellulase from *Trichoderma harzianum* interacts with roots and triggers induced systemic resistance to foliar disease in maize. Sci Rep 10:35543. https://doi.org/10.1038/srep35543

Saravanakumar K, Li Y, Yu C, Wang QQ, Wang M, Sun J, Chen J (2017) Effect of *Trichoderma harzianum* on maize rhizosphere microbiome and biocontrol of Fusarium Stalk rot. Sci Rep 7:1771. https://doi.org/10.1038/s41598-017-01680-w

Schouteden N, De Waele D, Panis B, Vos CM (2015) Arbuscular mycorrhizal fungi for the biocontrol of plant-parasitic nematodes: a review of the mechanisms involved. Front Microbiol 6:1280. https://doi.org/10.3389/fmicb.2015.01280

Schuster M, Kahmann R (2019) CRISPR-Cas9 genome editing approaches in filamentous fungi and Oomycetes. Fungal Genet Biol 130:43–53

Sengupta A, Gunri SK (2015) Microbial intervention in agriculture: an overview. Afr J Microbiol Res 9:1215–1226

Shafique HA, Sultana V, Ehteshamul-Haque S, Athar M (2016) Management of soilborne diseases of organic vegetables. J Plant Prot Res 56:221–230

Sharif M, Claassen N (2011) Action mechanisms of arbuscular mycorrhizal fungi in phosphorus uptake by *Capsicum annuum* L. Pedosphere **21**:502–511. https://doi.org/10.1016/S1002-0160(11)60152-5

Shoresh M, Harman GE, Mastouri F (2010) Induced systemic resistance and plant responses to fungal Biocontrol agents. Annu Rev Phytopathol 48:21–43

Singh A, Taylor LE 2nd, Vander Wall TA, Linger J, Himmel ME, Podkaminer K, Adney WS, Decker SR (2015) Heterologous protein expression in *Hypocrea jecorina*: a historical perspective and new developments. Biotechnol Adv 33:142–154. https://doi.org/10.1016/j.biotechadv.2014.11.009

Singh R, Parihar P, Singh M, Bajguz A, Kumar J, Singh S, Singh VP, Prasad SM (2017) Uncovering potential applications of Cyanobacteria and algal metabolites in biology, agriculture and medicine: current status and future prospects. Front Microbiol 8:515

Smith SE, Smith FA (2011) Roles of arbuscular mycorrhizas in plant nutrition and growth: new paradigms from cellular to ecosystem scales. Annu Rev Plant Biol 62:227–250

Soda N, Verma L, Giri J (2018) CRISPR-Cas9 based plant genome editing: significance, opportunities and recent advances. Plant Physiol Biochem 131:2–11. https://doi.org/10.1016/j.plaphy.2017.10.024

Sorek R, Lawrence CM, Wiedenheft B (2011) CRISPR-mediated adaptive immune systems in bacteria and archaea. Annu Rev Biochen 82:237–266

Spaepen S, Vanderleyden J (2011) Auxin and plant-microbe interactions. Cold Spring Harb Perspect Biol 3:a001438

Sperschneider J, Gardiner DM, Dodds PN, Tini F, Covarelli L, Singh KB, Manners JM, Taylor JM (2016) EffectorP: predicting fungal effector proteins from secretomes using machine learning. New Phytol 210:743–761

Srilatha B (2011) Nanotechnology in agriculture. J Nanomedic Nanotechnol 2:1–5

Téllez-Vargas J, Rodríguez-Monroy M, López-Meyer M, Montes-Belmont R, Sepúlveda-Jiménez G (2017) *Trichoderma asperellum* ameliorates phytotoxic effects of copper in onion (*Allium cepa* L.). Environ Exp Bot 136:85–93. https://doi.org/10.1016/j.envexpbot.2017.01.009

Thomazella DPDT, Brail Q, Dahlbeck D, Staskawicz BJ (2016) CRISPR-Cas9 mediated mutagenesis of a *DMR6* ortholog in tomato confers broad-spectrum disease resistance. bioRxiv.

Tripathi JN, Ntui VO, Ron M, Muiruri SK, Britt A, Tripathi L (2019) CRISPR/Cas9 editing of endogenous banana streak virus in the B genome of *Musa* spp. overcomes a major challenge in banana breeding. Commun Boil 2:46

Tyc O, Zweers H, De Boer W, Garbeva P (2015) Volatiles in inter-specific bacterial interactions. Front Microbiol 6:1412. https://doi.org/10.3389/fmicb.2015.01412

Vacheron J, Renoud S, Muller D, Babalola OO, Prigent-Combaret C (2015) Alleviation of abiotic and biotic stresses in plants by Azospirillum. In: Cassan FD, Okon Y, Creus C (eds) Handbook for Azospirillum: technical issues and protocols. Springer, Berlin, Heidelberg, pp 333–365

Van Der Heijden MGA, Bardgett RD, Van Straalen NM (2008) The unseen majority: soil microbes as drivers of plant diversity and productivity in terrestrial ecosystems. Ecol Lett 11:296–310. https://doi.org/10.1111/j.1461-0248.2007.01139.x

Viterbo A, Chet I (2010) *TasHyd1*, a new hydrophobin gene from the biocontrol agent *Trichoderma asperellum*, is involved in plant root colonization. Mol Plant Pathol 7:4. https://doi.org/10.1111/j.1364-3703.2006.00335.x

Vitorino LC, Bessa LA (2017) Technological microbiology: development and applications. Front Microbiol 8:827. https://doi.org/10.3389/fmicb.2017.00827

Vorholt JA (2012) Microbial life in the phyllosphere. Nat Rev Microbiol 10:828–840

Wang Q, Coleman JJ (2019) CRISPR/Cas9-mediated endogenous gene tagging in *Fusarium oxysporum*. Fungal Genet Boil 126:17–24

Wang L, Chen L, Li R, Zhao R, Yang M, Sheng J, Shen L (2017a) Reduced drought tolerance by CRISPR/Cas9-mediated SlMAPK3 mutagenesis in tomato plants. J Agric Food Chem 65:8674–8682. https://doi.org/10.1021/acs.jafc.7b02745

Wang W, Shi J, Xie Q, Jiang Y, Yu N, Wang E (2017b) Nutrient exchange and regulation in arbuscular mycorrhizal symbiosis. Mol Plant 10:1147–1158

Wang X, Tu M, Wang D, Liu J, Li Y, Li Z, Wang Y, Wang X (2019) CRISPR/Cas9-mediated efficient targeted mutagenesis in grape in the first generation. Plant Biotechnol J 16:844–855

Wensing A, Braun SD, Buttner P, Expert D, Volksch B, Ullrich MS, Weingart H (2010) Impact of siderophore production by *Pseudomonas syringae* pv. *Syringae* 22d/93 on epiphytic fitness and biocontrol activity against *Pseudomonas syringae* pv. *glycinea* 1a/96. Appl Environ Microbiol 76:2704–2711. https://doi.org/10.1128/AEM.02979-09

Werner S, Polle A, Brinkmann N (2016) Belowground communication: impacts of volatile organic compounds (VOCs) from soil fungi on other soil-inhabiting organisms. Appl Microbiol Biotechnol 100:8651–8665. https://doi.org/10.1007/s00253-016-7792-1

West SA, Griffin AS, Gardner A (2007) Evolutionary explanations for cooperation. Curr Biol 17:R661–R672

Williams TR, Moyne AL, Harris LJ, Marco ML (2013) Season, irrigation, leaf age, and *Escherichia coli* inoculation influence the bacterial diversity in the lettuce phyllosphere. PLoS One 8(7):e68642. https://doi.org/10.1371/journal.pone.0068642

Woo SL, Pepe O (2018) Microbial consortia: promising probiotics as plant biostimulants for sustainable agriculture. Front Plant Sci 9:1801. https://doi.org/10.3389/fpls.2018.01801

Xie K, Zhang J, Yang Y (2014) Genome-wide prediction of highly specific guide RNA spacers for CRISPR–Cas9-Mediated genome editing in model plants and major crops. Mol Plant 7:923–926

Yang D, Pomraning K, Kopchinskiy A, Karimi Aghcheh R, Atanasova L, Chenthamara K, Baker SE, Zhang R, Shen Q, Freitag M, Kubicek CP, Druzhinina IS (2015) Genome sequence and annotation of *Trichoderma parareesei*, the ancestor of the cellulase producer *Trichoderma reesei*. Genome Announc 3:e00885–e00815. https://doi.org/10.1128/genomeA.00885-15

Youssef MMA, Eissa MFM (2014) Biofertilizers and their role in management of plant parasitic nematodes. A review. E3 J Biotechnol Pharm Res 5:1–6

Yu X, Ai C, Xin L, Zhou G (2011) The siderophore producing bacterium, *Bacillus subtilis* CAS15, as a biocontrol effect on *Fusarium* wilt and promotes the growth of pepper. Eur J Soil Biol 47:138–145

Zhang C, Wohlhueter R, Zhang H (2016) Genetically modified foods: a critical review of their promise and problems. Food Sci Human Wellness 5:116–123

Zhang S, Liu Q, Han Y, Han J, Yan Z, Wang Y, Zhang X (2019) Nematophin, an antimicrobial dipeptide compound from *Xenorhabdus nematophila* YL001 as a potent biopesticide for *Rhizoctonia solani* control. Front Microbiol 10:1765. https://doi.org/10.3389/fmicb.2019.01765. eCollection 2019

Rhizobacteria Versus Chelating Agents: Tool for Phytoremediation

9

Charanjeet Kaur, Babli Bhandari, Alok Srivastava, and Vijai Pal Singh

Abstract

Many anthropogenic activities could magnify the concentrations of nonessential heavy metals (HMs) in soil, which, in turns, enter into the food chain and cause damage to plant, animal, and/or human population. The soil remediation is done in many ways such as conventional ones (physical and chemical methods), which are very expensive and damage the natural environment, and phytoremediation, which is quite affordable and is a green approach as compared to the conventional methods. Various chelating agents (organic and synthetic) are also used as amendments in phytoremediation of heavy metal-contaminated soil, which are very useful too. Although the chemical-assisted phytoremediation is useful, it has many risks/drawbacks, e.g., low efficiency, leaching of HM-chelator complex into the soil, and accumulation of HMs in plant parts. The microbe-assisted phytoremediation is an emerging and better tool for phytoremediation. The risks associated with this method are negligible as compared to chemical-assisted phytoremediation, and it augments the biological system of plants while removing HM. Hence, microbe-assisted phytoremediation is a better tool for phytoremediation.

Keywords

Phytoremediation · Heavy metal · Chelators · PGPR · AMF

C. Kaur · B. Bhandari · A. Srivastava · V. P. Singh (✉)
Department of Plant Science, M. J. P. Rohilkhand University, Bareilly, Uttar Pradesh, India

P. K. Arora (ed.), *Microbial Technology for Health and Environment*, Microorganisms for Sustainability 22,
https://doi.org/10.1007/978-981-15-2679-4_9

9.1 Introduction

There are different definitions of heavy metals in terms of metallurgy, physics, and biochemistry, where these could be defined by their densities (3.5 g/cm^3 to above 7 g/cm^3; Duffus 2002), higher atomic weight, and/or atomic number. In environmental chemistry, metals and metalloids (showing resemblance with metals) that have densities greater than 5 g/cm^3 are considered as heavy metals (HMs) (Järup 2003). However, HMs could be categorized into two forms on the basis of their biological roles: essential and nonessential HMs. Most of the essential elements from periodic table 4–6 are required in trace amount for certain biological processes such as the following: iron and copper (respiratory enzyme complex; Emsley 2011), cobalt (coenzyme syntheses and other cell metabolism; Emsley 2011), zinc (all enzyme classes), manganese (important component of photosystem II), chromium (glucose utilization), nickel (cell growth; Emsley 2011), and cadmium and molybdenum (catalysis of redox reactions; Emsley 2011). Some HMs such as arsenic (Uthus 1994), lead, and mercury are not required for biological functions of living organisms, hence termed as nonessential HMs. Not all HMs are toxic at traces, although they are assumed to be highly toxic and damaging to the environment at higher concentrations (Duffus 2002).

HMs may find their way into the soil through natural processes, e.g., volcanic eruptions and weathering of rocks, and anthropogenic activities such as mining, industrial processes, fossil fuels, and pesticides; higher concentration of these HMs can adversely affect the environment and biological systems of living organisms (Mani and Kumar 2014; Prasad and Strzalka 2002). Heavy metals are nonbiodegradable and can persist in soils for centuries, which cause a great concern to our ecosystem, ultimately leading a serious risk to living organisms (Iheanacho et al. 2017). A suitable approach is necessary to make soil reusable and metal-free. Low-cost and environment-friendly technologies are prime key for the remediation of HM-polluted sites (Mani and Kumar 2014).

Nowadays, various conventional (physical and chemical) methods have been used for the remediation of heavy metals, which are highly destructive, costly, and difficult to implicate (Gupta et al. 2008; Singh and Prasad 2015). Contrarily, phytoremediation is emerging as low-cost, eco-friendly, and easy method for heavy metal treatment (Gupta et al. 2008; Singh and Prasad 2015). The conceptual basis for phytoremediation, the use of plants to clean contaminated soils, came from identifying plants that accumulate metals in very high concentration (Ansari et al. 2014; Prasad 2004). Specific plants and wild species that accumulate increasing amounts of toxic HMs by their roots and transport/translocate them through various plant tissues where they can be metabolized, sequestered, and volatilized are used in this technique (Ansari et al. 2014; Prasad 2001). Phytoremediation can be done in different ways such as rhizofiltration, phytostabilization, phytovolatilization, phytoextraction, and phytodegradation (Fig. 9.1) (Ansari et al. 2014).

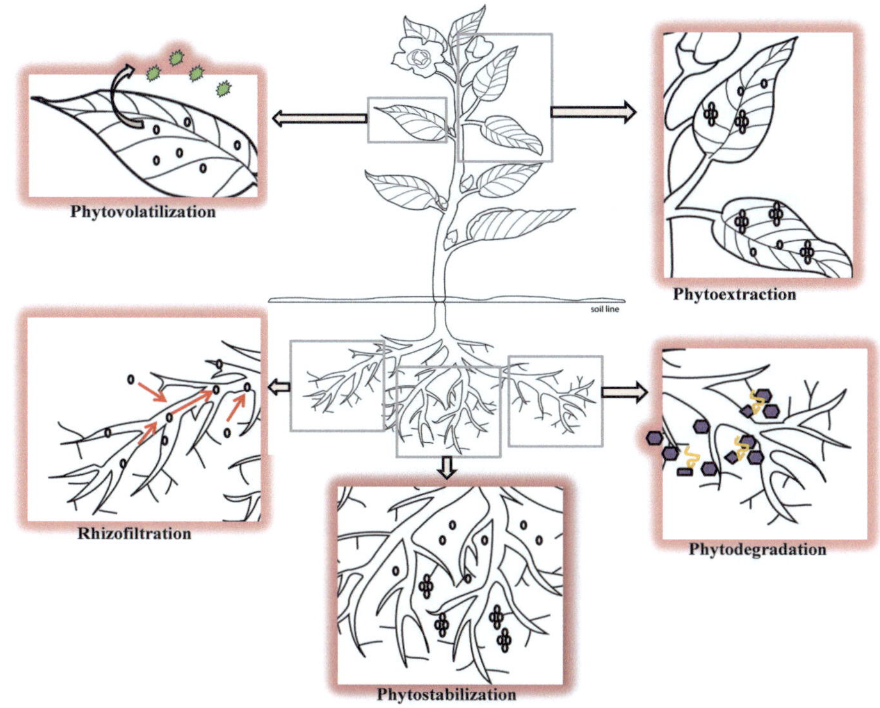

Fig. 9.1 Schematic representation of the processes involved in phytoremediation of heavy metals

9.1.1 Phytoextraction

Phytoextraction, also known as phytoaccumulation, refers to uptake of metals from soil by roots to aerial parts of the plants; certain plants accumulate higher concentration of heavy metals as compared to other plants, and such plants are called hyperaccumulators (Lone et al. 2008; Prasad and Freitas 2000). After grown in the contaminated soil, plants are harvested; while they have accumulated heavy metals in the vegetative parts, they are incinerated following harvesting (Garbisu and Alkorta 2001; Chen et al. 2003) (Fig. 9.1).

9.1.2 Phytostabilization

Involvement of a plant cover to reduce the mobility of various pollutants such as HMs is termed as phytostabilization. It primarily aims at confining pollutants to the soil surface by stabilizing pollutants in the root or rhizosphere to prevent exposure pathways to the new site (Lambrechts et al. 2014). It prevents migration of contaminants by wind, water erosion, and direct contact with animals or humans (Mitter et al. 2013).

It inhibits leaching of contaminants vertically, which relieves from the contamination of underground water (Etim 2012). Phytostabilization does all these by root absorption and chemical fixation by various soil amendments (Flathman and Lanza 1998; Berti and Cunningham 2000; Schnoor 2000).

9.1.3 Phytovolatilization

Phytovolatilization refers to the uptake of the contaminants by the plant in the soluble form and releasing them in the volatile form into the atmosphere by the aerial parts of the plants. It refers to the uptake and transpiration of contaminants, primary organic compounds, by plants (Kumar et al. 2017). The contaminant, present in the water and/or soil taken up by the plant, passes through the plant, is modified by the plant, and is released to the atmosphere (Gupta et al. 2016). The process may be enhanced by using transgenic plants with genes overexpressing the enzymes responsible for high transpiration rates and/or by transferring the genes for Se volatilization from hyper-accumulating plants to non-accumulating ones (Van Huysen et al. 2003; Le Duc et al. 2004; Gupta et al. 2016).

9.1.4 Rhizofiltration

Rhizofiltration is a type of phytoremediation that deploys aquatic and hydroponically cultivated plants for absorption and precipitation of toxic chemicals from polluted effluents (Schmoger et al. 2000). The water is collected from the contaminated site and plants are then grown in it, or plants are directly allowed to flourish in the site. The roots of Indian mustard (*Brassica juncea*), sunflower (*Helianthus annuus*), and various grasses effectively remove metal toxins like Pb, Cd, Cu, Ni, Zn, and Cr. *Arabidopsis halleri*, *Pistia stratiotes*, and *Thlaspi caerulescens* are some of the hyper-accumulators that are suitable for rhizofiltration.

Table 9.1 A brief description of different processes involved in phytoremediation technique

Process	Method
Rhizofiltration	Transfer of the pollutant from the soil and accumulationin the roots of the plants
Phytostabilization	Stabilization of heavy metals in soil/root surface and reductionof heavy metal mobility
Phytovolatilization	Transfer of pollutant from soil to the atmosphere
Phytoextraction	Transfer of pollutants from the soil and accumulationin the aboveground parts of the plant
Phytodegradation	Enhancement of the microbial community and increaseof biodegradation in the rhizosphere

9.1.5 Phytodegradation

Phytodegradation is also known as phyto-transformation; plant performing phytodegradation breaks down the contaminant either metabolically inside the plant tissue or outside the plant in the soil by secreting enzymes and root exudates (Arthur et al. 2005). It helps in the breakdown of certain organic compounds such as chlorinated solvents, ammunition wastes, and herbicides (Table 9.1).

A commonly accepted opinion for efficient phytoremediation of HM polluted sites is that it is essential to use plants having high biomass and fast growth rate, increased metal tolerance, and metal-accumulating capabilities and that are easily cultivable and harvestable (Ansari et al. 2014). Most of the commonly known plants recommended for phytoremediation belong to the families *Asteraceae*, *Brassicaceae*, *Caryophyllaceae*, *Cyperaceae*, *Cunouniaceae*, *Fabaceae*, *Flacourtiaceae*, *Lamiaceae*, *Poaceae*, *Violaceae*, and *Euphobiaceae* because a number of species belonging to these families are HM-tolerant metallophytes and hyper-accumulators (Prasad and Freitas 2003; Prasad et al. 2001). Among them, *Brassicaceae* has largest number of species that are hyper-accumulators of HMs (Prasad and Freitas 2003). However, the plant growth and metal uptake may be significantly inhibited in extremely polluted sites even for tolerant species (Chibuike and Obiora 2014; Ojuederie and Babalola 2017; Ayangbenro and Babalola 2017). There are basically four mechanisms through which HMs employ toxicities in plants (Singh et al. 2016), viz.,

(a) Competition with similar nutrient cations for absorption at root (e.g., competition of As and Cd with P and Zn, respectively; Chorom et al. 2013);
(b) Inactivation of plant proteins by HMs by directly interacting with their functional groups (e.g., sulfhydryl (SH) and phosphate (PO_4) groups) and rendering their activities (Kumar et al. 2017);
(c) Disruption of the function of specific enzymes by replacing cofactors in their prosthetic groups (Ayangbenro and Babalola 2017); and.
(d) Damage of the macromolecules by generating reactive oxygen species (Shahid et al. 2014; Keunen et al. 2011; Singh et al. 2016).

Due to the limited plant species with a high capacity to accumulate metals, especially metals with low bioavailability in soil, and to produce a large amount of biomass, two alternative approaches, using chelating agents (chelators/chelants) and implicating plant growth-promoting rhizobacteria (PGPR), have been used worldwide to improve the uptake of metals by high-biomass plants (Ali and Shamsuddin 2010; Jing et al. 2007). Bioavailability of metals in soil solutions could be determined by the soil properties and types of chelators applied (Singh and Prasad 2015; Qi et al. 2017)

9.2 Chelator-Assisted Phytoremediation

Phytoremediation can be made robust by complimenting with different amendments; one such amendment is the use of chelators (Wang et al. 2017). Chelators or chelating agents are the chemical compounds (a polydentate or multiple-bonded ligands) whose structure permits to form coordinate bonds with the central metal, thus forming stable complex that are soluble in water (Flora and Pachauri 2010). The chelators desorb metals in the solubilized form from the soil matrix, which move to the rhizosphere and are taken up by the plant roots (Sun et al. 2016). The soluble form of the metal can be removed by aerial parts of plants through phytoextraction (Lone et al. 2008).

Chelators could be inorganic or organic substances; all biochemical substances have the capability to form coordinate bonds with metals; hence, all proteins and polysaccharides are very good chelators and/or polydentate ligands for a variety of metal ions (Gupta and Diwan 2017). HM toxicity causes stress in plants; hence, they produce some amino acids (glutamic acid, glycine, histidine, proline, etc.; Irini et al. 2017; Jain and Chen 2018; Kishor et al. 2015; Sharma et al. 2014; Zemanová et al. 2013); phytochelators by glutathione, which is a metal-binding peptide (Farooq et al. 2016; Hossain et al. 2012; Sharma et al. 2017a, b); and amines (spermine, spermidine, putrescine, nicotianamine, etc.; Singh et al. 2016; Takahashi and Kakehi 2009; Wen et al. 2010), which solubilize metals and are known as natural chelators (Wen et al. 2010). Other low-molecular-weight organic acids, such as citric acid (CA), lactic acid, malic acid, malonic acid, oxalic acid, succinic acid, and tartaric acid, are also produced by plant root exudates and microbial residues, which are helpful in chelating heavy metals (Naidu and Harter 1998; Nascimento 2006; Wuana et al. 2010; Nworie et al. 2017; Chen et al. 2018). These low-molecular-weight organic acids and synthetic chelators (ethylenediaminetetraacetic acid (EDTA), nitrilotriacetic acid (NTA), diethylenetriaminepentaacetic acid (DTPA), ethylenediaminedisuccinic acid (EDDS)) could be amended to the plant roots for better remediation of metal-contaminated soil (Mujahid et al. 2013; Ebrahimi 2014; Nanthavong and Sampanpanish 2015; Yang et al. 2013).

9.2.1 Citric Acid

Citric acid ($C_6H_7O_8$) is the first stable compound of citric acid cycle being used as anticoagulant because of its calcium-binding property. The chelating feature of citric acid is employed to the heavy metal uptake by the plants as it has the capability of forming coordinate bonds with divalent ions of heavy metals (Wuana et al. 2010). Studies of citric acid to enhance the phytoaccumulation ability in *Brassica napus* and *Crotalaria juncea* have been reported previously (Ehsan et al. 2014; Alidoust et al. 2009). They analyzed that the application of CA in the spiked soil helped the plant to uptake higher concentration of the Cr in the plant with enhanced antioxidant activities. Studies were not only limited to *B. napus*; in *Helianthus annuus*, accumulation of metal was also reported as a result of application of citric acid in soil

(Turgut et al. 2004). Similar results had been obtained in *Tagetes erecta* by Sinhal et al. (2010), using citric acid and EDTA as amendments. It is not only applicable to contaminated soil, but sludge and wastewater can also be made free from heavy metals by using citric acid and other organic acids (Dacera and Babel 2006).

Although citric acid is biodegradable and nontoxic chelator, it is not a very efficient heavy metal removal agent as compared to synthetic chelators as reported earlier (Qu et al. 2011; Salt et al. 1995; Markovska et al. 2018).

9.2.2 Tartaric Acid

Tartaric acid ($C_4H_6O_6$) is a low-molecular-weight organic acid that helps in fortifying the HM accumulation by plants. Tartaric acid has been used as amendment along with malic and succinic acid in phytoremediation of chromium-contaminated soil by maize plants (Ling et al. 2011). There is another example of increasing the uptake of Cr(VI) in *Spirodela polyrhiza* by using tartaric acid, citric acid, and glycerol as amendments (Bala and Thukral 2011). It has also been reported to chelate many other HMs such as Zn, Cd, Cr, Pb, and Cu (Ding et al. 2014; Ke et al. 2006; Lin et al. 2009; Wuana et al. 2010).

Tartaric acid has been reported to increase the uptake of many HMs as mentioned (Bala and Thukral 2011; Ding et al. 2014; Ke et al. 2006; Lin et al. 2009; Ling et al. 2011; Wuana et al. 2010), but its effectiveness is very low, and as compared to citric acid, it is not very effective (Qu et al. 2011).

9.2.3 Oxalic Acid

Oxalic acid ($C_2H_2O_4$) is a reducing agent with conjugate base $C_2H_2O_{4-}{}^{2}$. Oxalic acid aids leaching of Pb-, Cu-, Zn-, and Cd-contaminated soils (Wuana et al. 2010; Ding et al. 2014). Oxalic acid provides self-defense to the plants in many aspects (Prasad and Shivay 2017), and apart from that, it is widely used as a chelating agent for the removal of HMs from contaminated soil (Prasad and Shivay 2017). It has been reported to increase the accumulation of various HMs in hyperaccumulator plants, e.g., *Beauveria caledonica*, *Leersia hexandra*, *Senecio coronatus*, *Sedum alfredii*, and *Zea mays* (Fomina et al. 2005; Nezami et al. 2016; Tao et al. 2016; Wang et al. 2012). The chelating property of oxalic acid lessened the chromium toxicity in *Hibiscus sabdariffa* L. seedlings, and the plants could mitigate the oxidative stress as compared to only chromium treatment (Ogunleye et al. 2016).

The efficiency of oxalic acid is not as high as compared to citric acid and synthetic chelators (Ogunleye et al. 2016; Tao et al. 2016).

9.2.4 Synthetic Chelators

Apart from natural chelators, a lot of studies have been done evincing the utility of synthetic chelators; heavy metal phytoextraction can be enhanced chemically by assisting with various types of synthetic chelators (Liu et al. 2008; Saifullah et al. 2008). The past couple of decades have been indulged in discerning the vanity of synthetic chelators to aggravate the metal uptake, especially lead, followed by cadmium and other heavy metals, from contaminated sources (Arabi et al. 2017). The synthetic chelators that happen to be of great importance in amending phytoextraction by hyper-accumulators as well as non-hyper-accumulators (Sheoran et al. 2010) are mentioned ahead.

9.2.5 Ethylenediaminetetraacetic Acid (EDTA)

EDTA is an amino-polycarboxylic acid and can form a maximum of six coordinate bond with the transition metal ion and main group ions (Beck 2009). It is the most widely used chelating agent because of its strong binding ability for different metals, consequently increasing the bioavailability of metals in the soil (Gupta et al. 2008; Liphadzi and Kirkham 2006; Sinhal et al. 2010). It has been used to chelate metallic ions and results in desorption of HMs, ultimately increasing the uptake of HMs by plants (Lawal and Sauban 2014; Lestan et al. 2008; Li et al. 2018; Shazia et al. 2014; Singh and Prasad 2015). EDTA is reported to improve phytoremediation better than citric acid applications as reported previously (Markovska et al. 2018; Turgut et al. 2004; Zhang et al. 2018).

EDTA is reported to be the most efficient chelator of HM; however, there are many concerns raised about using EDTA in phytoremediation technology (Saifullah et al. 2008). Its use at large scale is controversial due to the following reasons (Saifullah and Zia-Ur-Rehman 2015):

1. EDTA can adversely affect the composition of soil (physical and chemical properties also) and microbial activities in rhizosphere of plants, which may retard the growth of plants.
2. EDTA is a highly soluble compound and is not easily biodegradable; hence, it may leach to groundwater and persist in soil for months.
3. At higher concentrations, EDTA, applied in soil, can cause eutrophication due to excessive release of nitrogen.
4. It can intensify the stimulation of HMs leaching into groundwater.
5. EDTA can adversely affect soil nutrients because of unspecific co-mobilization of macro- and micronutrients.

9.2.6 Nitrilotriacetic Acid (NTA)

Nitrilotriacetic acid (NTA, $C_6H_9NO_6$) is a tertiary amino-polycarboxylic acid, which forms coordination compounds with metals to form soluble complexes; because of this property, it has been used as a chelating agent worldwide (Quartacci et al. 2006; Reinoso-Maset et al. 2013). NTA is reported to be highly biodegradable than EDTA (Quartacci et al. 2005; Ruley et al. 2006), but its degradability is not as high as oxalic and citric acid, as reported recently (Freitas and Nascimento 2016). NTA is found to increase the HM concentrations in shoots of Indian mustard that too at minimal leaching (Quartacci et al. 2006).

Although NTA is biodegradable, its rate of degradation is not as high as low-molecular-weight organic acids; hence, it can cause adverse ecological effects by leaching in the soil (Freitas and Nascimento 2016; Song et al. 2016).

9.2.7 Diethylenetriaminepentaacetic Acid (DTPA)

Diethylenetriaminepentaacetic acid (DTPA, $C_{14}H_{23}N_3O_{10}$), also known as pentetic acid, is used as a chelator for phytoremediation studies of HMs such as Cu, Cd, Hg, Ni, Pb, and Zn (Ghasemi et al. 2017; Liu et al. 2018; Pastor et al. 2007; Robinson et al. 1999).

DTPA is used as HM chelator; however, it is less effective than EDTA and costlier too (Ghasemi et al. 2017; Liu et al. 2018).

9.2.8 Ethylenediaminedisuccinic Acid (EDDS)

A stereoisomer (S, S) of ethylenediaminedisuccinic acid (EDDS, $C_{10}H_{16}N_2O_8$) is used as a biodegradable alternative to EDTA in phytoremediation studies of heavy metals (Ullmann et al. 2013; Fabbricino et al. 2013; Saifullah and Zia-Ur-Rehman 2015; Wang et al. 2012). It is found to be highly biodegradable, and it enhanced the solubilization and accumulation of various HMs such as Cd, Cu, Hg, Ni, Pb, and Zn by various plant varieties (Yan et al. 2010; Yang et al. 2013; Xu and Thomson 2007). Xu and Thomson (2007) found it to be a better chelator than EDTA and NTA for phytoextraction as well as ex situ washing of HMs.

EDDS is capable of improving phytoremediation of various HMs and highly biodegradable too; however, many workers have reported the risk of metal leaching to groundwater level (Fedje et al. 2013; Hauser et al. 2005; Hu et al. 2007; Wang et al. 2012; Yan et al. 2010)

9.3 Microbe-Assisted Phytoremediation

The interaction of plant root exudates and plant-associated microorganisms plays an important role in adaptation to metal-contaminated soil, which could be useful in phytoremediation technology (Ma et al. 2016; Mishra et al. 2017; Rajkumar et al. 2012). Many plant growth-promoting rhizobacteria (PGPR) and arbuscular mycorrhizal fungi (AMF) may enhance the biomass of plants and neutralize the detrimental effects of HMs on plant growth and nutrition through various mechanisms (Ayangbenro and Babalola 2017; Jambon et al. 2018; Safronova et al. 2011; Rajkumar et al. 2012). PGPR and AMF arouse plant growth by producing phytohormones (auxins, cytokinins, gibberellins) and diminish turbulences in the plant hormonal status. These microbes also secrete siderophores, phytochelators, organic acids, and amines (glutathione, glutamic acid, glycine, histidine, proline, spermine, spermidine, putrescine, nicotianamine, etc.), which in turn lessen the metal toxicity in plants and enhance metal bioavailability and phytostabilization of metals (Ali et al. 2013; Kamaludeen and Ramasamy 2008; Leskó and Simon-Sarkadi 2002; Farooq et al. 2016; Sharma et al. 2017a, b). Biogeochemical processes mediated by microbes (e.g., biological nitrogen fixation and bacterial phosphate solubilization or production of siderophores) can alleviate the HM toxicity by enhancing nutrient uptake and improving the plant transport systems (Alori et al. 2017; Khan et al. 2007; de Souza et al. 2015).

PGPR and AMF may ease the solubility and speciation of HMs in rhizosphere via various mechanisms, e.g., (a) intercellular sequestration of HMs by components of cell wall or by production of intracellular metal-binding substances (metallothioneins, phytochelatins, bacterial siderophores and catecholates, fungal siderophores, and hydroxamate siderophores) (Ojuederie and Babalola 2017); (b) blockage of HM uptake by altering biochemical pathways (Pal et al. 2018); (c) biosorption, precipitation, or bioaccumulation of HMs in external and intracellular spaces (Hrynkiewicz et al. 2014; Mosa et al. 2016); and (d) PGPR and AMF, which can alleviate the intracellular concentration of HMs by using specific plasmid-encoded efflux systems (Laetitia and Puchooa 2017; Roane and Pepper 2000). PGPR and AMF are more efficient in transforming, mobilizing, and solubilizing nutrients and, therefore, are the major driving forces for recycling of nutrients present in the soil, leading to increased fertility of the soil. However, the increased mobilization of HMs can also increase their phytoavailability and toxicity (Violante et al. 2010; Tangahu et al. 2011).

PGPR were initially used for increasing plant yield and growth, supporting nutrient uptake by plants, and providing tolerance to plant diseases. However, in recent years, the application of PGPR has been extended to assist phytoremediation of HMs also (Ma et al. 2016; Mishra et al. 2017; Laetitia and Puchooa 2017; Mosa et al. 2016; Alori et al. 2017). Recently, many workers (Xun et al. 2015; Dong et al. 2014) found that applying PGPR (*Serratia marcescens* BC-3) and AMF (*Glomus intraradices*), in pot experiments, could improve the biomass of plant, activities of antioxidant enzymes (superoxide dismutase, catalase, and peroxidase), and soil enzymes (urease, sucrase, and dehydrogenase) and degraded total petroleum hydrocarbon. Many PGPR and AMF isolates, such as *Bacillus cereus*, *B. subtilis*, *Planomicrobium chinense*, *Pseudomonas fluorescens*, *P. aeruginosa*, *P. stutzeri*, *Providencia vermicola, Rhizophagus irregularis, and Amanita strobiliformis*, had been used for phytoremediation of HMs (Ag, Cd, Cu, Ni, Pb, and Zn) worldwide (Zaidi et al. 2006; Khan and Bano 2016; Sharma et al. 2017a, b; González-Guerrero et al. 2008; Han et al. 2015; Hložková et al. 2016).

9.4 Conclusion

There are many ways to remediate heavy metal-contaminated soils, and among them, phytoremediation is a low-cost, green approach to save the environment. There are many drawbacks of using organic and synthetic chelators as mentioned earlier; however, the use of PGPR and AMF could augment the phytoremediation process without adversely affecting the plants. Hence, microbe-assisted phytoremediation is recommended over chelator-assisted one.

Acknowledgments V.P.S. acknowledges the University Grants Commission, New Delhi, India, for providing him UGC-BSR Faculty Fellowship. C.K. is also thankful to the University Grant Commission for providing her DSKPDF.

References

Ali H, Khan E, Anwar SM (2013) Phytoremediation of heavy metals-Concepts and applications. Chemosphere 91(7):869–881. https://doi.org/10.1016/j.chemosphere.2013.01.075

Ali K Shamsuddin ZH (2010) Phytoremediation of heavy metals with several efficiency enhancer methods. Afr J Biotechnol 9:3689–3698

Alidoust D, Suzuki S, Matsumura S, Yoshida M (2009) The role of citric acid in enhanced phytoextraction of heavy metals in an andosol by *Crotalaria juncea*. Fresenius Environ Bull 18:835–842

Alori ET, Glick BR, Babalola OO (2017) Microbial phosphorus solubilization and its potential for use in sustainable agriculture. Front Microbiol 8:971. https://doi.org/10.3389/fmicb.2017.00971

Ansari MK, Iqbal M, Ahmad A, Ansari MKA, Qureshi MI, Aref IM, Khan PR, Hegazy SS, El-Atta H, Husen A, Hakeem KR (2014) Improving the phytoextraction capacity of plants to scavenge metal (loid)-contaminated sites. Environ Rev 23:1–22. https://doi.org/10.1139/er-2014-0043

Arabi Z, Homaee M, Asadi ME, Kapourchal SA (2017) Cadmium removal from Cd-contaminated soils using some natural and synthetic chelates for enhancing phytoextraction. Chem Ecol 33(9):389–402

Arthur LE, Rice JP, Rice JP, Anderson T, Baladi MS, Henderson LDK, Coats J (2005) Phytoremediation–an overview. Crit Rev Plant Sci 24. https://doi.org/10.1080/07352680590952496

Ayangbenro AS, Babalola OO (2017) A new strategy for heavy metal polluted environments: a review of microbial biosorbents. Int J Environ Res Public Health 14(1):94. PMC5295344

Bala R, Thukral AK (2011) Phytoremediation of Cr(VI) by *Spirodela polyrrhiza* (l.) Schleiden employing reducing and chelating agents. Int J Phytoremediation 13(5):465–491. https://doi.org/10.1080/15226511003758861

Beck W (2009) Metal complexes of biologically important ligands, CLXXII [1]. Metal ions and metal complexes as protective groups of amino acids and peptides—reactions at coordinated amino acids. Z Naturforsch 64b:1221–1245

Berti WR, Cunningham SD (2000) Phytostabilization of metals. In: Raskin I, Ensley BD (eds) Phytoremediation of toxic metals: using plants to clean-up the environment. Wiley, New York, pp 71–88

Chen YX, Lin Q, Luo YM, He YF, Zhen SJ, Yu YL, Tian GM, Wong MH (2003) The role of citric acid on the phytoremediation of heavy metal contaminated soil. Chemosphere 50(6):807–811

Chen H, Dou J, Xu H (2018) The effect of low-molecular-weight organic-acids (LMWOAs) on treatment of chromium-contaminated soils by compost-phytoremediation: Kinetics of the chromium release and fractionation. J Environ Sci 70:45–53

Chibuike GU, Obiora SC (2014) Heavy metal polluted soils: effect on plants and bioremediation methods. Appl Environ Soil Sci 2014:752708. https://doi.org/10.1155/2014/752708

Chorom M, Mohammadzadeh RK, Kaviani B, Kalkhajeh YK (2013) Monometal and competitive adsorption of Cd, Ni, and Zn in soil treated with different contents of cow manure. Appl Environ Soil Sci 2013:510278. https://doi.org/10.1155/2013/510278

Dacera DDM, Babel S (2006) Use of citric acid for heavy metals extraction from contaminated sewage sludge for land application. Water Sci Technol 54(9):129–135

de Souza R, Ambrosini A, Passaglia LM (2015) Plant growth-promoting bacteria as inoculants in agricultural soils. Genet Mol Biol 38(4):401–419

Ding YZ, Song ZG, Feng RW, Guo JK (2014) Interaction of organic acids and pH on multi-heavy metal extraction from alkaline and acid mine soils. Int J Environ Sci Technol 11:33–42. https://doi.org/10.1007/s13762-013-0433-7

Dong R, Gu L, Guo C, Xun F, Liu J (2014) Effect of PGPR *Serratia marcescens* BC-3 and AMF *Glomus intraradices* on phytoremediation of petroleum contaminated soil. Ecotoxicology 23(4):674–680

Duffus JH (2002) 'Heavy metals'-a meaningless term? Pure Appl Chem 74(5):793–807. https://doi.org/10.1351/pac200274050793

Ebrahimi M (2014) Effect of EDTA and DTPA on phytoremediation of Pb-Zn contaminated soils by eucalyptus camaldulensis dehnh and effect on treatment time. Desert 19(1):65–73

Ehsan S, Ali S, Noureen S, Mahmood K, Farid M, Ishaque W, Shakoor MB, Rizwan M (2014) Citric acid assisted phytoremediation of cadmium by *Brassica napus* L. Ecotoxicol Environ Saf 106:164–172. https://doi.org/10.1016/j.ecoenv.2014.03.007

Emsley J (2011) Nature's building blocks, new edition. Oxford University Press, Oxford, ISBN: 978-0-19-960563-7

Etim EE (2012) Phytoremediation and its mechanisms: a review. Int J Environ Bioenerg 2(3):120–136

Fabbricino M, Ferraro A, Del Giudice G, d'Antonio L (2013) Current views on EDDS use for ex situ washing of potentially toxic metal contaminated soils. Rev Environ Sci Biotechnol 12(4):391–398. https://doi.org/10.1007/s11157-013-9309-z

Farooq MA, Ali B, Gill RA, Islam F, Cui P, Zhou W (2016) Chapter 2: Breeding oil crops for sustainable production: heavy metal tolerance. In: Gupta SK (ed) Breeding oilseed crops for sustainable production. Academic Press, New York, pp 19–31. ISBN: 9780128013090. https://doi.org/10.1016/B978-0-12-801309-0.00002-1

Fedje KK, Yillin L, Stromvall AM (2013) Remediation of metal polluted hotspot areas through enhanced soil washing—evaluation of leaching methods. J Environ Manag 128:489–496

Flathman PE, Lanza GR (1998) Phytoremediation: current views on an emerging green technology. J Soil Contamin 7(4):415–432

Flora SJS, Pachauri V (2010) Chelation in metal intoxication. Int J Environ Res Public Health 7(7):2745–2788. https://doi.org/10.3390/ijerph7072745

Fomina M, Hillier S, Charnock JM, Melville K, Alexander IJ, Gadd GM (2005) Role of oxalic acid over excretion in transformations of toxic metal minerals by *Beauveria caledonica*. Appl Environ Microbiol 71:371–381

Freitas E, Nascimento C (2016) Degradability of natural and synthetic chelating agents applied to a lead-contaminated soil. J Soils Sediments 17(5):1272–1278. https://doi.org/10.1007/s11368-015-1350-9

Garbisu C, Alkorta I (2001) Phytoextraction: a cost-effective plant-based technology for the removal of metals from the environment. Bioresour Technol 77(3):229–236

Ghasemi F, Ebrahimi M, Shirazi MP (2017) Lead phytoremediation capacity of *Puccinellia distans* (Jacq.) Parl. using EDTA and DTPA and associated potential leaching risk. Global NEST J 19:359–366

González-Guerrero M, Melville LH, Ferrol N, Lott JN, Azcon-Aguilar C, Peterson RL (2008) Ultrastructural localization of heavy metals in the extraradical mycelium and spores of the arbuscular mycorrhizal fungus *Glomus intraradices*. Can J Microbiol 54:103–110. https://doi.org/10.1139/w07-119

Gupta P, Diwan B (2017) Bacterial exopolysaccharide mediated heavy metal removal: a review on biosynthesis, mechanism and remediation strategies. Biotechnol Rep 13:58–71. https://doi.org/10.1016/j.btre.2016.12.006

Gupta D, Srivastava A, Singh VP (2008) EDTA enhances lead uptake and facilitates phytoremediation by vetiver grass. J Environ Biol 29:903–906

Gupta A, Joia J, Sood A, Sood R, Sidhu C, Kaur G (2016) Microbes as potential tool for remediation of heavy metals: a review. J Microb Biochem Technol 8:364–372. https://doi.org/10.4172/1948-5948.1000310

Han Y, Wang R, Yang Z, Zhan Y, Ma Y, Ping S et al (2015) 1-aminocyclopropane-1-carboxylate deaminase from *Pseudomonas stutzeri* a1501 facilitates the growth of rice in the presence of salt or heavy metals. J Microbiol Biotechnol 25:1119–1128. https://doi.org/10.4014/jmb.1412.12053

Hauser L, Tandy S, Schulin R, Nowack B (2005) Column extraction of heavy metals from soils using the biodegradable chelating agent EDDS. Environ Sci Technol 39:6819–6824

Hložková K, Matěnová M, Žáčková P, Strnad H, Hršelová H, Hroudová M et al (2016) Characterization of three distinct metallothionein genes of the Ag-hyperaccumulating ectomycorrhizal fungus *Amanita strobiliformis*. Fungal Biol 120:358–369. https://doi.org/10.1016/j.funbio.2015.11.007

Hossain MA, Piyatida P, da Teixeira SJA, Fujita M (2012) Molecular mechanism of heavy metal toxicity and tolerance in plants: central role of glutathione in detoxification of reactive oxygen species and methylglyoxal and in heavy metal chelation. Aust J Bot 2012:872875

Hrynkiewicz K, Złoch M, Kowalkowski T, Baum C, Niedojadło K, Buszewski B (2014) Strain-specific bioaccumulation and intracellular distribution of Cd^{2+} in bacteria isolated from the rhizosphere, ectomycorrhizae, and fruitbodies of ectomycorrhizal fungi. Environ Sci Pollut Res Int 22(4):3055–3067

Hu N, Luo Y, Wu L, Song J (2007) A field lysimeter study of heavy metal movement down the profile of soils with multiple metal pollution during chelate-enhanced phytoremediation. Int J Phytoremerned 9:257–268

Huysen T, Ghany SA, Hale KL, LeDuc D, Terry N, Smits EAH (2003) Overexpression of cystathionine synthase enhances selenium volatilization in *Brassica juncea*. Planta 218:71–78

Iheanacho EU, Ndulaka JC, Onuh CF (2017) Environmental pollution and heavy metals. Eur J Biotechnol Biosci 5(5):73–78

Irini NO, Panagiotis M, Kissoudis C, Voulgari G, Chronopoulou E, Tsaftaris A, Labrou NE (2017) Plant glutathione transferase-mediated stress tolerance: functions and biotechnological applications. Plant Cell Rep 36(6):791–805

Jain A, Chen WN (2018) Involvement of organic acids and amino acids in ameliorating Ni(II) toxicity induced cell cycle dysregulation in Caulobacter crescentus: a metabolomics analysis. Appl Microbiol Biotechnol 102(10):4563–4575. https://doi.org/10.1007/s00253-018-8938-0

Jambon I, Thijs S, Weyens N, Vangronsveld J (2018) Harnessing plant-bacteria-fungi interactions to improve plant growth and degradation of organic pollutants. J Plant Interact 13(1):119–130. https://doi.org/10.1080/17429145.2018.1441450

Järup L (2003) Hazards of heavy metal contamination. Br Med Bull 68(1):167–182. https://doi.org/10.1093/bmb/ldg032

Jing Y, He Z, Yang X (2007) Role of soil rhizobacteria in phytoremediation of heavy metal contaminated soils. J Zhejiang Univ (Sci) 8(3):192–207. https://doi.org/10.1631/jzus.2007.B0192

Kamaludeen SP, Ramasamy K (2008) Rhizoremediation of metals: harnessing microbial communities. Indian J Microbiol 48(1):80–88

Ke X, Li PJ, Zhou Q, Zhang Y, Sun TH (2006) Removal of heavy metals from a contaminated soil using tartaric acid. J Environ Sci (China) 18:727–733

Keunen E, Remans T, Bohler S, Vangronsveld J, Cuypers A (2011) Metal-induced oxidative stress and plant mitochondria. Int J Mol Sci 12(10):6894–6918

Khan N, Bano A (2016) Modulation of phytoremediation and plant growth by the treatment with PGPR, Ag nanoparticle and untreated municipal wastewater. Int J Phytoremediation 18(12):1258–1269

Khan MS, Zaidi A, Wani PA (2007) Role of phosphate-solubilizing microorganisms in sustainable agriculture—a review. Agron Sustain Dev 27(1):29–43

Kishor BPK, Kumari PH, Sunita MSL, Nese S (2015) Role of proline in cell wall synthesis and plant development and its implications in plant ontogeny. Front Plant Sci 6:544. https://doi.org/10.3389/fpls.2015.00544

Kumar B, Kumari S, Luis CF (2017) Plant mediated detoxification of mercury and lead. Arab J Chem 10(2):S2335–S2342. https://doi.org/10.1016/j.arabjc.2013.08.010

Laetitia H, Puchooa D (2017) Bioremediation of heavy metals from aquatic environment through microbial processes: a potential role for probiotics? J Appl Biol Biotechnol 5:14–23. https://doi.org/10.7324/JABB.2017.50603

Lambrechts T, Sébastien F, Stanley L, Muñoz-Carpena R, Charles B (2014) Impact of plant growth and morphology and of sediment concentration on sediment retention efficiency of vegetative filter strips: flume experiments and VFSMOD modeling. J Hydrol 511:800–810. https://doi.org/10.1016/j.jhydrol.2014.02.030

Lawal M, Sauban OO (2014) Ethylene diamine tetra-acetate (EDTA) chelate effects on pennisetum purpureum as heavy metal phytoextraction from contaminated soil in Mokwa/New-Bussa Highway Road, Niger State. J Trop Forest Res 30:127–136

Le Duc DL, Tarun AS, Montes-Bayon M, Meija J, Malit MF (2004) Overexpression of selenocysteine methyltransferase in Arabidopsis and Indian mustard increases selenium tolerance and accumulation. Plant Physiol 135:377–383

Leskó K, Simon-Sarkadi L (2002) Effect of cadmium stress on amino acid and polyamine content of wheat seedlings. Periodica Polytechnica Ser Chem Eng 46(1/2):65–71

Lestan D, Luo C, Li XD (2008) The use of chelating agents in the remediation of metal-contaminated soils: a review. Environ Pollut 153:3–13. https://doi.org/10.1016/j.envpol.2007.11.015

Li Y, Luo J, Yu J, Xia L, Zhou C, Cai L, Ma X (2018) Improvement of the phytoremediation efficiency of *Neyraudia reynaudiana* for lead-zinc mine-contaminated soil under the interactive effect of earthworms and EDTA. Sci Rep 8(1):6417. https://doi.org/10.1038/s41598-018-24715-2

Lin L, Li GD, Jing P, Yuan YF, Zhou NN, Jiao W (2009) Removal effect of EDTA and organic acids on Pb in soil. Health Environ Res Online (HERO) 9(4):51–53

Ling G, Cheng MZF, Wang QH (2011) Effects of natural organic acids on growth of maize and uptake of chromium by maize in chromium contaminated soil. Adv Mater Res 281:21–24. https://doi.org/10.4028/www.scientific.net/AMR.281.21

Liphadzi MS, Kirkham MB (2006) Availability and plant uptake of heavy metals in EDTA-assisted phytoremediation of soil and composted biosolids. South African J Bot 72(3):391–397

Liu D, Islam E, Li T, Yang X, Jin X, Mahmood Q (2008) Comparison of synthetic chelators and low molecular weight organic acids in enhancing phytoextraction of heavy metals by two ecotypes of Sedum alfredii Hance. J Hazard Mater 153:114–122

Liu Z, Wang L-A, Ding S, Xiao H (2018) Enhancer assisted-phytoremediation of mercury-contaminated soils by *Oxalis corniculata* L., and rhizosphere microorganism distribution of *Oxaliscorniculata* L. Ecotoxicol Environ Saf 160:171–177. https://doi.org/10.1016/j.ecoenv.2018.05.041

Lone MI, He Z, Stoffella PJ, Yang X (2008) Phytoremediation of heavy metal polluted soils and water: progresses and perspectives. J Zhejiang Univ (Sci) 9(3):210–220. https://doi.org/10.1631/jzus.B0710633

Ma Y, Oliveira RS, Freitas H, Zhang C (2016) Biochemical and molecular mechanisms of plant-microbe-metal interactions: relevance for phytoremediation. Front Plant Sci 7:918. https://doi.org/10.3389/fpls.2016.00918

Mani D, Kumar C (2014) Biotechnological advances in bioremediation of heavy metals contaminated ecosystems: an overview with special reference to phytoremediation. Int J Environ Sci Technol 11:843–872. https://doi.org/10.1007/s13762-013-0299-8

Markovska Y, Stancheva I, Petrov P, Georgieva T, Geneva M, Georgieva KM, Ivanova K (2018) EDTA and citrate impact on heavy metals phytoremediation using paulownia hybrids. Int J Environ Pollut 63:31. https://doi.org/10.1504/IJEP.2018.10014111

Mishra J, Singh R, Arora NK (2017) Alleviation of heavy metal stress in plants and remediation of soil by rhizosphere microorganisms. Front Microbiol 8:1706. https://doi.org/10.3389/fmicb.2017.01706

Mitter B, Günter B, Muhammad A, Stephane C, Muhammad N, Friederike T, Angela S (2013) Chapter seven-advances in elucidating beneficial interactions between plants, soil, and bacteria. In: Sparks DL (ed) Advances in agronomy, vol 121. Academic Press, New York, pp 381–445. https://doi.org/10.1016/B978-0-12-407685-3.00007-4

Mosa KA, Saadoun I, Kumar K, Helmy M, Dhankher OP (2016) Potential biotechnological strategies for the cleanup of heavy metals and metalloids. Front Plant Sci 7:303. https://doi.org/10.3389/fpls.2016.00303

Mujahid F, Ali S, Shakoor M, Bharwana S, Rizvi H, Ehsan S, Tauqeer HM, Iftikhar U, Hannan F (2013) EDTA assisted phytoremediation of Cadmium, Lead and Zinc. Int J Agron Plant Product 4:2833–2846

Naidu R, Harter RD (1998) Effect of different organic ligands on cadmium sorption and extractability from soils. Soil Sc Soc Am J 62(3):644–650

Nanthavong K, Sampanpanish P (2015) Effect of NTA and EDTA on arsenic uptake from contaminated soil by mimosa pudica. Mod Appl Sci 9(9):280–291

Nascimento CWA (2006) Organic acids effects on desorption of heavy metals from a contaminated soil. Sci Agric 63(3):276–280. https://doi.org/10.1590/S0103-90162006000300010

Nezami S, Mohammad MJ, Malakouti J, Samani AB, Maragheh MG (2016) Effect of low molecular weight organic acids on the uptake of ^{226}Ra by corn (*Zea mays* L.) in a region of high natural radioactivity in Ramsar-Iran. J Environ Radioact 164:145–150

Nworie OE, Qin J, Lin C (2017) Differential effects of low-molecular-weight organic acids on the mobilization of soil-borne arsenic and trace metals. Toxics 5(3):18. https://doi.org/10.3390/toxics5030018

Ogunleye JA, Dagari MS, Balogun OL, Maigari A (2016) Effect of oxalic acid on cr^{3+} ion uptake, accumulation and oxidative stress by sorrel (*Hibiscus sabdariffa* L.) seedlings in hydroponic solution. Int J Chem Mater Environ Res 3(3):49–55

Ojuederie OB, Babalola OO (2017) Microbial and plant-assisted bioremediation of heavy metal polluted environments: a review. Int J Environ Res Public Health 14:1504. https://doi.org/10.3390/ijerph14121504

Pal A, Chakraborty A, Sengupta C (2018) Differential effects of plant growth promoting rhizobacteria on chilli (*Capsicum annuum* L.) seedling under cadmium and lead stress. Plant Sci Today 5:182. https://doi.org/10.14719/pst.2018.5.4.419

Pastor J, Aparicio AM, Gutierrez-Maroto A, Hernández AJ (2007) Effects of two chelating agents (EDTA and DTPA) on the autochthonous vegetation of a soil polluted with Cu, Zn and Cd. Sci Total Environ 378:114–118. https://doi.org/10.1016/j.scitotenv.2007.01.022

Prasad MNV (2001) Metals in the environment: analysis by biodiversity. Marcel Dekker, New York, 487 p. ISBN: 0-58-540412-7. p. 504

Prasad MNV (2004) Phytoremediation of metals and radionuclides in the environment: the case for natural hyperaccumulators, metal transporters, soil-amending chelators and transgenic plants. In: Heavy metal stress in plants: from biomolecules to ecosystems, 2nd edn. Springer-Verlag, Heidelberg, pp 345–392

Prasad MNV, Freitas H (2000) Removal of toxic metals from the aqueous solution by the leaf, stem and root phytomass of *Quercus ilex* L. (Holly Oak). Environ Pollut 110(2):277–283

Prasad MNV, Freitas H (2003) Metal hyperaccumulation in plants—biodiversity prospecting for phytoremediation technology. Electron J Biotechnol 6(3). https://doi.org/10.2225/vol6-issue3-fulltext-6

Prasad R, Shivay YS (2017) Oxalic acid/oxalates in plants: from self-defence to phytoremediation. Curr Sci 112:1665–1667. https://doi.org/10.18520/cs/v112/i08/1665-1667

Prasad MNV, Greger M, Landberg T (2001) *Acacia nilotica* L. bark removes toxic metals from solution: Corroboration from toxicity bioassay using *Salix viminalis* L. in hydroponic system. Int J Phytoremediation 3(3):289–300

Prasad MNV, Strzalka K (2002) Physiology and biochemistry of metal toxicity and tolerance in plants. Kluwer Academic Publishers, Dordrecht, 432 p. ISBN: 1-40-200468-0

Qi F, Kuppusamy S, Naidu R, Bolan NS, Ok YS, Lamb D, Li Y, Yu L, Semple KT, Wang H (2017) Pyrogenic carbon and its role in contaminant immobilization in soils. Crit Rev Environ Sci Technol 47:795–876. https://doi.org/10.1080/10643389.2017.1328918

Qu J, Lou CQ, Yuan X, Wang XH, Cong HQ, Wang L (2011) The effect of sodium hydrogen phosphate/citric acid mixtures on phytoremediation by alfalfa & metals availability in soil. J Soil Sci Plant Nutr 11(2):85–95

Quartacci MF, Argilla A, Baker AJM, Navari-Izzo F (2006) Phytoextraction of metals from a multiply contaminated soil by Indian mustard. Chemosphere 63:918–925

Quartacci MF, Baker AJM, Navari-Izzo F (2005) Nitriloacetate- and citric acid-assisted phytoextraction of cadmium by Indian mustard (*Brassica juncea* (L.) Czernj, *Brassicaceae*). Chemosphere 59:1249–1255

Rajkumar M, Prasad MNV, Swaminathan S, Freitas H (2012) Perspectives of plant-associated microbes in heavy metal phytoremediation. Biotechnol Adv 30:1562–1574. https://doi.org/10.1016/j.biotechadv.2012.04.011

Reinoso-Maset E, Worsfold PJ, Keith-Roach MJ (2013) Effect of organic complexing agents on the interactions of Cs^{+}, Sr^{2+} and with silica and natural sand. Chemosphere 91:948–954

Roane TM, Pepper IL (2000) Microorganisms and metal pollution. In: Maier RM, Pepper IL, Gerba CB (eds) Environmental microbiology. Academic, London, p 55

Robinson BH, Brooks RR, Clothier BE (1999) Soil amendments affecting nickel and cobalt uptake by *Berkheya coddii*: potential use for phytomining and phytoremediation. Ann Bot 84(6):689–694. https://doi.org/10.1006/anbo.1999.0970

Ruley AT, Sharma NC, Sahi SV, Singh SR, Sajwan KS (2006) Effects of lead and chelators on growth, photosynthetic activity and Pb uptake in *Sesbania drummondii* grown in soil. Environ Pollut 144:11–18

Safronova VI, Piluzza G, Bullitta S, Belimov AA (2011) Use of legume-microbe symbioses for phytoremediation of heavy metal polluted soils: advantages and potential problems. In: Handbook of phytoremediation Editor: IA Golubev. ISBN: 978–1–61728-753-4

Saifullah U, Abdul G, Manzoor Q (2008) Lead phytoextraction by wheat in response to the EDTA application method. Int J Phytoremediation 11:268–282. https://doi.org/10.1080/15226510802432702

Saifullah, Shahid M., Zia-Ur-Rehman M., Sabir M., Ahmad H.R. (2015). Phytoremediation of Pb-contaminated soils using synthetic chelates soil remediation and plants: prospects and challenges. Hakeem K., Sabir M., Ozturk M., Murmet A. In Soil remediation and plants 1st Elsevier, Amsterdam. 397–414. https://doi.org/10.1016/B978-0-12-799937-1.00014-0

Salt DE, Blaylock M, Kumar NPBAV, Dushenkov BD, Ensley IC, Salt DE, Prince RC, Pickering IJ, Raskin I (1995) Mechanisms of cadmium mobility and accumulation in Indian mustard. Plant Physiol 109:1427–1433

Schmoger ME, Oven M, Grill E (2000) Detoxification of arsenic by phytochelatins in plants. Plant Physiol 122:793–801

Schnoor JL (2000) Phytostabilization of metals using hybrid poplar trees. In: Raskin I, Ensley BD (eds) Phytoremediation of toxic metals: using plants to clean-up the environment. Wiley, New York, pp 133–150

Shahid M, Pourrut B, Dumat C, Nadeem M, Aslam M, Pinelli E (2014) Heavy-metal-induced reactive oxygen species: phytotoxicity and physicochemical changes in plants. Rev Environ Contam Toxicol 232:1–44. https://doi.org/10.1007/978-3-319-06746-9_1

Sharma R, Bhardwaj R, Thukral AK, Bali S, Kohli S, Kaur P, Ohri P, Kaur R, Vig A, Sharma M (2017a) Role of phytochelators in land reclamation: recent progress and perspectives. Bentham Sci Curr Environ Eng 4(1):42–52. https://doi.org/10.2174/2212717804666161220160825

Sharma J, Shamim K, Dubey SK, Meena RM (2017b) Metallothionein assisted periplasmic lead sequestration as lead sulfite by *Providencia vermicola* strain SJ2A. Sci Total Environ 579:359–365. https://doi.org/10.1016/j.scitotenv.2016.11.089

Sharma R, Bhardwaj R, Thukral AK, Handa N, Kaur R, Kumar V (2014) Chapter 17: Osmolyte dynamics: new strategies for crop tolerance to abiotic stress signals. In: Ahmad P, Rasool S (eds) Emerging technologies and management of crop stress tolerance. Academic Press, New York, pp 405–430. ISBN 9780128008751. https://doi.org/10.1016/B978-0-12-800875-1.00017-X

Shazia A, Shazia I, Mahmood UH (2014) Effect of chelating agents on heavy metal extraction from contaminated soils. Res J Chem Sci 4(9):70–87

Sheoran V, Sheoran AS, Poonia P (2010) Role of hyperaccumulators in phytoextraction of metals from contaminated mining sites: a review. Crit Rev Environ Sci Technol 41(2):168–214. https://doi.org/10.1080/10643380902718418

Singh A, Prasad SM (2015) Remediation of heavy metal contaminated ecosystem: an overview on technology advancement. Int J Environ Sci Technol 12:353–366

Singh S, Parihar P, Singh R, Singh VP, Prasad SM (2016) Heavy metal tolerance in plants: role of transcriptomics, proteomics, metabolomics, and ionomics. Front Plant Sci 6:1143. https://doi.org/10.3389/fpls.2015.01143

Sinhal VK, Srivastava A, Singh VP (2010) EDTA and citric acid mediated phytoextraction of Zn, Cu, Pb and Cd through marigold (*Tagetes erecta*). J Environ Biol 31:255–259

Song Y, Ammami MT, Benamar A, Mezazigh S, Wang H (2016) Effect of EDTA, EDDS, NTA and citric acid on electrokinetic remediation of As, Cd, Cr, Cu, Ni, Pb and Zn contaminated dredged marine sediment. Environ Sci Pollut Res 23(11):10577–10586

Sun Y, Sun G, Xu Y, Liu W, Liang X, Wang L (2016) Evaluation of the effectiveness of sepiolite, bentonite, and phosphate amendments on the stabilization remediation of cadmium-contaminated soils. J Environ Manag 166:204–210. https://doi.org/10.1016/j.jenvman.2015.10.017

Takahashi T, Kakehi J (2009) Polyamines: ubiquitous polycations with unique roles in growth and stress responses. Ann Bot 105(1):1–6

Tangahu BV, Abdullah SRS, Basri H, Idris M, Anuar N, Mukhlisin M (2011) A review on heavy metals (As, Pb, and Hg) uptake by plants through phytoremediation. Int J Chem Eng 2011:939161. https://doi.org/10.1155/2011/939161

Tao Q, Hou D, Li T (2016) Oxalate secretion from the root apex of *Sedum alfredii* contributes to hyperaccumulation of cadmium. Plant Soil 398(1):139–152

Turgut C, Pepe MK, Cutright TJ (2004) The effect of EDTA and citric acid on phytoremediation of Cd, Cr, and Ni from soil using *Helianthus annuus*. Environ Pollut 131(1):147–154

Ullmann A, Brauner N, Vazana S, Katz Z, Goikhman R, Seemann B, Marom H, Gozin M (2013) New biodegradable organic-soluble chelating agents for simultaneous removal of heavy metals and organic pollutants from contaminated media. J Hazard Mater 260:676–688

Uthus EO (1994) Arsenic essentiality and factors affecting its importance. In: Chappell WR, Abernathy CO, Cothern CR (eds) Arsenic exposure and health. Science and Technology Letters, Northwood, UK, pp 199–208

Violante A, Cozzolino V, Perelomov L, Caporale AG, Pigna M (2010) Mobility and bioavailability of heavy metals and metalloids in soil environments. J Soil Sci Plant Nutr 10(3):268–292

Wang D, Zhang X, Liu J, Zhu Y, Zhang H, Zhang A, Jin X (2012) Oxalic acid enhances Cr tolerance in the accumulating plant *Leersia hexandra* Swartz. Int J Phytoremediation 14(10):966–977

Wang S, Wu W, Liu F, Liao R, Hu Y (2017) Accumulation of heavy metals in soil-crop systems: a review for wheat and corn. Environ Sci Pollut Res 24:15209–15225. https://doi.org/10.1007/s11356-017-8909-5

Wen XP, Ban Y, Inoue H, Matsuda N, Moriguch T (2010) Spermidine levels are implicated in heavy metal tolerance in a *spermidine synthase* overexpressing transgenic European pear by exerting antioxidant activities. Transgenic Res 19(1):91–103

Wuana RA, Okieimen FE, Imborvungu JA (2010) Removal of heavy metals from a contaminated soil using organic chelating acids. Int J Environ Sci Technol 7(3):485–496

Xu X, Thomson NR (2007) An evaluation of the green chelant EDDS to enhance the stability of hydrogen peroxide in the presence of aquifer solids. Chemosphere 69:755–762

Xun F, Xie B, Liu S, Guo C (2015) Effect of plant growth-promoting bacteria (PGPR) and arbuscular mycorrhizal fungi (AMF) inoculation on oats in saline-alkali soil contaminated by petroleum to enhance phytoremediation. Environ Sci Pollut Res 22(1):598–608

Yan DYS, Yui MMT, Yip TCM, Tsang DCW, Lo IMC (2010) Influence of EDDS-tometal molar ratio, solution pH, and soil-to-solution ratio on metal extraction under EDDS deficiency. J Hazard Mater 178:890–894

Yang L, Luo CL, Liu Y, Quan LT, Chen YH, Shen ZG (2013) Residual effects of EDDS leachates during EDDS-assisted phytoremediation oKf copper contaminated soil. Sci Total Environ 444:263–270

Zaidi S, Usmani S, Singh BR, Musarrat J (2006) Significance of *Bacillus subtilis* strain SJ 101 as a bioinoculant for concurrent plant growth promotion and nickel accumulation in *Brassica juncea*. Chemosphere 64:991–997. https://doi.org/10.1016/j.chemosphere.2005.12.057

Zemanová V, Pavlík M, Pavlíková D, Tlustoš P (2013) The changes of contents of selected free amino acids associated with cadmium stress in *Noccaea caerulescens* and *Arabidopsis halleri*. Plant Soil Environ 59(9):417–422

Zhang X, Zhong B, Shafi M, Guo J, Liu C, Guo H, Peng D, Wang Y, Liu D (2018) Effect of EDTA and citric acid on absorption of heavy metals and growth of Moso bamboo. Environ Sci Pollut Res 25(19):18846–18852

Effective and Sustainable Solid Waste Management in India: A Challenge

10

Sushila Saini, Meenakshi Nandal, and Geeta Dhania Bahamnia

Abstract

Rapid unplanned urbanization and population leads to the generation of large amount of waste in India. Municipal solid waste is generated from various human activities like domestic and industrial. Huge amount of waste generation due to lack of efficient and effective management causes various diseases and environmental contamination. Municipal solid waste management (SWM) is nowadays a big issue not only due to environment and health issue but also to generation of large quantities of waste. Developments of an integrated effective management system have the understanding of amount of waste generated, availability of resources, and environmental condition of society. SWM is a discipline that concerns in controlling the generation, storage, collection, transportation, disposal, and processing of solid waste in a way that has no effects on economy, health, and environment and suits with public attitudes.

Keywords

Municipal solid waste (MSW) · Solid waste management (SWM) · Incineration · Composting · Recycle · Swachh Bharat Mission (SBM) · Municipal solid waste management (MSWM)

S. Saini
Department of Botany, Janta Vidya Mandir Ganpat Rai Rasiwasia (JVMGRR) College, Charkhi Dadri, Haryana, India

M. Nandal · G. D. Bahamnia (✉)
Department of Environmental Science, Maharshi Dayanand (MD) University, Rohtak, Haryana, India

P. K. Arora (ed.), *Microbial Technology for Health and Environment*, Microorganisms for Sustainability 22,
https://doi.org/10.1007/978-981-15-2679-4_10

10.1 Introduction

Solid waste means any unwanted or useless solid materials that are generated by community activities, and about 70% of total solid waste is comprised by municipal solid waste (MSW). MSW is generated by everyday human activity in residential setting, educational institutions, and commercial complexes. It includes wet, dry, and hazardous household wastes. Rapid development, rising urbanization, and changing lifestyle are the main culprits behind the generation of enormous amount of biodegradable and nonbiodegradable waste. In spite of the production of huge quantity and great variability of waste, the practices of waste management are still outdated, and this has resulted in heaps of waste everywhere in cities and towns (Gupta et al. 2015). It is pertinent to note that in India only 22–28% of the collected MSW is processed and treated, while the rest is discarded at different dump sites. With time, these wastes lead to emission of greenhouse gases and leach in soil, causing groundwater, environmental, and health problems (Shazwin and Nakagoshi 2010; Srivastava et al. 2014). Segregation of waste at the generator site is of foremost important, but in our country, people often dumps mixed waste, which makes management more difficult. There is requirement of more sustainable practices as the present system includes only collection and dumping without treatment (Kumar et al. 2017). However, in the last 4 years, due to various programs by the Govt. of India, such as Swachh Bharat Mission (SBM) and concept of smart cities, MSWM had made people more aware about the magnitude of problem. These schemes had provided an atmosphere and also funds to concentrate more on the problem, but still, a large gap remains between policy and implementation (NITI Aayog 2015). In the year 2000 and then again in 2016, very strict rules were framed, but implementation on ground level is still lacking. The role of the informal sector has not been duly recognized, and there is requirement to incorporate them in recycling of waste by providing them better working condition. The need of the hour is to include integrated solid waste management strategy by the promotion of waste segregation, waste recycling, compost production, and waste to power generation (MoUD 2014; Nandan et al. 2017). This present study evaluates the current status and identifies challenges, barriers, and opportunities associated with improving waste management in India.

10.2 Generation of Municipal Solid Waste in India

10.2.1 Sources and Composition of Municipal Solid Waste

Municipal solid waste comprises highly bio degradable wastes (food waste, textiles, newspapers, garden waste, street sweepings, paper packaging material), moderately biodegradable waste (disposable napkins, disposable tableware, sanitary refuse), and nondegradable waste (rubber, plastic, metal, ceramics, glass, ash, electronic) from residential, institutional, commercial, and industrial sources (Fig. 10.1). It also includes hazardous household wastes such as paints, broken compact fluorescent

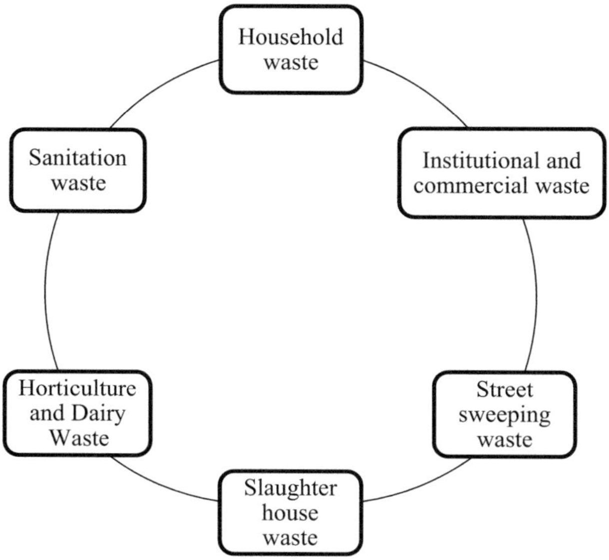

Fig. 10.1 Sources of municipal solid waste

Table 10.1 Change in the composition of municipal solid waste across the country

Year	Composition (%)							
	Biodegradable	Paper	Plastic/rubber	Metal	Glass	Rags	Others	Inert
1996	42.21	3.63	6.60	0.49	0.60	Nil	Nil	45.13
2005	47.43	8.13	9.22	0.50	1.01	4.49	4.016	25.16
2011	42.51	9.63	10.11	0.63	0.96	Nil	Nil	17.00

Source: Planning Commission Report (2014), Govt. of India

lamp (CFL) and bulbs, nail polish remover, thermometers, insecticides, and batteries. (Bhat et al. 2018). Composition of MSW depends on per capita income, degree of urbanization and industrialization, socioeconomic status, geographical region, and cultural habits (Kumar and Kaushal 2015).

In our country, MSW contain less hazardous and more organic material than developed countries like Canada, the USA, and other European countries. Also, in India, MSW contains less paper, plastic, and metal contents than western countries due to recycling of these materials. The constituents of MSW comprise mainly organic fraction (40–60%) having a lot of moisture and inert such as ash and sweepings (30–50%) followed by recyclables, viz., paper, plastics, glass, and metals. Food is the most important consumable item, and it contributes a lot to the organic fraction (Kaushal et al. 2012). Carbon/nitrogen (C/N) ratio in MSW ranges from 20 to 30 with calorific value of about 1700–1800 Kcal/Kg. During 1996–2011, constituents of MSW have changed considerably with increase in the proportion of high calorific value waste as shown in the Planning Commission Report 2014, Govt. of India. Major rise occur in plastic waste, which is of chief concern due to its nonbiodegradable nature (Table 10.1). Categorization of wastes is necessary to know changing trends in the composition of wastes as appropriate technologies could be selected for the treatment of waste on the basis of this (Pamnani and Srinivasarao 2014).

10.2.2 Municipal Solid Waste Generation Rate in Different Cities of India

Rapid explosion in population and urbanization had resulted in substantial increase in the quantity of waste generation (Table 10.2) and also resulted in change in its composition (Nguyen et al. 2011). After independence, there has been a marked change in the lifestyle of people which resulted in an eightfold increase in the generation of waste. From 1947 to 1997, waste generation has increased from 6 million tons to 48 million tons (Sharholy et al. 2006). According to Annepu (2012), waste generation in 2001 was 31.6 million tons that increased to 47.3 million tons in 2011 with an increase of 50% in 10 years. It has been estimated that in 2041 the waste generation will show a fivefold increase, and it will be 161 million ton (Table 10.3). Presently, there has been an increase in per capita waste generation from 1% to 1.33%. Depending upon economic status and density of population per capita generation, it varies from 200 g/day in small town and villages to 800 g/day in metro cities (Pattnaik and Reddy 2010; Siddiqqui 2018).

The Central Pollution Control Board (CPCB) along with the *Environment Protection Training and Research Institute* (EPTRI), National Environmental Engineering Research Institute (NEERI), and Central Institute of Plastics Engineering & Technology (CIPET) presented a report on production of waste from 1999 to 2011 by conducting a survey in 35 metro cities and 24 state capitals. According to this report in 2011, the national capital (New Delhi) produced maximum amount of solid waste (6800 ton/day) (Table 10.4) followed by Mumbai (6500 ton/day) and Chennai (4500 ton/day) (Kumar et al. 2009). Of the total waste produced, only 70% is collected, while the rest of the 30% spreads in every nook and corner of cities (Ghosh and Kansal 2014). The CPCB (2017) presented the recent status amount of waste generated, collected, processed, and landfilled in Indian states and UT after so much funding and campaigning under SBM (Table 10.5). In this report, Maharashtra topped the list with 21,860 tons of solid waste generated per day, followed by Uttar Pradesh (15,190 tons/day) and Delhi (9620 tons/day). Of

Table 10.2 Per head generation of waste in Indian cities

S. no.	Population	Per capita generation (g/day)
1	Cities with population less than 2 lakh	200–300
2	Cities with population of 2–5 lakh	300–350
3	Cities with population of 5–10 lakh	350–400
4	Cities with population more than 10 lakh	400–800

Source: CPCB (2012)

Table 10.3 Future calculation of waste generation up to 2041

S. no.	Year	Population (in millions)	Generation/individual (kg/day)	Waste generated (metric tons/day)
1	2001	197.3	0.439	31.63
2	2011	260.1	0.498	47.30
3	2021	342.8	0.569	71.15
4	2031	451.8	0.649	107.01
5	2036	518.6	0.693	131.24
6	2041	595.4	0.741	160.96

Source: Annepu (2012)

the total waste generated, only 20% is treated, while the rest goes in landfills or lost in urban environment.

10.3 Solid Waste Management Practice in India

Basic principle of SWM: 4Rs (refuse, reduce, recycle, reuse) (Fig. 10.2).

Refuse: Buy only those things that are needed and don't buy those things that are not needed.

Reduce: Minimize the amount of waste generated by yourself by changing your lifestyle.

Reuse: Use the things for maximum time and make secondary use of the things.

Recycle: Develop the habit of segregation by giving recyclable material to kabadiwallahs, and convert biodegradable waste into manure and other useful products for reuse.

10.3.1 Collection

Collection of waste from house to house is a very tedious process because of the varied behavior of the public. People use their own intelligence for the segregation, and it will create a problem to the professional. Waste is collected from different site, door to door, and transferred to disposal site for further processing. Collection of waste is a very complex and costly process. Waste separation at the point of source into three categories, i.e. biodegradable, recyclable, and nonrecyclable, reduced the cost. Proper planning and better management is effective in collection and reduction of cost.

10.3.2 Segregation

Segregation of waste is challenge for all, and it is unorganized (Advani and Somani 2018). The whole process becomes so easy once the process of segregation becomes part of each person's daily life.

Table 10.4 Waste generated (tons/day) by metro cities/state capital from 1999 to 2011 (CPCB 2012)

S. no.	Name of city	Municipal solid waste (tons/day)		
		1999–2000[a]	2004–2005[b]	2010–2011[c]
1	Agartala	–	77	102
2	Agra	–	654	520
3	Ahmedabad	1683	1302	2300
4	Aizawl	–	57	107
5	Allahabad	–	509	350
6	Amritsar	–	438	550
7	Asansol	–	207	210
8	Bangalore	2000	1669	3700
9	Bhopal	546	574	350
10	Bhubaneshwar	–	234	400
11	Chandigarh	–	326	264
12	Chennai	3124	3036	4500
13	Coimbatore	350	530	700
14	Daman	–	15	25
15	Dehradun	–	131	220
16	Delhi	4000	5922	6800
17	Dhanbad	–	77	150
18	Faridabad	–	448	700
19	Gandhinagar	–	44	97
20	Gangtok	–	13	26
21	Guwahati	–	166	204
22	Hyderabad	1566	2187	4200
23	Imphal	–	43	120
24	Indore	350	557	720
25	Itanagar	–	12	102
26	Jabalpur	–	216	400
27	Jaipur	580	904	310
28	Jammu	–	215	300
29	Jamshedpur	–	338	28
30	Kanpur	1200	1100	1600
31	Kavaratti	–	3	2
32	Kochi	347	400	15
33	Kohima	–	13	45
34	Kolkata	3692	2653	3670
35	Lucknow	1010	475	1200
36	Ludhiana	400	735	850
37	Madurai	370	275	450
38	Meerut	–	490	52
39	Mumbai	5355	5320	6500
40	Nagpur	443	504	650
41	Nashik	–	200	350
42	Panjim	330	32	25

(continued)

Table 10.4 (continued)

S. no.	Name of city	Municipal solid waste (tons/day)		
		1999–2000[a]	2004–2005[b]	2010–2011[c]
43	Patna	–	511	220
44	Pondicherry	–	130	250
45	Port Blair	700	76	45
46	Pune	–	1175	1300
47	Raipur	–	184	224
48	Rajkot	–	207	230
49	Ranchi	–	208	140
50	Shillong	–	45	97
51	Shimla	–	39	50
52	Silvassa	–	16	35
53	Srinagar	–	428	550
54	Surat	900	1000	1200
55	Thiruvananthapuram	–	171	250
56	Vadodara	400	357	600
57	Varanasi	412	425	450
58	Vijayawada	–	374	600
59	Visakhapatnam	300	584	334
	Total MSW	30,057	39,031	50,592

Municipal solid waste study conducted by CPCB through:
[a]EPTRI (1999–2000)
[b]NEERI-NAGPUR (2004–2005)
[c]CIPET during 2010–2011

10.3.3 Transportation

Solid wastes after collection are transported to the dumping/disposal site by open van and truck/compactor trucks. Industrial solid waste collection and transportation is done by private contractor to disposal site.

10.3.4 Disposal

10.3.4.1 Open Dumping

It is the easiest and commonly used methods by underdeveloped and developing countries like India. MSW are dumped outside the city in low-lying area without taking care of the environment. Open dumping does not need equipment and expertise and is of low cost. But once done, the remediation of that place is a costly affair. It is impossible to revert contamination of groundwater to healthy condition. Methane gas is released out from biodegradation in anaerobic condition, a major cause of global warming. Again, burning of waste at dump site creates respiratory problems due to release of fine particles. This unscientific way of disposal creates problem during rainy season and flooding, which causes various diseases and leads to contamination of groundwater.

Table 10.5 Solid waste generated, collected, and processed in Indian states and Union territories (UT) (CPCB (2017), CPCB (2015–2016) report)

S. no.	States and UT	Waste generated (TPD)	Collected (TPD)	Processed (TPD)	Landfilled (TPD)
1	Andaman and Nicobar Islands	70	70	5	–
2	Andhra Pradesh	6440	6331	500	143
3	Arunachal	13	11	Nil	Nil
4	Assam	7920	6336	200	Nil
5	Bihar	1670	0	0	No
6	Chandigarh	370	360	Nil	230
7	Chhattisgarh	2245.25	2036.97	828.18	1290.97
8	Daman and Diu	85	85	0	–
9	Delhi	9620	8300	3240	5060
10	Goa	450	400	182	–
11	Gujarat	10,480	10,480	2565	7730
12	Haryana	4837.35	3102.51	188	2163.18
13	Himachal Pradesh	276	207	125	150
14	Jharkhand	3570	3570	65	3505
15	Jammu and Kashmir	1634.5	1388.7	3.45	425
16	Karnataka	8842	7716	3584	3946
17	Kerala	1339	655	390	–
18	Nagaland	344	193	–	–
19	Lakshadweep	21	–	–	–
20	Madhya Pradesh	6678	Nil	Nil	Nil
21	Maharashtra	21867.27	21867.21	6993.2	14993.67
22	Manipur	176	125	–	–
23	Mizoram	552	276	0	–
24	Meghalaya	187	156	36	122
25	Orissa	2574.7	2283.9	30	–
26	Punjab	4456.2	4435	3.72	3214
27	Puducherry	513	513	10	503
28	Rajasthan	5037	2491	490	–
29	Sikkim	49	49	0.3	–
30	Tamil Nadu	230	210	–	207
31	Telangana	6628	6625	3175	3050
32	Tripura	414	368.2	250.4	164.4
33	Uttarakhand	917	917	No MSW treatment facility existing	No sanitary landfill site
34	Uttar Pradesh	15,192	11,394	1857	–
35	West Bengal	9500	8075	851	575
	Total	135198.27	111027.55	25572.25	47415.62

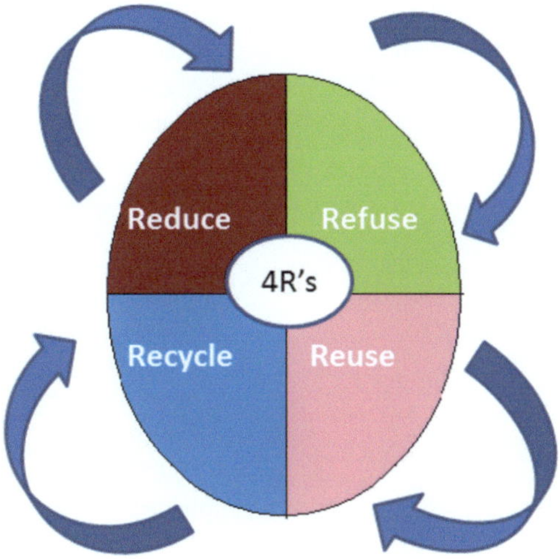

Fig. 10.2 4 Rs principles

10.3.4.2 Landfilling

Landfilling is an engineered structure designed for disposal of waste to avoid environment contamination and for public health. Landfill site should help in waste compaction, protection of environment, and control of public health. In this process, MSW is deposited in landfill where decomposition takes place by various physical, chemical, and biological processes in the absence of oxygen. Biogas and leachate are two main by-products of landfilling. Landfill site when designed and constructed scientifically with technological specification can reduce environmental pollution. They are also designed according to the type of waste generated (hazardous, non-hazardous, and inert waste) and checked periodically for their performance. An effective and safe landfill site has been planned by administrative and municipal solid waste management system. Major landfill sites present in various states of India are shown in Table 10.6.

10.3.4.3 Biological Treatment of Organic Waste

It is a fast process of decomposition of organic matter in warm moist by microorganism in aerobic and anaerobic condition. It is the simplest, eco-friendly, efficient, and cost-effective method of handling MSW. Composting can occur naturally, and it will take 6 months to degrade waste. This is known as passive composting. Compositing process is enhanced by heat, moisture, and temperature to provide a suitable condition, and then, it will complete the degradation process within 3 months. Compost (humus) is formed as end product of composting, which is rich in nutrients and used as slow release fertilizer. Using compost helps in soil erosion and increases the fertility of soil. CPCB Annual Report 2017 reported working composting plant all over India as shown in Table 10.7.

Table 10.6 Landfill site for disposal of solid waste in various states of India

States	Number of landfill sites constructed	Number of landfill sites working
Andhra Pradesh	01	01
Chandigarh	01	01
Goa	06	04
Gujarat	11	03
Haryana	00	10
Karnataka	52	157
Nagaland	01	01
Madhya Pradesh	10	03
Maharashtra	04	04
Tamil Nadu	12	11
West Bengal	06	07
Tripura	01	01

Source: CPCB (Annual Report 2016–2017)

Table 10.7 Number of composting/vermicomposting plants in India

State	Number of plants working	State	Number of plants working
Andaman and Nicobar	02	Karnataka	140
Andhra Pradesh*	39	Gujarat	23
Assam	01	Maharashtra	73
Chandigarh	01	Madhya Pradesh	20
Chhattisgarh	03	Punjab	01
Goa	09	Tamil Nadu	895
Haryana	04	Uttar Pradesh*	15
Uttarakhand	01	West Bengal	10
Sikkim	02	Puducherry	01

Source: CPCB (Annual Report 2016–2017)

10.3.4.3.1 Aerobic Compositing

It is a process of composting that occurs in the presence of air, humid, and warm environments with the presence of microorganisms.

10.3.4.3.2 Vermicomposting

It is a kind of compositing where the environment is suitable for survival and reproduction of red worm and earthworm. Earthworm feeds on semi-decomposed matter and excretes natural organic substance that looks like tiny soil, which is rich in nutrients. Vermicomposting is done in controlled environment, i.e., temperature, moisture, and types of organic matter (Kumar and Pandit 2013). It is found that compost formed is free of pathogen, but if the initial matter is rich in pathogen, some of the pathogen is left behind in the compost. It is the simplest technology with low cost that can be used for treatment of urban and rural waste. But segregation of organic waste is the prerequisite of vermicomposting.

Novel technology of waste management by black soldier flies not only reduces the volume of waste but also provides business to the fish-growing people. The larvae feed on organic matter and reduce the volume of dry mass by 40–50%. At prepupae stage, its body is rich in protein and fats, so it is a good feed for fishes (https://sswm.info/sites/default/files/reference_attachments/EAWAG%20SANDEC%202008%20Module%206%20Solid%20Waste%20Management%20Lecture.pdf).

10.3.4.3.3 Anaerobic Digestion

It is the formation of methane from organic matter by use of microorganism in the absence of air. Indian waste is rich in organic matter, and this process generates methane and manure enriched in nutrients. It is the most important and sustainable methods of treatment of biodegradable waste. It is commonly known as biomethanation. Biogas is stabilized and generated and is used as fuel for electricity generation and household activities. The Govt. of India in rural as well as in urban areas encourages people to utilize municipal, animal, and agricultural waste for biogas production.

10.3.4.4 Incineration

It is the process where energy is produced from waste and then processed in air at high temperature of 850 °C. Along with energy, carbon dioxide, incombustible material, and solid residue with water (bottom ash) are produced. Ash produced can be solidify and used to form concrete to control the migration of contaminants. It helps in reducing volume and weight of waste (Durgekar 2016). Energy released can be used for electricity generation, process steam, and hot water for public heater. Incineration is mostly done in hospital for medical waste where it helps in the breakdown of hazardous organic waste and nonmetallic waste and eradication of bacteria and virus. It is not successful as calorific value of Indian waste is low (low energy and high moisture) (Patel and Baredar 2016). Developing countries including India have many problems to start with full equipment-operated incineration system because of financial constraints such as infrastructure, maintenance, and pollution control equipment. In many countries like Singapore and Bangkok, fully functional incinerator plant used 90% of MSW generated (UNEP-IETC et al. 1996). Gas emission (CO_2, SO_2, dust, particulate matter, oxides of nitrogen, etc.) by incinerator creates major problems to human health and environment.

10.3.4.5 Thermal Treatment

The use of high temperature to degrade the waste is known as thermal treatment. It is done by incineration, gasification, and pyrolysis. Municipal solid wastes in India commonly contain large amount of organic matter with high moisture content and low calorific value that ranges 800–1100 kcal/kg, which makes its unsuitable for incineration (Kansal 2002).

Table 10.8 Number of biogas plants in India

State	Number of plants working	State	Number of plants working
Andhra Pradesh	10-RDF and 7-BG	Maharashtra	03-RDF, 34-BG
Chandigarh	01-RDF and 1-BG	Gujarat	1-BG
Goa	01	Karnataka	13-BG
Madhya Pradesh	01-RDF, 01-BG	Pondicherry	01-RDF, 02-BG
Tamil Nadu	07-RDF, 39-BG	Punjab	08
Uttar Pradesh	04-RDF	West Bengal	01

Source: CPCB (Annual Report 2016–2017); *RDF* refuse-derived fuels; *BG* biogas

1. Pyrolysis is a process where carbonaceous material is converted into gas, tar, ash, coke, and char by thermal process.
2. Gasification is used in the treatment of solid waste, which is able to decrease the pollution and increase the recovery of heat. It is done at temperature range of 900–1400 °C in limited supply of oxygen. Mostly, gasifiers are used to burn agro biomass, and only few number of gasifier is installed in India (see Table 10.8).

10.4 Roles of Public and Private Partnership

In our country, regulation of solid waste is considered as the sole accountability of urban local bodies (ULBs) due to the public and local nature of service. Due to increased urbanization, there is a marked rise in the magnitude of waste generated, and it also consists of lots of nonbiodegradable material. So, there is a gap between the requirement of infrastructure and services for management of waste and the capability of ULBs to provide the same. To overcome this shortfall, state and local government are increasingly adopting private contractor for collection, transportation, and disposal of waste. There are various forms of partnership among ULBs, private sector, and community at various places of country. These partnerships have been categorized mainly into four types. First category involves the management of MSW by ULBs alone as in cities like Jabalpur, Bokaro, and Tiruchirappalli. Second category includes partnership between ULBs and private sector for processing of waste, as in Hyderabad and Rajkot. In Guwahati, ULBs have engagement with both private and informal sector for the execution of waste management practices. In

Table 10.9 Forms of Public Private Partnership adopted by various ULBs for MSWM

	Forms of partnership			
	ULB (at their own)	ULB and private sector	ULB and community	ULB, private sector, and community
Cities involved	Bokaro, Trichy, Munger, Patna	Hyderabad, Rajkot, Chennai (from 1995), Bengaluru, Ahmedabad	Chennai (1989–1995), Namakkal, Trivandrum	Guwahati

Source: Collected from SWM tool kit, case studies, CDPs, etc.

some cities like Chennai and Trivandrum, local communities manage the waste of their own areas as well as of the vicinity. Here, ULBs along with small help group (SHG) or NGOs undertake the waste management activities. In Mumbai, Brihanmumbai Municipal Corporation has launched advanced locality management scheme in which members of locality are provided with subsidies and technical help for construction of composting facilities (Chatri and Aziz 2012) (Table 10.9).

In India, now, many private companies are in partnership with municipal bodies due to realization of business opportunity in MSWM, and under Public Private Partnership (PPP) mode, many projects are running. Some Indian companies involved are ESSEL Infra, Zen Global Finance Ltd., Hanjer Biotech, Excel Industries, Enkem Engineers Ltd., SELCO international Ltd., etc. Recently, some international companies have also jumped in Indian market for MSWM such as Lunde, TBW, and BTA of Germany, EISU of UK, Entac of Austria, and Nellemen and Nielsen of Denmark (Joshi and Ahmed 2016). Figure 10.3 shows the number of private projects undertaken by ULBs in some states of India. Karnataka has undertaken seven projects, while Rajasthan and Tamil Nadu have five and four projects, respectively, in partnership with private sector. Gujarat and Maharashtra has only one project each under PPP (Chatri and Aziz 2012). However, the concept of integrated solid waste management (Fig. 10.4) is still in nascent stage and is adopted by a few cities only. There is need for more states to come forward and adopt PPP mode for efficient management of MSW.

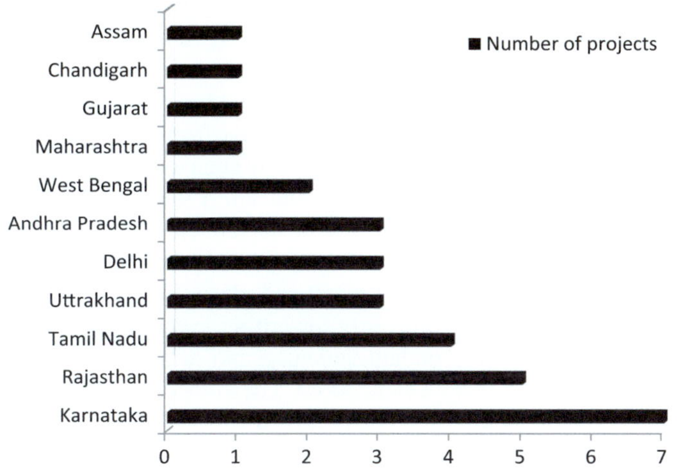

Fig. 10.3 SWM projects taken by some states under Public Private Partnership (PPP India Database, Chatri and Aziz 2012)

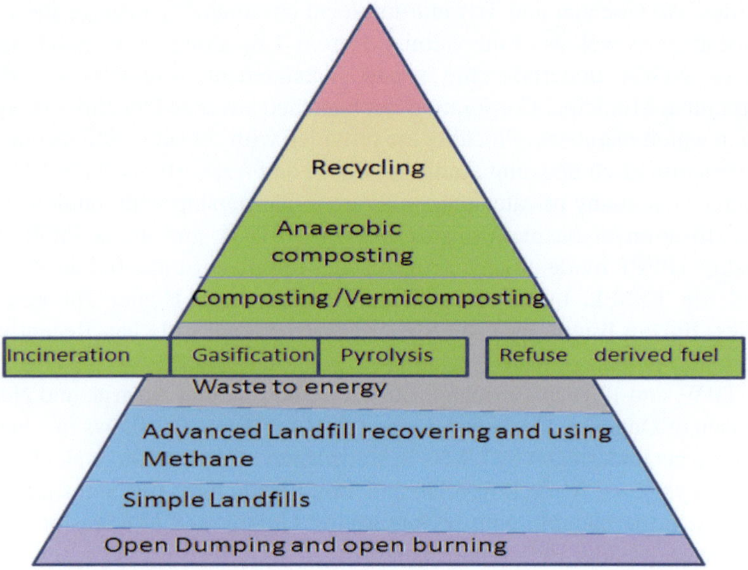

Fig. 10.4 Hierarchy of sustainable solid waste management

10.5 Role Played by Rag Pickers and Health Risks

In our country, the role played by rag pickers is very important due to apathy on the part of generator for the segregation of waste resulting in its dumping in outskirts of cities and towns. Most of the rag pickers are migrants from rural areas who have come in cities in search of employment. They live under extremely unhygienic conditions in the suburbs of cities. These people wander from one dump/landfill site to another for collecting, sorting, and recycling waste such as newspaper, plastic items, glass bottles, carton, gatta, and metal scrap and sell it to waste dealers to generate income (Kumar et al. 2004). An adult rag picker collects on an average 40 kg of waste per day, which includes 5–15 kg of plastic and 10–15 kg of paper and cardboards in addition to small amount of metal and glass (Syamala Devi et al. 2014). Thus, rag picker contributes significantly toward recovery of recyclable materials, saves about 14% of municipal budget annually, and decreases landfill load up to 20% (Chintan NGO report, Pappu et al. 2007). They make a significant contribution to the environmental management while putting their own health at risk. They suffer from many diseases such as tuberculosis, asthma, bronchitis, pneumonia, dysentery, and anemia (Syamala Devi et al. 2013). They often get infection by coming in contact with human and animal excreta, sputum, and dead animals. Cuts from sharp metal objects, syringes, blades, and broken bottles are more, common thus exposing them to tetanus and other infections. Injuries from medical waste are more dangerous because rag pickers may get infected by HIV, hepatitis B and C, and other bacterial infections by contaminated syringe and needles. They are often bitten by mosquitoes, snakes, and rodents. They also suffer poisoning from various chemicals, heavy metals, and pesticides (Sarkar 2003).

Considering the worth of their work steps should be taken to improve their living and working condition, efforts should be made to organize them in cooperatives with the help of NGOs. They should be allowed to collect waste directly from households instead of searching at dump sites (MoUD 2014). Besides giving them due recognition in society, it will help in providing better working condition along with reducing occupational health hazards. Cities like Pune, Rajkot, and Mysuru helped in organizing waste pickers with the help of NGOs, provided them with better working condition, and integrated their contribution with formal solid waste management system (Ahluwalia and Patel 2018). In Pune, rag pickers' cooperatives called SWaCH receives uniforms, identity cards, equipment, and also sheds for sorting recyclable from dry waste (SWaCH website 2012). Other Indian cities can adopt Pune's way of managing their waste by incorporating informal sector.

10.6 Swachh Bharat Mission

Four years ago, on second October 2014, Gandhi Jayanti Prime Minister Narendra Modi launched the Swachh Bharat Mission (SBM) for urban areas (NITI Aayog 2015). It is India's biggest cleanliness drive with the motive to eliminate open defecation, management of MSW in scientific manner, eradication of manual scavenging, and generation of health awareness. The scheme was started with a target of 80% of SWM with an increase of 2% per year by providing financial assistance and by ensuring information, education, and communication (IEC). "Smart City Mission" was initiated on 25th June 2015, and from 2016 onward, a city ranking inspection system for cities and towns called Swachh Survekshan was started to review city's performance under SBM. It surveys ULB documentation, independent observation, and verification along with citizen feedback. It is the first-ever largest survey in the world impacting around 40 core people. On fifth June 2017, "Har din do bin" campaign was launched to encourage 100% source segregation of waste into wet and dry in green and blue bins. Star rating framework was initiated on 20th January 2018 to ensure no visible garbage, reduction in waste generation, and enhancement in waste processing in all cities and towns. Compost Banao Compost Apnao scheme was started to encourage stakeholders to make compost from wet waste. Besides these, an ICT-based Grievance Redressal System called Swachhata

Table 10.10 Solid waste management scenario (SBM reporting, MoUD 2017)

MSW generation	1.45 lakh tons/day
MSW processed	33,215 tons/day (23%)
Total number of wards with 100% door to door collection	57,475 (68%)
Landfilled (crude dumping)	1.22 lakh tons/day (72%)
Waste to compost production	13.11 lakh TPA (tons/annum)
Waste to energy production	88.4 MW

Table 10.11 Waste to compost production (SBM reporting, MoUD 2017)

Waste to compost potential—54 lakh TPA	
Number of functional plants	145
Input capacity of functional plants (TPA)	62.3 lakh
Total production of city compost (TPA)	13.11 lakhs
Number of plants under construction	150
Input capacity of plant under construction (TPA)	33.48 lakh

App was launched to enable people to sort out their grievances by concerned Municipal Corporation (Singh 2018).

10.6.1 Achievements Under SBM

- Under SBM, 68% urban wards have 100% collection of MSW from the generator, and this collection increased to 83% till 2018.
- MSW processing capacity increased from 24% (in 2017) to 34% (in 2018) (Table 10.10).
- To manage city compost, new waste to compost plants are constructed. Presently, 145 plants are operating, while 150 plants are under construction (Table 10.11).
- As per SBM, 7 plants are operating under waste to energy generation, and 56 plants with a power generation capacity of 415 MW are under construction (Table 10.12).
- Under SBM, 7365 cores were allocated for SWM.
- Cleanliness has become a drive like never before.

10.6.2 Areas Lacking Behind Under SBM

- Segregation of waste is the biggest challenge, and this will be the game changer whenever implemented properly. Mostly, mixed waste ends in the dump site, and people are not segregating in true spirit. Under SWM 2016 Rules, it is mandatory to segregate waste into dry, wet, and domestic hazardous waste.

Table 10.12 Waste to energy production (SBM reporting, MoUD 2017)

Waste to energy potential—51 MW	
Number of functional plants	7
Production capacity of functional plants (MW)	88.4
Number of plants under construction	56
Production capacity of plant under construction (MW)	415

- Although so much campaign and incentives are given for production of city compost, neither the fertilizer companies nor the farmers are interested in city compost.
- Waste to energy plants can treat only nonbiodegradable, nonrecyclable, high calorific value waste, but only a limited amount of waste fit in this category as the waste coming from these plants is of mixed quality. So these plants have more capacity to treat waste than what they are doing.
- Out of allocated funds for SWM, only 2126.24 cores (28%) have been dispersed.
- Although so much investment has been done for public awareness, impact on ground level is very poor.
- To achieve better ranking under Swachh Survekshan, municipalities commence the activities a few months before the survey.
- SBM have focused much on prevention of open defecation, and attention on SWM is missed out.

Although the steps taken by the government under SBM to achieve cleanliness are laudable, people in India consider SWM as the job of municipal authorities only. SBM can be achieved only when all the stakeholders including the waste generators consider it their responsibility to manage waste.

10.7 Solid Waste Management Rules, 2016

In our country, before 2000, there is no specific rule regarding collection, segregation, transportation, processing, and disposal of waste. Often, waste was dumped in the periphery of cities where slums and unauthorized colonies got established for picking up recyclable waste that poses serious threat to public health. So, in the year 1996, a Public Interest Litigation was filed in the Supreme Court. This resulted in an appointment of a committee under MoEF to frame rules for waste management, and for the first time, Municipal Solid Waste (Management and Handling) Rules 2000 came into existence (MoEF 2000). But these rules did not bring improvement in solid waste management, so Solid Waste Management (SWM) Rules 2016 were proposed by MoEF&CC to improve the present scenario (MoEF&CC 2016).

10.7.1 Rules and Legal Provisions Under SWM Rules 2016 (MoEF&CC 2017, Yadav 2017)

- These rules for the first time clearly define the duties of MSW generator. These rules emphasize the duty of producer to separate out the waste into wet, dry, and special waste.
- The area under application of MSW rules was broadened by including residential and nonresidential properties, places of pilgrims, railways, airport, ports, defense, hospitals, hotels, educational institutions, sport complexes, etc. under its range.

- It introduced the concept of integrated solid waste management system by increasing PP partnership under Swachh Bharat Mission. Under this, concept of 5Rs was introduced such as reduce, reuse, recover, recycle, refine, and remanufacture for better management practices.
- It includes provisions of instant fines on those throwing waste in open places and not doing segregation.
- Street vendors have to keep bins for collecting the waste generated whole day and not to litter it on roads.
- Manufacturers of sanitary napkins and diapers should provide a wrapper for their disposal, and these must be disposed of in dry waste bin.
- All biodegradable waste must be treated by composting or biomethanation. Ministry of Chemical and Fertilizer was directed to provide assistance in developing market for compost.
- Ministry of New and Renewable Energy Resources should help in the improvement of infrastructure for developing new energy plants from waste by providing financial assistance.
- It is mandatory on the part of municipal authority to send annual reporting on MSW operation to the Ministry of Urban Development (MoUD).
- Secretary of State Urban Development, municipal administrator, local bodies, and village Panchayats have to make a waste management strategy by consulting all the stakeholders including rag pickers, self-help groups, and NGOs.
- The Ministry of Urban Development along with State Urban Development should prepare a state policy and provide technical guidelines and training to local bodies and other stakeholders.
- CPCB will annually review the execution of rules and then submit an Annual Report on performance of states and UT under these rules.

10.8 Conclusion

Segregation of waste, collection of waste from door to door, various treatment technologies, and limitation of land resource and lack of scientific method of waste disposal are major challenges for the management of solid waste. The first initiative is waste segregation/separation at source place, which reduced the cost of waste management system, and that amount can be used in other processing unit. Landfill site constructed should have longer life span and reduce environmental impacts. Law should be followed strictly so that every citizen can realize the responsibility of waste management. Municipal bin should collect dry and wet waste in separate bins. Strong transportation and setting of transfer station are also the requirement of today's SWM. Implementation of integrated solid waste management system as shown in Fig. 10.5 somehow helps in the management of this problem.

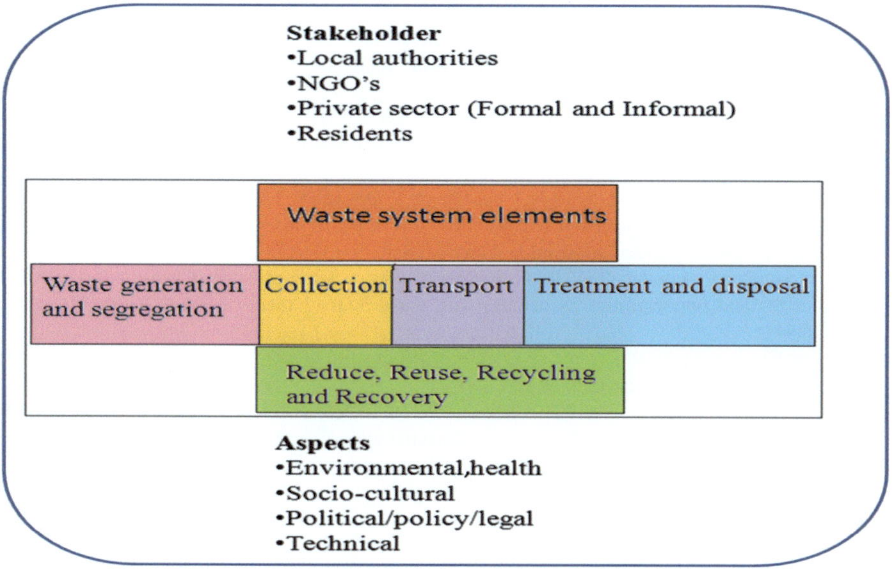

Fig. 10.5 Integrated solid waste management system

References

Advani P, Somani A (2018) Key challenge and deficiencies in the present solid waste management system for Indian cities. Int J Eng Res Mechan Civ Eng 3(2):20–24

Ahluwalia IJ, Patel U (2018) Solid waste management in india: an assessment of resource recovery and environmental impact. Working Paper No. 356 Indian Council for Research on International Economic Relations

Annepu RK(2012) Report on sustainable solid waste management in India. Waste-to-Energy Research and Technology Council (WTERT) 1–189

Bhat RA, Dar SA, Dar DA, Dar GA (2018) Municipal solid waste generation and current scenario of its management in India. Int J Adv Res Sci Eng 7(2):419–431

Chatri AK, Aziz A (2012) Public private partnerships in solid waste management: potential and strategies. Public Policy Team, Athena Infonomics

CPCB (2012) Consolidated annual review report on implementation of municipal solid wastes (management and handling) rules, 2000. Central Pollution Control Board, Delhi

CPCB (2017) Consolidated annual review report (2015–16) on implementation of municipal solid wastes (management and handling) rules, 2000. Central Pollution Control Board, Delhi

Durgekar V (2016) Towards sustainable waste management through technological innovations, effective policy, supply chain integration & participation. Procedia Environ Sci 35:140–149

Ghosh R, Kansal A (2014) Urban challenges in India and the mission for a sustainable habitat. Interdisciplina 2(2):281–304

Gupta N, Yadav KK, Kumar VA (2015) Review on current status of municipal solid waste management in India. J Environ Sci 37:206–217

Joshi R, Ahmed S (2016) Status and challenges of municipal solid waste management in India: a review. Cogent Environ Sci 2:1139434

Kansal A (2002) Solid waste management strategies for India. Indian J Environ Prot 22:444–448

Kaushal RK, Varghese GK, Chabukdhara M (2012) Municipal solid waste management in India-current state and future challenges: a review. Int J Eng Sci Technol 4(4):1473–1489

Kumar P, Kaushal RK (2015) Avenues of collection and disposal of municipal solid wastes management in India—a review. Int J Eng Sci Invent Res Develop 1(9):458–466
Kumar V, Pandit RK (2013) Problems of solid waste management in Indian cities. Int J Sci Res Publ 3(3):1–7
Kumar S, Mondal AN, Gaikwad SA, Devotta S, Singh RN (2004) Qualitative assessment of methane emission inventory from municipal solid waste disposal sites: a case study. Atmos Environ 38:4921–4929
Kumar S, Bhattacharyya JK, Vaidya AN, Chakrabarti T, Devotta S, Akolkar AB (2009) Assessment of the status of municipal solid waste management in metro cities, state capitals, class I cities, and class II towns in India: an insight. Waste Manag 29:883–895
Kumar S, Smith SR, Fowler G, Velis C, Kumar SJ, Arya SR, Kumar R, Cheeseman C (2017) Challenges and opportunities associated with waste management in India. R Soc Open Sci 4:160764
Ministry of Environment and Forests (2000) "Municipal solid wastes (management and handling) rules, 2000." Accessible at: moef.nic.in
Ministry of Urban Development (2014) Manual on municipal solid waste management. Central Public Health & Environment Engineering Organization: Government of India, New Delhi
MoEF&CC (2016) Solid wastes (handling and management) rules. Ministry of Environment, Forests and Climate Change, Government of India
Nandan A, Yadav BP, Baksi S, Bose D (2017) Recent scenario of solid waste management in India. World Scientific News 66:56–74
Nguyen PT, Yasuhiro M, Takeshi F (2011) Assessment of plastic waste generation and its potential recycling of household solid waste in Can Tho City, Vietnam. Environ Monit Assess 175:23–35
NITI Aayog (2015) Report of the Sub-Group of Chief Ministers on Swachh Bharat Abhiyaan, NITI Aayog (erstwhile Planning Commission), Government of India
Pamnani A, Srinivasarao M (2014) Municipal solid waste management in India: a review and some new results. Int J Civil Eng Technol 5(2):01–08
Pappu A, Saxena M, Asolekar SR (2007) Solid wastes generation in India and their recycling potential in building materials. Build Environ 42:2311–2320
Patel NK, Baredar P (2016) Opportunity and challenges of municipal waste disposal in India. International Journal of Innovative Research in. Sci Technol 3(2):253–264
Pattnaik S, Reddy MV (2010) Assessment of municipal solid waste management in Puducherry Pondicherry, India. Resour Conserv Recycl 54(8):512–520
Planning Commission Report (2014) Reports of the task force on waste to energy (Vol-I) (in the context of Integrated MSW management). Retrieved from http://planningcommission.nic.in/reports/genrep/rep_wte1205.Pdf
Sarkar P (2003) Solid waste management in Delhi—a social vulnerability study. In: Bunch MJ, Madha Suresh V, Vasantha Kumaran T (eds) Proceedings of the Third International Conference. Environment and Health, Chennai, India, pp 451–464
Sharholy M, Ahmad K, Mahmood G, Trivedi RC (2006) Development of prediction models for municipal solid waste generation for Delhi city. In: Proceedings of national conference of advanced in mechanicalengineering (AIME-2006). Jamia Millia Islamia, New Delhi, India, pp 1176–1186
Shazwin TM, Nakagoshi N (2010) Sustainable waste management through international cooperation: review of comprehensive waste management technique and training course. J Int Develop Cooperat 16(1):23–33
Siddiqqui MA (2018) Municipal solid waste management in India-status and challenges: an overview. Int J Biol Res 3(1):126–133
Singh N (2018) Rethinking waste valorization in fast growing middle-income cities in AFD-CPR Workshop on 'Valuing Waste or Wasting Value'. Solid Waste Management in the Developing Countries, Khulna, Bangladesh
Srivastava R, Krishna V, Sonkar I (2014) Characterization and management of municipal solid waste: a case study of Varanasi city. India Int J Curr Res Acad Rev 2:10–16
SWaCH Website (2012). Available at www.swachcoop.com

Syamala Devi K, Swamy AVVS, Hema Krishna R (2013) The potential adverse health effects on residents near hazardous municipal solid waste dump site -At Jawahar Nagar-Hyderabad. Glob J Appl Environ Sci 3(2):67-80

Syamala Devi K, Swamy AVVS, Hema Krishna R (2014) Studies on the solid waste collection by rag pickers at greater Hyderabad Municipal Corporation, India. Int Res J Environ Sci 3(1):13–22

UNEP-IETC and HIID (1996) International source book on environmentally sound technologies for municipal solid waste management. UNEP, International Environmental Technology Centre, 427 pp

Yadav M (2017) Municipal solid waste management: challenges and legislative framework In India. J Adv Sci Res Eng Trend 2(4):43–46

Rhizospheric Treatment of Hydrocarbons Containing Wastewater

11

Pankaj Kumar Gupta, Ajay Kumar, Lalit Goswami, and Basant Yadav

Abstract

Hydrocarbons have a global attention as some of harmful contaminants due to their potential in causing fatal disease to mankind. In India, their usage is being continuously increasing to meet the needs of growing population from last few decades. Hydrocarbon discharge from various anthropogenic activities (viz., petrochemical industries, gasification, incineration) are primarily causing the detrimental effect onto the soil health and groundwater. Therefore, several methodologies and hybrid technologies are being developed for the remediation of these hydrocarbons, including physical, chemical, and biological processes. But, the remediation processes employing microorganisms and plants have been considered as environmental friendly as well as cost-effective techniques. Moreover, several efforts have been made in improving the effectiveness of these technologies. This chapter provides an understanding of remediation techniques

P. K. Gupta (✉)
Remwasol Remediation Technologies Pvt. Ltd, Samastipur, India

Faculty of Environment, University of Waterloo, Waterloo, ON, Canada
e-mail: pgupta@hy.iitr.ac.in; Pankaj.gupta@remwasol.com

A. Kumar
Indian Institute of Technology Roorkee, Roorkee, Uttrakhand, India
e-mail: akumar1@hy.iitr.ac.in

L. Goswami
Indian Institute of Technology Guwahati, Guwahati, Assam, India
e-mail: g.lalit@iitg.ac.in

B. Yadav
Remwasol Remediation Technologies Pvt. Ltd, Samastipur, India

Canfield University, Bedford, UK
e-mail: Basant.Yadav@cranfield.ac.uk

P. K. Arora (ed.), *Microbial Technology for Health and Environment*, Microorganisms for Sustainability 22,
https://doi.org/10.1007/978-981-15-2679-4_11

by highlighting the multidisciplinary aspects. These approaches can be effectively deployed for the soil and groundwater remediation. Further, an approach of exploring the experimental outcomes in combination with the numerical modeling has been discussed which is a beneficial tool for making the technology transfer feasible from laboratory to field scale applications effectively as well as in cost-effective manner.

Keywords
Hydrocarbon pollutants · Plant-assisted bioremediation · Wastewater · Concurrent treatment

11.1 Introduction

With holding the second largest populated country in the world, still India is among the fast-growing economics of the globe. In India, agriculture is the primary supportive sector followed by the industry as secondary supportive sector for economic growth. With over a half of the country's population is living underneath the poverty level and lacking access to the basic facilities (Gupta and Sharma 2018), the policy makers and stakeholders are simultaneously adopting various schemes and policies to affirm the basic needs such as food, health, education, and livelihood to uphold the rising demand. Therefore, governments continuously encourage foreign direct investment (FDI) to boost their economy so that the provision of basic facilities could be ensured (Kuntluru et al. 2012). This scenario attracts the global manufacturer to relocate their industries in the land of the country. Currently, India imports 84% of the petroleum products, and in accordance with the Directorate General of Commercial Intelligence and Statistics (2015), during the financial year April 2018 to March 2019, the country imported around 46.6 million tons of crude oil. Several chemical industries have been established efficaciously all over the country that may release of numerous chemicals in the environment via transportation, processing, and storage. These contaminants generated via the leakage of petrochemicals further lead to the groundwater and soil pollution (Goswami et al. 2019a, b; Kushwaha et al. 2017; Seeger et al. 2011; Gupta 2020).

11.2 Hydrocarbon in Environment

Hydrocarbon contamination of soil and water is a ubiquitous problem all over the world, and remediation of these polluted resources is needed to eliminate risk to human and to the environment. Common anthropogenic sources of hydrocarbon contamination include the transportation and mishandling of petrochemical products and disposal and land application of petrochemicals from different sources and industrial sites (Goswami et al. 2017a, b; Sathe et al. 2020). According to Directorate General of Commercial Intelligence and Statistics (DGCIS), Government of India, there is increasing growth of the petrochemical industries and the exports and imports of the petrochemical products. Many researchers and agencies categorized hydrocarbon as a toxic and hazardous chemical for the ecosystem and human (Goswami et al. 2019c). United States Environmental Protection Agency (USEPA) refers hydrocarbons as toxic chemicals and are recommended for the bioremediation of all polluted sites (USEPA 1995). Similarly, USGS studies a long-term and interdisciplinary projects for the hydrocarbon-polluted sites. For examples, USGS sponsored a research project for crude oil-contaminated soil-water site near Bemidji, Minnesota (Delin et al. 1998).

When released to land, these contaminants can migrate downward through unsaturated zone, and consequently, light phase aqueous hydrocarbons float and move on top of the water table, while dense phase move downward through the water table and penetrate into the saturated zone (Dobson et al. 2007). The variable environment conditions like temperatures, soil moisture, nutrient supply, and water table fluctuation pose distribution pattern of the hydrocarbon plumes in the soil-water system. Therefore, the hydrocarbon spills present a significant threat to environment as they can result in extensive pollution from small spillages.

11.3 Decontamination Techniques

Over the past few decades, the hydrocarbon pollution is among the major problem, globally (Nedwell 1999; Goswami et al. 2018a). The current practice for remediating hydrocarbon-polluted sites relies heavily on encapsulation or isolation (capping, barriers); neither of which addresses the issue of decontamination. Cleaning these sites via immobilization or extraction by physiochemical techniques can be prohibitively expensive and is often appropriate only for small sites where rapid and complete removal is required (Kumar et al. 2019; Kushwaha et al. 2019; Bind et al. 2018; Goswami et al. 2017c). Costly methods, such as ex situ treatment and soil washing, have an adverse effect on the biological diversity (Gupta and Joshi 2017; Gupta and Yadav 2017c; Gupta et al. 2018d), soil structure, and fertility (Yadav and Hassanizadeh 2011).

For the safe drinking water production and the equilibrium of the natural resources with better ecosystem services, many technological approaches are applied in the last few decades (Gupt et al. 2018; Kumar et al. 2016; Kushwaha et al. 2015). The research studies are applied for the remediation of hydrocarbon-contaminated site by different process or integration of process such as pump-and-treat, in situ biodegradation, phytoremediation, soil washing, surfactant and co-solvent flushing, air stripping, and thermal entrapments (Bento et al. 2005). The physicochemical and other relative techniques are very economic and not feasible to its cleanup. Hence, amidst all the remediation techniques, bioremediation is the cost-efficient and sustainable technique for the eradicating the hydrocarbon contamination (Goswami et al. 2020). However, the devoid of operational facilities and interdisciplinary knowledge gaps on the research topics, literature seriously lacks the information of contamination sites particularly in India (Yadav et al. 2019; Gupta et al. 2020).

The other promising treatment options are through biological processes like bioremediation (Gupta et al. 2017), phytoremediation (Kushwaha et al. 2018; Susarla et al. 2002), and wetlands (Farhadian et al. 2008). Bioremediation is a developing cost-effective technique and causes no harm to the contaminated ecosystem as compared to the above-mentioned traditional chemical and physical methods since the biodegradation of hydrocarbons depends on the indigenous microorganisms stimulated by the pollutant (Borah et al. 2019; Goswami et al. 2018b). Various bioremediation techniques are developed to clean up residual BTEX from polluted soils, marine shorelines, and surface and groundwater systems under a broad range of environmental conditions (Gupta et al. 2018b). These techniques are readily utilized as a complementary polishing method after deploying the established techniques for the substantial removal of pure phase contamination. BTEX compounds get biodegraded in their aqueous phase by naturally occurring microorganisms in the subsurface environment, but the process is quite slow (Gupta et al. 2019). Therefore, engineered/enhanced bioremediation is practiced using additives to the natural environmental media. This involves the addition of seeded cultures, bioaugmentation or addition of nutrients, and biostimulation. The key role in the success of bioremediation in contaminated soil-water systems is played by microorganisms and various site-specific environmental parameters (Abhishek et al. 2018a, b). Use of plants may provide a multi-synchronous environment favorable for metabolism of microorganisms by increasing O_2 diffusion and root exudates, subsequently enhancing the rate of biodegradation in contaminated root zone (Gupta and Yadav 2017a, b; Gupta et al. 2018a, b, c; Gupta et al. 2019). Therefore, many researchers strongly recommended the urgent needs for knowledge development on the advance and interdisciplinary approaches of the remediation technology specially rhizoremediation/concurrent treatments (Goswami et al. 2018b; Ouyang 2002). To clean up by remediation using biological agents, three main strategies have been used: (a) stimulation of microorganism by providing the addition of substrate, (b) incubation of active organisms, and (c) integration with plant species. Rhizoremediation of petroleum contaminants is a phytoremediation process that depends on interactions among plants, microbes, and soils (Basu et al. 2015). During the rhizoremediation/plant-assisted biostimulation, some processes promote the remediation of a wide range of chemical at toxic site. Such processes are (1) modification of the physical and chemical properties of sites, (2) uses of nutrient organic carbon by releasing root exudates, (3) the aeration by transferring the oxygen to root zones, (4) retardation of the movements of chemicals by PRBs, (5) enhancement of the plant enzymatic transformation (Susarla et al. 2002).

The zone or electron acceptor-based application of the microbe diversity can result to more effective and efficiency rhizodegradation. The sub-surface zones represent the unique ecological niches due to their specific environment condition like separation of nutrient contents in different zones. Generally, the microbes are associated with thermodynamically favorable electron acceptors. Therefore, each zone process of electron acceptors inhibits the specific adaptive microorganisms. Various molecular surveys of microbial communities in various anoxic environmental, characteristic degrader ecotypes become evident for the particular environment and electron acceptor process (Kleinsteuber et al. 2008). The rhizosphere soil has 10–100 times more microbes than unvegetated (Imfeld et al. 2009). Plants influence xenobiotic biodegradation by increasing in microbial cell numbers and microbial activation that occurs in rhizospheres as a result of growth on carbon substrate provided by rhizodegradation. Plant microbes' interaction increased the mineralization process and immobilization process resulting in enzymatic enhancements. The deep fibrous root systems of plants may improve the aeration in soil by removing water through transports and by alternation of soil structure through agglomeration. The decay of dead root hair and fine root serves as an important source of the carbon for growth of rhizospheric microorganisms (Susarla et al. 2002). Plants also secrete surfactants which reduce the surface tension and solubilize contaminants in soil water. Therefore, bioavailability in different zones is increasing due to reduction in the toxicity. The constructed wetlands are the examples to stimulate the combined effects of the biostimulation, bioaugmentation, and phytoremediation. Some specialized plant species play a very important role in phytostimulation under wetland condition, which removes almost 100% of pollutants from soil-water system. Microbial growth kinetics meets to mass transfer kinetics and enzymatic kinetics which results as the ultimate biodegradation of substrate. Many researchers investigate batch experiment and column experiment using different plant species and reported different kinetics models. These are zero order kinetics models, first-order kinetics models, Monod's kinetics, etc. Mathematical modeling of plant-assisted bioremediation is helpful for the bioremediation technology, proposed schemes, policy, and managements of contaminant site (Narayanan et al. 1998a, b). Therefore, a better understanding is needed for plant-assisted bioremediation of hydrocarbon-contaminated soil-water system that is presented here with special emphasis on the rhizoremediation strategies and their kinetics and mathematical approaches for two-dimensional and three-dimensional modeling. That will help in the policy framework and recommendations for the cost-effective remediation technologies. It is also powerful for the common platform to address and respond to dialogue, priority setting, and policy formulation for the better managements of cleanup technology. Models in the rhizoremediation during fate and transports of hydrocarbons are summarized as follows with governing equation and mechanisms in Fig. 11.1.

11.4 Concurrent Treatment Facilities

This low-tech in situ approach of concurrent treatments is more attractive for bioremediation of hydrocarbon-polluted soils as it offers site restoration, partial decontamination, and maintenance of the biological activity, which is visually unobtrusive, and there is the possibility of clean water production (Van Gestel et al. 2003). Due to the enormous potential for its cost and environmental savings (Gupta et al. 2017), there is a significant interest in this technology that is in its early stage of development, and very little information is available related to site cleanup from start to finish. Effective implementation of concurrent treatments requires a thorough understanding of the soil-plant-atmospheric continuum processes which is currently

1. Moisture based geo-chemical mechanisms of the hydrocarbons transports

$$\frac{\partial \theta}{\partial t} - \nabla . D(\theta)\nabla\theta - \frac{\partial K}{\partial z} = 0$$

4. Plants Uptakesmechanisms

$$\frac{\partial C}{\partial t} = -q\, T_{scf} C_s$$

5. Biodegradation Kinetics mechanisms

2. Multi-dimensional advective-dispersive mechanisms

$$D_x \frac{\partial^2 C}{\partial x^2} + D_y \frac{\partial^2 C}{\partial y^2} + D_z \frac{\partial^2 C}{\partial z^2} - V_x \frac{\partial C}{\partial x} - \frac{r}{n} = \frac{\partial C}{\partial t} \qquad \frac{\partial C}{\partial t} = \mu_{max} C \frac{(C_o + X_o - C)}{(K_s + C)}$$

3. Sorptionmechanisms

$$\frac{\partial \phi C^i}{\partial t} = -r_{ads}{}^i + r_{des}{}^i$$

6. Contamination Evaporation (Volatilization) mechanisms

$$C_g^i = H^i C_{lig}^i$$

Fig. 11.1 Describes the summery of governing equation and the mechanisms of the rhizospheric treatments of the hydrocarbons polluted soil-water

poorly understood and makes this technology expensive and inefficient despite the tremendous potential mentioned above. Most of the current research deals with the effect of microorganisms on soil and water, wherein the hydrocarbon-contaminated effluent or sewage sludge is applied to the cropped soil and the plants grown on such sites are analyzed experimentally to determine their capacity to remove hydrocarbons from the root zone. These studies simply correlate the hydrocarbon contaminant concentration in growing media with its presence in the plant biomass for a particular soil-water-plant system without hypothesizing how the fate and transports actually take place. This lack of understanding hinders the efforts of researchers in their quest to develop concurrent treatments from contaminated soil-water system. Techniques are used separately for treating wastewater and contaminated water. Therefore, innovative concurrent method for treating both the resources in symbiosis way is urgently needed. Moreover, isolated experimental and modeling works are mostly performed for both experimental and numerical methods, which are needed to be studied together. Similarly, there is limited scientific information on the impacts of the rhizospheric treatment during the fate and transports of the hydrocarbon in the soil-water system. Other than this, one remaining hurdle for commercial implementation of such treatment has been the disposal of the produced contaminated biomass, which is addressed rarely by the researchers so far. Therefore, the focus of this chapter is to generate the interdisciplinary and multidisciplinary aspects of effectiveness and the mechanisms of the highly complex soil-water-plant-atmospheric continuum processes during the concurrent treatments of wastewater and hydrocarbon-polluted soil-water resources from start (laboratory investigation) to finish (modeling approaches and field application). The specific research topics include interdisciplinary aspects as listed in Box 11.1.

Box 11.1 Interdisciplinary and multidisciplinary aspects of treatment facilities

A. Contaminant hydrology and biochemical engineering.
  - A1—the mechanisms governing the concurrent treatment process.
  - A2—plants and chelating agent's for hydrocarbon removal from polluted sites.

B. Mass transfer (analytical tool) and mass balance (modeling tool).
  - B1—gas chromatography-mass spectrometry (GC-MS) for quantitative measurement of organic matter and metal dynamics in plant biomass and root zone.
  - B2—magnetic resonance imaging (MRI) for in situ measurement of plant growth and dynamic root density distribution.
  - B3—metal distribution in the soil solution and plant biomass using flame atomic absorption spectrophotometer.

C. Use of mathematical modeling in vadose zone processes.
  - C1—simulating water and contaminant dynamics in vadose zone and their uptake by plant.
  - Biomass using a realistic approach.
  - C2—validating of the developed model using experimental data obtained in B1, B2, and B3.

D. Technology comparison in cleanup processing.
  - D1—valorization of plant-enhanced decontamination during rhizospheric treatment for different techniques.
  - D2—economic evaluation of rhizospheric treatment against traditional techniques.

11.5 Numerical Modeling

The modeling involves the simultaneous movement of soil water and hydrocarbons through soil-water-plant system by coupling of the moisture flow equation with the contaminant transport equation in the presence of sink terms as mentioned in Eqs. (1) and (2). Similarly, the degradation pattern of the hydrocarbons can be used to numerically simulate the movement of water and hydrocarbon transport through the heterogeneous variably saturated zone (Gupta and Yadav 2019). The modeling involves the simultaneous movement of soil water and pollutant movement through soil-water-plant system by coupling of the moisture flow equation with the contaminant transport equation in the presence of sink terms. The transient moisture dynamics in variably saturated porous media is expressed by a parabolic partial differential equation popularly known as Richards' equation which is derived by integrating the Darcy's law with the equation of continuity. This equation is in its three-dimensional mixed form which is coupled by non-uniform sink function for water by plants:

$$\frac{\partial \theta}{\partial h}\left(\frac{\partial h}{\partial t}\right)=\frac{\partial}{\partial x}K_x(h)\frac{\partial h}{\partial x}+\frac{\partial}{\partial y}K_y(h)\frac{\partial h}{\partial y}+\frac{\partial}{\partial z}K_z(h)\frac{\partial h}{\partial z}+K(h)-S(t,h) \quad (1)$$

where θ is the volumetric water content defined in the volume of water per unit volume of soil and h is the pressure head. $S(t, h)$ is a sink function that represents the water extraction by surface vegetation, z is the depth of root zone measured positive upwards, K is the hydraulic conductivity of the soil, and t is the time. This equation is highly nonlinear for unsaturated flow, since hydraulic conductivity K and the volumetric water content are nonlinear functions of the dependent variable h, the soil moisture pressure head. To solve this equation, explicit expressions for the soil constructive relationship between the dependent variable h and the nonlinear terms K and θ are required.

The classical convection dispersion equation is used for contaminant transport in multidimensions taking the contaminant extraction term (Kumari et al. 2019). The Fick's law coupled with the mass balance equation yields a modified form of advective-dispersive equation.

$$\begin{aligned}\frac{\partial(\rho_s S_D)}{\partial t}+\frac{\partial(\theta C)}{\partial t}&=\frac{\partial}{\partial x}\left[D_{xx}\theta\frac{\partial C}{\partial x}+D_{xy}\theta\frac{\partial C}{\partial y}+D_{xz}\theta\frac{\partial C}{\partial z}\right]+\\ &\frac{\partial}{\partial y}\left[D_{yx}\theta\frac{\partial C}{\partial x}+D_{yy}\theta\frac{\partial C}{\partial y}+D_{yz}\theta\frac{\partial C}{\partial z}\right]+\frac{\partial}{\partial z}\left[D_{zx}\theta\frac{\partial C}{\partial x}+D_{zy}\theta\frac{\partial C}{\partial y}+D_{zz}\theta\frac{\partial C}{\partial z}\right]\\ &-q_x C-q_y C-q_z C+S(t,C)\end{aligned} \tag{2}$$

Yadav and Hassanizadeh (2011) described the general expression for the solute biodegradation in soil-water system, in which only microbial densities and the contaminant concentration determine the degradation kinetics.

$$\frac{\partial C}{\partial t}=\mu_{\max}C\frac{(C_0+X_0-C)}{(K_s+C)} \tag{3}$$

where $\mu_{\max}$ is the maximum growth rate, C is the contamination concentration at time t, C_0 is the initial contamination concentration, X_0 is the contamination required to produce initial microbial density, and K_s is the half saturation constant.

The equilibrium adsorption isotherms founded in the case of hydrocarbons in soil-water system is mostly Langmuir equilibrium adsorption isotherms. The Langmuir equation is

$$S_{\text{eq}}=\frac{S_{\max}K_L C_{\text{eq}}}{1+K_L C_{\text{eq}}} \tag{4}$$

where S_{eq} is the concentration of adsorbed viruses and C_{eq} is the concentration of free viruses after apparent equilibrium has been reached. $S_{\max}$ is the maximum adsorbed concentration when all active surface sites are occupied; K_L is a constant related to the bonding energy. The movement of water and hydrocarbons in soils is generally better described with multidimensional non-equilibrium models than with more commonly used one-dimensional and/or equilibrium models. Furthermore, such equations are solved for the validation of the different field data set. Therefore, it plays very important role for the contamination fate and transport modeling including hydrocarbon contamination in the vadose zone and/or also saturated zone.

11.6 Rhizospheric Treatment Facilities: Solar/Wind-Based Design

Rhizosphere deliberates as the "ecological remediation unit" for treating contaminated soils, possessing huge amount of microbes particularly bacteria, fungi, and rhizobacteria (symbiont with the plant roots). This approach gives how and in what extent the extreme environmental variations of soil moisture content, temperature, and water table dynamics could affect the biodegradation of hydrocarbon contaminants in variably saturated soils. Direct practical importance for remediating hydrocarbon-polluted natural resources is very high for Indian climatic conditions. A successful transformation of this cost-effective technology of bioremediation from laboratory to the field would have a significant impact on science and industrial application, not only in India but also for countries having the similar environmental conditions. An improved understanding of bioremediation processes that control biodegradation of organic contaminants is required to effectively implement this environmental-friendly technology for decontaminating the polluted sites (Mustapha et al. 2018). Such remediation technologies are convenient for the polluted site where the handling of petrochemical substances is established such as oil refineries and port and costal area. The produced database and knowledge gained in this chapter can be used to encourage petrochemical and hydrocarbon production industries and other environmental agencies for remediating hydrocarbon-contaminated soil-water systems. If the application of bioremediation from the lab to the field proves to be efficient, this would have a positive impact on sustainability and the marketing of petrochemical of the country. At the same time, soil-water systems are referring under vulnerability due to such activities. Therefore, the technological supports to the commercial or industrial activities become the millstones

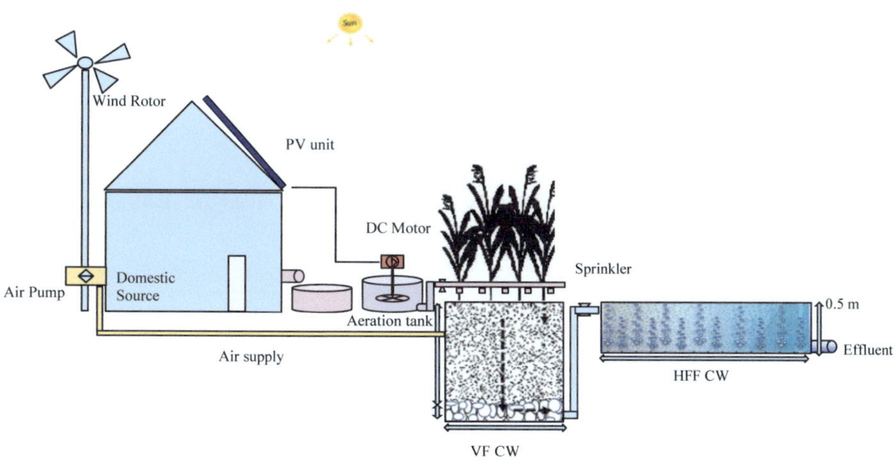

Fig. 11.2 Rhizospheric treatment: solar/wind-driven aeration of hybrid CW (VF + HFF)

for the sustainable developments. In this regard, a new treatment facility (Fig. 11.2) has been designed for treatment of wastewater containing hydrocarbon pollutants. The treatment system consisted of influent tank having a capacity of 200 L made of high-density polyethylene plastic containers located between source and aeration tank. The aeration tank having the capacity of 400 L is located next to influent tank and peristaltic pumps used to pump the influent into the hybrid CWs. The hybrid CW was composed of two parts: (1) a vertical flow (VF) CW planted with *Canna generalis* and (2) a horizontal flow filter (HFF) CW. Aeration of root zone enhances the biodegradation of hydrocarbon in VF chamber where motor of aeration can be driven by solar/wind system.

11.7 Summary and Recommendations

In India, the increase in the demand of hydrocarbons and its utilization cause the devastating effects to the ecosystems due to occurrence of mishandling episodes and lack of infrastructure, which in turn requires the development of engineered technologies in their remediation. Rhizospheric treatment alone with the conventional methods seems to be effective and reliable in this respect. However, appropriate innovations are needed to upgrade the literature for direct practical implication of the technique according to Indian climatic conditions. Some recommendations are as follows:

1. Use of nano-biomaterials and biochar to enhance the rhizospheric degradation of hydrocarbon is a new direction of research and application (Ranjan et al. 2018).
2. Solar/wind-driven aeration may accelerate the aerobic biodegradation of petrochemical in root zone; however to maintain optimal aeration, it is important to investigate the other operational parameters.
3. In situ aerobic heating, i.e., providing optimal heated water using PV system, can be an effective approach.
4. Modeling of root zone mechanisms is needed to understand the accurate root uptake and pollutant distribution in subsurface.

Acknowledgments Supports from Remwasol Remediation Technologies Pvt. Ltd. is well acknowledged.

References

Abhishek A, Gupta PK, Yadav BK, Amandeep A, Tomar AS, Kataria S, Kumar S (2018a) Phytoremediation of toluene polluted groundwater under nutrient loading using constructed wetlands. Poster presentation (B33G-2766) in AGU Fall Meeting 2018 held in Washington D.C. USA during 10–14 Dec., 2018

Abhishek A, Yadav BK, Gupta PK (2018b) Morphological variations in unsaturated porous media due to LNAPL contamination. Poster in Japan Geoscience Union (JpGU) Chiba-city, Japan, May 20–24 2018

Basu S, Yadav BK, Mathur S (2015) Enhanced bioremediation of BTEX contaminated groundwater in pot-scale wetlands. Environ Sci Pollut Res 22(24):20041–20049

Bento FM, Camargo FAO, Okeke BC, Frankenberger WT (2005) Comparative bioremediation of soils contaminated with diesel oil by natural attenuation, biostimulation and bioaugmentation. Bioresour Technol 96(9):1049–1055. https://doi.org/10.1016/j.biortech.2004.09.008

Bind A, Goswami L, Prakash V (2018) Comparative analysis of floating and submerged macrophytes for heavy metal (copper, chromium, arsenic and lead) removal: sorbent preparation, characterization, regeneration and cost estimation. Geol Ecol Landscapes 2(2):61–72

Borah SN, Sen S, Goswami L, Bora A, Pakshirajan K, Deka S (2019) Rice based distillers dried grains with solubles as a low cost substrate for the production of a novel rhamnolipid biosurfactant having anti-biofilm activity against *Candida tropicalis*. Colloids Surf B: Biointerfaces 182:110358

Delin GN, Essaid HI, Cozzarelli IM, Lahvis MH, Bekins BA (1998) Ground water contamination by crude oil near Bemidji, Minnesota (No. 084-98). US Geological Survey

Directorate General of Commercial Intelligence and Statistics. Government of India. Ministry of Commerce and Industry (2015) Annual report

Dobson R, Schroth MH, Zeyer J (2007) Effect of water-table fluctuation on dissolution and biodegradation of a multi-component, light nonaqueous-phase liquid. J Contam Hydrol 94:235–248

Farhadian M, Vachelard C, Duchez D, Larroche C (2008) In situ bioremediation of monoaromatic pollutants in groundwater: a review. Bioresour Technol 99(13):5296–5308. https://doi.org/10.1016/j.biortech.2007.10.025

Goswami L, Kumar RV, Manikandan NA, Pakshirajan K, Pugazhenthi G (2017a) Simultaneous polycyclic aromatic hydrocarbon degradation and lipid accumulation by *Rhodococcus opacus* for potential biodiesel production. J Water Process Eng 17:1–10

Goswami L, Manikandan NA, Pakshirajan K, Pugazhenthi G (2017b) Simultaneous heavy metal removal and anthracene biodegradation by the oleaginous bacteria *Rhodococcus opacus*. 3 Biotech 7(1):37

Goswami L, Namboodiri MT, Kumar RV, Pakshirajan K, Pugazhenthi G (2017c) Biodiesel production potential of oleaginous *Rhodococcus opacus* grown on biomass gasification wastewater. Renew Energy 105:400–406

Goswami L, Kumar RV, Borah SN, Manikandan NA, Pakshirajan K, Pugazhenthi G (2018a) Membrane bioreactor and integrated membrane bioreactor systems for micropollutant removal from wastewater: a review. J Water Process Eng 26:314–328

Goswami L, Manikandan NA, Dolman B, Pakshirajan K, Pugazhenthi G (2018b) Biological treatment of wastewater containing a mixture of polycyclic aromatic hydrocarbons using the oleaginous bacterium *Rhodococcus opacus*. J Clean Prod 196:1282–1291

Goswami L, Kumar RV, Arul Manikandan N, Pakshirajan K, Pugazhenthi G (2019a) Anthracene biodegradation by Oleaginous *Rhodococcus opacus* for biodiesel production and its characterization. Polycycl Aromat Compd 39(3):207–219

Goswami L, Kumar RV, Pakshirajan K, Pugazhenthi G (2019b) A novel integrated biodegradation-microfiltration system for sustainable wastewater treatment and energy recovery. J Hazard Mater 365:707–715

Goswami L, Manikandan NA, Taube JCR, Pakshirajan K, Pugazhenthi G (2019c) Novel waste-derived biochar from biomass gasification effluent: preparation, characterization, cost estimation, and application in polycyclic aromatic hydrocarbon biodegradation and lipid accumulation by *Rhodococcus opacus*. Environ Sci Pollut Res 26(24):25154–25166

Goswami L, Pakshirajan K, Pugazhenthi G (2020) Utilization of biochar immobilized polyurethane foam as the potential packing material in an up-flow packed bed bioreactor for enhanced biological treatment of biomass gasification wastewater by *Rhodococcus opacus*. J Clean Prod

Gupt CB, Yamsani SK, Prakash A, Medhi CR, Sreedeep S (2018) Appropriate liquid-to-solid ratio for sorption studies of bentonite. J Environ Eng 145(2):04018138

Gupta PK (2020) Fate, transport, and bioremediation of biodiesel and blended biodiesel in subsurface environment: a review. J Environ Eng 146(1):03119001
Gupta PK, Joshi P (2017) Assessing groundwater resource vulnerability by coupling GIS based DRASTIC and solute transport model in Ajmer District, Rajasthan. J Geol S India (Springer). https://doi.org/10.1007/s12594-018-0958-y
Gupta PK, Sharma D (2018) Assessments of hydrological and hydro-chemical vulnerability of groundwater in semi-arid regions of Rajasthan, India. Sustain Water Res Manag:1–15. https://doi.org/10.1007/s40899-018-0260-6
Gupta PK, Yadav BK (2017a) Bioremediation of non-aqueous phase liquids (NAPLS) polluted soil and water resources. Chapter 8, environmental pollutants and their bioremediation approaches, ISBN 9781138628892. CRC Press, Taylor and Francis Group, Florida, USA
Gupta PK, Yadav BK (2017b). Role of climatic variability on fate and transport of LNAPL pollutants in subsurface. Session H060: Groundwater Response to Climate Change and Variability, AGU Fall Meeting 2017, New Orleans, USA
Gupta PK, Yadav BK (2017c) Effects of climatic variation on dissolution of LNAPL pollutants in subsurface environment. In: Chapter 8: Climate change resource conservation and sustainability strategies, ISBN 9789384871086. DBH Publishers and Distributors, New Delhi
Gupta PK, Yadav BK (2019) Subsurface processes controlling reuse potential of treated wastewater under climate change conditions. In: Water conservation, recycling and reuse: issues and challenges. Springer, Singapore, pp 147–170
Gupta PK, Yadav BK, Hassanizadeh SM (2017) Engineered bioremediation of LNAPL polluted soil-water resources under changing climatic conditions. Proceedings of International Conference on Modeling of environmental and water resources systems (ICMEWRS-2017), HBTU Kanpur, 24–26th March, 2017 (ISBN 978-93-85926-53-2)
Gupta PK, Abhishek, Yadav BK (2018a) Impact of hydrocarbon pollutants on partially saturated soil media in batch system: morphological analysis using SEM techniques. Chapter 5, Water Quality Management; Water Science and Technology Library, ISBN: 978-981-10-5794-6, Vol. 79, Springer
Gupta PK, Ranjan S, Kumar D (2018b) Groundwater pollution by emerging industrial pollutants and its remediation techniques. Chapter 2, In Recent advances in environmental management, CRC Press Taylor & Francis Group, ISBN 9780815383147, Vol 1
Gupta P.K., Ranjan S., Kumar D., (2018c). Groundwater pollution by emerging industrial pollutants and its remediation techniques. Chapter 2, In recent advances in environmental management, CRC Press Taylor & Francis Group: Routledge/Boca Raton, ISBN 9780815383147, Vol 1
Gupta PK, Yadav B, Yadav BK (2018d) Transport of LNAPL and biofilm growth in subsurface under dynamic groundwater conditions. C001723-Oral presentation in Japan Geoscience Union (JpGU) Chiba-city, Japan, May 20–24 2018
Gupta PK, Yadav B, Yadav BK (2019) Assessment of LNAPL in subsurface under fluctuating groundwater table using 2D sand tank experiments. ASCE J Environ Eng 145. https://doi.org/10.1061/(ASCE)EE.1943-7870.0001560
Gupta PK, Yadav B, Kumar A, Singh RP (2020) India's Major subsurface pollutants under future climatic scenarios: challenges and remedial solutions. In: Contemporary environmental issues and challenges in era of climate change. Springer, Singapore, pp 119–140
Imfeld G, Braeckevelt M, Kuschk P, Richnow HH (2009) Monitoring and assessing processes of organic chemicals removal in constructed wetlands. Chemosphere 74(3):349–362. https://doi.org/10.1016/j.chemosphere.2008.09.062
Kleinsteuber S, Schleinitz KM, Breitfeld J, Harms H, Richnow HH, Vogt C (2008) Molecular characterization of bacterial communities mineralizing benzene under sulfate-reducing conditions. FEMS Microbiol Ecol 66:143–157
Kumar RV, Goswami L, Pakshirajan K, Pugazhenthi G (2016) Dairy wastewater treatment using a novel low cost tubular ceramic membrane and membrane fouling mechanism using pore blocking models. J Water Process Eng 13:168–175

Kumar M, Goswami L, Singh AK, Sikandar M (2019) Valorization of coal fired-fly ash for potential heavy metal removal from the single and multi-contaminated system. Heliyon 5(10):e02562

Kumari B, Gupta PK, Kumar D (2019) In-situ observation and nitrate-N load assessment in Madhubani District, Bihar, India. J Geol Soc India (Springer) 93(1):113–118. https://doi.org/10.1007/s12594-019-1130-z

Kuntluru S, Muppani VR, Khan MAA (2012) Foreign direct investment and export performance of pharmaceutical firms in India: An empirical approach. Int J Econ Financ 4(5):216–226

Kushwaha A, Rani R, Kumar S, Gautam A (2015) Heavy metal detoxification and tolerance mechanisms in plants: implications for phytoremediation. Environ Rev 24(1):39–51

Kushwaha A, Rani R, Kumar S, Thomas T, David AA, Ahmed M (2017) A new insight to adsorption and accumulation of high lead concentration by exopolymer and whole cells of lead-resistant bacterium *Acinetobacter junii* L. Pb1 isolated from coal mine dump. Environ Sci Pollut Res 24(11):10652–10661

Kushwaha A, Hans N, Kumar S, Rani R (2018) A critical review on speciation, mobilization and toxicity of lead in soil-microbe-plant system and bioremediation strategies. Ecotoxicol Environ Saf 147:1035–1045

Kushwaha A, Rani R, Patra JK (2019) Adsorption kinetics and molecular interactions of lead [Pb (II)] with natural clay and humic acid. Int J Environ Sci Technol:1–12

Mustapha IH, Gupta PK, Yadav BK, van Bruggen JJA, Lens PNL (2018) Performance evaluation of duplex constructed wetlands for the treatment of diesel contaminated wastewater. Chemosphere 205:166–177. https://doi.org/10.1016/j.chemosphere.2018.04.036

Narayanan M, Tracy JC, Davis LC, Erickson LE (1998a) Modeling the fate of toluene in a chamber with alfalfa plants 1. Theory and modeling concepts. J Hazardous Substance Res 1:1–30

Narayanan M, Tracy JC, Davis LC, Erickson LE (1998b) Modeling the fate of toluene in a chamber with alfalfa plants 2. Numerical results and comparison study. J Hazardous Substance Res 1

Nedwell DB (1999) Effect of low temperature on microbial growth lowered affinity for substrates limits growth at low temperature. FEMS Microbiol Ecol 30:101–111

Ouyang Y (2002) Phytoremediation: modeling plant uptake and contaminant transport in the soil–plant–atmosphere continuum. J Hydrol 266(1-2):66–82. https://doi.org/10.1016/S0022-1694(02)00116-6

Ranjan S, Gupta PK, Yadav BK (2018) Application of nano-materials in subsurface remediation techniques—challenges and future prospects. In: Chapter 6, Recent advances in environmental management, CRC Press Taylor & Francis Group, ISBN 9780815383147, Vol 1

Sathe SS, Goswami L, Mahanta C, Devi LM (2020) Integrated factors controlling arsenic mobilization in an alluvial floodplain. Environ Technol Innov 17:100525

Seeger EM, Kuschk P, Fazekas H, Grathwohl P, Kaestner M (2011) Bioremediation of benzene-, MTBE- and ammonia-contaminated groundwater with pilot-scale constructed wetlands. Environ Pollut 159(12):3769–3776. https://doi.org/10.1016/j.envpol.2011.07.019

Susarla S, Medina VF, McCutcheon SC (2002) Phytoremediation: an ecological solution to organic chemical contamination. Ecol Eng 18(5):647–658. https://doi.org/10.1016/S0925-8574(02)00026-5

USEPA (U.S. Environmental Protection Agency) (1995) Light nonaqueous phase liquids. Office of solid waste and emergency response, Washington, DC. EPA/540/S-95/500

Van Gestel K, Mergaert J, Swings J, Coosemans J, Ryckeboer J (2003) Bioremediation of diesel oil-contaminated soil by composting with biowaste. Environ Pollut 125(3):361–368. https://doi.org/10.1016/S0269-7491(03)00109-X

Yadav BK, Hassanizadeh SM (2011) An overview of biodegradation of LNAPLs in coastal (semi)-arid environment. Water Air Soil Pollut 220:225–239. https://doi.org/10.1007/s11270-011-0749-1

Yadav B, Gupta PK, Patidar N, Himanshu SK (2019) Ensemble modelling framework for groundwater level prediction in urban areas of India. Sci Total Environ:135539

12 Metabolism of Nitroaromatic Compounds by Microbes and Study of Chemotaxis Toward These Compounds

Debarati Paul

Abstract

Nitroaromatic compounds are mainly man-made compounds having diverse functions in industry and otherwise. These are toxic compounds, and their complete mineralization by natural or engineered microbes is desirable via aerobic, anaerobic, or dual pathways. Bacterial chemotaxis has been shown to improve degradation rates and also result in biofilm formation, which in turn assists breakdown of the toxic compounds. These properties may be harnessed for engineering bugs for enhanced and varied degradation of NACs. The microbial diversity of unculturable microbes may be tapped for discovering "new" genes for mineralization of xenobiotic and persistent/recalcitrant compounds.

Keywords

Nitroaromatic · Xenobiotic · Chemotaxis · Bioavailability · Breakdown

Abbreviations

HMX Octahydro-1,3,5,7-tetranitro-1,3,5,7-tetrazine
NACs Nitroaromatic compounds
RDX Cyclotrimethylenetrinitramine
TNB 1,3,5-Trinitrobenzene
TNT 2,4,6-Trinitrotoluene

D. Paul (✉)
Amity Institute of Biotechnology, Amity University, Noida, Uttar Pradesh, India
e-mail: dpaul@amity.edu

P. K. Arora (ed.), *Microbial Technology for Health and Environment*, Microorganisms for Sustainability 22,
https://doi.org/10.1007/978-981-15-2679-4_12

12.1 Nitroaromatic Compounds: Synthesis and Applications

Nitroaromatic compounds (NACs) are mainly man-made and applied for the manufacture of explosives (TNT, RDX, etc.), as pesticides/insecticides (Ju and Parales 2010), and in industries, e.g., tannery, polyurethane foams, rubber photographic chemicals, azo dyes, varnishes, and pharmaceuticals (derivatives of phenothiazines, substituted nitrobenzenes, chloromycetin). The natural formation of nitroaromatic compounds may occur in both air and water conditions. In cities or small towns, hydrocarbons are released due to complete burning and incomplete burning of fossil fuels and thereafter become substrates for generating nitrobenzene(s), nitrotoluene(s), and nitro-polyaromatic hydrocarbons (nitro-PAHs) after nitration with nitrogen dioxide (Ju and Parales 2010). In aquatic environment, nitration and halogenations are sunlight catalyzed with the formation of 2- and 4-nitrophenol, chlorophenol, and bromophenol (Ju and Parales 2010).

Currently, two mechanisms are known for the production of biogenic nitroaromatic compounds (NACs). Oxygenases and haloperoxidases (under unnatural/stress) are known to catalyze the addition of nitro moieties to aromatic compounds. The other mechanism of formation of biogenic nitroaromatic compounds is via an electrophilic interaction of a nitronium cation that may directly help in attaching a nitro group to the aromatic ring (Ju and Parales 2010). NACs are also biologically active metabolites existing in plants and fungal species (e.g., alkaloids) although the reason for their presence is largely unknown. Table 12.1 includes some of the naturally and artificially produced NACs and their applications in various areas.

Man-made or synthetic NACs include picric acid, TNT, lidocaine, and dinoseb and are primarily produced by nitration at *para*, *meta*, and/or *ortho* positions of the aromatic ring. The Zincke nitration is where sodium nitrite reacts with phenols to replace Br with a NO_2 group (Raiford and LeRosen 1944), and Wolfenstein-Böters reaction is where benzene can be converted to 1,3,5-trinitrobenzene (Davis et al. 1921). TNT was generally synthesized via sequential nitrification of toluene (Ju and Parales 2010). Aromatic amines, e.g., anilines used by the chemical industry, are manufactured by catalytically reducing NACs (Ju and Parales 2010).

12.2 Toxicity and Health Issues of Nitroaromatic Compounds

Several of the NACs are considered as "priority pollutants" and listed by the US Environmental Protection Agency (USEPA), and most of them are toxic and even mutagenic and capable of causing cancers on long exposures. The properties of NACs that are preferable for use as pesticides or other industrial application make them dangerous to mankind and animals. Ames test using *Salmonella* and *E. coli* tester strains has been popularly used to detect the potential of NACs to cause mutation that consequently leads to DNA damage through deletions, transversions, and transitions (Purohit and Basu 2000).

Several nitroaromatic compounds pose a threat to the environment and to all living beings, e.g., benzene, naphthalene, polycyclic aromatic hydrocarbons, and biphenyls (Kovacic and Somanathan 2014). The toxicity is mainly due to formation

Table 12.1 List of synthetic or naturally occurring nitroaromatic and chloro-nitroaromatic compounds, their applications, and effects on human beings

Name of compound	Application	Natural/ synthetic	Effects
TNT, DNT (tri-, dinitrotoluene)	Explosive	Synthetic	Adverse
RDX (cyclotrimethylenetrinitramine)	Explosive	Synthetic	Adverse
HMX (cyclotetramethylenetetranitramine)	Explosive	Synthetic	Adverse
p-Nitrophenol	Tannery	Natural	Mineralized by microbes but otherwise adverse effect
4-Nitrocatechol	Useful metabolic marker for the presence of functional cytochrome P450 2E1 in mammalian cell microsomes	Natural	Mineralized by microbes but otherwise adverse effect
Nitrobenzoates	Dye industry	Synthetic	Allergic skin reaction, eye damage
3-Methyl-4-nitrophenol, *m*-nitrophenol	Diesel exhaust particles, pesticides	Synthetic	Adverse
Nitrobenzene or halonitrobenzenes	Pharmaceuticals	Synthetic	Cancer inducing
Chloronitrobenzenes	Precursor for useful compounds	Synthetic	Hazardous
Lidocaine	Local anesthetic	Synthetic	Useful
Anilines	Drugs, rubber, polyurethane foams, azo dyes, photographic chemicals, varnishes	Synthetic	Disorientation, dizziness
4-Nitropyrene	Diesel and gasoline engine exhausts	Natural	Carcinogenicity
p-Nitrochlorobenzene	Pesticides and dyes	Synthetic	Coughing and wheezing
1-Nitronaphthalene	Industrially important chemical	Synthetic	Not significant
Chloramphenicol	Antibiotic	Natural	Beneficial to human

of electron transfer, reactive oxygen species, and oxidative stress. Different classes of nitroaromatic compounds are known to affect human population via various mechanisms, and this has been briefly described below.

12.2.1 Nitrobenzenes

Nitrobenzenes, commonly used as pesticides, drugs, ammunition, or explosives, intermediates of chemical synthesis of industrial products, are known to be potential carcinogens due to nitro group reduction via two or more mechanisms leading to formation of reactive oxygen species or causing oxidative stress. When mice were exposed for a long time (2 years) to *o*-nitrotoluene, alterations in ras, p53, and β-catenin genes were observed in hemangiosarcomas leading to mutagenesis (Hong et al. 2003). The toxicity of 2,4,6-trinitrotoluene (TNT), a well-known explosive, covalently binds proteins and DNA in its reduced forms and also perturbs enzymatic redox cycling and/or serves as redox-cycling substrates for single ET as shown experimentally (Šarlauskas et al. 2004). Upon chronic exposure to TNT, there has been DNA damage in testes in rats and reduced semen secretion in Chinese workers (Homma-Takeda et al. 2002; Li et al. 1993)

12.2.2 Nitrobenzanthrones (NBA)

Nitrobenzanthrones (NBA), consisting of four fused aromatic rings, e.g., 3-nitrobenzanthrone (3-NBA), occur in diesel exhausts and in airborne particles that exhibit significant mutagenic activity and serve as potent carcinogens, which cause tumors in the lungs of rodents and cause damage via H_2O_2 formation in human cells (Murata et al. 2006).

4-Nitrobiphenyl (NBP) is a bladder carcinogen of dogs and has mutagenic properties. When enzymatically reduced, the intermediate *N*-hydroxylaminobiphenyl is also reported as a mutagen, and the major product 4-aminobiphenyl (ABP) is a bladder carcinogen (Culp et al. 1997; Kovacic and Somanathan 2014).

Nitrated derivatives of polyaromatic hydrocarbons (PAHs) are reported as airborne pollutants (e.g., 1-nitropyrene and 1,3- and 1,8-dinitropyrene) that mainly arise from combustion of diesel in engines. Diesel engine emission poses as an important contaminant and the exhaust releases particles that are potent air pollutants. Oxidative attack of some nitroaromatic molecules on DNA and formation of DNA adduct molecules are important in cancer formation, for example, 3-NBA and nitropyrenes (Kovacic and Somanathan 2014).

12.3 Degradation of Nitroaromatics

Enormous use of NACs in explosives, dyes, agricides, etc. and their release in the environment via groundwater, soil, and streams/water bodies have flamed up strong criticism and aroused concerns due to their potential health hazards. The costs for conventional cleanup have been estimated to be enormous and might not be sustainable and environment friendly either, and therefore, biological means of cleanup are being considered and researched upon. Nitroaromatic compounds are comparatively more recalcitrant to biodegradation than their analogs, which are not nitrated (Alexander and Lustigman 1966). To understand biodegradation, a few concepts are very important, i.e., **mineralization**, **co-metabolism,** and **transformation**. *Mineralization* (complete degradation) refers to catabolism of the pollutant/substrate to its elements and is the preferred over the other types of degradation for developing bioremediation technologies. Once a compound is mineralized, it yields energy in the biological system and may be incorporated into various biomolecules to increase cellular biomass (Alexander 1981). Since energy is generated during the catabolism of NACs, the reaction is a self-sustainable process and proceeds continuously when the contaminant exists in proper concentration, also providing a selective pressure to promote the proliferation and growth of the degrading organism over others. In contrast to mineralization, there is another process called *co-metabolism*, where enzymes involved in the breakdown of some growth-inducing substrate (primary substrate) nonspecifically transform another contaminant/substrate (Alexander 1981). Few co-metabolized compounds may provide nutrition and energy, but only as long as the primary substrate exists and therefore, (i) need for some primary contaminant and (ii) lack of selective pressure created by primary substrate,entails co-metabolic bioremediation more expensive and labor intensiveas compared to mineralization. The stark difference between mineralization (complete degradation) and *transformation* is the difference in the products, which in the former case are harmless minerals and biomass, whereas, in the latter case, the products are essentially organic derivatives of the contaminant that may be more toxic than before or nontoxic (preferred for bioremediation). Polynitroaromatic compounds are partially biodegraded or transformed to generate amino-nitro products and are not mineralized further (Kaplan 1992); however, para-nitrophenol is degraded by several bacteria and is easily mineralized (Prakash et al. 1996;

Samanta et al. 2000; Chauhan et al. 2000; Pandey et al. 2002). Table 12.1 lists nitroaromatics that undergo mineralization, co-metabolization, or transformation via biological agents.

12.3.1 Aerobic or Anaerobic Degradation Pathways

Microbes evolved various strategies for dispensing nitro group(s) during conversion of NACs to simplified forms. Nitro group(s) present in the compound may be converted to NO_2^- after dioxygenation of its aromatic ring to an intermediate (dihydroxy compound) and monooxygenation to another intermediate (epoxide and/or hydride-Meisenheimer complex) (Nishino et al. 2000). Partial reduction to hydroxylaminobenzenes generates ammonia. The hydroxylaminobenzenes are processed by mutases and rearranged to *o*-aminophenols by few microbes. Alternately, hydroxylamino intermediate(s) may be transformed to catechol upon the release of the ammonium moiety. Few common interpretations about biodegradation of different types of NACs have been outlined below:

Mononitrophenols, e.g., (2-nitrophenol (2NP), 4-nitrophenol (4NP), 4-chloro-2-nitrophenol (4C2NP)) get hydroxylated to replace the NO_2 group consequently releasing (NO_2^-) resulting in *ortho*–/*para*-dihydroxybenzene (Nishino et al. 2000) (Fig. 12.1). Few organ phosphate pesticides, such as parathion and methyl-parathion, are transformed to *p*-nitrophenol and then hydroxylated. *Para*-nitroanisole is *O*-demethylated to *p*-nitrophenol and then hydroxylated. **Flavoprotein monooxygenases** may be involved in monooxygenation and all the monooxygenation reactions in some cases giving rise toquinone(s) (Kadiyala and Spain 1998).

Sometimes, microbial enzymes catalyze **dioxygenation** reaction to produce dihydroxy intermediates, which further attacked dioxygenases as in 2-nitrotoluene

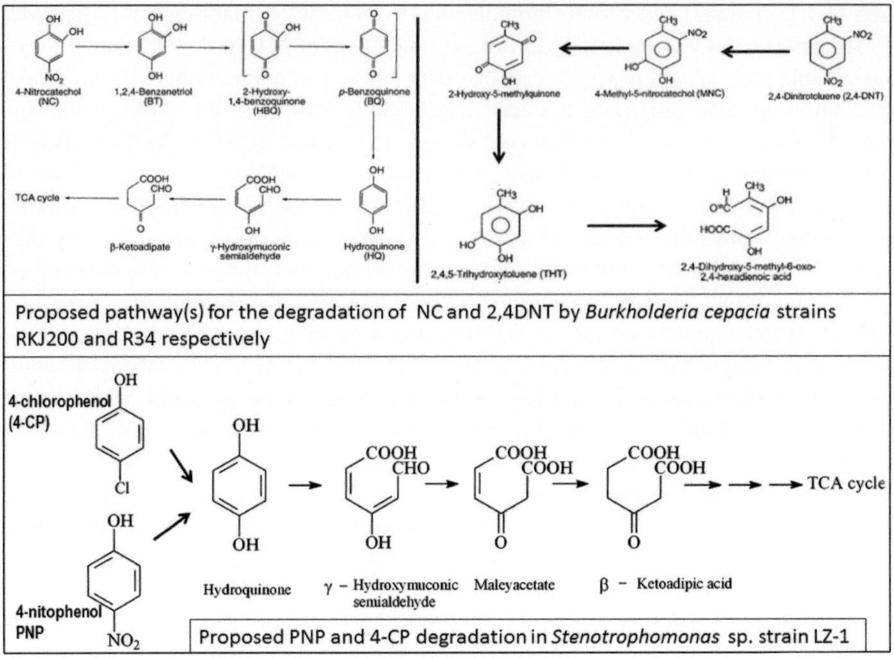

Fig. 12.1 The aerobic degradation of few NACs following various pathways. (Adapted from Chauhan et al. (2000), Johnson et al. (2002), and Liu et al. (2009))

(2NT), NB, 2,6-dinitrotoluene (2,6-DNT), 2,4-dinitrotoluene (2,4-DNT), 2,6-dinitrophenol, 1,3-dinitrobenzene, and 3-nitrobenzoic acid (3-NBA) (Nishino et al. 2000). The **dihydroxy-nitro cyclohexadienes** (intermediates) formed upon dioxygenase activity of nitro groups are unstable and re-aromatize after elimination of nitrites to produce **catechols**. Microbes aerobically convert 2,4-dinitrophenol and 2,4,6-trinitrophenol to unstable hydride-Meisenheimer complex and release the first nitrite (Nishino et al. 2000; Behrend and Heesche-Wagner 1999) (Fig. 12.1).

Anaerobic pathways include microaerobic or partial aerobic conditions. *Rhodobacter capsulatus* converts 2,4-dinitrophenol to *o*-amino-*p*-nitrophenol under anaerobic and/or microaerobic environment in the presence of light using suitable nitro reductases. Subsequently, *o*-amino-*p*-nitrophenol is degraded via a constitutive activity requiring light, O_2, and other sources of carbon and nitrogen. Nitro group is partially reduced to NO_2^- via the hydroxylamino derivative following a well-known chemical reaction.

Hydroxylamino moieties may be transformed by **hydroxylaminolyase** to corresponding **catechols** and dissemination of ammonium moiety. **Mutase**s catalyze the intramolecular rearrangement of hydroxylaminophenol to *o*-aminophenol, e.g., 2-chloro-5-nitrophenol, *p*-chloronitrobenzene, and *m*-nitrophenol (Nishino et al. 2000; Blasco and Castillo 1997; Meulenberg and de Bont 1995).

12.3.2 Anaerobic-Aerobic Dual Systems

Biodegradation of nitrobenzene has been carried out using a dual system consisting of aerobic microbes, followed by anaerobic ones (Dickel et al. 1993). Nitrobenzene may be completely degraded by aerobic microbial processes, but there are problems that may be ameliorated via an anaerobic process by reducing nitrobenzene to aniline and subsequently converting it via aerobic reactions. Anaerobic phase would be using glucose as carbon source (C source) and hydrogen donor. For the intermediates released during TNT biodegradation, a similar two-stage system has been tried (Fig. 12.2). Hydroxytoluenes or amino toluenes may be eliminated quickly under aerobic conditions (Funk et al. 1993; Rieger and Knackmuss 1995).

12.3.3 Fungal Degradation, Phytoremediation, and Composting of NACs

The degradation mechanism of several microorganisms is based on the breakdown of any aromatic nitro group to an amino moiety. However, white-rot fungi produce extracellular ligninolytic enzymes that oxidatively transform and/or mineralize xenobiotics such as TNT, polychlorinated biphenyls (PCBs), polyaromatic hydrocarbons (PAHs), chlorinated phenols, and pesticides, e.g., DDT. This process is an

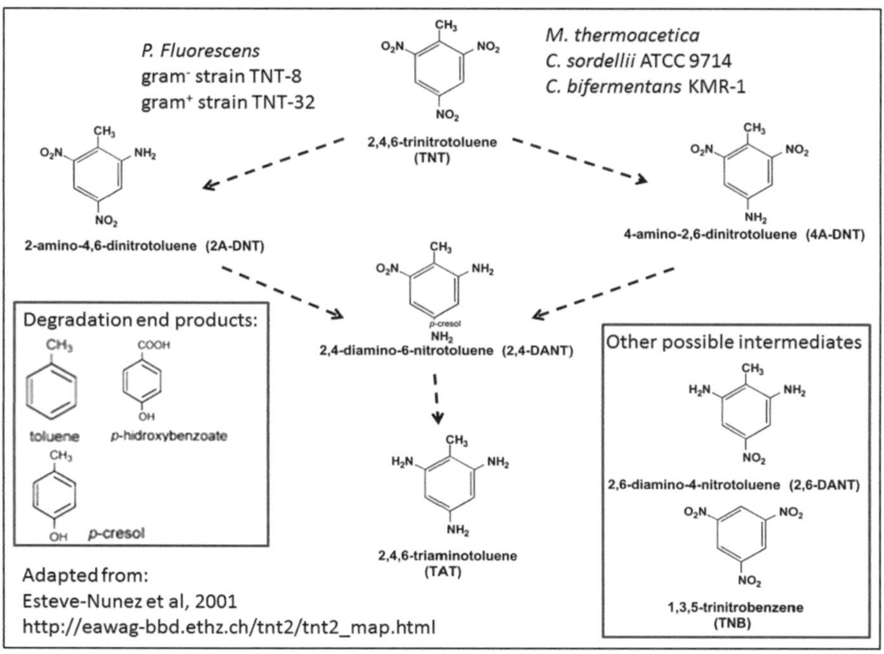

Fig. 12.2 Anaerobic degradation of TNT following various pathways

example of co-metabolism where carbohydrates are utilized as growth-promoting substrates for the fungi. Therefore, white-rot fungi have been fruitfully used as tools for bioremediation of persistent toxicants from contaminated sites. TNT and related compounds have been successfully treated by *Phanerochaete chrysosporium* (syn. *Sporotrichum pulverulentum*) and is a favorite model for research (Alexander and Lustigman 1966; Alexander 1981; An et al. 1994).

TNT is first reduced to mono-amino-dinitrotoluene and then transformed to azoxy, azo, and hydrazo intermediates. Primary metabolites (Hydroxy-Azo-DiNitro Toluenes and Azo DiNitro Toluenes) are rearranged to aminophenols (Bamberger rearrangement). A condensation reaction of nitroso and hydroxylamino intermediates generates azoxy compounds. Other researchers have proposed a combination of fungi with bacteria in bioremediation systems in which the fungi detoxify or modify the xenobiotic compound such that the bacterial population that may be mineralized by bacteria (Barr and Aust 1994).

The rapid disappearance of contaminants, e.g., TNT in aquatic environment dominated by specific plant species, has provoked researchers to consider phytoremediation as an option for removal of NACs using water and terrestrial plant systems. The researchers focused on deciphering plant metabolism and learning about mechanisms for phytoremediation including remediation by microbe in the phyllosphere and rhizosphere and alterations made by the plant body rendering the

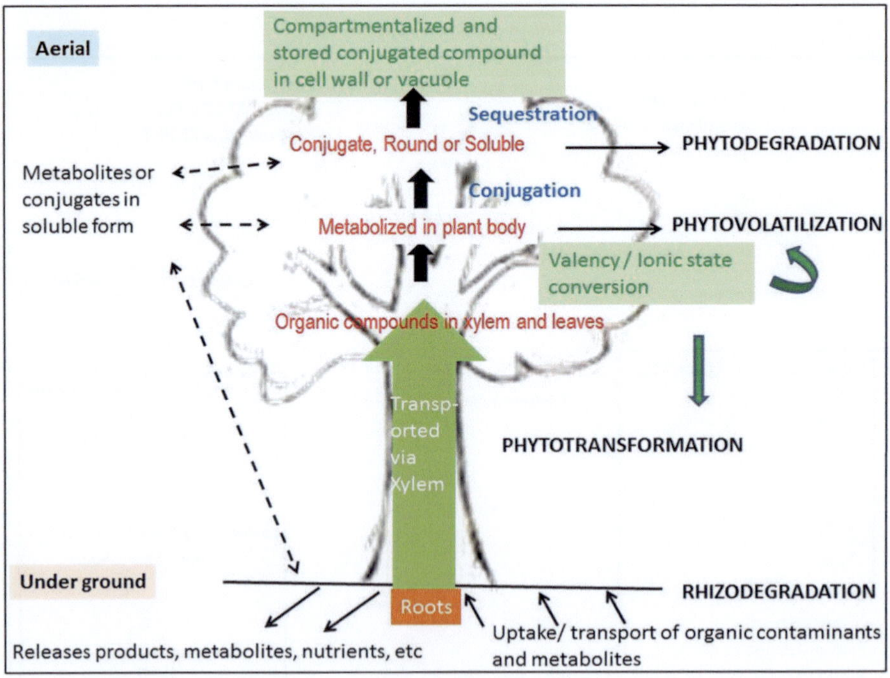

Fig. 12.3 The "green liver" model of **phytoremediation** for removal of xenobiotics in three stages: transformation, conjugation, and sequestration

environment conducive for decontamination (pH, redox changes, etc.). For certain cases, indirect mechanisms are more significant for decontamination, and therefore, one might use "plant-assisted" remediation. The green liver model (Fig. 12.3) is often used to correctly justifies the treatment of xenobiotics as, unlike microbes, plants do not exhibit a vast range of enzymatic pathways to break down given metabolites; instead, it recognizes foreign matter as toxins and degrades or transforms them, and this detoxification mechanism is common to the human liver; hence, the jargon "green liver" prevails (Klein and Scheunert 1982; Sandermann 1994). The theory was introduced because it was observed that plants exposed to herbicides readily metabolized it in three stages: transformation, conjugation, and sequestration (Fig. 12.3).

(a) Transformation: It includes reduction, oxidation, and hydrolysis reactions. Oxygenases or hydroxylases are involved.
(b) Conjugation: Compounds released after transformation are conjugated with plants' organic molecule (glucose, malonate), which leads to reduction in toxicity (Singh and Jain 2003).
(c) Sequestration: The storage of conjugate in plant organelles such as vacuoles, or are "bound" as xenobiotic conjugates and incorporated into biopolymers such as lignin where they are no longer capable of interfering with cell function.

Composting (humification) has been of great interest for environmental waste management as it greatly reduces weight and volume and results in less hazardous and often a useful biofertilizer. In the case of NACs, ex situ physicochemical treatment of incineration has proven to be costly and undesirable although it is accepted in case of TNT contamination. However, the ex situ technique of composting at lab and pilot scale has been successfully applied to various contaminants and xenobiotics (herbicides, chlorophenols, etc.). Synthetic musk fragrances (fragrances in cosmetics, soaps, lotions, washing powder, etc.) contain NACs that have significant environmental consequences and are classified as persistent organic pollutants (POPs), e.g., musk xylene, which is similar to TNT, and musk ketone, which is similar to 2,4-DNT. The presence of such toxicants in human milk, adipose tissue, and food items has been alarmingly recognized especially as musk xylene is suspected to be a carcinogen. They accumulate in agricultural systems via sewage sludge, so controlled composting is believed to be a possible treatment for decontamination. Soil/sludge is mixed with degradable organic matter and sufficient moisture to start composting (Sandermann 1994; Williams et al. 1992). Composting proceeds via the following phases in which the microbial community also changes sequentially: mesophilic phase (up to 45 °C), thermophilic phase (45–60 °C, <80 °C), cooling phase (45–20 °C), and maturation span (~20 °C). A highly diverse microbial population is active during the mesophilic phase, followed by the second phase (shortest), in which spore-forming bacteria, e.g., *Bacillus* sp., and thermophilic fungi dominate and highest amount of degradation occurs during this phase. Finally, fungi emerge as dominant species because spores withstand high temperatures and fungal enzymes degrade lignin-cellulosic parts to derive energy. Rise in temperature to about 80 °C deactivates microbial processes and inhibits composting (Garg et al. 1991, Griest et al. 1993). Important factors influencing composting include aeration, pH, moisture, and C/N content of the substrate.

Although composting is an aerobic press, in the case of TNT contamination, anaerobic/aerobic composting systems have been shown to be highly efficient. Aerating the compost by injecting air or by turning over the compost pile favors composting. Structure of composting material mainly assists aeration and may sometimes create anaerobic micro-pockets. Light material, such as wood chips and straw, inhibits compression of the compost (Bruns-Nagel et al. 2000).

The compost pH varies from 5.5 to 8.0; however, in case of anaerobic pre-phase, pH drops below 4.0 but returns to optimal range once aerobic treatment commences. The optimal moisture tolerated by bacteria in compost is ~50–60%. Moisture over this limit creates anaerobic conditions. The C/N ratio for an efficient composting process is 26–35.32 Higher N_2 concentrations give rise to ammonia that elevates pH, but lower nitrogen slows cellular growth and increases organic acids to elevate pH, and both conditions impede composting. Composting is of five major types: (1) in-vessel static piles, (2) static piles, (3) mechanically agitated in-vessel (MAIV) composting, (4) windrow composting, and (5) anaerobic/aerobic composting.

Different contaminants show varying rates of decomposition, e.g., some petroleum hydrocarbons can be mineralized, and some co-metabolized to less dangerous forms and are incorporated into humus. Doyle et al. (1986) tried to compost radio-labeled[14] C-TNT,[14] C-RDX,[14] and C-HMX using different conditions and achieved varying degrees of degradation by increasing the amount of soil. Craig et al. (1995) showed better results on windrow composting of TNT, RDX, and HMX, proving that composting is successful for removing dangerous NACs from soil and may be successfully applied in case of significantly polluted soils because it may be used as an ex situ or in situ technique and not very cost-intensive or labor-intensive. On the other hand, phytoremediation may be applied to mildly polluted large areas that are abandoned due to pollution and can be used to reclaim them, although weather conditions and risks due to introduction of non-native species persist. This technique is not cost- or labor-intensive like composting, unlike using bioreactors for treatment.

12.4 Evolution of New Pathways Via Genetic Changes

Although synthetic NACs have been introduced in nature for a short period of time, bacteria capable of breaking down these chemicals have been isolated from polluted sites suggesting that they are capable of adaptation by evolving new enzymes and pathways to endure and survive the selective stress created by the contaminants (Kivisaar 2009). The contaminated sites are therefore good resources to fish out strains with excellent properties for bioremediation.

It has been observed that genes encoding enzymes for catabolizing aromatic compounds may be associated with transposons carried by plasmids showing conjugation and genomic islands, to promote dissemination via horizontal transfer (Nojiri et al. 2004; Juhas et al. 2009). Bacteria oppose stress of starvation, desiccation, and unsuitable pH and/or temperature, where stress and starvation lead to (1) genetic changes and further transposition of mobile elements (Kivisaar 2003; Robleto et al. 2007) and (2) errors in DNA synthesis and mechanism of duplication (Kivisaar 2009; Tark et al. 2005).

Error-prone DNA polymerase or homologues of Pol-V may be coded by naturally occurring conjugative plasmids, and they may also have a contribution for manifestation of bacterial metabolic diversity (Permina et al. 2002; Tark et al. 2005). Few NACs, e.g., nitrobenzene, dinitrotoluenes, and nitrophenols, are powerful carcinogens (Kulkarni and Chaudhari 2007; Kivisaar 2009) and have the potential to induce mutations at higher concentrations (badly contaminated environment).

Bacterial evolution to use NACs as nutrients is exemplified by the 2,4-dinitrotoluene (2,4-DNT) pathway of *Burkholderia cepacia* R34. In this organism, the sequence of genes contained in the 27-kb DNA segment coding for 2,4-DNT pathway indicates that pathway gene(s) depict three points of ancestry (Johnson et al. 2002): (a) initial dioxygenase (DNT dioxygenase), catalyzing the first denitrification of the intermediate, that may have originated from naphthalene catabolic pathway; (b) second denitrification (by 4-methyl-5-nitrocatechol monooxygenase) from chloroaromatic degradation pathway; (c) third gene from amino acid pathway. Open reading frames having unknown function in 2,4-DNT degradation in *Burkholderia cepacia* R34 and the presence of several point mutations and transposon in the region advocate an intermediate phase during evolution of the pathway by incorporating genetic material with divergent ancestry (Johnson et al. 2002; Kivisaar 2009) via horizontal transfer and movement of transposable elements.

In a recently hypothesized concept about evolution depending on the proximity of proteins (O'Loughlin et al. 2006; Tokuriki and Tawfik 2009), it is suggested that protein evolves by directed evolution to perform a new role via "enzyme promiscuity" resulting in unique enzymes that break down newly introduced synthetic compounds (Aharoni et al. 2005; Afriat et al. 2006). The concept of "promiscuity" also plays significant roles during the development of novel regulators that respond to effectors (Cases and de Lorenzo 2005). XylR, a transcriptional activator in the toluene degradation pathway, is encoded by TOL plasmid pWW0 and acquires a new role by responding to both 2,4-DNT and its monosubstituted precursor molecules and to the dissimilar isomeric compound *m*-nitrotoluene and various chlorophenols (Galvão et al. 2007). The mutations leading to such a change were based on amino acid substitutions at the surface of proteins leading to conformational shifts affecting binding of effector and modulating transmission of signal between XylR domains (Galvão et al. 2007).

Ju et al. (2009) demonstrated that the functioning of the regulator NtdR (nitrotoluene responsive) was contributed by five mutations in NagR-like ancestor regulator. NtdR and NagR differed by 5 amino acids, located at position numbers 74, 169,

189, 227, and 232 (Lessner et al. 2003); however, NagR recognized 5 out of the 63 tested compounds, namely, salicylate, gentisate, 4-hydroxybenzoate, 4-isopropylbenzoate, and methyl salicylate, and, especially, does not interact with the NACs. On the other hand, NtdR, could activate 2-nitrotoluene degradation pathway genes in the presence of NACs, and a broad range of aromatic acids and their analogues, that may not be metabolized by *Acidovorax* sp. strain JS42 (exhibiting 2-nitrotoluene pathway).

12.5 Bacterial Metabolism of Chloro-Nitroaromatics

Chlorinated nitroaromatic compounds (CNAs) such as chloronitrobenzenes, chloronitrophenols, and chloronitrobenzoic acids enter our environment via agricultural practices, industrial discharges, or improper waste disposal and are known to be hematotoxic, immunotoxic, splenotoxic, genotoxic, hepatotoxic, nephrotoxic, and carcinogenic (Travlos et al. 1996; Li et al. 1999). 4-Chloronitrobenzene undergoes three types of transformation in mammals: (a) nitro group reduction, (b) glutathione conjugation for chloride displacement, and (c) hydroxylation of ring. The toxic nature renders many of the CNAs as priority pollutants as listed by the USEPA.

There are several methods including advanced oxidation processes (AOPs) to remove CNAs from industrial wastewater, which relies on nonspecific oxidation by hydroxyl radical (OH∗) (Vilhunen and Sillanpää 2010). AOPs also treat wastewater using UV rays, H_2O_2, UV-H_2O_2, photo-Fenton, ozonation, catalytic ozonation (Vilhunen and Sillanpää 2010), and the combination of these techniques. However, the main drawback of physicochemical techniques is their unsuitability for in situ biological application, and they are not cost-effective. CNA degradation is significantly affected by the position of chloro- and nitro-substitution on the benzene ring. The compounds, which have nitro groups at the *ortho* or *meta* positions, are predicted to be more difficult to degrade, as compared to the compounds having nitro groups at *para* positions (Arora et al. 2012). 4C2NP is more difficult to degrade than 2-chloro-4-nitrophenol (2C4NP) (Arora et al. 2012). Bacteria that utilize 2C4NP could not catabolize 4C2NP due to altered position of Cl and NO_2 groups (Pandey et al. 2011). The underlying reason for this phenomenon is that the enzymes, which catalyze reaction at the *para* positions, do not function at the *meta* or *ortho* position and vice versa (Arora et al. 2012). Degradation of CNAs with Cl and NO_2 moieties at different positions is acted upon by specified set of enzymes. The nitro group makes the benzene ring more recalcitrant than chloro group (Arora et al. 2012).

The **enzymes** mediating 2C4NP degradation in strain OCNB-1 were determined to be *aniline dioxygenase*, *nitrobenzene reductase*, and *catechol-1,2-dioxygenase*. A 100-kb plasmid carried gene(s) for degrading 4C2NP in *P. putida* strain ZWL73 and *Comamonas* sp. CNB-1. Some CNAs showed multiple routes of degradation, e.g., pentachloronitrobenzene (PCNB), which gives rise to metabolites with or without sulfur, many of which can be detected in soils contaminated with PCNB. Several reports on bioremediation of CNAs have proven the utility of rhizoremediation (*Comamonas* sp. CNB-1 associated with alfalfa roots), bioaugmentation (*P. putida*

ZWL73, *Rhodococcus imtechensis* RKJ300), and decontamination via membrane bioreactor followed by bacterial growth.

12.6 Bioavailability and Biodegradation via Microbes

Chemotaxis (chemo, chemical; taxis, movement) is the swimming of microbes with the aid of their flagella toward or away from a particular chemical. If microbial cells migrate toward a substance, the progression is called positive chemotaxis, while cellular motion in reverse direction is known as negative chemotaxis (Pandey and Jain 2002). Many microbial isolates have been applied for degrading NPs and CNPs, out of which few are capable of moving and show chemotaxis positive (Arora et al. 2012, 2014). Such strains are exemplified by *Burkholderia* sp. SJ98, which utilizes 4-nitrophenol, 2-chloro-4-nitrophenol, and 3-methyl-4-nitrophenol as the only C source and energy resource, and exhibited positive chemotaxis toward the above-said compounds (Bhushan et al. 2000, Samanta et al. 2000, Pandey et al. 2012). *Bacillus subtilis* PA-2 exhibited chemotaxis away from 4-chloro-2-nitrophenol, 4-nitrophenol, and 2,6-dichloro-4-nitrophenol (Arora et al. 2015).

The optimal environment for a bacterium is one in which energy generation is maximum, such as when there is a balance between the amount of electron donor and electron acceptor available (Taylor and Zhulin 1998). Since the chemoeffectors that attract bacteria (chemo-attractants) are often electron donors bacteria consume, metabolism of the chemo-attractant in the cell produces a gradient of electron acceptors to which the bacteria can also respond.

Positive chemotaxis helps microbes to sense a chemical that it can metabolize (either as nutrient) or co-metabolize by moving up a gradient of concentration of the particular chemical (here, NACs or CNAs) such that it avoids toxic levels and at the same time can continue feeding until the chemical persists in the environment. However, negative chemotaxis is a survival strategy shown when the organism feels engendered by toxic levels of the chemical in question and avoids it by moving away (Pandey et al. 2002, Arora et al. 2015).

The first step in bioremediation is the bioavailability of the substrate to the microbes, expedited by positive chemotaxis. Non-bioavailability or unavailability of organic pollutants for microbial cells significantly limits the efficiency of bioremediation of contaminated areas (Head 1998; Stelmack et al. 1999). Soil from contaminated sites possesses a separate or nonaqueous-phase liquid such as drops or liquid films on the surface of soil. Biodegradation readily occurs when target substrates are soluble in any aqueous phase (Stelmack et al. 1999, Pandey and Jain 2002; Law and Aitken 2003). Several pollutants, specially hydrophobic compounds, are nearly insoluble and persist superficially adsorbed on the nonaqueous-phase liquid (Head 1998; Stelmack et al. 1999; Parales and Haddock 2004). For onset and progress of biodegradation, bacteria must be able to access target compounds, by either dissolving it in liquid phase or by adhering directly to the interface of water and nonaqueous-phase liquid. For gaining access to adsorbed substances, pollutant-degrading bacteria may attach to surfaces possibly by forming biofilms.

12.6.1 Response of Microbes

Molecular mechanisms behind chemotaxis have been explained by three routes: (a) Signaling by chemoreceptors across the cell membrane (transmembrane) is the superior-most method acquired by bacteria where a ligand binds an external domain of a particular receptor that spans the membrane, thereby producing a signal to modulate the activity of kinase inside the cell, which results in chemotaxis (Pandey et al. 2002; Falke and Hazelbauer 2001). This transmembrane signaling mechanism does not depend on catabolism of the pollutant or its non-catabolizable analogues. (b) Chemotaxis therefore only depends on alteration in the cells' energy affected by complete metabolism of substrate (Alexandre and Zhulin 2001). This chemotactic response is referred to as metabolism dependent, and it is observed in many microbes, e.g., *Escherichia coli*, *Rhodobacter sphaeroides*, and *Azospirillum brasilense* (Pandey et al. 2002; Alexandre et al. 2000).

(c) The third mechanism suggests that chemotaxis signals are generated in collaboration with transport of effectors into cells. For example, chemotactic movement of *Bacillus subtilis* and *E. coli* toward carbohydrates and sugars is attached to the transport of the chemo-attractant. In *P. putida*, the gene pcaK codes for an unessential transporter protein for carrying 4-hydroxybenzoate and is essential for chemotaxis. Similarly, a permease protein coded by tfdK enables *Ralstonia eutropha* JMP123 to move toward very low levels of 2,4-dichlorophenoxyacetate (2,4-D); however, in this case, this gene is not responsible for entering the cells (Hawkins and Harwood 2002; Harwood et al. 1994).

12.7 Chemotaxis in Semisolid Medium TNT Chemotaxis Recent

There are several methods to study the chemotaxis of bacteria toward and away from the cells. The most commonly describes ones are (a) drop plate assay, (b) swarm plate assay, and (c) capillary assay.

To study negative chemotaxis, researchers have used various techniques, namely, (1) the chemical-in-plug method, (2) chemical-in-pond method, (3) chemical-in-plate method, (4) test-tube method, and (5) high-throughput micro-well method. To demonstrate negative chemotaxis, the chemical-in-plug method is widely acceptable (Arora et al. 2015; Tso and Adler 1974).

Here, we describe the most prevalent techniques with suitable diagrammatic representations.

For drop plate assay, degradative microbes would be cultured on a suitable medium trypticase soy broth, collected at middle logarithmic phase and resuspended in minimal medium containing 0.3% Bacto agar to prepare petri plates. Some crystals of NACs/CNAs placed in the center of the petri plate would act as chemo-repellant, and the samples should be incubated at 30°C. The response should be observed after 6 h of incubation. Chemotaxis would be suggested by the ring

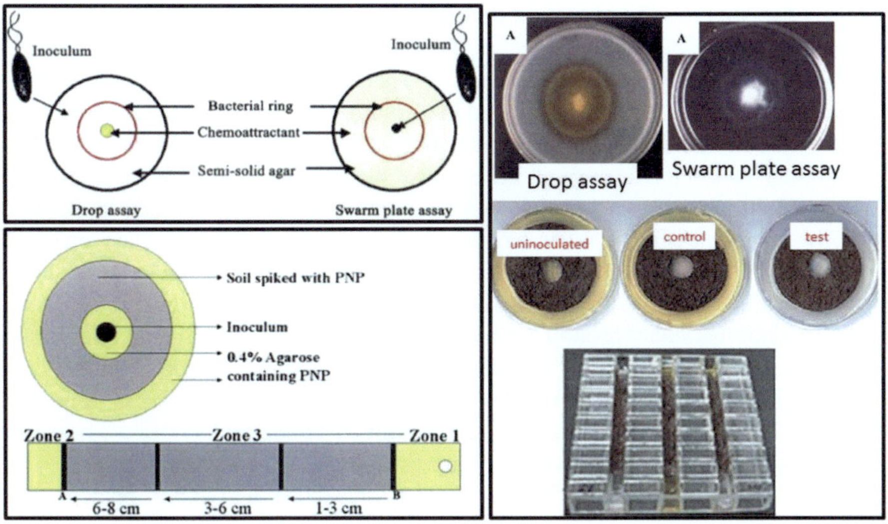

Fig. 12.4 Assays of chemotaxis (diagrammatic and actual experiment) against coloured nitroaromatic compounds in semisolid medium (drop and swarm plate assay) and in soil using plate and tray assays (Adapted from: Pandey et al. 2002; Paul et al. 2006)

formation (clearing zone) around the crystal as cells move away from it; heat-killed cells of the same strain may be used as control.

To perform swarm plate assay, *p*-nitrophenol (conc. 0.3 mM) should be dissolved in swarm plate agar medium (minimal medium having 0.16% Bacto agar) prior to casting plates. Around 75–100 μL induced and washed cells suspended in minimal medium should be gently poured in a petri dish and incubated at around 25 °C. One millimolar glucose may be provided to the cell in suspension for energizing them. Ring development was viewed 12–16 h after incubation (Fig. 12.4).

To perform capillary assays, special capillary tubes (Drummond Scientific, USA) of 1 μL volumetric capacity should be used. The appropriate NAC should be added to chemotaxis buffer consisting of 100 mM pot. Phosphate buffer (pH 7.2) and 20 mM EDTA to attain concentrations of 10, 20, and 200 mM. Aspartate may be used as positive substrate control. Capillaries loaded with the above were inserted in cell suspension (10^6–10^7 cells/mL in buffer for chemotaxis) taken on a shallow cover glass. After 30 min incubation time, cell suspension from the capillaries was diluted serially and plated on nutrient agar (NA). Bacterial colonies obtained on the plates were counted after overnight cultivation at 30 °C. Suitable positive and negative controls were maintained. Chemotaxis index (CI) was calculated as follows:

Number of bacterial cells accumulated in the test capillary containing NAC divided by that of the control.

12.7.1 Negative Chemotaxis Assays

1. Chemical-in-capillary method. Positive chemotaxis is assayed by inserting a capillary that was previously dipped in a solution of chemo-attractant into a suspension (the "pond") of bacterial cells that are motile. Microbes move toward the capillary and enter through it. After a given time, the total number of bacteria entered into the capillary is detected by counting colony-forming units (CFUs). If the chemical is inert, some bacteria nevertheless enter the capillary owing to random motility of the bacteria This is the "background" number and is deducted from reading as noise. If the given compound behaves as a chemo-repellent, fewer than this "background" number will be present inside the capillary. This capillary method is helpful in determining chemo-attractants and/or repellents, although repelling effects may be too insignificant for detection. Inhibition of motility might end up giving the same result as negative chemotaxis; however, this may be cross-checked via "motility assay."
2. Chemical-in-pond method. The repellent is present in the pond of bacteria, and none is put into the capillary. The bacteria then escape into the capillary for "refuge," and the number accumulated in the capillary is determined as before. Without any repellent, some bacteria enter the capillary by random swimming; with repellent, the accumulation is larger, and hence, this assay provides a positive result. (The values are never as strikingly above background as for positive chemotaxis in the chemical-in-capillary method.) There is a threshold value for repulsion. At high concentrations, loss of motility or viability, or saturation of the chemotactic apparatus. The negative chemotactic assay against 4C2NP was reported by Arora et al. (2015) (using *Bacillus subtilis* strain PA2) using chemical-in-plug method. For this bacteria, suspension was prepared as described before (10^6–10^7 cells/mL chemotaxis buffer). The bacterial solution was poured around hard agar plugs composed of minimal media, 2% Bacto agar, and 4C2NP (100 mM) or 4NP (100 mM) or dichloro nitrophenol (DCNP) (100 mM). After solidifying, the plates were incubated at 30∘C for 6 h, at which time they were evaluated for chemotactic response.

A convenient agarose-in-plug bridge method was used to demonstrate chemotaxis in the Archaeon *Halobacterium salinarum* (Yu and Alam 1997). Hot liquid agarose solution with chemo-effectors was placed in the center of a microscope slide with a bridge that is built by using two plastic strips 16 mm apart. A coverslip should be placed on the molten agarose immediately. The agarose plug gets encircled quickly by the bacterial cell suspension. This method has been tried for amino acids, but not for NACs.

Leungsakul et al. (2005) observed the chemotaxis toward 2,4,6-trinitrotoluene (TNT), 2,5-DNT, 2,3-dinitrotoluene (2,3-DNT), 2,6-DNT, 2,4-DNT, 2-nitrotoluene (2NT), 4NT, and 4-methyl-5-nitrocatechol (4-M-5NC). They used drop assay for their studies and found that, although there are cases where the substrate is not the source of C and nitrogen for the organism, they still can be a chemo-attractant. For example, 2,4-DNT in case of *B. cepacia* R34 and *Burkholderia* sp. DNT is the C and nitrogen source, but not a good chemo-attractant as 2,5-DNT (not a C/N source). Also TNT, 2,3-DNT, 2,5-DNT, 2NT, and 4NT are chemo-attractants, but not carbon and energy sources. This implies that the chemotactic machinery of the above strains works for other NACs as well apart from the compounds that they degrade or serve as intermediates in degradation pathway. The results also suggest that degradation pathway gene(s) are not associated with chemotaxis, so their presence/absence has no impact.

Organophosphates (OP) degrading *Pseudomonas* BUR11 isolated from an agricultural site utilized parathion and chlorpyrifos or their intermediates as sole sources of carbon along with being positively chemotactic toward them (Pailan and Saha 2015).

12.7.2 Chemotaxis Through Soil

Chemotaxis of *Ralstonia* sp. SJ98 toward *p*-nitrophenol was demonstrated in laboratory using various assays in semisolid medium; two assays were designed for demonstrating chemotaxis in soil microcosm, i.e., a small-scale qualitative assay in petri plate and another assay was done using a specially designed tray giving quantifiable results on a larger scale. For such experimental strategies, bacterial suspension was prepared in PO_4 buffer saline (PBS) in order to attain an approximated density of 10^{10} cells/mL.

1. Plate assay: Soil spiked with para-nitrophenol (PNP) was used to prepare concentric zones of soil and Bacto agar in a petri plate such that the organism travels radially out of the center, through the soil and agar zones (Fig. 12.4). Bright-yellow color of PNP changes to colorless indicating depletion of PNP. Positive and negative controls were maintained. Moisture content of soil was maintained at 40–50% of soil's water holding capacity (WHC) by sprinkling water whenever necessary.
2. Tray assay: For quantitative soil chemotaxis, a glass tray with three parallel lanes was fabricated, containing markings at 1 cm interval (Fig. 12.4). Glass stoppers prevented the mixing of agar and soil. As indicated in the diagram, the zones 1 and 2 constituted of 0.5 mM *p*-nitrophenol suspended in 0.4% semisolid Bacto agar and zone 3 had PNP-spiked soil in all three lanes. The first lane was inoculated with cell suspension of chemotaxis-positive strain *Ralstonia* sp. SJ98, second lane was inoculated with non-chemotactic *Burkholderia* sp. *strain* RKJ200, and the third lane was kept uninoculated. Both types of bacteria were capable of degrading PNP. The yellow color of PNP in zone 1 disappeared, indicating that bacteria travelled through agar zone and reached the soil.

CFUs of soil bacteria were determined by spreading proper dilutions of soil taken at different time points from each lane (Fig. 12.4). For determining residual PNP in samples collected from the start and end at various zones of the soil, high-performance liquid chromatography (HPLC) was performed by a method as reported earlier (Labana et al. 2005).

12.7.3 Application of Chemotaxis in Biofilms

Flagella are required for attaching cells to various surfaces and facilitating biofilm development (Pratt and Kolter 1999). In addition, biofilm-forming bacteria show chemotaxis to move along the surface to grow and spread (Stelmack et al. 1999). These biofilms may be useful for degradation of CNAs and NACs. For removal of 2,4-dichlorophenol (DCP) from wastewater, Kargey and Ekker made use of a perforated rotating tube biofilm reactor comprising of a mixed biomass of microbes originating from an activated sludge and was supplemented with DCP-degrading *P. putida*. It was observed that DCP was completely mineralized. Similarly, bacteria capable of adhering to polyaromatic hydrocarbons (PAHs) often expedite breakdown of PAHs (Wolfaardt et al. 1995). This phenomenon is exemplified by the two-ring herbicide called diclofop-methyl, methyl 2-[4-(2,4-dichlorophenoxy)phenoxy] pyruvate, which adsorbs onto biofilms formed by microbial exopolymers. The microbial community of the biofilm catabolized the diclofop-methyl during a period of starvation. Nitroaromatic compounds fall under another group of xenobiotics having multiple uses during the synthesis of pharmaceuticals, foams, pesticides, and explosives. Due to the nitro side group, these compounds are resistant to biodegradation; microbial transformation might lead to harmful metabolic intermediates (Lendenmann et al. 1998). Lendenmann et al. (1998) used a consortium that degraded a mixture of dinitrotoluene (DNT) using fluidized-bed biofilm reactor containing 2,4-DNT (40 mg/L) and 2,6-DNT (10 mg/L). Efficiency of degradation was more than 98% for 2,4-DNT and ~94% for 2,6-DNT. Degradation of 4,6-dinitro-ortho-cresol (an old synthetic pesticides) was reported using batch cultures in a fermenter called fixed-bed column reactor.

12.8 Microbial Diversity, Microbial Evolution, and Bioremediation Strategies

The unexplored "unculturable" microbial wealth holds a tremendous promise for various resources. About 1 g of soil might hold approximately 1000–10,000 unknown species belonging to prokaryotes, and further diversity is expected within each species (Torsvik and Ovreas 2002). Phylogenetically, the "unculturable" microorganisms may show some or 100% similarity to the culturable ones while possessing a unique physiological conformation making them recalcitrant to standard culturing techniques (Rondon et al. 1999). Another probability could be that the unyielding remaining microbial population might represent novel lineages phylogenetically dissimilar in nature and therefore cannot be grown via standard lab techniques (Rondon et al. 1999). Various techniques showed that significant dissimilarities were observed on comparing the community of contaminated to uncontaminated sites, especially with respect to unculturable organisms. In an example, experimental plots where oil was spilt (to mimic contamination) were compared to that of control sites (uncontaminated) via techniques such as phospholipid fatty acid analysis (PLFA) along with denaturing gradient gel electrophoresis (DDGE) analysis (MacNaughton et al. 1999). These studies suggested that culturing-based method elucidated only small fractions of the entire soil microbial diversity. Therefore, soil metagenome continues to be an unexploited reservoir of novel gene(s) and/or gene cluster(s) for bioremediation and other applications. The biosynthetic diversity of microbes from different environments has been accessed using "metagenomics" where operons or genes responsible for the degradation of pollutants are acquired from the metagenome. Large-scale projects, e.g., http://www.tigr.org/tdb/MBMO/, led by "The Institute for Genomic Research" and "Monterey Bay Coastal Ocean Microbial Observatory" are databases that make metagenomic information available on many unexplored metabolic processes exhibited by microbes. Fang et al. (2014) used metagenomics study using six data sets (16 Gb data) to report diversity, abundance, and potential biodegradation genes (BDGs) and metabolic pathways of recalcitrant pollutants of freshwater and marine sediments, e.g., dichloro-diphenyl-trichloro-ethane (DDT), hexachloro-cyclo-hexane (HCH), and atrazine (ATZ). Nearly complete catabolic pathways for breakdown of DDT and ATZ were found in the sediments.

Such recent approaches help in constructing effective "designer biocatalysts" for various biotechnological applications including bioremediation. Exploring the microbial diversity leads to the elucidation of (a) effective or new pathways for improved catabolism of pollutants, (b) chemotactic or biosurfactant-producing strains for better or faster access and solubilization of sparingly soluble or "aged" chemicals for degradation, and (c) regulatory operons and gene(s) for construction of containment systems for regulated bioremediation. It is the responsibility of biotechnologists to understand the ethical responsibilities before application of novel techniques for biological applications. Also, there should be complete understanding of genetic modifications, and confinement of the introduced constructs should be ascertained before releasing designer bugs for bioremediation of contaminated sites.

References

Afriat L, Roodveldt C, Manco G, Tawfik DS (2006) The latent promiscuity of newly identified microbial lactonases is linked to a recently diverged phosphotriesterase. Biochemistry 45(46):13677–13686

Aharoni A, Gaidukov L, Khersonsky O, Gould SM, Roodveldt C, Tawfik DS (2005) The 'evolvability' of promiscuous protein functions. Nat Genet 37(1):73

Alexander M (1981) Biodegradation of chemicals of environmental concern. Science 211(4478):132–138

Alexander M, Lustigman BK (1966) Effect of chemical structure on microbial degradation of substituted benzenes. J Agric Food Chem 14(4):410–413

Alexandre G, Zhulin IB (2001) More than one way to sense chemicals. J Bacteriol 183(16):4681–4686

Alexandre G, Greer SE, Zhulin IB (2000) Energy taxis is the dominant behavior in Azospirillum brasilense. J Bacteriol 182(21):6042–6048

An D, Gibson DT, Spain JC (1994) Oxidative release of nitrite from 2-nitrotoluene by a three-component enzyme system from Pseudomonas sp. strain JS42. J Bacteriol 176(24):7462–7467

Arora PK, Sasikala C, Ramana CV (2012) Degradation of chlorinated nitroaromatic compounds. Appl Microbiol Biotechnol 93(6):2265–2277

Arora PK, Srivastava A, Singh VP (2014) Bacterial degradation of nitrophenols and their derivatives. J Hazard Mater 266:42–59

Arora PK, Jeong MJ, Bae H (2015) Chemotaxis away from 4-chloro-2-nitrophenol, 4-nitrophenol, and 2,6-dichloro-4-nitrophenol by Bacillus subtilis PA-2. J Chemother 2015:1

Barr DP, Aust SD (1994) Mechanisms white rot fungi use to degrade pollutants. Environ Sci Technol 28(2):78A–87A

Behrend C, Heesche-Wagner K (1999) Formation of hydride-Meisenheimer complexes of picric acid (2,4,6-trinitrophenol) and 2,4-dinitrophenol during mineralization of picric acid by Nocardioides sp. strain CB 22–2. Appl Environ Microbiol 65(4):1372–1377

Bhushan B, Samanta SK, Chauhan A, Chakraborti AK, Jain RK (2000) Chemotaxis and biodegradation of 3-methyl-4-nitrophenol by Ralstonia sp. SJ98. Biochem Biophys Res Commun 275(1):129–133

Blasco R, Castillo F (1997) Characterization of 2,4-Dinitrophenol uptake byRhodobactercapsulatus. Pestic Biochem Physiol 58(1):1–6

Bruns-Nagel D, Steinbach K, Gemsa D, von Löw E (2000) Composting (humification) of nitroaromatic compounds. Biodegradation of nitroaromatic compounds and explosives. Lewis Publishers, Boca Raton, FL, pp 357–393

Cases I, De Lorenzo V (2005) Promoters in the environment: transcriptional regulation in its natural context. Nat Rev Microbiol 3(2):105

Chauhan A, Samanta SK, Jain RK (2000) Degradation of 4-nitrocatechol by Burkholderia cepacia: a plasmid-encoded novel pathway. J Appl Microbiol 88(5):764–772

Craig HD, Sisk WE, Nelson MD, Dana WH (1995) Bioremediation of explosives-contaminated soils: a status review. In: Proceedings of the 10th annual conference on hazardous waste research. Kansas State University, Manhattan, NY, USA, pp 164–179

Culp SJ, Roberts DW, Talaska G, Lang NP, Fu PP, Lay JO Jr, Teitel CH, Snawder JE, Von Tungeln LS, Kadlubar FF (1997) Immunochemical, 32P-postlabeling, and GC/MS detection of 4-aminobiphenyl–DNA adducts in human peripheral lung in relation to metabolic activation pathways involving pulmonary N-oxidation, conjugation, and peroxidation. Mutat Res 378(1–2):97–112

Davis TL, Worral DE, Drake NL, Helmkamp RW, Young AM (1921) The role of mercuric nitrate in the "catalyzed" nitration of aromatic substances. J Am Chem Soc 43:594–607

Dickel O, Haug W, Knackmuss HJ (1993) Biodegradation of nitrobenzene by a sequential anaerobic-aerobic process. Biodegradation 4(3):187–194

Doyle RC, Isbister JD, Anspach GL, Kitchens JF (1986) Composting explosives/organics contaminated soils. Atlantic Research Corp, Alexandria, VA

Falke JJ, Hazelbauer GL (2001) Transmembrane signaling in bacterial chemoreceptors. Trends Biochem Sci 26(4):257–265

Fang H, Cai L, Yang Y, Ju F, Li X, Yu Y, Zhang T (2014) Metagenomic analysis reveals potential biodegradation pathways of persistent pesticides in freshwater and marine sediments. Sci Total Environ 470:983–992

Funk SB, Roberts DJ, Crawford DL, Crawford RL (1993) Initial-phase optimization for bioremediation of munition compound-contaminated soils. Appl Environ Microbiol 59(7):2171–2177

Galvão TC, Mencía M, De Lorenzo V (2007) Emergence of novel functions in transcriptional regulators by regression to stem protein types. Mol Microbiol 65(4):907–919

Garg R, Grasso D, Hoag G (1991) Treatment of explosives contaminated lagoon sludge. Hazard Waste Hazard Mater 8(4):319–340

Griest WH, Stewart AJ, Tyndall RL, Caton JE, Ho CH, Ironside KS, Caldwell WM, Tan E (1993) Chemical and toxicological testing of composted explosives-contaminated soil. Environ Toxicol Chem 12(6):1105–1116

Harwood CS, Nichols NN, Kim MK, Ditty JL, Parales RE (1994) Identification of the pcaRKF gene cluster from Pseudomonas putida: involvement in chemotaxis, biodegradation, and transport of 4-hydroxybenzoate. J Bacteriol 176(21):6479–6488

Hawkins AC, Harwood CS (2002) Chemotaxis of Ralstonia eutropha JMP134 (pJP4) to the herbicide 2,4-dichlorophenoxyacetate. Appl Environ Microbiol 68(2):968–972

Head IM (1998) Bioremediation: towards a credible technology. Microbiology 144(3):599–608

Homma-Takeda S, Hiraku Y, Ohkuma Y, Oikawa S, Murata M, Ogawa K, Iwamuro T, Li S, Sun GF, Kumagai Y, Shimojo N (2002) 2,4,6-trinitrotoluene-induced reproductive toxicity via oxidative DNA damage by its metabolite. Free Radic Res 36(5):555–566

Hong HL, Ton TV, Devereux TR, Moomaw C, Clayton N, Chan P, Dunnick JK, Sills RC (2003) Chemical-specific alterations in ras, p53, and β-catenin genes in hemangiosarcomas from B6C3F1 mice exposed to o-nitrotoluene or riddelliine for 2 years. Toxicol Appl Pharmacol 191(3):227–234

Johnson GR, Jain RK, Spain JC (2002) Origins of the 2,4-dinitrotoluene pathway. J Bacteriol 184(15):4219–4232

Ju KS, Parales RE (2010) Nitroaromatic compounds, from synthesis to biodegradation. Microbiol Mol Biol Rev 74(2):250–272

Ju KS, Parales JV, Parales RE (2009) Reconstructing the evolutionary history of nitrotoluene detection in the transcriptional regulator NtdR. Mol Microbiol 74(4):826–843

Juhas M, Van Der Meer JR, Gaillard M, Harding RM, Hood DW, Crook DW (2009) Genomic islands: tools of bacterial horizontal gene transfer and evolution. FEMS Microbiol Rev 33(2):376–393

Kadiyala V, Spain JC (1998) A two-component monooxygenase catalyzes both the hydroxylation of p-nitrophenol and the oxidative release of nitrite from 4-nitrocatechol in Bacillus sphaericus JS905. Appl Environ Microbiol 64(7):2479–2484

Kaplan DL (1992) Biological degradation of explosives and chemical agents. Curr Opin Biotechnol 3(3):253–260

Kivisaar M (2003) Stationary phase mutagenesis: mechanisms that accelerate adaptation of microbial populations under environmental stress. Environ Microbiol 5(10):814–827

Kivisaar M (2009) Degradation of nitroaromatic compounds: a model to study evolution of metabolic pathways. Mol Microbiol 74(4):777–781

Klein W, Scheunert I (1982) Bound pesticide residues in soil, plants and food with particular emphasis on the application of nuclear techniques. In Agrochemicals: fate in food and the environment

Kovacic P, Somanathan R (2014) Nitroaromatic compounds: environmental toxicity, carcinogenicity, mutagenicity, therapy and mechanism. J Appl Toxicol 34(8):810–824

Kulkarni M, Chaudhari A (2007) Microbial remediation of nitro-aromatic compounds: an overview. J Environ Manag 85(2):496–512

Labana S, Pandey G, Paul D, Sharma NK, Basu A, Jain RK (2005) Pot and field studies on bioremediation of p-nitrophenol contaminated soil using Arthrobacter protophormiae RKJ100. Environ Sci Technol 39(9):3330–3337

Law AM, Aitken MD (2003) Bacterial chemotaxis to naphthalene desorbing from a nonaqueous liquid. Appl Environ Microbiol 69(10):5968–5973

Lendenmann U, Spain JC, Smets BF (1998) Simultaneous biodegradation of 2,4-dinitrotoluene and 2,6-dinitrotoluene in an aerobic fluidized-bed biofilm reactor. Environ Sci Technol 32(1):82–87

Lessner DJ, Parales RE, Narayan S, Gibson DT (2003) Expression of the nitroarene dioxygenase genes in Comamonas sp. strain JS765 and Acidovorax sp. strain JS42 is induced by multiple aromatic compounds. J Bacteriol 185(13):3895–3904

Leungsakul T, Keenan BG, Smets BF, Wood TK (2005) TNT and nitroaromatic compounds are chemoattractants for Burkholderia cepacia R34 and Burkholderia sp. strain DNT. Appl Microbiol Biotechnol 69(3):321–325

Li Y, Jiang QG, Yao SQ, Liu W, Tian GJ, Cui JW (1993) Effects of exposure to trinitrotoluene on male reproduction. Biomed Environ Sci 6(2):154–160

Li Q, Minami M, Hanaoka T, Yamamura Y (1999) Acute immunotoxicity of p-chloronitrobenzene in mice: II. Effect of p-chloronitrobenzene on the immunophenotype of murine splenocytes determined by flow cytometry. Toxicology 137(1):35–45

Liu Z, Yang C, Jiang H, Mulchandani A, Chen W, Qiao C (2009) Simultaneous degradation of organophosphates and 4-substituted phenols by Stenotrophomonas species LZ-1 with surface-displayed organophosphorus hydrolase. J Agric Food Chem 57(14):6171–6177

MacNaughton SJ, Stephen JR, Venosa AD, Davis GA, Chang YJ, White DC (1999) Microbial population changes during bioremediation of an experimental oil spill. Appl Environ Microbiol 65:3566–3574

Meulenberg R, de Bont JA (1995) Microbial production of catechols from nitroaromatic compounds. In: Biodegradation of nitroaromatic compounds. Springer, Boston, MA, pp 37–52

Murata M, Tezuka T, Ohnishi S, Takamura-Enya T, Hisamatsu Y, Kawanishi S (2006) Carcinogenic 3-nitrobenzanthrone induces oxidative damage to isolated and cellular DNA. Free Radic Biol Med 40(7):1242–1249

Nishino SF, Spain JC, He Z (2000) Strategies for aerobic degradation of nitroaromatic compounds by bacteria: process discovery to field application. Biodegrad Nitroaromat Compound Explos 7:61–65

Nojiri H, Shintani M, Omori T (2004) Divergence of mobile genetic elements involved in the distribution of xenobiotic-catabolic capacity. Appl Microbiol Biotechnol 64(2):154–174

O'Loughlin TL, Patrick WM, Matsumura I (2006) Natural history as a predictor of protein evolvability. Protein Eng Des Sel 19(10):439–442

Pailan S, Saha P (2015) Chemotaxis and degradation of organophosphate compound by a novel moderately thermo-halo tolerant Pseudomonas sp. strain BUR11: evidence for possible existence of two pathways for degradation. PeerJ 3:e1378

Pandey G, Jain RK (2002) Bacterial chemotaxis toward environmental pollutants: role in bioremediation. Appl Environ Microbiol 68(12):5789–5795

Pandey G, Chauhan A, Samanta SK, Jain RK (2002) Chemotaxis of a Ralstonia sp. SJ98 toward co-metabolizable nitroaromatic compounds. Biochem Biophys Res Commun 299(3):404–409

Pandey J, Heipieper HJ, Chauhan A, Arora PK, Prakash D, Takeo M, Jain RK (2011) Reductive dehalogenation mediated initiation of aerobic degradation of 2-chloro-4-nitrophenol (2C4NP) by Burkholderia sp. strain SJ98. Appl Microbiol Biotechnol 92(3):597–607

Pandey J, Sharma NK, Khan F, Ghosh A, Oakeshott JG, Jain RK, Pandey G (2012) Chemotaxis of Burkholderia sp. strain SJ98 towards chloronitroaromatic compounds that it can metabolise. BMC Microbiol 12(1):19

Parales RE, Haddock JD (2004) Biocatalytic degradation of pollutants. Curr Opin Biotechnol 15(4):374–379

Paul D, Singh R, Jain RK (2006) Chemotaxis of Ralstonia sp. SJ98 towards p-nitrophenol in soil. Environ Microbiol 8(10):1797–1804

Permina EA, Mironov AA, Gelfand MS (2002) Damage-repair error-prone polymerases of eubacteria: association with mobile genome elements. Gene 293(1–2):133–140

Prakash D, Chauhan A, Jain RK (1996) Plasmid-encoded degradation of p-Nitrophenol by Pseudomonas cepacia. Biochem Biophys Res Commun 224(2):375–381

Pratt LA, Kolter R (1999) Genetic analyses of bacterial biofilm formation. Curr Opin Microbiol 2(6):598–603

Purohit V, Basu AK (2000) Mutagenicity of nitroaromatic compounds. Chem Res Toxicol 13(8):673–692

Raiford LC, LeRosen AL (1944) The nitration of brominated flurophenols by the Zincke method. J Am Chem Soc 66:1872–1873

Rieger PG, Knackmuss HJ (1995) Basic knowledge and perspectives on biodegradation of 2,4,6-trinitrotoluene and related nitroaromatic compounds in contaminated soil. In: Biodegradation of nitroaromatic compounds. Springer, Boston, MA, pp 1–18

Robleto EA, Yasbin R, Ross C, Pedraza-Reyes M (2007) Stationary phase mutagenesis in *B. subtilis*: a paradigm to study genetic diversity programs in cells under stress. Crit Rev Biochem Mol Biol 42(5):327–339

Rondon MR, Goodman RM, Handelsman J (1999) The Earth's bounty: assessing and accessing soil microbial diversity. Trends Biotechnol 17(10):403–409
Samanta SK, Bhushan B, Chauhan A, Jain RK (2000) Chemotaxis of a Ralstonia sp. SJ98 toward different nitroaromatic compounds and their degradation. Biochem Biophys Res Commun 269(1):117–123
Sandermann JH (1994) Higher plant metabolism of xenobiotics: the 'green liver' concept. Pharmacogenetics 4(5):225–241
Šarlauskas J, Nemeikaite-Č A, Anusevičius Ž, Misevičien L, Julvez MM, Medina M, Gomez-Moreno C, Narimantas Č (2004) Flavoenzyme-catalyzed redox cycling of hydroxylamino-and amino metabolites of 2,4,6-trinitrotoluene: implications for their cytotoxicity. Arch Biochem Biophys 425(2):184–192
Singh OV, Jain RK (2003) Phytoremediation of toxic aromatic pollutants from soil. Appl Microbiol Biotechnol 63(2):128–3519
Stelmack PL, Gray MR, Pickard MA (1999) Bacterial adhesion to soil contaminants in the presence of surfactants. Appl Environ Microbiol 65(1):163–168
Tark M, Tover A, Tarassova K, Tegova R, Kivi G, Horak R, Kivisaar M (2005) A DNA polymerase V homologue encoded by TOL plasmid pWW0 confers evolutionary fitness on Pseudomonas putida under conditions of environmental stress. J Bacteriol 187(15):5203–5213
Taylor BL, Zhulin IB (1998) In search of higher energy: metabolism-dependent behaviour in bacteria. Mol Microbiol 28(4):683–690
Tokuriki N, Tawfik DS (2009) Protein dynamism and evolvability. Science 324(5924):203–207
Torsvik V, Ovreas L (2002) Microbial diversity and function in soil: from genes to ecosystems. Curr Opin Microbiol 5:240–245
Travlos GS, Mahler J, Ragan HA, Chou BJ, Bucher JR (1996) Thirteen-week inhalation toxicity of 2-and 4-chloronitrobenzene in F344/N rats and B6C3F1 mice. Fundam Appl Toxicol 30(1):75–92
Tso WW, Adler J (1974) Negative chemotaxis in Escherichia coli. J Bacteriol 118(2):560–576
Vilhunen S, Sillanpää M (2010) Recent developments in photochemical and chemical AOPs in water treatment: a mini-review. Rev Environ Sci Biotechnol 9(4):323–330
Williams RT, Ziegenfuss PS, Sisk WE (1992) Composting of explosives and propellant contaminated soils under thermophilic and mesophilic conditions. J Ind Microbiol 9(2):137–144
Wolfaardt GM, Lawrence JR, Robarts RD, Caldwell DE (1995) Bioaccumulation of the herbicide diclofop in extracellular polymers and its utilization by a biofilm community during starvation. Appl Environ Microbiol 61(1):152–158
Yu HS, Alam M (1997) An agarose-in-plug bridge method to study chemotaxis in the archaeon Halobacterium salinarum. FEMS Microbiol Lett 156(2):265–269

Potential of Thallophytes in Degradation of Dyes in Industrial Effluents

13

Saroj Kumar Pradhan and Rohita Singla

Abstract

Diverse groups of microorganisms have inhabited this earth, which use different types of sources for energy and growth. Industries revolutionize the lifestyle of humankind, which affects negatively the ecosystem. Synthetic dyes impart fabulous colors to cloth, food, paper, and cosmetics. Due to their xenobiotic nature, they are mostly insurmountable for degradation and also toxic. Most of them are washed off during the various processes and mixed in the industrial effluents. Microorganisms have enzymatic system for the decolorization of dyes or simply they can adsorb them on their surface. Several genera of algae, bacteria, and fungi have developed a system to use these unwanted compounds in the water. They can also biotransform or degrade them into non-toxic products. Degradation of the dyes depends upon their toxicity and chemical structure and the type of strain used. Some species were found to be efficient against a variety of dyes at a high concentration level. The present review describes the diversity of three genera *Chlorella*, *Pseudomonas*, and *Aspergillus* of thallophytes for the degradation and decolorization of various dyes in industrial effluents and also the use of integrated approach of different consortia or other treatments for their application in wastewater treatment plants.

Keywords

Xenobiotic compounds · Industrial effluents · Azo dyes · Decolorization · *Chlorella* · *Pseudomonas* · *Aspergillus*

S. K. Pradhan (✉)
PG Department of Botany, Dev Samaj College for Women, Ferozepur, Punjab, India

Department of Botany, Punjabi University, Patiala, India

R. Singla
PG Department of Botany, Dev Samaj College for Women, Ferozepur, Punjab, India

P. K. Arora (ed.), *Microbial Technology for Health and Environment*,
Microorganisms for Sustainability 22,
https://doi.org/10.1007/978-981-15-2679-4_13

13.1 Introduction

Dyes are synthetic or natural compounds used to color or change the shade of any substance. From the beginning natural dyes from plants were used, but the invention of synthetic dyes by the British chemist William Perkin (1856) from coal tar revolutionized the chemical industry. During the next few decades, production of synthetic dyes has been popularized due to their use in every sector of industries. Dyes are used in food products, paper and textile industry, tanning, cosmetics, pharmaceutical, etc. Commercial products use colors to attract the customers. Due to their high usage, they are concentrated in our environment as xenobiotic compounds. The major share of production goes to textile industry which uses more than 10,000 types of dyes, and most are used as excess levels with 1000 tonnes per annum. About 10–25% is lost at some stage in the dyeing process, and approximately 2–20% is discharged as effluents in water and soil (Carmen and Daniel 2012). They are highly toxic, if not disposed properly as most of them are washed off in the effluents of these industries and reach the water and soil bodies. Dyes and by-products cause environmental, esthetic, and health problems. Dyes can be categorized as disperse, basic, acid, direct, and reactive dyes (Asgher 2012). The breakdown of chromophore groups (azo or anthraquinone) from dyes leads to the formation of toxic compounds (Katheresan et al. 2018). They break down in the form of several carcinogenic or mutagenic forms (aromatic compounds, benzidine, naphthalene, etc.) and cause serious health problems in the food chain. With the time, xenobiotic compounds accumulate in Mother Nature and become problematic for every type of organism. They are mostly degraded or adsorbed by microorganisms, but sometimes become recalcitrant in nature because of insolubility, absence of transporting enzymes, and non-accessibility as substrates (Godheja et al. 2016).

The thallophytes are a group of non-mobile organisms which included algae, bacteria, fungi, and lichens. This group of organisms inhabited the earth in almost all types of conditions like hot springs, volcanoes, and Arctic and Antarctic regions. A variety of microorganisms can tolerate these conditions as well as adapt themselves for their survival. The xenobiotics or industrial effluents make the natural water bodies more acidic and also disturb the growth of biota. Some species of the group were found capable of removing the color from industrial effluents by adsorption or biodegradation or biotransformation or mineralization (Chang et al. 2001a). As compared to chemicophysical treatments, biological degradation of dyes is always cost-effective and also can remove the toxic amines in the effluents, and further the combination of both treatments can produce better results (Hai et al. 2007). The exploration of the diversity and deciphering the underlying mechanism of adaptability will be helpful to make the positive planning to transform the worst environmental conditions (Rampelotto 2010). In the present chapter, we have summarized three different genera, *Chlorella* (algae), *Pseudomonas* (bacteria), and *Aspergillus* (fungi), implicated in the natural degradation of dyes in industrial effluents and the underlying mechanism of decolorization.

13.2 Algae

Algae are a group of aquatic microorganisms having photosynthetic machinery and ca. 50,000 species adapted to various ecological conditions (Xu et al. 2006). They come under the group of thallophytes as due to undifferentiated roots, stems, and leaves. The major commercially available groups of microalga are Chlorophyta, Dinophyta, Haptophyta, Rhodophyta, and Stramenopiles (Heimann and Huerlimann 2015). The microalgal genera studied for the biotreatment of industrial wastewater are *Spirogyra*, *Oscillatoria*, *Spirulina*, *Scenedesmus*, *Cosmarium*, etc. (Fazal et al. 2018) Among these groups, *Chlorella* taxa have been majorly investigated for the treatment of various types of industrial effluents (Banat et al. 1997; Munoza and Guieysse 2006; Safi et al. 2014).

13.2.1 *Chlorella*

The genus is spherical shaped single cell green algae. It is widely used in the field of productions of biofuels, cosmetics, food, and pigments and wastewater treatments (de Andrade and de Andrade 2017). Industrial wastewater contains dyes and nutrients used by algal community for their growth, which can be used as a sustainable approach for biodiesel production and bioremediation (Fazal et al. 2018). The two species, i.e., *C. vulgaris* and *C. pyrenoidosa*, were well documented by various authors for the treatment of effluents of textile industry (Table 13.1).

The first report of degradation of azo dyes by *Chlorella* was given by Jinqi and Houtian (1992). They tested 30 azo compounds for the decolorization process and found removal percentage in the range of 5–100%. The most easily degradable dye was Direct Blue 71 (100%), and Methyl Red was not decolorized from the medium. The azoreductase enzyme was found to be responsible for the bioconversion of aniline intermediate into carbon dioxide. The same type of degradation product was confirmed by Acuner and Dilek (2004) while studying *C. vulgaris* for the decolorization of Tectilon Yellow 2G. Sinha et al. (2016) reported the degradation of many industrial pollutants by *C. pyrenoidosa* NCIM 2738-based photobioreactor. The organism was able to decolorize the dye completely within 2.16 days and also improved the water quality.

The dyes can be degraded into simpler products, or simply they can be adsorbed by the microalgae. Adsorption capacity of microalgae can vary for different dyes and their initial concentration (Aksu and Tezer 2005). The initial pH of the solution was a determining factor for the proper biosorption of the dyes, and it can also vary with the specific dyes. Aksu and Tezer (2005) found that the highest uptake of vinyl sulphone-type reactive dyes occurred at pH 2.0 by dried *C. vulgaris*, while Daneshvar et al. (2007) demonstrated that basic pH was more favorable for the decolorization of Malachite Green. Similar results were observed by Tsai and Chen (2010) by altering the pH from 3.0 to 11.0. To attain the highest uptake of cationic dyes, the surface should acquire more negative charge which is only possible at this pH. The functional groups, i.e., hydroxyl and carbonyl groups, present on the surface of

Table 13.1 Removal of different dyes by *Chlorella* species

Sr. no.	*Chlorella* species	Dyes (concentration)	Mechanism (enzyme(s))	Removal time (percentage)	By-product	References
1.	*C. pyrenoidosa* *C. vulgaris*	Azo dyes	Biodegradation (azoreductase) –	(5–100%)	Aromatic amines, CO_2	Jinqi and Houtian (1992)
2.	*C. ellipsoidea* *C. kessleri* *C. vulgaris*	Tartrazine and Ponceau (5–20 ppm)		6 days (40–55%)	Aromatic amines	Hanan (2008)
3.	*C. vulgaris*	Tectilon Yellow 2G (400 mg L^{-1})	Bioconversion	200 h (83%)	Aniline, CO_2	Acuner and Dilek (2004)
		Remazol Golden Yellow (200 mg L^{-1}); Remazol Red and Black B (800 mg L^{-1});	Biosorption	–	–	Aksu and Tezer (2005)
		G-Red, Orange II, and Methyl Red (20 ppm); basic cationic (10 ppm); basic fuchsin (5 ppm)	Biosorption and Biodegradation (azoreductase)	7 days (4–91%)	Aromatic amines	El-Sheekh et al. (2009)
		Malachite Green (6 mg L^{-1})	Biosorption	90 min (91.61%)	–	Kousha et al. (2013)
		Congo Red (5–25 mg L^{-1})	Biosorption and Biodegradation (azoreductase)	96 h (83 and 58 %)	–	Hernández-Zamora et al. (2015)
4.	*C. vulgaris* UMACC 001	Lanaset Red 2GA (7.25 mg L^{-1})	Biosorption	10 days (48.7%)	–	Chu et al. (2009)
		Supranol Red 3BW (20 mg L^{-1})		10 days (50%)		Lim et al. (2010)
5.	*C.* sp.	Malachite Green (5 ppm)	Decolorization	2.5 h (80.7%)	–	Daneshvar et al. (2007)
6.	*C. pyrenoidosa*	Thioflavin T and Malachite Green	Biosorption	–	–	Horník et al. (2013)

(continued)

Table 13.1 (continued)

Sr. no.	*Chlorella* species	Dyes (concentration)	Mechanism (enzyme(s))	Removal time (percentage)	By-product	References
		Textile wastewater and Methylene Blue dye (10–60 mg L^{-1})		60 min (40–90%)	–	Pathak et al. (2015)
		Methylene Blue dye (100 mg L^{-1})		(98.20%)	–	Lebron et al. (2018)
7.	*C. pyrenoidosa* NCIM 2738	Direct Red-31 dye (40 mg L^{-1})	Biodegradation (azoreductase)	2.16 days (100%)	Aromatic amines	Sinha et al. (2016)

microalgae help them for the biosorption of dyes (Horník et al. 2013). The optimal temperature range for the dye uptake by *Chlorella* lies between 25 and 35 °C; however, a wide range has little effect on the biosorption (Tsai and Chen 2010).

The continuous lighting conditions used in the case of mixed culture of algae (13 taxa including *Chlorella*) removed 80% color within 30 days as compared to 60% after 60 days of exposure under simulated field lighting conditions from the pulping effluent (Dilek et al. 1999). El-Sheekh et al. (2009) tested *C. vulgaris* among five taxa of microalgae for the removal of basic fuchsin, basic cationic, G-Red, Methyl Red, and Orange II. The most susceptible dyes were basic cationic and basic fuchsin. *C. vulgaris* removed 43.7 and 59.12% of Orange II and G-Red dyes. The G-Red dye acts as an inducer of the azoreductase enzyme and increases the activity up to 72.25%. Kousha et al. (2013) compared the biosorption activity for Malachite Green of the same species against *Scenedesmus quadricauda*. They considered the different parameters like dye concentration, contact time, algae amount, and pH. The maximum dye removal was done by *C. vulgaris* (91.61%) as compared to the latter one (73.49%). Similarly, Lebron et al. (2018) recorded maximum elimination of Methylene Blue by *C. vulgaris* (98.20%) as compared to *Spirulina maxima* (94.19%). Recently, Zhao et al. (2018) evaluated the effectiveness of wastewater treatment by *C. vulgaris*, *C. zofingiensis*, and *Scenedesmus* sp. in terms of the activity of photosystem II, nutrient loading, and lipid productivity. *C. zofingiensis* shows higher absorption capability, productivity, and efficiency as compared to the other two species, even in worse environmental conditions.

The immobilized form of microalgae has more advantages over the free cell suspension for the elimination of heavy metals and xenobiotics in wastewater (Luan et al. 2006). Chu et al. (2009) investigated the immobilized *C. vulgaris* UMACC 001 (1% κ-carrageenan and 2% sodium alginate) for the treatment of three dyes and textile wastewater. The algae immobilized on 2% sodium alginate has higher color removal efficiency for the textile wastewater and dyes. The immobilized form is more stable, easy to harvest, and protected from the direct exposure to toxicity as compared to free cells. Later, Gao et al. (2011) also found the same results for the removal of nonylphenol using the same type of matrix. Horník et al. (2013) investigated the biosorption capacity of dried biomass of *C. pyrenoidosa* immobilized in polyurethane foam. The process of sorption of cationic dyes (Thioflavin T and Malachite Green) depends upon the preliminary concentration of dyes, flow rate of solution through the column, bed height, and biomass concentration. The simple or modified polyurethane-based adsorbent has been reported as an efficient sorbent for the elimination of dyes from wastewater (Sultan 2017).

Apart from the treatment of dyes, the genus has been also directly tested for the exclusion of xenobiotics directly from the textile wastewater. The organism utilizes textile wastewater for its growth and also removes the color in the range of 41.8–50.0% as reported by Lim et al. (2010). It also reduces phosphate, nitrate content, BOD, and COD from the effluents. The dried biomass was found more efficient as a biosorbent than wet algal biomass, due to its high binding affinity and large surface area. It can be cultured in the wastewater for color and COD removal and biomass production (El-Kassas and Mohamed 2014; Pathak et al. 2015; Tao et al. 2017). The integrated approach for the treatment of wastewater and production of biomass, lipids, biofuels, bioelectricity, etc. is the promising application of *Chlorella* in the industry (Logroño et al. 2017; Wang et al. 2017; Fazal et al. 2018). Malla et al. (2015) tested *C. minutissima* for biodiesel production and nutrient removal from primary and tertiary treated wastewater. The species removed TDS (90–98%), N (70–80%), P (60–70%), and K (45–50%) from the wastewater within 12 days. Zheng et al. (2017) demonstrated the enhanced production of biofuel by using kelp waste extracts combined with acetate in *C. sorokiniana*.

Seo et al. (2015) used oxidized dye wastewater composed of Methylene Blue and Methyl Orange for the harvesting of algae. The exposed amine groups of oxidized dyes act as amine-based coagulants. Daneshvar et al. (2018) investigated the feasibility of cultivation of *C. vulgaris* in a combination of aquaculture and pulp effluents. The carbohydrate, lipid, and protein percentage was very much high in the microalgae from the wastewater as compared to Bold's Basal Medium (BBM) solution. Another aspect of the use of microalgae and textile dyeing sludge was proved by Peng et al. (2015), as the combination of the duo improved char catalytic effect and increased the combustion process for the decomposition of textile dyeing sludge residue at high temperature (530–800 °C).

Undoubtedly, the discharge of the dyes into the aquatic ecosystem causes serious threats for the growth of many microorganisms. Toxicity studies of many dyes on *Chlorella* have been done by many workers (Hanan 2008; Qian et al. 2008; Hernández-Zamora et al. 2014; Kanhere et al. 2014; Xu et al. 2015). The deteriorated metabolic activity, growth rate, respiration, and photosynthesis efficiency of *C. vulgaris* were observed due to the direct exposure of Congo Red (Hernández-Zamora et al. 2014). After the bioremoval of the effluents by the species, the influents were less toxic to the primary consumer (*Daphnia magna*) of the aquatic ecosystem (Hernández-Zamora et al. 2015). Kanhere et al. (2014) observed genotoxic and cytotoxic effects of Malachite Green on *C. pyrenoidosa* in the form of altered cell morphology, high oxidative stress, DNA damage, and cell death. The growth was inhibited in a dosage-dependent manner, and *D. magna* ingest the dye even at very low concentrations. Thus, there would be the same type of negative effects on the other aquatic organisms.

13.3 Bacteria

The prominent genera of bacteria explored by different workers are *Aeromonas*, *Bacillus*, *Escherichia*, *Eubacterium*, *Citrobacter*, *Pseudomonas*, *Sphingomonas*, and *Staphylococcus* (Rafii et al. 1990; Bumpus 1995; Banat et al. 1997; Keck et al. 1997; Sugiura et al. 1999; Nakanishi et al. 2001; Coughlin et al. 2003). Several anaerobic bacteria produce azoreductase for the degradation of dyes and produced metabolites. Biochemical and molecular characterization has shown that the enzyme presumably a flavin reductase or FMN-dependent NADH-azoreductase or tetrameric NADPH-dependent flavoprotein, as found from *Sphingomonas*, *Escherichia*, and *Staphylococcus*, respectively (Nakanishi et al. 2001; Suzuki et al. 2001; Chen et al. 2005). Bacteria can degrade the xenobiotic compounds in either aerobic or anaerobic or both conditions. Many strains of *Pseudomonas* have degraded them into non-hazardous products and simultaneously utilized the dyes for their growth (Pandey and Upadhyay 2006). The next section of the chapter reviews the diversity of different species/strains of *Pseudomonas* capable of degrading dyes in industrial effluents (Table 13.2).

13.3.1 *Pseudomonas*

Several workers have isolated the azoreductase enzyme from different species of bacteria implicated in the deterioration of azo dyes (Michaels and Lewis 1985; Zhipei and Huifang 1991; Yatome et al. 1990; Hu 1994; Bumpus 1995; Banat et al. 1997). The bacteria utilize them as a source of carbon and nitrogen. However, in the case of RP_2B dye, it only acts as an inducer rather than as a growth substrate in the case of *P. luteola* (Hu 1998). The enzyme was found to be substrate specific, and the susceptibility of the bacterial attack depends on the substitution of the chemical and charged group at specific positions (Zimmermann et al. 1982; Yatome et al. 1990; Ben Mansour et al. 2009a). The degradation reaction of azo dyes into aromatic amines was fully catalyzed by the enzyme under anaerobic conditions, but to produce complete inorganic compounds, aerobic conditions are needed (Zhipei and Huifang 1991; Idaka et al. 1987a, b).

Zimmermann et al. (1982) isolated oxygen-insensitive azoreductase from *Pseudomonas* KF46, able to degrade the aromatic amines and complete mineralization of carboxy-Orange II. Nachiyar and Rajkumar (2004, 2005) proposed the mechanism of systematic elimination of Navitan Fast Blue S5R by the oxygen-insensitive enzyme, purified from *P. aeruginosa*. The intermediate metabolites of the dye may have undergone further oxidative deamination/decarboxylation and further enter the TCA cycle to release carbon dioxide. One of the intermediates

Table 13.2 Removal of different dyes by *Pseudomonas* species

Sr. no.	*Pseudomonas* species	Dyes (concentration)	Mechanism (enzyme(s))	Removal time and percentage	By-product	References
1.	*P. sp.* KF46	Orange II	Orange II azoreductase	–	4-Aminobenzoate and sulfanilic acid	Zimmermann et al. (1982)
2.	*P.* sp. GM3	Indigo Carmine, Reactive Blue 2, Acid Red 183, Acid Green 27, Acid Violet 7 (100 mg L^{-1})	Decolorization	72 h (18–97%)	–	Yu et al. (2001)
3.	*P.* sp. PR41-1	Methyl Red	Biodegradation (azoreductase)	–	Dimethyl *p*-phenylenediamine and *o*-aminobenzoic acid	Sugiura et al. (1999)
4.	*P.* sp. S-42	Diamira Brilliant Orange RR, Direct Brown M, Eriochrome Brown R	Biodegradation (azoreductase)	(70.5–95.3%)	Aromatic amines	Zhipei and Huifang (1991)
5.	*P.* sp.	Direct Black 38 and Congo Red (100 mg L^{-1})	Degradation	5 days (83–100 %)	Benzidine	Işik and Sponza (2003)
6.	*P.* sp. OX1	Acid Orange 7 (800 mg L^{-1})	Biodegradation	–	–	Lodato et al. (2007)
7.	*P.* sp. SUK1	Red BLI, Reactive Navy Blue RX, Reactive Red M5B, Reactive Red 6BI, Reactive Red HE, Reactive Red HE3B, Reactive Orange HE2R, and Reactive Orange M2R (50 mg L^{-1})	Biotransformation (aminopyrine *N*-demethylase and NADH-DCIP reductase)	80–360 min. (85.33–99.28%)	Nitroso compound, oximes, and imines	Kalyani et al. (2008)
		Reactive Red 2 (5 g L^{-1})	Biodegradation (lignin peroxidase and azoreductase)	6 h	2-Naphthol	Kalyani et al. (2009)
8.	*P.* sp. DY1	Acid Black 172 (100–300 mg L^{-1})	Decolorization	25 h (12.2–79.6%)	–	Du et al. (2010)

Sr. no.	*Pseudomonas* species	Dyes (concentration)	Mechanism (enzyme(s))	Removal time and percentage	By-product	References
9.	*P.* sp. LBC1	Direct Brown MR (100 mg L^{-1})	Biodegradation	14–42 h (90%)	*p*-Dihydroperoxybenzene, 2-hydroxy-7-aminonaphthol-3-sulfonic acid, and 3,6-dihydroxybenzoic acid	Telke et al. (2012)
10.	*P. aeruginosa*	Navitan Fast Blue S5R	Azoreductase	48 h	Metanilic acid, peri acid, aniline, and β-ketoadipic acid	Nachiyar and Rajkumar (2004, 2005)
11.	*P. aeruginosa* CR-25	Remazol Black B (50–500 mg L^{-1})	Decolorization	24 h (67–96%)	–	Joe et al. (2011)
12.	*P. aeruginosa* KY284155	Remazol Black B (200 mg L^{-1})	Degradation (azoreductase)	32 h (100%)	–	Hashem et al. (2018)
13.	*P. aeruginosa* NCIM 2074	Malachite Green (50 mg L^{-1})	Biodegradation (MG reductase, laccase, and aminopyrine *N*-demethylase)	5 h (100%)	Benzophenone	Kalyani et al. (2012)
14.	*P. aeruginosa* 23N1	Reactive Red 21 (50, 150 mg L^{-1})	Decolorization	48 h (91.5–93.5%)	–	Mishra and Maiti (2018)
15.	*P. desmolyticum* NCIM 2112	Red HE7B (100 mg L^{-1})	Biodegradation (lignin peroxidase and aminopyrine *N*-demethylase)		2-Hydroxyl-6-oxalyl-benzoic acid and 8-amino-naphthalene-1,3,6,7-tetraol	Kalme et al. (2007a)
		Direct Blue 6 (100 mg L^{-1})	Biodegradation (lignin peroxidase, laccase, and tyrosinase)	72 h (100%)	Aminonaphthalenesulfonic acid and 4-amino naphthalene	Kalme et al. (2007b)

(continued)

Table 13.2 (continued)

Sr. no.	*Pseudomonas* species	Dyes (concentration)	Mechanism (enzyme(s))	Removal time and percentage	By-product	References
16.	*P. fluorescens*	Acid Yellow-9	Biodegradation	4 days	4-Aamino-2-hydroxybenzene sulfonic acid sodium salt, 2-amino-4-hydroxy-benzene sulfonic acid sodium salt, and 2,4-dihydroxybenzene sulfonic acid sodium salt	Pandey and Upadhyay (2006)
17.	*P. luteola*	Azo Dye RP_2B	Azoreductase	5 days (95%)	Orthanilic acid	Hu (1998)
		Reactive Acid Yellow (30–200 mg L^{-1}), Reactive Black B (29–252 mg L^{-1}); Reactive Red 22 (61–353 mg L^{-1})	Biosorption	–	–	Chen (2002)
		Reactive Red 22 (200 ppm)	Decolorization	(80–98%)	–	Chen and Lin (2007)
		Congo Red (100, 210 ppm), Eriochrome Black T (100, 230 ppm), Methyl Orange (100–400 ppm), and Methyl Red (100–450 ppm)	Decolorization	20–25 h (100 %)	–	Hsueh and Chen (2007)
		Reactive Red 22 (200 mg L^{-1})	Degradation (azoreductase)	60 h (75–80%)	Aromatic amines	Chang et al. (2001a, b)
18.	*P. mendocina* PM2	Malachite Green (50–1800 mg L^{-1})	Biodegradation (MG reductase, lignin peroxidase, and manganese peroxidase)	24 h (73.5–99.5%)	–	Chaturvedi et al. (2013)
19.	*P. nitroreducens*	Methyl Red	Decolorization	15 h (80%)	–	Adedayo et al. (2004)

Sr. no.	*Pseudomonas* species	Dyes (concentration)	Mechanism (enzyme(s))	Removal time and percentage	By-product	References
20.	*P. otitidis* WL-13	Triphenylmethane dyes (Malachite Green, Brilliant Green, and Crystal Violet) (500 μmol/L)	Adsorption	12 h (13–95%)	–	Wu et al. (2009)
21.	*P. putida*	Crystal Violet (60 μmol/L)	Biodegradation	7 days (80%)	*N*-Demethylation intermediates and pararosaniline	Chen et al. (2007)
22.	*P. putida* SKG-1	Orange II (100 mg L^{-1})	Biodegradation	96 h (92.8%)	Sulfanilic acid and 1,2-naphthoquinone	Kumar Garg et al. (2012)
23.	*P. putida* mt-2	Acid Yellow 17, Violet 7, and Orange 52 (100 mg L^{-1})	Azoreduction and oxygen-dependent metabolization	60 h (100%)	Sulfanilic acid, *N*,*N*-dimethyl-*p*-phenylenediamine, and 4′-aminoacetanilide	Ben Mansour et al. (2007)
		Acid Violet 7 (200 mg L^{-1})	Biodegradation (azoreductase)		4-Aminoacetanilide	Ben Mansour et al. (2009a, b)
24.	*P. putida* MTCC 4910	Basic Violet 3 and Acid Blue 93 (250 mg L^{-1})	Biosorption	8 h (50–100%)	–	Arunarani et al. (2013)
		Direct Red 28 (10–250 mg L^{-1})	Biosorption	1 h (45–85%)	–	Deepa et al. (2013)
25.	*P. stutzeri* IAM 12097	4′-Dimethylaminoazobenzene-2-carboxylic acid (4.5×10^{-5} mol dm^{-3})	Biodegradation	(90%)	*o*-Aminobenzoic acid, *N*,*N*-dimethyl-*p*-phenylenediamine, and catechol	Yatome et al. (1993)
26.	*P. alcaligenes* *P. mendocina* *P. putida* *P. stutzeri*	Methyl Violet (0.9–16.5 mg L^{-1})	Decolorization	7 days (33.3–86.7%)	–	Sarnaik and Kanekar (1995)
27.	*P. aeruginosa* *P. fluorescens* *P. putida*	Navitan Fast Blue S5R (100–1200 mg L^{-1})	Biodegradation (azoreductase)	72 h (72–92 %)	Metanilic acid	Nachiyar and Rajkumar (2003)

formed in this study, i.e., metalinic acid, was further degraded into aniline and β-ketoadipic acid (Nachiyar et al. 2007). Işik and Sponza (2003) used aerobic and anaerobic conditions to study the color removal efficiency of *Pseudomonas* sp. They found that decolorization of Direct Black 38 and Congo Red was 83% and 100% under anaerobic incubation while 74% and 76% under microaerophilic conditions. The aerobic degradation occurs by the action of lignin peroxidase, tyrosinase, and laccase as reported by Kalme et al. (2007b) in *P. desmolyticum* NCIM 2112. Further, they purified laccase enzyme from the species and demonstrated the asymmetric breakdown of azo bond and that the specificity depends on the position of amino, hydroxyl, and sulfonic group in a dye. The decolorization rate is less when hydroxyl group and sulfonic group are at *meta* position or charged carboxyl group at *ortho* position to the azo bond (Nigam et al. 1996; Chen 2006; Kalme et al. 2007b, 2009). The presence of electron-withdrawing groups or absence of charged groups also enhances the rate of decolorization as stated by Hsueh and Chen (2007, 2008) in *P. luteola*. The toxicity of dyes depends on the type of azo bond, molecular structure, functional groups, and types of intermediates or degraded products. The lesser the toxicity of the dye, the easier will be the decolorization. Chen (2002) tested the toxicity of three reactive dyes against *P. luteola* (Acid Yellow, Black B, and Red 22). The Reactive Red 22 was easily decolorized, while Reactive Black B was highly toxic as it contains two azo bonds. As in this study decolorization is not growth-associated, the viability of the cells is the important criterion for the metabolism and expression of enzymes. Alternatively the cells can go for biosorption rather than decolorization.

Various authors have also isolated the laccase enzyme from different strains/species of *Pseudomonas* and showed its applicability in the elimination of synthetic dyes in industrial effluents (Telke et al. 2009; Kuddus et al. 2013; Wang et al. 2012). Phugare et al. (2011) purified a highly active enzyme, i.e., veratryl alcohol oxidase, from *P. aeruginosa* BCH. The enzyme has specificity for wide varieties of substrates and decolorizes seven dyes (Methyl Orange, Rubine 3GP, Congo Red, Remazol Black, Red HE7B, Red HE8B, and Red HE3B) in the range of 85–100%. One of the dyes, i.e., Remazol Black, was decolorized completely within 6 h and degraded into 7-diazenyl-naphathalene-1-ol and naphthalene-1,2,7-triol. Kalyani et al. (2011) reported a heme-containing peroxidase enzyme isolated from *Pseudomonas* sp. for the symmetric cleavage of Methyl Orange into *N*,*N*-dimethyl-1,4-benzenediamine and an intermediate 4-aminobenzenesulfonic acid. The intermediate formed was further degraded into aniline.

Toxicity analysis of the decolorized dyes should be done either by elucidating the structure of the degraded products by FTIR, GC-MS, HPLC, and NMR techniques or by using different organisms or cell lines. Several authors have checked the genotoxicity/cytoxicity/mutagenic potential of the metabolites formed by *Pseudomonas* during the remediation of industrial effluents (Adedayo et al. 2004; Pandey and Upadhyay 2006; Kalme et al. 2007a; Kalyani et al. 2009). Perei et al. (2001) isolated an aerobic bacterium called *P. paucimobilis* from the contaminated sites for the effective degradation of mutagenic metabolite sulfanilic acid. During the degradation of Orange 52, Violet 7, and Acid Yellow 17 by *P. putida* mt-2, genotoxic metabolites were found high in static cultures as compared to shaken conditions (Ben Mansour et al. 2007). Later on the authors demonstrated that the amines were mutagenic formed under static conditions, which later on vanished during shaken incubation. Further, the metabolite 4′-aminoacetanilide exhibited maximum mutagenicity, while 5-acetamido-2-amino-1-hydroxy-3,6-naphthalene disulfonic acid shows less effect due to presence of sulfonic groups (Ben Mansour et al. 2009b). Telke et al. (2012) tested the toxicity assays of *p*-dihydroperoxybenzene, 2-hydroxy-7-aminonaphthol-3-sulfonic acid, and 3,6-dihydroxy benzoic acid, metabolites formed during biodegradation of Direct Brown MR by *Pseudomonas* sp. LBC1. The textile effluents and the dye were more toxic to *Vigna radiata* and *Sorghum bicolor* as compared to the biodegraded metabolites.

In the case of Methyl Orange, there wasn't any kind of removal under aerobic conditions by *P. putida* mt-2 (Thao et al. 2013). So an immobilized bacterial system can solve the problem for oxygen-sensitive decolorization by creating miniature anoxic environment and complementarily increasing the biomass concentration and providing mechanical strength, feasibility of continuous processing, low-cost recovery, and reusability of biocatalyst (Stormo and Crawford 1992; Park and Chang 2000; Chang et al. 2001a). Puvaneshwari et al. (2002) studied the effective role of immobilized *P. fluorescens* on sodium alginate for the degradation of Direct Blue (71%) and Direct Red (82%). Chen and Lin (2007) used silicate/alginate sol-gel beads of *P. luteola* for the decolorization of Reactive Red 22. The rate of decolorization of the free cells decreased, while the immobilized system was static after five repeated batch cycles. Tuttolomondo et al. (2014) reported the biodegradation of Methyl Orange, Benzyl Orange, and Remazol Black by immobilized *Pseudomonas* sp. in sol-gel silica matrices due to higher expression of extracellular enzymes. The encapsulation directly protects the bacteria from toxic conditions and consecutively increases the production of enzymes involved in degradation. *Pseudomonas* sp. DY1 immobilized in the fungi (*A. oryzae*) cellular mass shows 96% decolorization in the batch cycle, still after 16 days. Inhibition test confirmed that the activity of the pellets was mainly due to the bacteria, demonstrating their stable and long-term usability for the dye treatment (Yang et al. 2011a, b). Recently, Roy et al. (2018) used immobilized *Pseudomonas* sp. in fly ash for the biodegradation of Reactive Yellow. The highest removal percentage (98.72%) was recorded in *Pseudomonas* sp. on fly ash as compared to sorption by fly ash (88.51%) and degradation by species (92.62%).

The activated carbon in combination with *P. luteola* was found to be very much effective for the adsorption and biodegradation of Reactive Red 22 (Lin and Leu 2008). Selvakumar et al. (2010) use electro-oxidation and bio-oxidation by *P. aeruginosa* for the removal of color from textile effluent having Procion Blue 2G dye. Later the treated effluents have been treated with photo-oxidation to remove the bacteria, so that water can be recycled. Similarly, Srinivasan et al. (2011) combined the sonolysis pretreatment with post-biological treatment by the mutant strain of *P. putida* in the case of Tectilon Yellow 2G.

The studies on the optimization of the conditions like temperature, pH, presence of organic compounds, carbon and nitrogen source, concentration range of dyes, and aerobic or anaerobic or both conditions are very much necessary, depending on the nature of the dye to be treated by *Pseudomonas*. Yu et al. (2001) observed that presence of nitrate at concentration 1000 mg/L inhibits the process completely, while increase in the temperature from 10 to 35 °C enhances the decolorization rate of *Pseudomonas* strain GM3. Chang et al. (2001b) found that tryptone and yeast extract enhances the decolorization process of Reactive Red 22, while retarded by the added glucose concentration and dissolved oxygen. The activity of azoreductase enzyme isolated from cell-free extract also depends upon the growth phase of bacteria. Lodato et al. (2007) proved that depletion of dye can be achieved irrespective of the initial concentration by changing the aerobic-anaerobic operating conditions. In the aerobic conditions, growth of *Pseudomonas* sp. OX1 can be achieved, while in the anaerobic conditions, depletion of dye takes place. Similarly, Lin et al. (2010) observed complete mineralization of Reactive Blue 13 by *Pseudomonas* sp. L1 in the same conditions. Joe et al. (2011) investigated the optimal conditions needed for Remazol Black B dye by *P. aeruginosa* CR-25. The maximum rate of removal occurs at 37 °C, pH7 with supplementation of peptone, yeast extract, glucose and fructose as nitrogen and carbon sources under static conditions. The same results have been observed under the above-said conditions by other workers using different species of *Pseudomonas* (Kalyani et al. 2008; Telke et al. 2009; Thao et al. 2013). Kumar Garg et al. (2012) showed that supplementation of ammonium sulfate (0.1%, w/v) and glucose (0.4% w/v) improved the decolorization of Orange II. Mishra and Maiti (2018) demonstrated that yeast extract has positive effect, while peptone and glucose have negative effect on the decolorization of Reactive Red 21 by *P. aeruginosa* 23N1. This may be due to the fact that species must have utilized peptone and glucose as primary sources of nitrogen and carbon rather than the dye molecule. Recently, Hashem et al. (2018) isolated a pH-tolerant *P. aeruginosa* KY284155 with high decolorization rate for Remazol Black B. With the addition of iron, magnesium, and yeast extract in the medium, the degradation rate was further accelerated. The heavy metals and salts at high concentrations in the medium have inhibitory effects on the decolorization of dyes (Gopinath et al. 2011). Some strains of *P. aeruginosa* were very effective in the degradation of reactive azo dyes even in the presence of heavy metals like lead, zinc, cadmium, and chromium (Maqbool et al. 2016; Hafeez et al. 2018).

The majority of the studies done in *Pseudomonas* were related to biodegradation of the dyes, but few authors have also studied the adsorption phenomena for the management of industrial effluents. Du et al. (2012) compared the adsorption capacity of live and heat-treated *Pseudomonas* sp. strain DY1 biomass for Acid Black 172. The heat-treated cells have high adsorption due to increased permeability and denatured intracellular proteins. Deepa et al. (2013) showed that 4 to 9 pH and 1 to 1000 mM NaCl concentrations have insignificant effect on the adsorption rate of Direct Red by *P. putida*. Later on, Arunarani et al. (2013) proved the same type of effect on the adsorption of Acid Blue 93 and Basic Violet 3 by the same taxa due to pH and salts. Liu et al. (2017) extracted a biosurfactant from *P. taiwanensis* L1011 and utilized it to accelerate the chemical and biological decolorization of Congo Red and Amaranth, respectively. Recently, Iqbal et al. (2018) developed a novel biosorbent using *P. aeruginosa* USM-AR2 cells immobilized on mesoporous rice husk ash silica (RHA-SiO2).

There is a lot of variability for the potential of degradation of dyes within the different genera of bacteria. Hu (1996) compared the adsorption efficiency of *Aeromonas*, *Bacillus*, *Escherichia*, *Pseudomonas*, and *Staphylococcus* for four reactive azo dyes. The dead biomass of the three genera exhibits higher adsorption capacity in the order of *Aeromonas* > *Pseudomonas* > *Escherichia*. Nachiyar and Rajkumar (2003) tested three species (*P. aeruginosa*, *P. fluorescens*, and *P. putida*) for the decolorization of Navitan Fast Blue S5R and found that *P. aeruginosa* exhibited maximum efficiency (72–92%) within 72 h. Silveira et al. (2009) compared 4 species (*P. oleovorans*, *P. putida*, *P. cepacia, and P. aeruginosa*) for the efficiency of decolorization of 14 commercial textile dyes. Among them, *P. aeruginosa* and *P. oleovorans* were more capable to decolorize ten textile dyes. The mixed consortia of *Pseudomonas*, *Acinetobacter*, *Escherichia*, *Enterobacter*, *Aspergillus*, and *Actinobacteria* were also found to significantly decolorize or degrade different kinds of azo dyes (Kadam et al. 2011; Yang et al. 2011a, b; Patel et al. 2012; Khan et al. 2014; Isaac et al. 2015; Kuppusamy et al. 2017; Sathishkumar et al. 2017).

Pseudomonas genus was also studied for the biotreatment of triphenylmethane dyes, used extensively as biological or dermatological agent, and in various processes in the food, medical, and textile industry (Sarnaik and Kanekar 1995, 1999; Yatome et al. 1981, 1990; Lin et al. 2004; Wu et al. 2009). Malachite Green and Crystal Violet dyes were extensively studied by several researchers (El-Naggar et al. 2004; Chen et al. 2007; Li et al. 2009; Huan et al. 2010; Kalyani et al. 2012; Chaturvedi et al. 2013). Enhancement of degradation of triphenylmethane dyes can be attained by adding glucose and sucrose as cosubstrates and heavy metals in the medium (Oranusi and Ogugbue 2005). Kalyani et al. (2012) showed that aminopyrine *N*-demethylase, MG reductase, and laccase enzymes were induced in *P. aeruginosa* NCIM 2074 and degraded Malachite Green into a non-toxic product. The same category of enzymes was also found to degrade heavy amounts of the dye (1800 mg/L) in *P. mendocina* (Chaturvedi et al. 2013). Li et al. (2009) isolated a strain of *Pseudomonas* sp. MDB-1 from water of an aquatic hatchery, capable of degrading various triphenylmethane dyes. Later on, *tmr2* gene encoding the enzyme (triphenylmethane reductase) was also fully characterized responsible for the biodegradation (Huan et al. 2010; Li et al. 2009). Zabłocka-Godlewska et al. (2014) compared SDz3 and Sz6 strains of *P. fluorescens* for the biodegradation of mixture containing triphenylmethane (Brilliant Green) and azo (Evans Blue) dyes. The strain Sz6 was able to degrade the dyes faster in shaken/semistatic conditions, and maximum removal (95.4%) was achieved in the case of Brilliant Green.

Various species of *Pseudomonas* were also reported for the removal of other xenobiotic compounds used for the preparation of dyes. The compounds include phenol by *P. putida* DSM 548, *Pseudomonas* CF600, and *P. stutzeri* (Sá and Boaventura 2001; Moharikar and Purohit 2003; Pazarlioğlu and Telefoncu 2005; Nowak and Mrozik 2018; Singh et al. 2018); 4-aminophenol by *Pseudomonas* ST-4 (Afzal Khan et al. 2006); pyridine by *Pseudomonas* sp. PI2 (Mohan et al. 2003); naphthalene and *p*-cresol by *P. putida* and *P. gessardii* LZ-E (Huang et al. 2016a, b;

Izmalkova et al. 2013; Surkatti and El-Naas 2014); chloroanilines by *P. putida* T57 (Nitisakulkan et al. 2014); polycyclic aromatic hydrocarbons by *P. stutzeri* (Álvarez et al. 2015); polynuclear aromatic hydrocarbons by *P. plecoglossicida* PB1 and *Pseudomonas* sp. PB2 (Nwinyi et al. 2016); and phenanthrene by *P. stutzeri* JP1 and P. *mendocina* NR802 (Mangwani et al. 2014; Kong et al. 2017).

13.4 Fungi

Many genera of fungi were also explored for the color removal from industrial effluents, especially actinomycetes and basidiomycetes (Chivukula and Renganathan 1995; McMullan et al. 2001). These organisms produce extracellular enzymes (laccase, peroxidases, and azoreductase) to catalyze dealkylation, oxidation, and hydroxylation reactions for the metabolism of dyes (Goszczynski et al. 1994). Most of the work was done for white rot fungus (*Phanerochaete*), as they are capable to degrade the majority of the azo dyes (Bumpus 1995; Banat et al. 1997; Cripps et al. 1990). The other fungal genera reported for the biodegradation of xenobiotic compounds are *Streptomyces*, *Lenzites*, *Coriolopsis*, *Neurospora*, *Penicillium*, *Pleurotus*, *Trichoderma*, and *Trametes* (Paszczynski et al. 1992; Chao and Lee 1994; Knapp and Newby 1999; Saparrat et al. 2014; He et al. 2018; Naraian et al. 2018; Pandey et al. 2018). The brown rot fungus (*Aspergillus*) has also shown potential to biodegrade a variety of toxic xenobiotic compounds and for the biotreatment of wastewater (Ali et al. 2010; Abd El-Rahim et al. 2017; Gomaa et al. 2011). Recently, Ning et al. (2018) reported biodegradation of 15 dyes by *Aspergillus flavus* A5p1 in a range of 61.7–100.0%. So there is always a need to explore the different strains/species of the *Aspergillus* for the degradations of the wide varieties of dyes (Table 13.3).

13.4.1 *Aspergillus*

The genus is composed of 340 species, widespread in diverse habitats, and reported as a pathogen, spoils food materials, and produces mycotoxins (Bennett and Klich 2003; Houbraken et al. 2016). They reproduce by asexual reproduction via conidiophores. The key to identify or classify various species of the genus is based on the size, color, and arrangement of asexual spores of conidiophores. Some species are associated with serious health problems like allergic bronchopulmonary aspergillosis, liver cancer (consumption of food containing mycotoxins), etc. (Hedayati et al. 2007). Most of the species are also used to produce beneficial products (enzymes, food fermenters, antibiotics, etc.) in biotechnology industry (Samson et al. 2014). To mention some of the species with beneficial/harmful effects are *A. flavus* (aflatoxin), *A. fumigatus* (cellulose, xylanase), *A. niger* (homologous or heterologous proteins), *A. oryzae*, *A. sojae* (food fermentation), *A. tamari* (Japanese soya sauces), and *A. terreus* (lovastatin, terrein) (Park et al. 2017). The present section reviews the diversity found within the *Aspergillus* species for the elimination of hazardous dyes from the industrial effluents (Table 13.3).

Initial studies for the wastewater treatment were mainly focused on the white rot fungus group, as they have lignin-degrading enzymes for the oxidation of organic compounds (Bumpus and Aust 1987). *Aspergillus* genus (brown rot fungi) was also explored for the removal of dyes in the industrial effluents. Ryu and Weon (1992) analyzed four species of *Aspergillus* (six strains) and one species of *Phanerochaete* (two strains) for the biodegradation of three azo dyes and stated that the former genus was much more effective in the process. Mainly two processes for the treatment of dyes in the solution or synthetic effluents were studied extensively, either biosorption or biodegradation (Conatao and Corso 1996; Fu and Viraraghavan 2000, 2002a; Sumathi and Manju 2000; Zope et al. 2007; Esmaeili and Kalantari 2011; Almeida and Corso 2014). The biosorption of dyes was influenced by their chemical structure and functional group on the surface of fungus (Fu and Viraraghavan 2002b, 2003). Parshetti et al. (2007) observed faster adsorption rate in *A. ochraceus* in the shaking conditions. The treatment of *Aspergillus* species with immobilization beads, autoclaving, and specific compounds also accelerates the process of decolorization (Wang and Hu 2007; Wang et al. 2008; Patel and Suresh 2008). Yang et al. (2011a, b) demonstrated higher biosorption capacity in the CDAB (cetyldimethylammonium bromide) modified biomass of *A. oryzae*. The same type of result was seen by Huang et al. (2016a, b) while investigating the effect of heavy salts, metals, and SDS on the adsorption kinetics of chemically modified (cetyltrimethylammonium bromide) *A. versicolor*. They found a close relationship between low pH (2.0) and heavy metals on the biosorption rate. The chemical modification increases the surface area and functional groups. Naskar and Majumder (2017) used response surface methodology for *A. niger* and demonstrated that adsorption rate depends upon the concentration of biomass, temperature, and pH of the solution. Further, they also revealed that amine and carboxyl groups play an important role in dye sorption along with electrostatic interactions. The same type of phenomena was observed by the authors using different dyes and the same species (Xiong et al. 2010; Mahmoud et al. 2017). The high temperature and low pH range (1–3) in the solution speed up the uptake of the dyes, as the biosorption is mostly endothermic (Akar et al. 2009). This type of condition increases the kinetic energy and diffusion rate (Ramya et al. 2007; Aksu and Karabayır 2008; Abdallah and Taha 2012). Contradictory to this, other authors reported optimal temperature (28–30 °C) and pH (5) as much more favorable condition for the biodegradation of azo dyes (Ali et al. 2007a, b; Ameen and Alshehrei 2017; Sharma et al. 2009) by four *Aspergillus* spp. The nutritional condition needs to be standardized as sources of nitrogen and carbon in the medium, as they are also a detrimental factor for the rate of dye removal (Kaushik and Malik 2010, 2011). Gomaa et al. (2017) demonstrated the role of calcium chloride as stress response in *A. niger* and high removal efficiency for commercial dye Malachite Green.

The live fungal strains were extensively studied for the decolorization of dyes from industrial effluents; however, some workers used pellets and dead biomass for the process and found promising results as compared to the living strains (Abdallah and Taha 2012; Abdel Ghany and Al Abboud 2014; Lu et al. 2017). The formation of bioflocculants and silver and zinc oxide nanoparticles using different *Aspergillus*

Table 13.3 Removal of different dyes by *Aspergillus* species

Sr. no.	*Aspergillus* species	Dyes (concentration)	Mechanism (enzyme(s))	Removal time/ percentage	By-product	References
1.	*A.* sp. CB-TKL-1	Methyl Violet (5–30 mg L^{-1})	Biosorption and biodegradation	24 h (100%)	*N*-Demethylated compounds	Kumar et al. (2011, 2012)
		Brilliant Green (5–20 mg L^{-1})		72 h (99.24%)		
2.	*A.* sp. TS-A CGMCC 12,964	Mordant Yellow 1	Degradation (lignin oxidases)	1 h	–	Kang et al. (2017)
3.	*A.* sp.	Reactive Red, Yellow, Black, Blue, Coloron Violet, and Black	Decolorization	24 h (9–99%)	–	Ramya et al. (2007)
4.	*A. carbonarius* M333	Congo Red (25–125 mg L^{-1})	Biosorption	–	–	Bouras et al. (2017)
5.	*A. flavus SA2*	Drimarene Blue K2RL (50 mg L^{-1})	Biosorption and biodegradation	24 h (71.3%)	1,4-Dihydroxyanthraquinone, 2,3-dihydro-9,10-dihydroxy-1,4-anthracenedione, phthalic acid, and benzoic acid	Andleeb et al. (2012)
6.	*A. flavus*	Reactive Red 198 (25–100 ppm)	Decolorization	24 h (84.96%)	–	Esmaeili and Kalantari (2011)
7.	*A. flavus* A5p1	Direct Blue 71, Direct Blue 86, and Reactive Blue (100–1000 mg L^{-1})	Biosorption and biodegradation	(61.7–100.0%)		
8.	*A. foetidus*	Drimarene Red BF F3B1, Drimarene Navy Blue BF F2G1, and Drimarene Black HFGR1 (50, 100 mg L^{-1})	Decolorization	72 h (85–95%)	–	Sumathi and Manju (2000)
		Drimarene Red (50 mg L^{-1})	Decolorization	24 h (50%)	–	Bidisha et al. (2006)

Sr. no.	*Aspergillus* species	Dyes (concentration)	Mechanism (enzyme(s))	Removal time/ percentage	By-product	References
		Reactive Black 5 (100 mg L^{-1})	Biosorption	2 h (97%)	–	Patel and Suresh (2008)
9.	*A. fumigatus*	Methylene Blue (5 mg L^{-1})	Biosorption	90 min. (80%)	–	Abdallah and Taha (2012)
		Reactive Brilliant Red (96.6 mg L^{-1})	Biosorption	48 h	–	Wang et al. (2008)
10.	*A. fumigatus fresenius*	Acid Red 151 (100–200 mg L^{-1})	Biosorption	5 days (41.62–84.77%)	–	Sharma et al. (2009)
11.	*A. lentulus* FJ172995	Acid Blue 120 (100–300 mg L^{-1})	Decolorization	24 h (84.48–99.97%)	–	Kaushik and Malik (2011)
12.	*A. niger*	Basic fuchsin, nigrosin, and Malachite Green	Biosorption	6 days	—	Rani et al. (2014)
		Procion Blue MX-G (100 μg mL^{-1})	Biosorption	120 min. (99%)	–	Conatao and Corso (1996)
		Basic Blue 9	Biosorption	30 h	–	Fu and Viraraghavan (2000)
		Congo Red (50 mg L^{-1})	Biosorption	42 h	–	Fu and Viraraghavan (2002a)
		Acid Red 27, Acid Red 151, Reactive Blue 19, and Reactive Blue 81 (20–100 mg L^{-1})	Biosorption and biodegradation	9 days (70–80%)	–	Ali and El-Mohamedy (2012)

(continued)

Table 13.3 (continued)

Sr. no.	*Aspergillus* species	Dyes (concentration)	Mechanism (enzyme(s))	Removal time/ percentage	By-product	References
		Direct Blue 199 (400 mg L^{-1})	Biosorption	4 h	–	Xiong et al. (2010)
		Direct Red (100–1000 mg L^{-1})	Biosorption and biodegradation	7 days (58.6–99.69%)	–	Mahmoud et al. (2017)
		Synozol Red HF6BN and Synozol Yellow HF2GR	Biosorption	18 h (88%)	–	Khalaf (2008)
13.	*A. niger grp*	Reactive Red 195 and Reactive Green 11 (50–200 ppm)	Decolorization	72 h (75.4, 62.9%)	–	Zope et al. (2007)
14.	*A. niger* SA1	Acid Red 151, Orange II, Drimarene Blue K2RL, and Sulfur Black	Biosorption and Biodegradation	8 days (9.33–68.64%)	–	Ali et al. (2007a, b)
15.	*A. niger* ZJUBE-1	Congo Red	Biosorption and biodegradation	(>98.5%)	–	Lu et al. (2017)
16.	*A. ochraceus*	Reactive Blue 25 (100 mg L^{-1})	Biosorption and biodegradation (lignin peroxidase, laccase, and tyrosinase)	7 h (100%)	Phthalimide and diisobutyl phthalate	Parshetti et al. (2007)
17.	*A. parasiticus*	Reactive Red 198 (100–300 mg L^{-1})	Biosorption	50 min. (98.57%)	–	Akar et al. (2009)
18.	*A. sojae* B-10	Amaranth, Sudan III, and Congo Red	Decolorization	4–8 days (100%)	–	Ryu and Weon (1992)
19.	*A. versicolor*	Reactive Black 5 (200 mg L^{-1})	Biosorption	420 min. (98%)	–	Huang et al. (2016a, b)
20.	*A. wentii*	Brilliant Blue G (119.3–544.8 mg L^{-1})	Biosorption	3 h	–	Khambhaty et al. (2012)

Sr. no.	*Aspergillus* species	Dyes (concentration)	Mechanism (enzyme(s))	Removal time/ percentage	By-product	References
21.	*A. flavus* *A. terreus*	Acid Red 151 and Orange II (20 mg L^{-1})	Biosorption and biodegradation	96 h (48–97%)	α-Naphthol, sulfanilic acid, and aniline	Ali et al. (2010)
22.	*A. lentulus* *A. fumigatus*	Reactive Remazol Red, Reactive Blue, and Reactive Yellow dyes	Biosorption and bioaccumulation	4 h	–	Mathur et al. (2018)
23.	*A. niger* *A. terreus* *A. oryzae* *A. flavus* *A. fumigatus* *A. alabamensis*	Reactive Red, tartrazine, Direct Blue, Naphthol Blue Black, Direct Red, Trypan Blue, Direct Violet, Janus Green, Reactive Blue, Alizarin Yellow, Reactive Orange, Evans Blue, Fast Green, Brilliant Green, Methyl Red, Safranin, Crystal Violet, pararosaniline, Allura Red, and Ponceau S (100 μg/mL)	Biosorption and biodegradation	5 days (0.2–93.3%)	–	Abd El-Rahim et al. (2017)
24.	*A. niger* *A. terreus*	Procion Red MX-5B (200 μg/mL)	Biosorption and biodegradation (azoreductase)	336 h (98%)	Primary amines	Almeida and Corso (2014)
25.	*A. niger* *A. terreus* *A. flavus* *A. fumigatus*	Reactive Red 120 (100 ppm)	Biodegradation	7 days (84–86%)	Sodium 2-aminobenzenesulfonate	Ameen and Alshehrei (2017)
26.	*A. flavus* *A. fumigatus* *A. ochraceus* *A. puniceus* *A. sulphureus* *A. versicolor*	Reactive Blue 19, Poly R-478, Poly S-119, Acid Blue 113, Acid Blue 225, Acid Red 111, Reactive Blue 214, Reactive Blue 41, Reactive Blue 49, Reactive Red 243, Direct Blue 81, and Direct Red 80 (200, 1000 ppm)	Biosorption	2 h (90%)	–	Anastasi et al. (2009)

spp. has also the potential for the color removal from industrial effluents (Deng et al. 2005; Muthu Kumara Pandian et al. 2016; Kalpana et al. 2018a, b). Copete-Pertuz et al. (2019) demonstrated that *A. terreus* in combination with *Trichoderma viride* can act as a co-inducer for *Leptosphaerulina* sp. ligninolytic enzyme activity and improved removal of Reactive Black 5 dye.

Survey of literature reveals that most of the studies were related to the biosorption mechanism rather than the degradation. The metabolites formed during degradation process are shown in Table 13.3. The enzymes involved in the biodegradation were laccase, manganese peroxides, and lignin-modifying enzymes, which mineralize synthetic lignin of dyes (Ali and El-Mohamedy 2012; Hasanin et al. 2019). Azoreductase is one of the key enzymes found in the degradation pathways of the organism. Ameen and Alshehrei (2017) found laccase and azoreductase to be involved in the degradation of Reactive Red 120 into sodium 2-aminobenzenesulfonate. Tamayo-Ramos et al. (2012) characterized three forms of laccase-like multicopper oxidase enzymes having high catalytic activity for several phenolic compounds and synthetic dyes. The optimization process for the high production and activity of laccase enzyme has been done for several *Aspergillus* species. The factors associated are pH, temperature, carbon and nitrogen sources, inoculum size, etc. (Jin and Ning 2013; Benghazi et al. 2013; Kumar et al. 2016). Recently, Abd El-Rahim et al. (2017) isolated 18 strains belonging to 6 species from the wastewater sample and evaluated them against 20 azo dyes. The most resistant dye was Fast Green azo dye, and easily degradable dyes were Direct Violet and Methyl Red. The decolorization process was enhanced by glucose supplementation, and the limiting factor was a nitrogen source, as in its absence the strains were unable to produce lignin peroxidase enzyme. The high pH has been also shown to be related to the low formation of residual products (Ali et al. 2007a, b).

The different *Aspergillus* species have shown very much diversity in the biodegradation of various dyes. Anastasi et al. (2009) compared five species of mitosporic fungi (*Penicillium*, *Cladosporium*, and *Aspergillus*) for the removal of nine industrial and two model dyes. They found that *A. ochraceus* and *A. flavus* were efficient for the decolorization of all the dyes tested and one species, i.e., *A. ochraceus*, causes over 90% decolorization against simulated effluents. Similarly, other workers found the maximum potential of *Aspergillus* as compared to *Penicillium* (Ali et al. 2010; Gomaa et al. 2011; Ali and El-Mohamedy 2012). Khalaf (2008) tested the effectiveness of *Spirogyra* sp. (green algae) and *A. niger* against the reactive dye (Synozol) in textile wastewater. The autoclaved biomass of the both species exhibited 88% and 85% dye removal, respectively. Some species have higher absorption capacity, but still they lack the ability to degrade them into non-toxic metabolites (Almeida and Corso 2014).

The degraded products should be checked for the toxicity assays, as decolorization does not always lead to the absence of toxicity, rather forming incomplete toxic metabolites (Almeida and Corso 2014). The extracellular enzymes were found to degrade triphenylmethane dye by stepwise demethylation into non-toxic *N*-demethylated products (Kumar et al. 2011, 2012). Andleeb et al. (2012)

investigated the toxicity of degraded products formed during biodegradation of Drimarene Blue dye by *A. flavus*. As compared to dye treatment, the germination and morphological characteristics in *Lolium perenne* were somewhat near to the untreated. Similarly, Parshetti et al. (2007) observed that germination of *Phaseolus mungo* was high or near to control in comparison to the Malachite Green treatment.

13.5 Conclusion

The treatment of industrial effluents with cost-effective methods is the urgent need of the society. The literature shows that aerobic and anaerobic conditions were well utilized by algae, bacteria, and fungi for the management of dyes. The effluents also serve as a growth substrate or also can be used to extract biomass. The integrated approach of remediation as successive treatment along with extraction of enzymes, lipids, and biofuels seems to be the best practice for sustainable development. The mixed consortium of best strains of algae, bacteria, and fungi should be tested for the degradation of toxic dyes. Genetically engineered strains may be used for the degradation of toxic amines in the severe environmental conditions. Toxicity assays clearly show which strain is best for the future applications to clear the water for recycling.

References

Abd El-Rahim WM, Moawad H, Abdel Azeiz AZ et al (2017) Optimization of conditions for decolorization of azo-based textile dyes by multiple fungal species. J Biotechnol 260:11–17

Abdallah R, Taha S (2012) Biosorption of methylene blue from aqueous solution by nonviable *Aspergillus fumigatus*. Chem Eng J 195-196:69–76

Abdel Ghany TM, Al Abboud MA (2014) Capacity of growing, live and dead fungal biomass for safranin dye decolourization and their impact on fungal metabolites. Aus J Basic Appl Sci 8:489–499

Acuner E, Dilek F (2004) Treatment of tectilon yellow 2G by *Chlorella vulgaris*. Process Biochem 39:623–631

Adedayo O, Javadpour S, Taylor C et al (2004) Decolourization and detoxification of methyl red by aerobic bacteria from a wastewater treatment plant. World J Microbiol Biotechnol 20:545–550

Afzal Khan S, Hamayun M, Ahmed S (2006) Degradation of 4-aminophenol by newly isolated *Pseudomonas* sp. strain ST-4. Enzym Microb Technol 38:10–13

Akar ST, Akar T, Cabu A (2009) Decolorization of a textile dye, RR198 by Aspergillus parasiticus fungal biosorbent. J Chem Eng 2:399–405

Aksu Z, Karabayır G (2008) Comparison of biosorption properties of different kinds of fungi for the removal of Gryfalan Black RL metal-complex dye. Bioresour Technol 99:7730–7741

Aksu Z, Tezer S (2005) Biosorption of reactive dyes on the green alga *Chlorella vulgaris*. Process Biochem 40:1347–1361

Ali NF, El-Mohamedy RSR (2012) Microbial decolourization of textile waste water. J Saudi Chem Soc 16(2):117–123

Ali N, Hameed A, Ahmed S, Khan AG (2007a) Decolorization of structurally different textile dyes by Aspergillus niger SA1. World J Microbiol Biotechnol 24(7):1067–1072

Ali N, Ikramullah, Lutfullah G et al (2007b) Decolorization of acid red 151 by *Aspergillus niger* SA1 under different physicochemical conditions. World J Microbiol Biotechnol 24:1099–1105

Ali N, Hameed A, Ahmed S (2010) Role of brown-rot fungi in the bioremoval of azo dyes under different conditions. J Microbiol 4:907–915

Almeida EJR, Corso CR (2014) Comparative study of toxicity of azo dye Procion Red MX-5B following biosorption and biodegradation treatments with the fungi *Aspergillus niger* and *Aspergillus terreus*. Chemosphere 112:317–322

Álvarez MS, Rodríguez A, Sanromán MÁ et al (2015) Simultaneous biotreatment of polycyclic aromatic hydrocarbons and dyes in a one-step bioreaction by an acclimated *Pseudomonas* strain. Bioresour Technol 198:181–188

Ameen F, Alshehrei F (2017) Biodegradation optimization and metabolite elucidation of Reactive Red 120 by four different *Aspergillus* species isolated from soil contaminated with industrial effluent. Ann Microbiol 67:303–312

Anastasi A, Prigione V, Casieri L et al (2009) Decolourisation of model and industrial dyes by mitosporic fungi in different culture conditions. World J Microbiol Biotechnol 25:1363–1374

Andleeb S, Atiq N, Robson GD, Ahmed S (2012) An investigation of anthraquinone dye biodegradation by immobilized Aspergillus flavus in fluidized bed bioreactor. Environ Sci Pollut Res 19(5):1728–1737

Arunarani A, Chandran P, Ranganathan BV et al (2013) Bioremoval of basic violet 3 and Acid Blue 93 by *Pseudomonas putida* and its adsorption isotherms and kinetics. Colloids Surf B: Biointerfaces 102:379–384

Asgher M (2012) Biosorption of reactive dyes: a review. Water Air Soil Pollut 223:2417. https://doi.org/10.1007/s11270-011-1034-z

Banat IM, Nigam P, McMullan G et al (1997) The isolation of thermophilic bacterial cultures capable of textile dyes decolorization. Environ Int 23:547–551

Ben Mansour H, Corroler D, Barillier D et al (2007) Evaluation of genotoxicity and pro-oxidant effect of the azo dyes: acids yellow 17, violet 7 and orange 52, and of their degradation products by *Pseudomonas putida* mt-2. Food Chem Toxicol 45:1670–1677

Ben Mansour H, Corroler D, Barillier D et al (2009a) Influence of the chemical structure on the biodegradability of acids yellow 17, violet 7 and orange 52 by *Pseudomonas putida*. Ann Microbiol 59:9–15

Ben Mansour H, Mosrati R, Corroler D et al (2009b) *In vitro* mutagenicity of Acid Violet 7 and its degradation products by *Pseudomonas putida* mt-2: correlation with chemical structures. Environ Toxicol Pharmacol 27:231–236

Benghazi L, Record E, Suárez A, Gomez-Vidal JA, Martínez J, de la Rubia T (2013) Production of the Phanerochaete flavido-alba laccase in *Aspergillus niger* for synthetic dyes decolorization and biotransformation. World J Microbiol Biotechnol 30(1):201–211

Bennett JW, Klich M (2003) Mycotoxins. Clin Microbiol Rev 16:497–516

Bidisha C, Sreeranjani R, Shaik A et al (2006) Bioaccumulation and biosorption of drimarene red dye by *Aspergillus foetidus*. Int J Environ Pollut 28:517–533

Bouras HD, Yeddou AR, Bouras N (2017) Biosorption of Congo red dye by *Aspergillus carbonarius* M333 and *Penicillium glabrum* Pg1: kinetics, equilibrium and thermodynamic studies. J Taiwan Inst Chem E 80:915–923

Bumpus JA (1995) Microbial degradation of azo dyes. In: Singh VP (ed) Biotransformations: microbial degradation of health risk compounds. Elsevier Science, Amsterdam, pp 157–176

Bumpus JA, Aust SD (1987) Biodegradation of environmental pollutants by the white rot fungus *Phanerochaete chrysosporium*: involvement of the lignin degrading system. Bio Essays 6:166–170

Carmen Z, Daniel S (2012) Textile organic dyes—characteristics, polluting effects and separation/elimination procedures from industrial effluents—a critical overview, organic pollutants ten years after the Stockholm convention, Tomasz Puzyn and Aleksandra Mostrag-Szlichtyng, IntechOpen, doi: 10.5772/32373. Available from: https://www.intechopen.com/books/organic-pollutants-ten-years-after-the-stockholm-convention-environmental-and-analytical-update/textile-organic-dyes-characteristics-polluting-effects-and-separation-elimination-procedures-from-in

Chang JS, Chou C, Chen SY (2001a) Decolorization of azo dyes with immobilized *Pseudomonas luteola*. Process Biochem 36:757–763
Chang JS, Chou C, Lin YC et al (2001b) Kinetic characteristics of bacterial azo-dye decolorization by *Pseudomonas luteola*. Water Res 35:2841–2850
Chao WL, Lee SL (1994) Decoloration of azo dyes by three white rot fungi: influence of carbon source. World J Microbiol Biotechnol 10:556–559
Chaturvedi V, Bhange K, Bhatt R et al (2013) Biodetoxification of high amounts of malachite green by a multifunctional strain of *Pseudomonas mendocina* and its ability to metabolize dye adsorbed chicken feathers. J Environ Chem Eng 1:1205–1213
Chen BY (2002) Understanding decolorization characteristics of reactive azo dyes by *Pseudomonas luteola*: toxicity and kinetics. Process Biochem 38:437–446
Chen BY (2006) Toxicity assessment of aromatic amines to *Pseudomonas luteola*: chemostat pulse technique and dose–response analysis. Process Biochem 41:1529–1538
Chen JP, Lin YS (2007) Decolorization of azo dye by immobilized *Pseudomonas luteola* entrapped in alginate–silicate sol–gel beads. Process Biochem 42:934–942
Chen H, Hopper SL, Cerniglia CE (2005) Biochemical and molecular characterization of an azoreductase from *Staphylococcus aureus*, a tetrameric NADPH—dependent flavoprotein. Microbiology 151:1433–1441
Chen CC, Liao HJ, Cheng CY et al (2007) Biodegradation of Crystal Violet by *Pseudomonas putida*. Biotechnol Lett 29:391–396
Chivukula M, Renganathan V (1995) Phenolic azo dye oxidation by laccase from *Pyricularia oryzae*. Appl Environ Microbiol 61:4374–4377
Chu WL, See YC, Phang SM (2009) Use of immobilised *Chlorella vulgaris* for the removal of colour from textile dyes. J Appl Phycol 21:641. https://doi.org/10.1007/s10811-008-9396-3
Conatao M, Corso CR (1996) Studies of adsorptive interaction between *Aspergillus niger* and the reactive azo dye procion blue MX-G. Ecletica Quim 21:97–102
Copete-Pertuz LS, Alandete-Novoa F, al PJ (2019) Enhancement of ligninolytic enzymes production and decolourising activity in *Leptosphaerulina* sp. by co–cultivation with *Trichoderma viride* and *Aspergillus terreus*. Sci Total Environ 646:1536–1545
Coughlin MF, Kinkle BK, Bishop PL (2003) High performance degradation of azo dye acid orange 7 and sulfanilic acid in a laboratory scale reactor after seeding with cultured bacterial strains. Water Res 37:2757–2763
Cripps C, Bumpus JA, Aust SD (1990) Biodegradation of azo and heterocyclic dyes by *Phanerochaete chrysosporium*. Appl Environ Microbiol 56:1114–1118
Daneshvar N, Khataee AR, Rasoulifard MH et al (2007) Biodegradation of dye solution containing malachite green: optimization of effective parameters using Taguchi method. J Hazard Mater 143:214–219
Daneshvar E, Antikainen L, Koutra E et al (2018) Investigation on the feasibility of *Chlorella vulgaris* cultivation in a mixture of pulp and aquaculture effluents: treatment of wastewater and lipid extraction. Bioresour Technol 255:104–110
de Andrade CJ, de Andrade LM (2017) An overview on the application of genus Chlorella in biotechnological processes
Deepa K, Chandran P, Sudheer Khan S (2013) Bioremoval of Direct Red from aqueous solution by *Pseudomonas putida* and its adsorption isotherms and kinetics. Ecol Eng 58:207–213
Deng S, Yu G, Ting YP (2005) Production of a bioflocculant by *Aspergillus parasiticus* and its application in dye removal. Colloids Surf B: Biointerfaces 44:179–186
Dilek FB, Taplamacioglu HM, Tarlan E (1999) Colour and AOX removal from pulping effluents by algae. Appl Microbiol Biotechnol 52:585–591
Du LN, Yang YY, Li G et al (2010) Optimization of heavy metal-containing dye Acid Black 172 decolorization by *Pseudomonas* sp. DY1 using statistical designs. Int Biodeterior Biodegrad 64:566–573
Du LN, Wang B, Li G et al (2012) Biosorption of the metal-complex dye Acid Black 172 by live and heat-treated biomass of *Pseudomonas* sp. strain DY1: kinetics and sorption mechanisms. J Hazard Mater 205-206:47–54

El-Kassas HY, Mohamed LA (2014) Bioremediation of the textile waste effluent by *Chlorella vulgaris*. Egypt J Aquat Res 40:301–308

El-Naggar MA, El-Aasar SA, Barakat KI (2004) Bioremediation of crystal violet using air bubble bioreactor packed with *Pseudomonas aeruginosa*. Water Res 38:4313–4322

El-Sheekh MM, Gharieb MM, Abou-El-Souod GW (2009) Biodegradation of dyes by some green algae and cyanobacteria. Int Biodeterior Biodegradation 63:699–704

Esmaeili A, Kalantari M (2011) Bioremoval of an azo textile dye, Reactive Red 198, by *Aspergillus flavus*. World J Microbiol Biotechnol 28:1125–1131

Fazal T, Mushtaq A, Rehman F et al (2018) Bioremediation of textile wastewater and successive biodiesel production using microalgae. Renew Sustain Energy Rev 82:3107–3126

Fu YZ, Viraraghavan T (2000) Removal of a dye from aqueous solution by the fungus *Aspergillus niger*. Water Qual Res J Can 35:95–111

Fu Y, Viraraghavan T (2002a) Removal of Congo Red from an aqueous solution by fungus Aspergillus niger. Adv Environ Res 7:239–247

Fu Y, Viraraghavan T (2002b) Dye biosorption sites in *Aspergillus niger*. Bioresour Technol 82:139–145

Fu Y, Viraraghavan T (2003) Column studies for biosorption of dyes from aqueous solutions on immobilized *Aspergillus niger* fungal biomass. Water South Africa 29:465–472

Gao QT, Wong YS, Tam NFY (2011) Removal and biodegradation of nonylphenol by immobilized *Chlorella vulgaris*. Bioresour Technol 102:10230–10238

Godheja J, Shekhar SK, Siddiqui SA et al (2016) Xenobiotic compounds present in soil and water: a review on remediation strategies. J Environ Anal Toxicol 6:5. https://doi.org/10.4172/2161-0525.1000392

Gomaa OM, Momtaz OA, Kareem HAE et al (2011) Isolation, identification, and biochemical characterization of a brown rot fungus capable of textile dye decolorization. World J Microbiol Biotechnol 27:1641–1648

Gomaa OM, Selim NS, Wee J et al (2017) RNA Seq analysis of the role of calcium chloride stress and electron transport in mitochondria for malachite green decolorization by *Aspergillus niger*. Fungal Genet Biol 105:1–7

Gopinath KP, Kathiravan MN, Srinivasan R et al (2011) Evaluation and elimination of inhibitory effects of salts and heavy metal ions on biodegradation of Congo red by *Pseudomonas* sp. mutant. Bioresour Technol 102:3687–3693

Goszczynski S, Paszczynski A, Pasti-Grigsby MB et al (1994) New pathway for degradation of sulfonated azo dyes by microbial peroxidases of by *Phanerochaete chrysosporium* and *Streptomyces chromofuscus*. J Bacteriol 176:1339–1347

Hafeez F, Farheen H, Mahmood F et al (2018) Isolation and characterization of a lead (Pb) tolerant *Pseudomonas aeruginosa* strain HF5 for decolorization of reactive red-120 and other azo dyes. Ann Microbiol 68:943–952

Hai FI, Yamamoto K, Fukushi K (2007) Hybrid treatment systems for dye wastewater. Crit Rev Environ Sci Technol 37:315–377

Hanan HO (2008) Algal decolorization and degradation of monoazo and diazo dyes. Pak J Biol Sci 11:1310–1316

Hasanin MS, Darwesh OM, Matter IA et al (2019) Isolation and characterization of non-cellulolytic *Aspergillus flavus* EGYPTA5 exhibiting selective ligninolytic potential. Biocatal Agri Biotechnol 17:160–167

Hashem RA, Samir R, Essam TM et al (2018) Optimization and enhancement of textile reactive Remazol black B decolorization and detoxification by environmentally isolated pH tolerant *Pseudomonas aeruginosa* KY284155. AMB Express 8:83. https://doi.org/10.1186/s13568-018-0616-1

He X, Song C, Li Y et al (2018) Efficient degradation of Azo dyes by a newly isolated fungus *Trichoderma tomentosum* under non-sterile conditions. Ecotox Environ Safety 150:232–239

Hedayati MT, Pasqualotto AC, Warn PA et al (2007) *Aspergillus flavus*: human pathogen, allergen and mycotoxin producer. Microbiology 153:1677–1692

Heimann K, Huerlimann R (2015) Microalgal classification: major classes and genera of commercial microalgal species. In: Se-Kwon K (ed) Handbook of marine microalgae: biotechnolgy advances. Academic Press, London, UK, pp 25–41
Hernández-Zamora M, Perales-Vela HV, Flores-Ortíz CM et al (2014) Physiological and biochemical responses of *Chlorella vulgaris* to Congo red. Ecotoxicol Environ Saf 108:72–77
Hernández-Zamora M, Cristiani-Urbina E, Martínez-Jerónimo F et al (2015) Bioremoval of the azo dye Congo Red by the microalga *Chlorella vulgaris*. Environ Sci Pollut Res Int 22:10811–10823
Horník M, Šuňovská A, Partelová D et al (2013) Continuous sorption of synthetic dyes on dried biomass of microalga *Chlorella pyrenoidosa*. Chem Pap 67:254–264
Houbraken J, Samson RA, Yilmaz N (2016) Taxonomy of *Aspergillus*, *Penicillium* and *Talaromyces* and its significance for biotechnology. In de Vries RP, Gelber IB, Andersen MR (eds), *Aspergillus* and *Penicillium* in the post-genomic era (pp. 1-16). Caister, UK, Academic Press
Hsueh CC, Chen BY (2007) Comparative study on reaction selectivity of azo dye decolorization by *Pseudomonas luteola*. J Hazard Mater 141:842–849
Hsueh CC, Chen BY (2008) Exploring effects of chemical structure on azo dye decolorization characteristics by *Pseudomonas luteola*. J Hazard Mater 154:703–710
Hu TL (1994) Decolourization of reactive azo dyes by transformation of *Pseudomonas luteola*. Bioresour Technol 49:47–51
Hu TL (1996) Removal of reactive dyes from aqueous solution by different bacterial genera. Water Sci Technol 34:89–95
Hu TL (1998) Degradation of azo dye RP_2B by *Pseudomonas luteola*. Water Sci Technol 38:229–306
Huan M, Lian-Tai L, Cai-Fang Y et al (2010) Biodegradation of malachite green by strain Pseudomonas sp. K9 and cloning of the tmr2 gene associated with an ISPpu12. World J Microbiol Biotechnol 27:1323–1329
Huang H, Wu K, Khan A et al (2016a) A novel *Pseudomonas gessardii* strain LZ-E simultaneously degrades naphthalene and reduces hexavalent chromium. Bioresour Technol 207:370–378
Huang J, Liu D, Lu J et al (2016b) Biosorption of reactive black 5 by modified Aspergillus versicolor biomass: kinetics, capacity and mechanism studies. Colloids Surf A Physicochem Eng Aspect 492:242–248
Idaka E, Ogawa T, Horitsu H (1987a) Reductive metabolism of aminoazobenzenes by *Pseudomonas cepacia*. Bull Environ Contam Toxicol 39:100–107
Idaka E, Ogawa T, Horitsu H (1987b) Oxidative pathway after reduction of p-aminoazobenzene by *Pseudomonas cepacia*. Bull Environ Contam Toxicol 39:108–113
Iqbal A, Sabar S, Mun-Yee MK et al (2018) *Pseudomonas aeruginosa* USM-AR2/SiO 2 biosorbent for the adsorption of methylene blue. J Environ Chem Eng 6:4908–4916
Isaac P, Martínez FL, Bourguignon N et al (2015) Improved PAHs removal performance by a defined bacterial consortium of indigenous *Pseudomonas* and actinobacteria from Patagonia, Argentina. Int Biodeterior Biodegradation 101:23–31
Işik M, Sponza DT (2003) Effect of oxygen on decolorization of azo dyes by *Escherichia coli* and *Pseudomonas* sp. and fate of aromatic amines. Process Biochem 38:1183–1192
Izmalkova TY, Sazonova OI, Nagornih MO, Sokolov SL, Kosheleva IA, Boronin AM (2013) The organization of naphthalene degradation genes in Pseudomonas putida strain AK5. Res Microbiol 164(3):244–253
Jin X, Ning Y (2013) Laccase production optimization by response surface methodology with Aspergillus fumigatus AF1 in unique inexpensive medium and decolorization of different dyes with the crude enzyme or fungal pellets. J Hazard Mater 262:870–877
Jinqi L, Houtian L (1992) Degradation of azo dyes by algae. Environ Pollut 75:273–278
Joe J, Kothari RK, Raval CM, Kothari CR (2011) Decolourization of textile dye Remazol black B by Pseudomonas aeruginosa CR-25 isolated from the common effluent treatment plant. J Bioremed Biodegrade 2:118. https://doi.org/10.4172/2155-6199.1000118

Kadam AA, Telke AA, Jagtap SS et al (2011) Decolorization of adsorbed textile dyes by developed consortium of *Pseudomonas* sp. SUK1 and *Aspergillus ochraceus* NCIM-1146 under solid state fermentation. J Hazard Mater 189:486–494

Kalme S, Ghodake G, Govindwar S (2007a) Red HE7B degradation using desulfonation by *Pseudomonas desmolyticum* NCIM 2112. Int Biodeterior Biodegrad 60:327–333

Kalme SD, Parshetti GK, Jadhav SU et al (2007b) Biodegradation of benzidine based dye Direct Blue-6 by *Pseudomonas desmolyticum* NCIM 2112. Bioresour Technol 98:1405–1410

Kalme S, Jadhav S, Jadhav M et al (2009) Textile dye degrading laccase from *Pseudomonas desmolyticum* NCIM 2112. Enzym Microb Technol 44:65–71

Kalpana VN, Kataru BAS, Sravani N et al (2018a) Biosynthesis of zinc oxide nanoparticles using culture filtrates of *Aspergillus niger*: antimicrobial textiles and dye degradation studies. Open Nano 3:48–55

Kalpana VN, Kataru BAS, Sravani N (2018b) Biosynthesis of zinc oxide nanoparticles using culture filtrates of *Aspergillus niger*: antimicrobial textiles and dye degradation studies. OpenNano 3:48–55

Kalyani DC, Patil PS, Jadhav JP et al (2008) Biodegradation of reactive textile dye Red BLI by an isolated bacterium *Pseudomonas* sp. SUK1. Bioresour Technol 99:4635–4641

Kalyani DC, Telke AA, Jadhav JP et al (2009) Ecofriendly biodegradation and detoxification of Reactive Red 2 textile dye by newly isolated *Pseudomonas* sp. SUK1. J Hazard Mater 163:735–742

Kalyani DC, Phugare SS, Shedbalkar UU et al (2011) Purification and characterization of a bacterial peroxidase from the isolated strain *Pseudomonas* sp. SUK1 and its application for textile dye decolorization. Ann Microbiol 61:483–491

Kalyani DC, Telke AA, Surwase SN et al (2012) Effectual decolorization and detoxification of triphenylmethane dye malachite green (MG) by *Pseudomonas aeruginosa* NCIM 2074 and its enzyme system. Clean Techn Environ Policy 14:989–1001

Kang Y, Xu X, Pan H, Tian J, Tang W, Liu S (2017) Decolorization of mordant yellow 1 using. TS-A CGMCC 12964 by biosorption and biodegradation. Bioengineered 9(1):222–232

Kanhere J, Gopinathan R, Banerjee J (2014) Cytotoxicity and genotoxicity of malachite green on non-target aquatic organisms: *Chlorella pyrenoidosa* and *Daphnia magna*. Water Air Soil Pollut 225:2134. https://doi.org/10.1007/s11270-014-2134-3

Katheresan V, Kansedo J, Lau SY (2018) Efficiency of various recent wastewater dye removal methods: a review. J Environ Chem Eng 6:4676–4697

Kaushik P, Malik A (2010) Effect of nutritional conditions on dye removal from textile effluent by Aspergillus lentulus. World J Microbiol Biotechnol 26(11):1957–1964

Kaushik P, Malik A (2011) Process optimization for efficient dye removal by Aspergillus lentulus FJ172995. J Hazard Mater 185(2–3):837–843

Keck A, Klein J, Kudlich M et al (1997) Reduction of azo dyes by redox mediators originating in the naphthalene sulfonic acid degradation pathway of *Sphingomonas* ssp. Strain BN6. Appl Environ Microbiol 63:3684–3690

Khalaf MA (2008) Biosorption of reactive dye from textile wastewater by non-viable biomass of *Aspergillus niger* and *Spirogyra* sp. Bioresour Technol 99:6631–6634

Khambhaty Y, Mody K, Basha S (2012) Efficient removal of Brilliant Blue G (BBG) from aqueous solutions by marine *Aspergillus wentii*: kinetics, equilibrium and process design. Ecol Eng 41:74–83

Khan Z, Jain K, Soni A et al (2014) Microaerophilic degradation of sulphonated azo dye- Reactive Red 195 by bacterial consortium AR1 through co-metabolism. Int Biodeterior Biodegradation 94:167–175

Knapp JS, Newby PS (1999) The decolourisation of a chemical industry effluent by white rot fungi. Water Res 33:575–577

Kong J, Wang H, Liang L et al (2017) Phenanthrene degradation by the bacterium *Pseudomonas stutzeri* JP1 under low oxygen condition. Int Biodeterior Biodegradation 123:121–126

Kousha M, Farhadian O, Dorafshan S et al (2013) Optimization of malachite green biosorption by green microalgae—*Scenedesmus quadricauda* and *Chlorella vulgaris*: application of response surface methodology. J Taiwan Inst Chemical E 44:291–294

Kuddus M, Joseph B, Wasudev Ramteke P (2013) Production of laccase from newly isolated *Pseudomonas putida* and its application in bioremediation of synthetic dyes and industrial effluents. Biocat Agri Biotechnol 2:333–338

Kumar Garg S, Tripathi M, Singh SK et al (2012) Biodecolorization of textile dye effluent by *Pseudomonas putida* SKG-1 (MTCC 10510) under the conditions optimized for monoazo dye orange II color removal in simulated minimal salt medium. Int Biodeterior Biodegrad 74:24–35

Kumar CG, Mongolla P, Sheik AB et al (2011) Decolorization and biotransformation of triphenylmethane dye, methyl violet, by *Aspergillus* sp. isolated from Ladakh, India. J Microbiol Biotechnol 21:267–273

Kumar CG, Mongolla P, Joseph J, Sarma VUM (2012) Decolorization and biodegradation of triphenylmethane dye, brilliant green, by *Aspergillus* sp. isolated from Ladakh, India. Process Biochem 47(9):1388–1394

Kumar R, Kaur J, Jain S, Kumar A (2016) Optimization of laccase production from Aspergillus flavus by design of experiment technique: partial purification and characterization. J Genet Eng Biotechnol 14(1):125–131

Kuppusamy S, Sethurajan M, Kadarkarai M et al (2017) Biodecolourization of textile dyes by novel, indigenous *Pseudomonas stutzeri* MN1 and *Acinetobacter baumannii* MN3. J Environ Chem Eng 5:716–724

Lebron YAR, Moreira VR, Santos LVS et al (2018) Remediation of methylene blue from aqueous solution by *Chlorella pyrenoidosa* and *Spirulina maxima* biosorption: equilibrium, kinetics, thermodynamics and optimization studies. J Environ Chem Eng 6:6680–6690

Li L, Hong Q, Yan X et al (2009) Isolation of a malachite green-degrading *Pseudomonas* sp. MDB-1 strain and cloning of the tmr2 gene. Biodegradation 20:769–776

Lim SL, Chu WL, Phang SM (2010) Use of *Chlorella vulgaris* for bioremediation of textile wastewater. Bioresour Technol 101:7314–7322

Lin YH, Leu JY (2008) Kinetics of reactive azo-dye decolorization by *Pseudomonas luteola* in a biological activated carbon process. Biochem Eng J 39:457–467

Lin SF, Yu P, Lin YM (2004) Study on decolorization of malachite green by a *Pseudomonas aeruginosa*. J Fujian Norm Univ 20:72–75

Lin J, Zhang X, Li Z et al (2010) Biodegradation of Reactive blue 13 in a two-stage anaerobic/aerobic fluidized beds system with a *Pseudomonas* sp. isolate. Bioresour Technol 101:34–40

Liu C, You Y, Zhao R et al (2017) Biosurfactant production from Pseudomonas taiwanensis L1011 and its application in accelerating the chemical and biological decolorization of azo dyes. Ecotoxicol Environ Saf 145:8–15

Lodato A, Alfieri F, Olivieri G et al (2007) Azo-dye conversion by means of *Pseudomonas* sp. OX1. Enzym Microb Technol 41:646–652

Logroño W, Pérez M, Urquizo G et al (2017) Single chamber microbial fuel cell (SCMFC) with a cathodic microalgal biofilm: a preliminary assessment of the generation of bioelectricity and biodegradation of real dye textile wastewater. Chemosphere 176:378–388

Lu T, Zhang Q, Yao S (2017) Efficient decolorization of dye-containing wastewater using mycelial pellets formed of marine-derived *Aspergillus niger*. Chin J Chem Eng 25:330–337

Luan TG, Jin J, Chan SMN et al (2006) Biosorption and biodegradation of tributyltin (TBT) by alginate immobilized *Chlorella vulgaris* beads in several treatment cycles. Process Biochem 41:1560–1565

Mahmoud MS, Mostafa MK, Mohamed SA (2017) Bioremediation of red azo dye from aqueous solutions by *Aspergillus niger* strain isolated from textile wastewater. J Environ Chem Eng 5:547–554

Malla FA, Khan SA, Rashmi et al (2015) Phycoremediation potential of *Chlorella minutissima* on primary and tertiary treated wastewater for nutrient removal and biodiesel production. Ecol Eng 75:343–349

Mangwani N, Shukla SK, Rao TS (2014) Calcium-mediated modulation of *Pseudomonas mendocina* NR802 biofilm influences the phenanthrene degradation. Colloids Surf B: Biointerfaces 114:301–309

Maqbool Z, Hussain S, Ahmad T et al (2016) Use of RSM modeling for optimizing decolorization of simulated textile wastewater by *Pseudomonas aeruginosa* strain ZM130 capable of simultaneous removal of reactive dyes and hexavalent chromium. Environ Sci Pollut Res 23:11224–11239

Mathur M, Gola D, Panja R, Malik A, Ahammad SZ (2018) Performance evaluation of two Aspergillus spp. for the decolourization of reactive dyes by bioaccumulation and biosorption. Environ Sci Pollut Res 25(1):345–352

McMullan G, Meehan C, Conneely A et al (2001) Microbial decolourisation and degradation of textile dyes. Appl Microbiol Biotechnol 56:81–87

Michaels GB, Lewis DL (1985) Sorption and toxicity of azo and triphenylmethane dyes to aquatic microbial populations. Environ Toxicol Chem 4:45–50

Mishra S, Maiti A (2018) Optimization of process parameters to enhance the bio-decolorization of Reactive Red 21 by *Pseudomonas aeruginosa* 23N1. Int J Environ Sci Technol 16:6685–6698. https://doi.org/10.1007/s13762-018-2023-1

Mohan SV, Sistla S, Guru RK et al (2003) Microbial degradation of pyridine using *Pseudomonas* sp. and isolation of plasmid responsible for degradation. Waste Manag 23:167–171

Moharikar A, Purohit HJ (2003) Specific ratio and survival of *Pseudomonas* CF600 as co-culture for phenol degradation in continuous cultivation. Int Biodeterior Biodegrad 52:255–260

Munoza R, Guieysse B (2006) Algal–bacterial processes for the treatment of hazardous contaminants: a review. Water Res 40:2799–2815

Muthu Kumara Pandian A, Karthikeyan C, Rajasimman M (2016) Isotherm and kinetic studies on nano-sorption of malachite green onto Aspergillus flavus mediated synthesis of silver nano particles. Environ Nanotechnol Monitor Manag 6:139–151

Nachiyar CV, Rajkumar GS (2003) Degradation of a tannery and textile dye, Navitan Fast Blue S5R by *Pseudomonas aeruginosa*. World J Microbiol Biotechnol 19:609–614

Nachiyar CV, Rajkumar GS (2004) Mechanism of Navitan fast Blue S5R degradation by *Pseudomonas aeruginosa*. Chemosphere 57:165–169

Nachiyar CV, Rajkumar GS (2005) Purification and characterization of an oxygen insensitive azoreductase from *Pseudomonas aeruginosa*. Enzym Microb Technol 36:503–509

Nachiyar CV, Vijayalakshmi K, Muralidharan D et al (2007) Mineralization of metanilic acid by *Pseudomonas aeruginosa* CLRI BL22. World J Microbiol Biotechnol 23:1733–1738

Nakanishi M, Yatome C, Ishida N et al (2001) Putative ACP phosphodiesterase gene encodes an azoreductase. J Biol Chem 49:46394–46399

Naraian R, Kumari S, Gautam RL (2018) Biodecolorization of brilliant green carpet industry dye using three distinct *Pleurotus* spp. Environ Sustain 1:141–148

Naskar A, Majumder R (2017) Understanding the adsorption behaviour of acid yellow 99 on *Aspergillus niger* biomass. J Mol Liq 242:892–899

Nigam P, Banat IM, Singh D et al (1996) Microbial process for the decolorization of textile effluent containing azo, diazo and reactive dyes. Process Biochem 31:435–442

Ning C, Qingyun L, Aixing T et al (2018) Decolorization of a variety of dyes by *Aspergillus flavus* A5p1. Bioprocess Biosyst Eng 41:511–518

Nitisakulkan T, Oku S, Kudo D et al (2014) Degradation of chloroanilines by toluene dioxygenase from *Pseudomonas putida* T57. J Biosci Bioeng 117:292–297

Nowak A, Mrozik A (2018) Degradation of 4-chlorophenol and microbial diversity in soil inoculated with single *Pseudomonas* sp. CF600 and *Stenotrophomonas maltophilia* KB2. J Environ Manag 215:216–229

Nwinyi OC, Ajayi OO, Amund OO (2016) Degradation of polynuclear aromatic hydrocarbons by two strains of *Pseudomonas*. Braz J Microbiol 47:551–562

Oranusi NA, Ogugbue CJ (2005) Effect of cosubstrates on primary biodegradation of triphenylmethane dyes by *Pseudomonas* sp. Afr J Appl Zool Environ Biol 7:38–44

Pandey BV, Upadhyay RS (2006) Spectroscopic characterization and identification of *Pseudomonas fluorescens* mediated metabolic products of Acid Yellow-9. Microbiol Res 161:311–315
Pandey RK, Tewari S, Tewari L (2018) Lignolytic mushroom *Lenzites elegans* WDP2: laccase production, characterization, and bioremediation of synthetic dyes. Ecotox Environ Safety 158:50–58
Park JK, Chang HN (2000) Microencapsulation of microbial cells. Biotechnol Adv 18:303–319
Park HS, Jun SC, Han KH et al (2017) Diversity, application, and synthetic biology of industrially important *Aspergillus* fungi. Adv Appl Microbiol 100:161–202
Parshetti GK, Kalme SD, Gomare SS (2007) Biodegradation of reactive blue-25 by *Aspergillus ochraceus* NCIM-1146. J Biotechnol 98:3638–3642
Paszczynski A, Pasti-Grigsby MB, Goszczynski S et al (1992) Mineralization of sulfonated azo dyes and sulfanilic acid by *Phanerochaete chrysosporium* and *Streptomyces chromofuscust*. Appl Environ Microbiol 58:3598–3604
Patel R, Suresh S (2008) Kinetic and equilibrium studies on the biosorption of reactive black 5 dye by *Aspergillus foetidus*. Bioresour Technol 99:51–58
Patel Y, Mehta C, Gupte A (2012) Assessment of biological decolorization and degradation of sulfonated di-azo dye Acid Maroon V by isolated bacterial consortium EDPA. Int Biodeterior Biodegrad 75:187–193
Pathak VV, Kothari R, Chopra A et al (2015) Experimental and kinetic studies for phycoremediation and dye removal by *Chlorella pyrenoidosa* from textile wastewater. J Environ Manag 163:270–277
Pazarlioğlu NK, Telefoncu A (2005) Biodegradation of phenol by *Pseudomonas putida* immobilized on activated pumice particles. Process Biochem 40:1807–1814
Peng X, Ma X, Xu Z (2015) Thermogravimetric analysis of co-combustion between microalgae and textile dyeing sludge. Bioresour Technol 180:288–295
Perei K, Rakhely G, Kiss I et al (2001) Biodegradation of sulfanilic acid by *Pseudomonas paucimobilis*. Appl Microbiol Biotechnol 55:101–107
Phugare SS, Waghmare SR, Jadhav JP (2011) Purification and characterization of dye degrading of veratryl alcohol oxidase from *Pseudomonas aeruginosa* strain BCH. World J Microbiol Biotechnol 27:2415–2423
Puvaneshwari N, Muthukrishnan J, Gunasekaran P et al (2002) Biodegradation of benzidine based azodyes direct red and direct blue by the immobilized cells of *Pseudomonas fluorescens* D41. Indian J Exp Biol 40:1131–1136
Qian HF, Chen W, Sheng GD et al (2008) Effects of glufosinate on antioxidant enzymes, subcellular structure, and gene expression in the unicellular green alga *Chlorella vulgaris*. Aquat Toxicol 88:301–307
Rafii F, Franklin W, Cerniglia CE (1990) Azoreductase activity of anaerobic bacteria isolated from human intestinal microflora. Appl Environ Microbiol 56:2146–2151
Rampelotto PH (2010) Resistance of microorganisms to extreme environmental conditions and its contribution to astrobiology. Sustain For 2:1602–1623
Ramya M, Anusha B, Kalavathy S et al (2007) Biodecolorization and biodegradation of Reactive Blue by *Aspergillus* spp. Afr J Biotechnol 6:1441–1445
Rani B, Kumar V, Singh J, Bisht S, Teotia P, Sharma S, Kela R (2014) Bioremediation of dyes by fungi isolated from contaminated dye effluent sites for bio-usability. Braz J Microbiol 45(3):1055–1063
Roy U, Sengupta S, Banerjee P et al (2018) Assessment on the decolourization of textile dye (Reactive Yellow) using *Pseudomonas* sp. immobilized on fly ash: response surface methodology optimization and toxicity evaluation. J Environ Manag 223:185–195
Ryu BH, Weon YD (1992) Decolorization of Azo Dyes by *Aspergillus sojae* B-10. J Microbiol Biotechnol 2:215–219
Sá CS, Boaventura RA (2001) Biodegradation of phenol by *Pseudomonas putida* DSM 548 in a trickling bed reactor. Biochem Eng J 9:211–219
Safi C, Zebib B, Merah O et al (2014) Morphology, composition, production, processing and applications of *Chlorella vulgaris*: a review. Renew Sust Energ Rev 35:265–278

Samson RA, Visagie CM, Houbraken J et al (2014) Phylogeny, identification and nomenclature of the genus *Aspergillus*. Studies Myco 78:141–173

Saparrat MCN, Balatti PA, Arambarri AM et al (2014) *Coriolopsis rigida*, a potential model of white-rot fungi that produce extracellular laccases. J Ind Microbiol Biotechnol 41:607–617

Sarnaik S, Kanekar P (1995) Bioremediation of colour of methyl violet and phenol from a dye-industry waste effluent using *Pseudomonas* spp. isolated from factory soil. J Appl Bacteriol 79:459–469

Sarnaik S, Kanekar P (1999) Biodegradation of methyl violet by *Pseudomonas mendocina* MCM B-402. Appl Microbiol Biotechnol 52:251–254

Sathishkumar K, Sathiyaraj S, Parthipan P et al (2017) Electrochemical decolorization of methyl red by RuO_2 -IrO_2 -TiO_2 electrode and biodegradation with *Pseudomonas stutzeri* MN1 and *Acinetobacter baumannii* MN3: an integrated approach. Chemosphere 183:204–211

Selvakumar KV, Basha CA, Prabhu HJ et al (2010) The potential of free cells of *Pseudomonas aeruginosa* on textile dye degradation. Bioresour Technol 101:2678–2684

Seo YH, Park D, Oh YK et al (2015) Harvesting of microalgae cell using oxidized dye wastewater. Bioresour Technol 192:802–806

Sharma P, Singh L, Dilbaghi N (2009) Response surface methodological approach for the decolorization of simulated dye effluent using *Aspergillus fumigatus* fresenius. J Hazard Mater 161:1081–1086

Silveira E, Marques PP, Silva SS et al (2009) Selection of *Pseudomonas* for industrial textile dyes decolourization. Int Biodeterior Biodegrad 63:230–235

Singh U, Arora NK, Sachan P (2018) Simultaneous biodegradation of phenol and cyanide present in coke-oven effluent using immobilized *Pseudomonas putida* and *Pseudomonas stutzeri*. Braz J Microbiol 49:38–44

Sinha S, Singh R, Chaurasia AK et al (2016) Self-sustainable *Chlorella pyrenoidosa* strain NCIM 2738 based photobioreactor for removal of Direct Red-31 dye along with other industrial pollutants to improve the water-quality. J Hazard Mater 306:386–394

Srinivasan R, Kathiravan MN, Gopinath KP (2011) Degradation of Tectilon Yellow 2G by hybrid technique: combination of sonolysis and biodegradation using mutant *Pseudomonas putida*. Bioresour Technol 102:2242–2247

Stormo KE, Crawford RL (1992) Preparation of encapsulated microbial cells for environmental applications. Appl Environ Microbiol 58:727–730

Sugiura W, Miyashita T, Yokoyama T et al (1999) Isolation of azo-dye degrading microorganisms and their application to white discharge printing of fabric. J Biosci Bioeng 88:577–581

Sultan M (2017) Polyurethane for removal of organic dyes from textile wastewater. Environ Chem Lett 15:347. https://doi.org/10.1007/s10311-016-0597-8

Sumathi S, Manju B (2000) Uptake of reactive textile dyes by *Aspergillus foetidus*. Enzym Microb Technol 27:347–355

Surkatti R, El-Naas MH (2014) Biological treatment of wastewater contaminated with p-cresol using *Pseudomonas putida* immobilized in polyvinyl alcohol (PVA) gel. J Water. Process Eng 1:84–90

Suzuki Y, Yoda T, Ruhul A et al (2001) Molecular cloning and characterization of the gene encoding azoreductase from *Bacillus* sp. OY 1-2isolated from soil. J Biol Chem 246:9059–9065

Tamayo-Ramos JA, van Berkel WJ, de Graaff LH (2012) Biocatalytic potential of laccase-like multicopper oxidases from *Aspergillus niger*. Microb Cell Factories 11:165. https://doi.org/10.1186/1475-2859-11-165

Tao R, Kinnunen V, Praveenkumar R et al (2017) Comparison of *Scenedesmus acuminatus* and *Chlorella vulgaris* cultivation in liquid digestates from anaerobic digestion of pulp and paper industry and municipal wastewater treatment sludge. J Appl Phycol 29:2845–2856

Telke AA, Kalyani DC, Jadhav UU et al (2009) Purification and characterization of an extracellular laccase from a *Pseudomonas* sp. LBC1 and its application for the removal of bisphenol A. J Mol Cata B: Enzymatic 61:252–260

Telke AA, Kim SW, Govindwar SP (2012) Significant reduction in toxicity, BOD, and COD of textile dyes and textile industry effluent by a novel bacterium *Pseudomonas* sp. LBC1. Folia Microbiol 57:115–122

Thao TP, Kao HC, Juang RS et al (2013) Kinetic characteristics of biodegradation of methyl orange by *Pseudomonas putida* mt2 in suspended and immobilized cell systems. J Taiwan Inst Chem Eng 44:780–785

Tsai WT, Chen HR (2010) Removal of malachite green from aqueous solution using low-cost chlorella-based biomass. J Hazard Mater 175:844–849

Tuttolomondo MV, Alvarez GS, Desimone MF et al (2014) Removal of azo dyes from water by sol–gel immobilized *Pseudomonas* sp. J Environ Chem Eng 2:131–136

Wang B, Hu Y (2007) Comparison of four supports for adsorption of reactive dyes by immobilized *Aspergillus fumigatus* beads. J Environ Sci 19:451–457

Wang BE, Hu YY, Xie L et al (2008) Biosorption behavior of azo dye by inactive CMC immobilized *Aspergillus fumigatus* beads. Bioresour Technol 99:794–800

Wang W, Zhang Z, Ni H et al (2012) Decolorization of industrial synthetic dyes using engineered *Pseudomonas putida* cells with surface-immobilized bacterial laccase. Microb Cell Factories 11:75. https://doi.org/10.1186/1475-2859-11-75

Wang L, Chen X, Wang H et al (2017) *Chlorella vulgaris* cultivation in sludge extracts from 2,4,6-TCP wastewater treatment for toxicity removal and utilization. J Environ Manag 187:146–153

Wu J, Jung BG, Kim KS et al (2009) Isolation and characterization of *Pseudomonas otitidis* WL-13 and its capacity to decolorize triphenylmethane dyes. J Environ Sci 21:960–964

Xiong XJ, Meng XJ, Zheng TL (2010) Biosorption of C.I Direct Blue 199 from aqueous solution by nonviable *Aspergillus niger*. J Hazard Mater 175:241–246

Xu H, Miao X, Wu Q (2006) High quality biodiesel production from a microalga *Chlorella* protothecoides by heterotrophic growth in fermenters. J Biotechnol 126:499–507

Xu C, Wang R, Zhang YF et al (2015) Stress response of *Chlorella pyrenoidosa* to nitro-aromatic compounds. Environ Sci Pollut Res 22:3784–3793

Yang Y, Hu H, Wang G et al (2011a) Removal of malachite green from aqueous solution by immobilized *Pseudomonas* sp. DY1 with *Aspergillus oryzae*. Int Biodeterior Biodegrad 65:429–434

Yang Y, Jin D, Wang G et al (2011b) Competitive biosorption of Acid Blue 25 and Acid Red 337 onto unmodified and CDAB-modified biomass of *Aspergillus oryzae*. Bioresour Technol 102:7429–7436

Yatome C, Ogawa T, Koga D et al (1981) Biodegradability of azo and triphenylmethane dyes by *Pseudomonas pseudomallei* 13 NA. J Soc Dye Colour 97:166–169

Yatome C, Ogaw T, Hishida H et al (1990) Degradation of azo dyes by cell-free extract from *Pseudomonas stutzeri*. J Soc Dye Colour 106:280–283

Yatome C, Matsufuru H, Taguchi T et al (1993) Degradation of 4-dimethylaminoazobenzene-2-carboxylic acid by *Pseudomonas stutzeri*. Appl Microbiol Biotechnol 39:778–781

Yu J, Wang X, Yue PL (2001) Optimal decolorization and kinetic modeling of synthetic dyes by *Pseudomonas* strains. Water Res 35:3579–3586

Zabłocka-Godlewska E, Przystaś W, Grabińska-Sota E (2014) Decolourisation of different dyes by two *Pseudomonas* strains under various growth conditions. Water Air Soil Pollut 225:1846. https://doi.org/10.1007/s11270-013-1846-0

Zhao W, Sun H, Ren Y et al (2018) *Chlorella zofingiensis* as a promising strain in wastewater treatment. Bioresour Technol 268:286–291

Zheng S, He M, Sui Y et al (2017) Kelp waste extracts combined with acetate enhances the biofuel characteristics of *Chlorella sorokiniana*. Bioresour Technol 225:142–150

Zhipei L, Huifang Y (1991) Decolorization and biodegradation metabolism of azo dyes Pseudomonas S-42. J Environ Sci 3:89–102

Zimmermann T, Kulla GH, Leisinger T (1982) Properties of purified orange II azoreductase, the enzyme initiating azo dye degradation by *Pseudomonas* KF46. Eur J Biochem 29:197–203

Zope V, Kulkarni M, Chavan M (2007) Biodegradation of synthetic textile dyes Reactive Red 195 and Reactive Green 11 by *Aspergillus niger grp*: an alternative approach. J Sci Ind Res 66:411–414

Microbial Metabolism of Organophosphates: Key for Developing Smart Bioremediation Process of Next Generation

14

Santanu Pailan, Kriti Sengupta, and Pradipta Saha

Abstract

Currently organophosphate compounds constitute one of the largest families of chemical compounds that are used for pest control, mainly for better crop yield worldwide. Due to their toxicity, persistence, and adverse effects, some organophosphates (like parathion and methyl parathion) were classified and registered as extremely hazardous by the World Health Organization (WHO) and US EPA (US Environmental Protection agency) and have been banned in many countries. Some of the hydrolysis intermediates (such as 4-nitrophenol and trichloropyridinol) of these organophosphates are more toxic and environmentally mobile (due to greater water solubility) and therefore more dangerous. However, existing reports suggest their illegal, extensive use and application without proper technical know-how (especially by illiterate farmers in underdeveloped/developing countries). Their indiscriminate and extensive application and use are responsible for possible contamination of several ecosystems and groundwater. Continuous and excessive use of organophosphates has been reported to be responsible for various ever-ending global problems such as contamination of air, water, and terrestrial ecosystems, decline in diversity of productive soil microflora, disruption of biogeochemical cycles, and death of nontarget macroscopic life forms. Organophosphates have been documented as neurotoxic and are potent inhibitors of acetylcholinesterase. They are responsible for serious adverse effect on the nervous, excretion, endocrine, reproductive, cardiovascular, and respiratory systems of target as well as nontarget organisms including humans. Moreover, these compounds are one of the major causes of accidental and suicidal deaths in rural population of the world. The situation therefore is of huge public interest, and hence, suitable cost-effective bioremediation technique

S. Pailan · K. Sengupta · P. Saha (✉)
Department of Microbiology, Burdwan University, Golapbag, Burdwan, West Bengal, India
e-mail: psaha@microbio.buruniv.ac.in

P. K. Arora (ed.), *Microbial Technology for Health and Environment*,
Microorganisms for Sustainability 22,
https://doi.org/10.1007/978-981-15-2679-4_14

must be developed for the restoration of organophosphate-contaminated environmental niches. Bioremediation of pollutants by biological system has emerged as the most effective method for clean up the contaminated sites. In order to implement bioremediation approach, proper understanding of microbial metabolism of these organophosphates compounds is of extreme importance. Microbial metabolism of OP compounds can be carried out catabolically (with organophosphates serving either as a sole source for C, N, or P) or co-metabolically (in the presence of other compounds, mainly carbohydrates). The metabolic conversion of organophosphates to CO_2 and H_2O (i.e., complete mineralization) is carried out through three main processes such as degradation, conjugation, and rearrangements that involves reactions like oxidation, hydrolysis, and reduction, all mediated through the enzyme-mediated pathways. The main enzymes that are involved in hydrolysis are phosphotriesterases (PTE) and phosphatase. The three major types of PTE are reported so far, such as organophosphate hydrolase (OPH), methyl parathion hydrolase (MPH), and organophosphorus acid anhydrolase (OPAA) encoded by *opd*, *mpd*, and *opaA* genes, which are either located on plasmid or on chromosomal DNA. Since most of the organophosphates are less soluble to make it physiologically available for microbes, solubilization is carried out either through the secretion of organic acid or by biosurfactants by the microbial cells. This is followed by adsorption and or uptake. Most of these adsorption and uptake mechanisms remain largely unknown. However, being lipophilic and small in size, these organophosphates can be transported to the periplasmic space where the metabolic transformation starts. The metabolic transformation involves either an initial oxidation or reduction followed by hydrolysis to release the toxic functional group and phosphate group. This hydrolysis step is most critical as it reduces the toxicity of organophosphates. The metabolic transformation of the toxic functional group is most well-studied and reported in literature. This is followed by a series of reactions that involves interconversion ultimately leading to ring cleavage reaction that opens up the molecule. Further reactions then convert these intermediates into a product that can act as suitable metabolite to be entered into the TCA cycle. The end products released from the TCA cycle are CO_2 and H_2O. Most of initial reactions are mediated in the periplasmic space of the bacterial cell. The interconversion of much less toxic metabolites occurs in the cytoplasm. Although many facets of organophosphates biodegradation have been excavated, still there remain many lacunas. Understanding microbial diversity, ecological aspects, and adaptation strategies might cater better prospects to hope for smart technologies.

Keywords
Organophosphates; 4-Nitrophenol · Parathion · Methyl parathion

14.1 Introduction

The population of human is probably going through zenith phase of development and to cater its need steady food supply for all is an absolute requirement. The latter is dependent on the continuous increase in food production. Unfortunately, nearly 15–20% (sometimes up to 33%) of the agricultural production are lost due to pest infestation (Puri et al. 2013). For tropical countries, products are damaged due to high humidity, temperature, and several conditions that provide highly favorable environment for the multiplication of insect pests (Lakshmi 1993; Abhilash and Singh 2009). Thus, to protect crops and food from insect attack, insecticides were introduced (Kannan et al. 1997). Initially, organochlorine (OC) insecticides were used; however, due to their high toxicity, long persistence in the environment, bioaccumulation, biomagnifications, and devastatingly ill ecological effects, the majority has been replaced by organophosphate insecticides (Aktar et al. 2009). Some common organophosphate insecticides used worldwide along with their chemical structure, mode of action, year of introduction, half-life, and toxicity are illustrated in Table 14.1.

14.2 Introduction of Organophosphate: Historical Perspectives and Current Scenario

The first organophosphate insecticide to be commercialized was Bladan, which contained tetraethyl pyrophosphate (TEPP) and was formulated by German chemist Gerhard Schrader in 1937 (Gallo and Lawryk 1991; Kanekar et al. 2004; Ghosh 2010). Parathion was synthesized in 1944 by same chemist-scientist (Gallo and Lawryk 1991) and was introduced in 1947; later on its methyl derivative, methyl parathion, was introduced in 1949 (Singh and Walker 2006). Chlorpyrifos was introduced in 1965 as acaricide and insecticide (Singh and Walker 2006). Due to its broad-spectrum nature, chlorpyrifos was used throughout the world to control a variety of chewing and sucking insect pests and mites on a range of economically important crops, including citrus fruit, bananas, vegetables, potatoes, coffee, cocoa, tea, cotton, wheat, and rice (Thengodkar and Sivakami 2010, Chen et al. 2012).

Currently, more than 140 organophosphates are reported to be used worldwide as insecticides, fertilizers, fungicides, weedicides, plant growth factors, and other agrochemicals for better crops yield and chemical warfare agents like soman and sarin. These organophosphates are used as a component of 100 different types of commercially available insecticides, and it has also been estimated that more than 1500 different types of organophosphates have been synthesized during the period of the last century. Presently, organophosphates represent the largest group of chemical insecticides used in plant protection throughout the world after the prohibition on use of organochlorine insecticides (Bhagobaty and Malik 2008; Ortiz-Hernandez and Sanchez-Salinas 2010).

Table 14.1 Some commonly used organophosphate compounds

Name of OP insecticides	Structure	Mode of action	Year of introduction	Half-life in soil (days)
Parathion	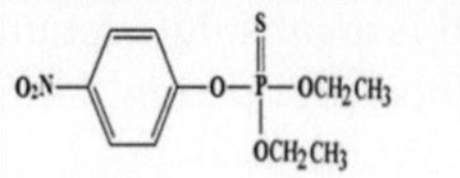	Insecticides	1947	30–180
Methyl parathion	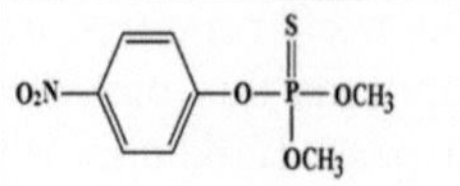	Insecticides	1949	25–130
Chlorpyrifos	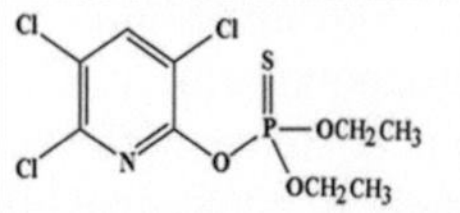	Acaricide/ insecticide	1965	10–120
Malathion	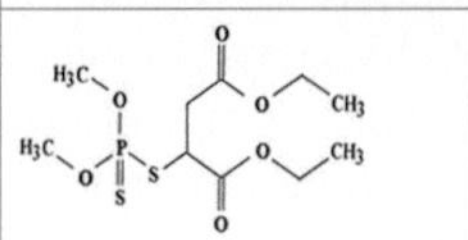	Insecticides	1950	1–25
Dimethoate	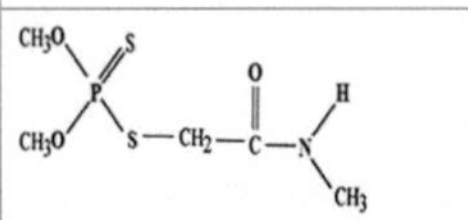	Insecticides	1955	2–40
Monocrotophos	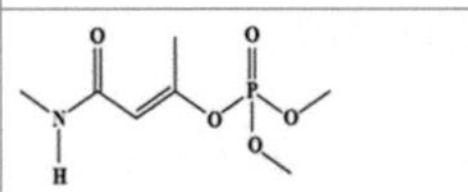	Insecticides	1965	40–60
Coumaphos	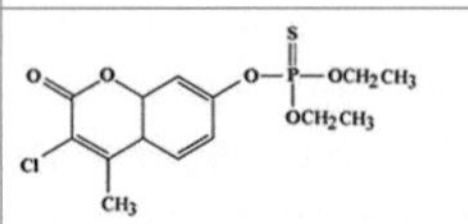	Insecticides	1952	24–1400

Data taken from Singh and Walker (2006); Kanekar et al. (2004)

14.2.1 Usage of Organophosphates

Historically, organophosphates were used as chemical warfare agents such as Sarin, Soman, and VX. About 200,000 tons of these extremely toxic organophosphates chemical warfare agents were manufactured and are stored. As per Chemical Weapons Convention (CWC) of 1993, these stocks must be destroyed within 10 years of ratification by the member states (Singh and Walker 2006).

Abhilash and Singh (2009) categorically pointed out the following six sectors where organophosphates insecticides are used extensively:

1. Agriculture—for control of weeds, insects, pests, and rodents mainly
2. Public health—for control of insect (mainly mosquito and others) vectors that spread various diseases (malaria, filariasis, dengue fever, Japanese encephalitis, etc.)
3. Domestic—for controlling insects (mosquitos, louse, etc.), flies that are common in houses and gardens (insects such as spiders that affect ornamental plants), ectoparasites (scab mites, blowfly, ticks, and lice) of domestic farmhouse cattle

4. Personal—applied in clothing or body for controlling head and body lice, mites, other small insects, etc.
5. Material building—incorporated into paints, plastics, wood (furniture, etc.), and other materials as well in building foundation, to prevent insect infestation
6. Others—control of vegetation in forests and factory sites, fumigation of buildings, and ships

14.3 Toxic Organophosphates: A Global Threat of Huge Public Interest

Organophosphates act as neurotoxic agents (Shimazu et al. 2001; Ghosh et al. 2010) and are mainly potent inhibitors of acetyl cholinesterase (Tago et al. 2006. Chao et al. 2008; Ortiz-Hernandez and Sanchez-Salinas 2010). Acetylcholine is a neurotransmitter and acetylcholinesterase constitutes a key enzyme of the nervous system. Generally, after completion of nerve impulse transmission, the function of acetylcholinesterase is to hydrolyze acetylcholine (neurotransmitter) into choline and acetyl-CoA (inactive components), so that these become available for further function. Upon irreversible binding of organophosphate to acetylcholinesterase, it loses its normal hydrolysis function. This results into accumulation of acetylcholine at the junction of the synaptic cleft. Eventually, overstimulation occurs that ultimately leads to paralysis and, under extreme condition, death (Kumar et al. 2010; Theriot and Grunden 2011; Chaudhry et al. 1988; Cho et al. 2004; Bhagobaty and Malik 2008; Ortiz-Hernandez and Sanchez-Salinas 2010). The failure of nerve impulse transmission, due to the organophosphate pesticide poisoning, causes health problems such as weakness, headache, excessive sweating, salivation, nausea, vomiting, diarrhea, abdominal pain, and paralysis which can ultimately lead to death (under extreme condition) (Kanekar et al. 2004). Some other health disorders reported due to organophosphate poisoning are malfunctioning of the endocrinal, respiratory, excretory, and cardiovascular systems as well as miscarriage during pregnancy, abnormal/retarded fetus development, etc. (Kumar et al. 2010).

Approximately, two million tons of organophosphate pesticides are used per year throughout the world. The major consumers are Europe (45%) followed by the USA (24%) and the rest of the world (25%). Herbicides are the main category of pesticide used globally followed by insecticides and fungicides (Gupta 2004).

14.4 Microbial Bioremediation: Best for Effective Environmental Cleanup of Organophosphates

Although several chemical, physical, and physicochemical methods have been developed for the removal of these toxic chemicals from its contaminated sites, bioremediation is considered to be the best. It is the green process of cleaning the environment by using different biological means (i.e., with the help of plants, animals, and microorganisms). It offers a more effective, cheap, eco-friendly, and safer alternative process toward cleaning up of toxic and hazardous contaminants/pollutants (Chen et al. 2012, 2014). Bioremediation using microorganisms has received huge attention in the last one decade. Organophosphate-hydrolyzing enzymes of bacterial origin are considered for detoxification (and bioremediation) due to broader substrate specificities and better kinetics (Dumas et al. 1989; Cheng et al. 1993).

The organophosphate-degrading microorganisms may be used for systematic investigation toward development of suitable technology for bioremediation of these toxic organophosphate agrochemicals from the contaminated agricultural fields (and other adjoining niches). This is a strong need and demand of the day toward greener and clean tomorrow.

14.5 Hunting Bacteria for Organophosphates: Key for Developing Bioremediation Process

Research works carried out over the past three decades have shown microorganisms as the major component of biological diversity on our planet earth with the representation of 10^{30} cells. These huge number of microbes are fundamental components toward the successful execution of biogeochemical cycles and all other processes that take care of the health of our planet earth (Whitman et al. 1998). Several studies has now unequivocally proven that a successful existence and survival of most of the other life forms (including macroscopic plants and animals) depends on the proper functioning and interaction of the very basic normal microbiota that varies from one living system to another (Berg et al. 2014).

Therefore, to understand the fate of organophosphate compounds in the ecosystems, its metabolic transformation must be properly investigated in the laboratory under precisely controlled conditions (Fig. 14.1). Since the diversity of bacteria is considered huge, lot being unknown and unexplored, this group is supposed to serve as the major reservoir of novel gene pool to hunt for. Since less than 1% of the total diversity is known, it is best to explore more. Bacterial systems are less complicated compared to eukaryotic ones (fungal and plants), and their genetic regulation has been well explored and better understood and thus can be better manipulated for biotechnological applications and bioremediation purposes. In general bacterial enzymes are given more importance than the same from other (plants and animals) sources due to the following reasons (Dumas et al. 1989; Cheng et al. 1993; Chen et al. 2011; Cycon' et al. 2011; Arora et al. 2012; Chen et al. 2014):

- They are generally cheaper to produce.
- Their enzyme contents are more predictable and controllable.
- Reliable supplies of raw material of constant composition are more easily arranged.
- Plant and animal tissues contain more potentially harmful materials than microbes, including phenolic compounds (from plants), endogenous enzyme inhibitors, and proteases.

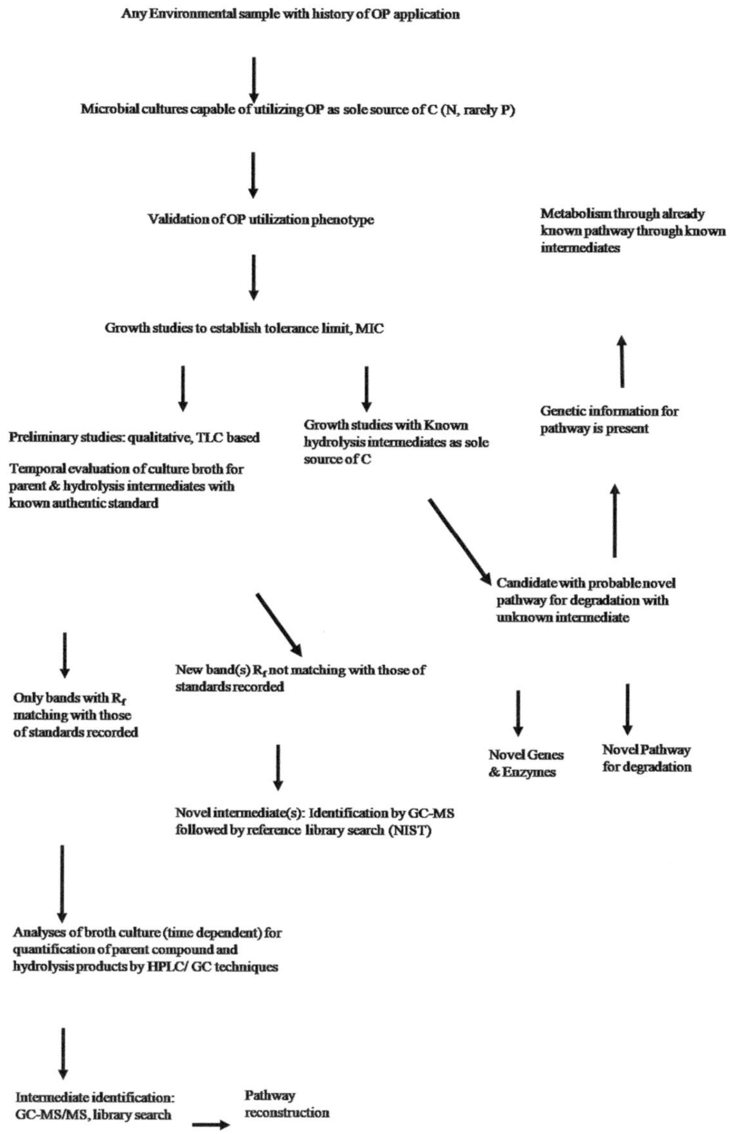

Fig. 14.1 A brief overview of the current trends in the study of organophosphate (OP) metabolism (catabolic) in microorganism, from isolation and hydrolysis product identification to pathway reconstruction

- Their enzyme-based biodegradation and bioremediation are more cost-effective and eco-friendly.
- Their enzymes have broad substrate specificity.
- Their enzyme can be used easily with bead-based remediation of toxic pollutant.

Although many organophosphate hydrolytic enzymes have been reported, considering the huge estimated diversity of the microbial world, these represent only the tip of hidden, unknown iceberg. From the rich collection, such as organophosphate-degrading microbes, much has been excavated in terms of microbial metabolism, biodegradation pathways, evolution, genetic, and molecular mechanisms. Still, in order to realize the full potential of organophosphate-degrading bacteria, their applications, and development of better strategies for bioremediation of contaminated sites, more intensive research is required. This involves isolation of organophosphate-degrading microorganisms from different ecological habitat (extreme habitats), understanding the detail molecular events of degradation and signaling pathways that initiate/activate the organophosphate-degrading genes, and development of modern technologies for better field applications (Singh 2009).

14.6 Study of Microbial Metabolism of Organophosphate Compounds

In general, the study of microbial metabolism of organophosphate compounds was started by Sethunathan and Yoshida (1973), when they reported a bacterial strain *Flavobacterium* sp. ATCC 27551 (now reclassified as (*Sphingobium fuliginis*), which could degrade and utilize diazinon and parathion as the sole carbon source and degrade chlorpyrifos co-metabolically followed by *Bacillus* sp. and *Pseudomonas* sp. (Siddaramappa et al. 1973); *Xanthomonas* (Rosenberg and Alexander 1979); *Arthrobacter* sp. (Nelson 1982); and *Pseudomonas diminuta* MG (Serdar et al. 1982; Mulbry et al. 1986). Singh et al. (2004) for first time reported the degradation of chlorpyrifos as the sole carbon source by *Enterobacter asburiae* strain B-4, which was followed by *Alcaligenes faecalis* (Yang et al. 2005); *Stenotrophomonas* sp. YC-1 (Yang et al. 2006); and *Sphingomonas* sp. DSP-2 (Li et al. 2007a, b). The overall general methodologies followed toward their studies are summarized in Fig. 14.2. So far, many bacterial strains have been reported to degrade parathion, chlorpyrifos, and other organophosphate compounds either catabolically or co-metabolically. A thorough and extensive list of bacterial spp. reported to be involved in the degradation of organophosphate compounds (mainly parathion and/or chlorpyrifos) is documented in Table 14.2.

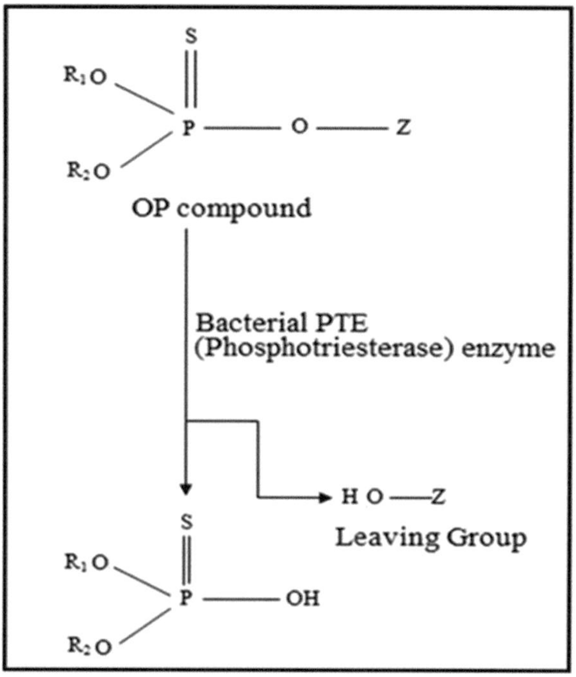

Fig. 14.2 General pathway for biodegradation of organophosphate compounds (Singh 2009)

14.7 General Trend for Organophosphate Metabolism in Microorganisms

The process of microbial metabolism of organophosphate compounds takes place through multistep pathway each being catalyzed by an enzyme. In most of the cases, the general reactions involved are hydrolysis and oxidation and rarely reduction.

All the organophosphate compounds share a similar general pattern for their degradation (Fig. 14.2). There are usually three ester bonds and breakdown of any one reduces toxicity of the compound. The most important step is the breakdown of ester bond with the main group (Z in Fig. 14.2) is bonded. This releases the group [4-NP in case of parathion and methyl parathion; 3,5,6-trichloro-2-pyridinol (TCP) in case of chlorpyrifos] to be metabolized further through enzyme catalyzed multiple steps. Finally, the ultimate end product enters into the TCA cycle for complete metabolic utilization (Singh 2009; Singh and Walker 2006).

14.8 Microbial Metabolism of Organophosphate: A Potential Source of C, P, and N for Growing Cells

Most of the studies related to understanding of microbial metabolism of organophosphate compounds started with isolation and degradation of organophosphate compounds by microorganisms. Two categories for metabolism studies have been

Table 14.2 List of organophosphate-degrading microorganisms

Name of strain	Organophosphate compound (Cat/ Co-Met utilization as C/P source)	Isolation (from) site	References
Flavobacterium sp. (ATCC 27551), reclassified as *Sphingobium fuliginis*	Par, Couma (Cat, C) Chlp (Co-met, C)	Paddy field water, Philippines	Sethunathan and Yoshida (1973); Kawahara et al. (2010)
Pseudomonas sp.	Par, 4-NP (Cat, C)	Parathion-amended soil	Siddaramappa et al. (1973)
4 species of *Pseudomonas* sp. (mixed culture)	Par (Co-met C)	Agri. wastes	Munnecke and Hsieh (1974)
Pseudomonas stutzeri	Par (Co-met, C)	–	Daughton and Hsieh (1977)
Pseudomonas sp.	Par (Cat, P)	Soil and sewage	Rosenberg and Alexander (1979)
Xanthomonas sp.	Par (Cat, C)	Soil and sewage	Rosenberg and Alexander (1979)
Pseudomonas diminuta MG	Par, chlp (Cat)	American isolate	Serdar et al. (1982), Mulbry et al. (1986)
Arthrobacter sp.	Par (Co-met, C)	Par-treated soil (Gilat, Israel)	Nelson (1982)
Bacillus sp.	Par (Co-met)		
Pseudomonas sp. (mixed culture)	Par, MPar (Co-met, C)	MPar-treated soil of farmland	Chaudhry et al. (1988)
Arthrobacter sp.	Chlp (Co-met)	Flooded soil treated with MPar	Misra et al. (1992)
Pseudomonas putida	MPar (Cat, C, and P)	–	Rani and Lalithakumari (1994)
Flavobacterium balustinum	MPar	Agri. soils (Anantapur, AP, India)	Somara and Siddavattam (1995)
Pseudomonas sp. A3	MPar (Cat, C, and P)	Rice field soil	Ramanathan and Lalithakumari (1996, 1999)
Micrococcus sp. (M-36 and AG-43)	Chlp (Cat)	Soil	Guha et al. (1997)
Bacillus sp.	MPar (Cat)	Cotton field soil (Guntur, AP, India)	Sreenivasulu and Aparna (2001)
Plesiomonas sp. strain M6	MPar (Co-met)	(Nanjing, Jiangsu, China)	Zhongli et al. (2001)

(continued)

Table 14.2 (continued)

Name of strain	Organophosphate compound (Cat/ Co-Met utilization as C/P source)	Isolation (from) site	References
Burkholderia cepacia, Bacillus sp.	MPar	Agri. soil	Keprasertsupa et al. (2001)
Agrobacterium radiobacter P230	MPar, Par	Soil, domestic yard (Brisbane, Australia)	Horne et al. (2002a)
Pseudomonas putida KT2442	Par (Cat)	–	Walker and Keasling (2002)
Enterobacter, Enterobacter asburiae strain B-4 (AJ564997 and AJ564998)[#]	Chlp (Co-met and Cat, C)	Soils of the UK and Australia	Singh et al. (2003, 2004)
Pseudomonas pseudoalcaligenes	MPar (Co-met)	Organophosphate-treated soil	Ningfeng et al. (2004)
Pseudomonas sp. strain WBC-3	MPar, 4-NP, Mala, Fen, Diazin (Cat, C, and N)	–	Liu et al. (2005)
	Chlp, TCP (Cat, C)	Soils (che. factory)	Yang et al. (2005)
7 bacterial species (*Pseudaminobacter* sp., *Achromobacter* sp., *Brucella* sp., *Ochrobactrum* sp.) (AY627033 to AY627039)[#]	MPar	MPar-contam. soil	Zhang et al. (2005, 2006a, b)
Ochrobactrum sp. B2 (AY661464)[#]	MPar (Co-met)	MPar-polluted soil	Qiu et al. (2006)
Stenotrophomonas sp. YC-1 (DQ537219)[#]	Chlp (Cat, C, and P)	Sludge (WW, OP pest. manuf.)	Yang et al. (2006)
Bacillus laterosporus strain DSP	Chlp	–	Wang et al. (2006); Zhang et al. (2012a, b)
Sphingomonas sp. DSP-2 (AY994060)[#]	Chlp (Cat, C)	Poll. water (chlp manuf. indust., Nantong, China	Li et al. (2007b)
Klebsiella sp.	Chlp	Acti. sludge (Damascus WW Treatment Plant, Syria)	Ghanem et al. (2007)
Serratia sp. (EF070125)[#]	Chp (Cat, C)	Acti. sludge (Tiancheng pesti. Co., Shandong, China)	Xu et al. (2007)
Bacillus sp. DM-1 (DQ201643)[#]	MPar (Co-met)	Organophosphate-polluted soil	Yang et al. (2007)

(continued)

Table 14.2 (continued)

Name of strain	Organophosphate compound (Cat/ Co-Met utilization as C/P source)	Isolation (from) site	References
Acinetobacter radioresistens USB-04	MPar, Par (Cat, C)	Sedi., WW treat., pesti., Shandong, China	Fang-Yao et al. (2007)
Burkholderia sp. JBA3	Par (Cat)	Agri. soil (Korea)	Kim et al. (2007)
Serratia sp. (AM050059)[#]	MPar, 4-NP (Cat, C)	Agri. soil (Anantapur, AP, India)	Pakala et al. (2007)
Delftia sp. XSP-1	MPar, chlp, Fen, Phoxim	Sludge collected from a pesti. manuf.	Shen et al. (2007)
Bacillus firmus strain BY6	Chlp (Cat, C)	Coral was collected from Teluk Awur North Java Sea, Indonesia	Sabdono (2007)
Pseudomonas stutzeri strain HS-D36	Me-Par (Cat, C)	Acti. sludge water treat. pond pesti. facto. in Hubei, China	Wang et al. (2008)
Arthrobacter sp. L1	MPar (Cat, C, and N)	Acti. sludge, enrich. tech.	Li et al. (2008a, b)
Brachybacterium sp., *Kytococcus* sp., *Brevibacterium* sp., *Chromobacterium* sp., *Oceanobacillus* sp., *Bacillus* sp. (AB449753, AB449754, AB449755, AB449757, AB449758, AB449765)	Chlp, Diazin, Ethn (Cat, C)	Coral surface (Teluk Awur, N. Java Sea, Indonesia	Sabdono and Radjasa (2008)
Paracoccus sp. strain TRP (EF070124)[#]	Chlp/TCP (Cat, C)	Acti. sludge (pesti. manuf., Shandong, China)	Xu et al. (2008)
Pseudomonas aeruginosa (NCIM 2074)	Chlp (Cat, C)	From NCIM, Pune, India	Fulekar and Geetha (2008)
Providencia stuartii (MTCC 8099)	Chlp (Cat, C)	Agri. soil (Chittoor, AP, India	Rani et al. (2008)
Pseudomonas sp. DSP-1 (DQ482656), DSP-3 (DQ482655), and DSP-5 (DQ115539), *Sphingomonas* sp. DSP-2 (AY994060), *Stenotrophomonas* sp. DSP-4 (DQ482654), *Bacillus* sp. DSP-6 (DQ237947), *Brevundimonas* sp. DSP-7 (DQ676936)[#]	Chlp (Cat, C)	Water sample of chlp indust. Pt. (Nan Tong, Jiangsu and soil agri. field Nanjing, China)	Li et al. (2008a, b)

(continued)

Table 14.2 (continued)

Name of strain	Organophosphate compound (Cat/ Co-Met utilization as C/P source)	Isolation (from) site	References
Bacillus pumilus C2A1	Chlp (Cat, C)	Soil sample from cotton fields at NIBGE, Jhang Road, Faisalabad, Pakistan	Anwar et al. (2009)
Pseudomonas aeruginosa	Chlp and TCP (Cat, C)	Pesti.-contam. soils (Punjab, India)	Lakshmi et al. (2009)
Pseudomonas sp., Burkholderia, *Arthrobacter, Pseudomonas, Variovorax, Ensifer*	Par, Fen, 4-NP, MPar (Cat, C)	Rice field soils	Min-Kyeong et al. (2009)
P. fluorescens, Brucella melitensis, Bacillus subtilis, Bacillus cereus, Klebsiella sp., *Serratia* sp., *P. aeruginosa* (consortium)	Chlp (Cat, C)	Pesti.-contam. soils of Punjab	Lakshmi et al. (2009)
Burkholderia sp. strain KR100 (HM101281)[#]	Chlp-Me, TCP (Cat, C)	Korean rice paddy soil	Kim and Ahn (2009)
Bacillus sp. and *Pseudomonas* sp.	Chlp, MPar, phorate, dichlorvos	Soil sample	Madhuri and Rangaswamy (2009)
Pseudomonas aeruginosa	MPar, Mono	MTCC, Chandigarh, India	Balamurugan et al. (2010)
Stenotrophomonas sp. SMSP-1 (EU312979)[#]	Par, MPar, Fen, Phoxim –	Sludge of a WW of pesticide manuf.	Shen et al. (2010a, b)
Bacillus licheniformis ZHU-1 (KC197213)[#]	Chlp (Cat, C)	Soil sample from Wuqi Farm in Shanghai, China	Zhu et al. (2010)
Sinorhizobium sp., *Pseudoxanthomonas* sp., *Streptomyces iakyrus, Microbacterium takaoensis, Isoptericola dokdonensis* (GU902282 to GU902303)[#]	Par (Cat, C)	Soil sample	Fodale et al. (2010)
Spirulina platensis (cyanobacteria)	Chlp	Obtained from Indian Agricultural Research Institute, Delhi, India	Thengodkar and Sivakami (2010)
Pseudomonas sp. (*aeruginosa/ putida*)	Paraoxon (Cat)	Soil samples Houston, Texas, Alvin Texas, League City, Texas Sealy, Texas Katy, Texas	Iyer et al. (2011)

(continued)

Table 14.2 (continued)

Name of strain	Organophosphate compound (Cat/Co-Met utilization as C/P source)	Isolation (from) site	References
4 species of *Pseudomonas* sp., 2 species of *Agrobacterium* sp. and *Bacillus* sp. (GQ149502-GQ149508)[#]	Chlp (Cat, C)	Soil sample from agri. farm of Banaras Hindu University, Varanasi, India	Maya et al. (2011)
Synechocystis sp. strain PUPCCC 64 (GQ907237)[#]	Chlp	Rice field of the village Dera Bassi of Patiala district of Punjab state, India	Singh et al. (2011)
Pseudomonas sp. strains RCC-2, *Staphylococcus* sp. GCC-1, *Flavobacterium* sp. GCC-3, and *Streptococcus* sp. JCC-3	Chlp	Soil samples from cultivated fields of Rajkot, Gujarat, India	Kumar (2011a, b)
Acinetobacter sp., *Pseudomonas putida, Bacillus* sp., *Pseudomonas aeruginosa, Citrobacter freundii, Stenotrophomonas* sp., *Flavobacterium* sp., *Proteus vulgaris, Pseudomonas* sp., *Acinetobacter* sp., *Klebsiella* sp., *Proteus* sp., and *Pseudomonas* sp. (consortium)	Chlp, MPar (Co-Met), 4-NP	Contam. garbage dump of Moravia, Medellin	Pino et al. (2011); Pino and Peñuela (2011)
Pseudomonas stutzeri, Pseudomonas Pseudoalcaligenes, Pseudomonas maltophilia, Pseudomonas vesicularis	Chlp (Cat, C)	Pest.-contaminated soil in Egypt	Awad et al. (2011)
Agrobacterium sp. strain Yw12 (DQ468100)[#]	MPar (Cat, C, and P)	OP-contaminated sludge Huayang pesti. manuf., Shandong, China	Wang et al. (2012)
Enterobacter sp. strain Cons002	Par, MPar, phorate (Co-met)	Agri. soil	Concepcio'n et al. (2012)
Bacillus pumilus W1	MPar	OP-contaminated soil of Khairpur, N. Sindh, Pakistan	Ali et al. (2012)

(continued)

Table 14.2 (continued)

Name of strain	Organophosphate compound (Cat/ Co-Met utilization as C/P source)	Isolation (from) site	References
Klebsiella sp., (NII 1118), *Pseudomonas putida* (NII 1117), *Pseudomonas stutzeri* (NII 1119), *Pseudomonas aeruginosa* (NII 1120) (consortium) (HM135446, HM135447, HM135448, HM135449)#	Chlp (Cat)	Chlp-contam. soil sample paddy field, Kancheepuram, Tamil Nadu, India	Sasikala et al. (2012)
Pseudomonas putida	Chlp (Co-Met)	Soil samples collected from different sites in and around Bangalore, India, having a history of repeated application of chlp	Vijayalakshmi and Usha (2012)
5 species of *Pseudomonas* sp. (individually)	Chlp (Cat, C, and P)	Efflu. storage pools of facto. producing pesti. and from soil moisture around them	Latifi et al. (2012)
Pseudomonas fluorescens, Bacillus subtilis, Klebsiella sp.	Chlp, Mono (Co-Met)	Pesti.-contam. soil of paddy field, Annamalai Nagar, Tamil Nadu, India	KaviKarunya and Reetha (2012)
Bacillus stearothermophilus, B. circulans, B. macerans	Chlp (Co-Met)	Soil from cabbage cultivated private agri. farm, Bangalore, India	Savitha and Raman (2012)
Bacillus cereus	Chlp,TCP (Cat N)	Soil from Jiangsu Jinghong Chemical Co., Ltd, China	Liu et al. (2012)
Four species of *Actinobacteria* (*Streptomyces* sp.) (JQ289350-JQ289353)#	Chlp (Co-Met)	Chlp-contam. agri. soil from blueberry field, Gorbea City in southern Chile	Briceño et al. (2012)
Stenotrophomonas maltophilia strain MHF ENV 20 and MHF ENV (HM625746, HQ661376)#	Chlp/TCP	Soil from banks of Surya River, Palghar (100 km away from Mumbai)	Dubey and Fulekar (2012)
Pseudomonas putida MAS-1	Chlp (Co-Met)	Indigenous agri. soil of Karachi, Pakistan	Ajaz et al. (2012)
Pseudomonas sp. WW5	Chlp (Co-Met)	–	Farhan et al. (2012)

(continued)

Table 14.2 (continued)

Name of strain	Organophosphate compound (Cat/Co-Met utilization as C/P source)	Isolation (from) site	References
Pseudomonas diminuta (EMP11c), *P. putida* (EMP12a), *P. aeruginosa* (EMP12b)	OP (Cat, C)	Agri. soil from Gwalior, Madhya Pradesh, India	Sharma et al. (2013)
Pseudomonas putida POXN01	MPar	Soil sample collected from rice field of Harlingen (Cameron Country, Texas)	Iyer et al. (2013)
Sphingobacterium sp. JAS3 (JQ514560)#	Chlp (Cat, C)	Soil collected from a paddy field in Vellore district, Tamil Nadu state, India	Abraham and Silambarasan (2013)
Naxibacter sp. strain CY6 (JX987079)#	Chlp, Par, MPar (Cat, C, P)	Soil samples from pesti.-contam. soil of a greenhouse	Kim et al. (2013)
Cupriavidus sp. DT-1 (JQ750642)#	Chlp,TCP (Cat, C)	Sludge collected from a chlp manuf. site in Changzhou, Jiangsu Province, China	Lu et al. (2013)
Kocuria sp.	Chlp	Agri. soil of West Godavari district of AP, India	Neti and Zakkula (2013)
Acinetobacter radioresistens, Pseudomonas frederiksbergensis, Bacillus pumilus, Serratia liquefaciens, Serratia marcescens, Burkholderia gladioli	Chlp, MPar, Diazin, Mala, Dime	Agri. soil of Beed district, Maharashtra, India	Hussaini et al. (2013)
Nocardia mediterranei	Chlp, MPar (Co-Met)	–	Sukirtha and Usharani (2013)
Pseudomonas aeruginosa, Bacillus megaterium, Staphylococcus aureus	MPar	Rhizos. soil MP-treated agri. res. farm, guava orchad. SHIATS and comm. farm, Jhunsi, Allahabad	Peter et al. (2014)
Bacillus subtilis strain C5 (JN942155)#	MPar	Marine sludge (China Bohai Sea)	Hao et al. (2014)

(continued)

Table 14.2 (continued)

Name of strain	Organophosphate compound (Cat/Co-Met utilization as C/P source)	Isolation (from) site	References
Pseudomonas aeruginosa, Serratia marcescens, and Klebsiella oxytoca	Chlp	Rice fields in Anaku, Omor, and Igbakwu towns in Ayamelum Local Govt. Area of Anambra State, Nigeria	Ifediegwu et al. (2015)
Bacillus cereus strain LR5 (JX966388)[#]	Chlp	Soil (treated with chlp) was collected from Zhejiang Academy of Agri. Sciences, Hangzhou, China	Chen et al. (2014)
Pseudomonas sp. strain YF-5 (KF584917)[#]	MPar, chlp (Cat, C)	Sludge (China)	Liu et al. (2014)
Pseudomonas sp. BF1–3 (KJ849233)[#]	Chlp	Balloon flower root	Barman et al. (2014)
Paenibacillus (*Bacillus*) *polymyxa* and *Azospirillum lipoferum*	Chlp, chlp-Me, Mala	–	Romeh and Hendawi (2014)
Stenotrophomonas sp. G1 (JN688160)[#]	Par, chlp, MPar, Diazin	Sludge, drain outlet (chlorpyrifos manufac. Plant, China)	Deng et al. (2015)
Achromobacter sp. C1	MPar (Cat, C)	Agri. soil, Jabalpur, India	Mishra (2015)
Mesorhizobium sp. HN3 (JN119831)[#]	Chlp, TCP (Cat, C)	Chlp-contam. agri. soil samples	Jabeen et al. (2015)
Cupriavidus taiwanensis (JN688161)[#]	Chlp	Sludge from outlet of a chlp manuf. in Jiangsu Province, China	Wang et al. (2015)
Bacillus aerius	Chlp	Soil samples from locations of the Nandimandalam village of YSR district Kadapa, AP, India	Jayasri et al. (2015)
Bacillus thuringiensis strain BRC-HZM2 (GQ140344)[#]	Chlp	Samples were collected from a facto (Fujian Sannong che. and pest. facto.), manuf. OP pesti., Sanming, Fujian Province, China	Wu et al. (2015)

(continued)

Table 14.2 (continued)

Name of strain	Organophosphate compound (Cat/ Co-Met utilization as C/P source)	Isolation (from) site	References
Bacillus aryabhattai SanPs1	MPar (Cat, C)	Rhizosphere soil of paddy field. Burdwan, India	Pailan et al. (2015)
Pseudomonas sp. BUR11	MPar (Cat, C)	Rhizosphere soil of paddy field. Burdwan, WB, India	Pailan and Saha (2015)
Acinetobacter sp. MemCl4	Chlp (Cat, C)	Rhizosphere soil of paddy field. Memari, WB, India	Pailan et al. (2016)
Pseudomonas putida X3	MPar (Cat, C)	–	Zhang et al. (2016)
Pseudomonas sp. *R1, R2, and R3*	Mpar (Cat)	Agri. soil, Visakhapatnam, AP, India	Begum and Arundhati (2016)
Cupriavidus nantongensis X1	Chlp	Isolated from sludge collected at drain outlet of a chlorpyrifos manuf. plant	Fang et al. (2016)
Staphylococcus warneri (CPI2), *Pseudomonas putida* (CPI 9), *and Stenotrophomonas maltophilia* (CPI 15) (consortium)	Chlp	Soil from different agric. areas in Kerala, India	John et al. (2016)
Xanthomonas sp. 4R3-M1, Pseudomonas sp. 4H1-M3, and Rhizobium sp. 4H1-M1	Chlp (catabolically as a sole source of C and N)	Sugarcane farms in the Mackay, Burdekin, and Tully areas in Queensland, Australia	Rayu et al. (2018)
Fungi			
Penicillium waksmani	Par	Flooded sulfate soil	Rao and Sethunathan (1974)
Trichoderma harzianum, Penicillium vermiculatum, and *Mucor* sp.	Chlp	Forest sample	Jones and Hastings (1981)
Phanerochaete chrysosporium	Chlp (Cat, N)	US Dept. of agri. Forest Products Laboratory, Madison, WI	Bumpus et al. (1993)
Aspergillus terreus, Trichoderma harzianum	Chlp	A clay soil taken from the Botanical Garden of Assiut University, Assiut, Egypt	Omar (1998)

(continued)

Table 14.2 (continued)

Name of strain	Organophosphate compound (Cat/ Co-Met utilization as C/P source)	Isolation (from) site	References
Coriolus versicolor, Hypholoma fasciculare	Chlp	–	Bending et al. (2002)
Aspergillus sp., *Trichoderma* sp.	Chlp	Soil pre-treated with chlp, China	Liu et al. (2003)
Fusarium sp. LK (WZ-I)	Chlp	–	Wang et al. (2005); Xie et al. (2010)
Verticillium sp. (DQ153250)[##]	Chlp (Cat, C)	Samples from farm soil, tree rhizos. soil, sedi. of a sewer, sludge, and piggery soil from Huajiachi Campus, Zhejiang University, Hangzhou, China	Yu et al. (2006)
Trichosporon sp. (EF091819)[##]	Chlp,TCP	Acti. sludge from Tiancheng pesti. Co., Shandong, China	Xu et al. (2007)
Verticillium sp. DSP	Chlp	Soil samples collected from farm field at Huajiachi Campus, Zhejiang University, Hangzhou, China	Fang et al. (2008)
Trichoderma viride	MPar	MTCC, Chandigarh, India	Balamurugan et al. (2010)
Aspergillus niger AN400	MPar (Co-Met, C)	–	Marinho et al. (2011)
Acremonium sp. strain GFRC-1	Chlp (Cat, C)	From agri. soils	Kulshrestha and Kumari (2011)
Cladosporium cladosporioides Hu-01	Chlp (Cat, C)	–	Chen et al. (2012)
Aspergillus terreus JAS1 (JQ361749)[##]	Chlp (Co-Met, C)	Paddy field chlp-contam. soil sample from Vellore, Tamil Nadu, India	Silambarasan and Abraham (2013)
Aspergillus sp. F1 (JQ898687), *Penicillium* sp. F2 and F3 (JQ898688, JQ898689), *Eurotium* sp. F4 (JQ898690), and *Emericella* sp. F5 (JQ898691)[##]	Chlp, TCP	Soil of Agri. farm of Banaras Hindu University, Varanasi (25° 18′ N, 83° 3′ E)	Maya et al. (2012)

(continued)

Table 14.2 (continued)

Name of strain	Organophosphate compound (Cat/Co-Met utilization as C/P source)	Isolation (from) site	References
Trichoderma harzianum, *Rhizopus nodosus*	Chlp, Ethn (Cat, C)	Chlp- and Ethn-contam. soil	Harish et al. (2013)
Fusarium sp. CR10 (JX915255); *Fusarium oxysporum* CR9 (JX915246); *Fusarium* sp. GR4 and CR13 (JX915247); *Gibberella moniliformis* CR11, GR1, GR3, and CR4 (JX915252, JX915251, and JX915250); *Dipodascaceae* sp. GR2 and CR12 (JX915245); *Chaetomium globosum* CR1 and CR14 (JX915254)##	Chlp	Soil (treated with chlp) was collected from Zhejiang Academy of Agri. Sciences, Hangzhou, China	Chen et al. (2014)
Isaria farinosa	Chlp	Chlp-contam. soil samples from Idukki, Kerala, India	Karolin et al. (2015)
Penicillium citrinum, *Fusarium proliferatum*	MPar	Isolated from the ascidian *Didemnum ligulum*	Rodrigues et al. (2016)

Abbreviation: *Acti* activated, *agri* agriculture, *Che* chemical, *Cat* catabolic, *C* carbon, *chlp* chlorpyrifos, *Couma* coumaphos, *Co-met* co-metabolic, *contam* contaminated, *Diazin* diazinon, *Efflu* effluent, *Ethn* ethion, *facto* factory, *Fen* fenitrothion, *Mpar* methyl parathion, *Par* parathion, *pesti* pesticides, *Mala* malathion, *manuf* manufacturer, *Mono* monocrotophos, *N* nitrogen, *P* phosphorus, *poll* polluted, *res* research, *rhizos* rhizosphere, *sedi* sediment, *WW* wastewater, # 16S rRNA gene sequence, ## 18S rRNA gene sequence

addressed in literature. This includes the following: the first includes the catabolic utilization/biodegradation studies, where, organophosphate compound has been used as the sole source of C, and the second includes co-metabolic utilization/biodegradation studies, where another C compound (along with organophosphate compound) has been used as sources of C for growth (Singh 2009). The metabolic conversion of organophosphate compounds has been proposed to occur through pathways, each having multiple steps. In this chapter, parathion has been considered as a representative compound.

Till date, three different pathways for metabolic conversion of parathion have been reported as shown in Fig. 14.3 (Singh and Walker 2006). The first pathway involves an initial oxidative step to generate paraoxon which is hydrolyzed to generate 4-NP and diethyl thiophosphoric acid (DETP). For the second pathway, the first step is hydrolysis, leading to the formation of 4-NP and DETP. While the third pathway is reductive one facilitated under anaerobic condition [although some oxygen-insensitive reductase from *Bacillus* (Yang et al. 2007) and *Anabaena* sp. PCC7120 (Barton et al. 2004) has been reported]. The reactions involve reduction

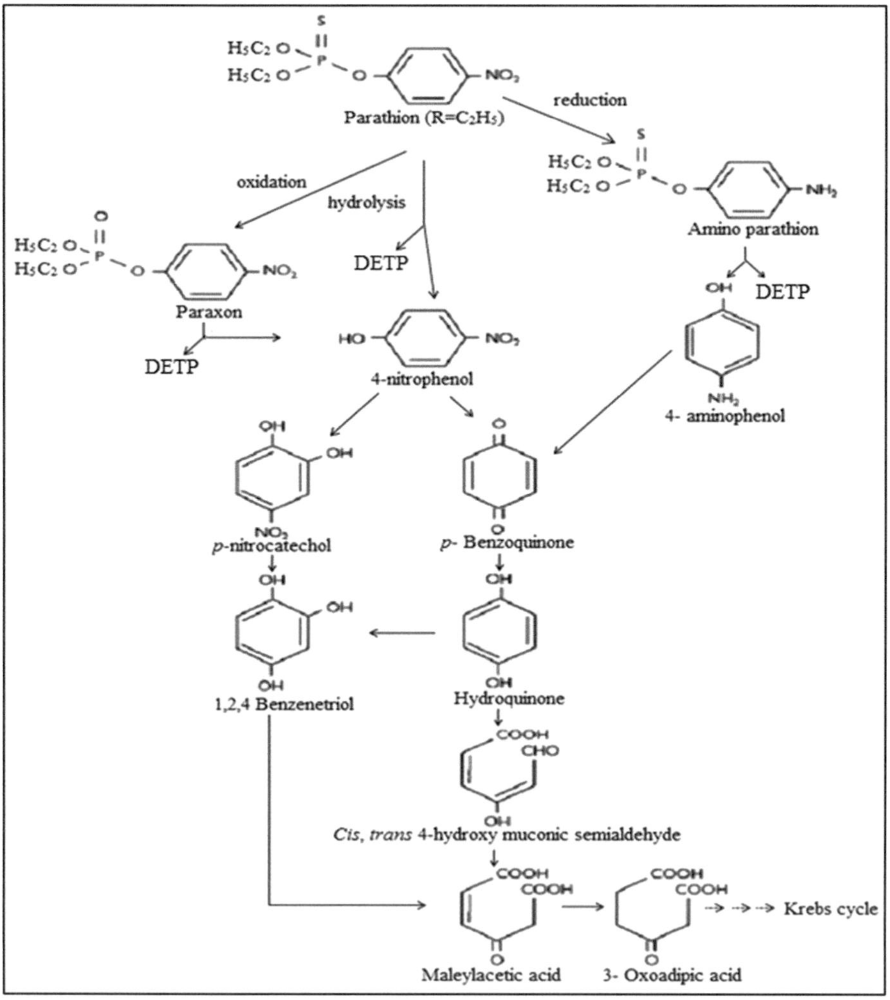

Fig. 14.3 Pathway of parathion biodegradation (Singh and Walker 2006)

of nitro group to amine (leading to formation of 4-aminoparathion), which up on further hydrolysis releases 4-aminophenol (4-AP) and DETP. In most of the literatures, the metabolisms of the main functional leaving groups are discussed. The fate of DETP, being common to all, is not followed.

It is clear from available literature that the second pathway (the hydrolysis one) is the most widely reported one. The 4-NP that is generated is reported to be utilized via two pathways: one that operates through formation of 4-NC and BT is more prevalent among Gram-positive bacteria [*Bacillus sphaericus* JS905 (Kadiyala and Spain 1998) and *Rhodococcus opacus* SAO101 (Kitagawa et al. 2004)], while the second that operates through formation of PBQ and HQ is more common among the Gram negatives [*Moraxella* sp. (Spain and Gibson 1991) and *Pseudomonas* sp.

strain WBC-3 (Zhang et al. 2009)]. However, in *Pseudomonas* sp. 1–7, both the pathways have been reported to be operative (Zhang et al. 2012b).

Very few reports on the degradation of parathion to paraoxon before hydrolysis of phosphotriester bond (i.e., the first pathway) were reported, except that from a mixed bacterial culture (Mastumura and Boush 1968; Tomlin 2000).

The third pathway has mainly been reported from a mixed bacterial consortium (by Munnecke and Hsieh 1976) under anaerobic environment. This pathway was also reported from aerobically growing *Bacillus* sp. (Sharmila et al. 1989 and Yang et al. 2007) and *Anabaena* sp. PCC7120 (Barton et al. 2004). The presence of possible involvement of oxygen-insensitive reductases is suggested for conversion in the aerobic bacteria (Barton et al. 2004). Very recently, Pailan and Saha (2015) reported evidence of two possible pathways (first, through 4-NP formation and, second, through 4-aminoparathion and 4-aminophenol) operative in *Pseudomonas* sp. strain BUR11. Through analytical techniques and growth-dependent experimental evidences, they reported on this aspect.

14.9 Overall Process of Organophosphate Metabolism

Several enzymes are reported to participate in the process of metabolism of organophosphate compounds. These can be broadly categorized into two major groups, namely, phase I and phase II enzymes. Phase I enzymes participate in reactions that makes the molecule much more polar, water-soluble, and amenable for enzymes of phase II to act. It may also be pointed that microbes can solubilize organophosphate by organic acid secretion and also by biosurfactants (Monteiro et al. 2007). In general, increase in solubility reduces half-life of the compounds rapidly. The major processes involved in metabolism are biodegradation, conjugation, and rearrangements. These include many chemical reaction types such as oxidation, reduction, dealkylation, ring cleavage, oxygenase, and peroxidize mechanisms.

Interaction of toxic organophosphate compound with microorganisms can proceed through three processes:

1. Transformation reaction leading to detoxification of parent organophosphate compound
2. Direct degradation and mineralization through catabolic pathway
3. Maintenance of cellular homeostasis

These three processes can occur together or in isolation depending up on what kind of genetic information the organism is equipped with.

Most of the literature has worked up on the second issue (Singh and Walker 2006; Pailan and Saha 2015). While, Longkumar et al. (2014) showed existence glutathione *S*-transferase mediated detoxification system in *Acinetobacter baumannii* strain DS002. The enzyme was reported to be involved in a dealkylation reaction that eventually reduced the toxicity of parent methyl parathion. There is a huge lacuna as far as the third issue is concerned. This issue is particularly true for those strain that do not have the capacity to degrade organophosphates but can tolerate them.

14.10 Quantitative Study of Microbial Metabolism

Most of the studies in literature have addressed the quantitative aspect of metabolism study by any one of the following two ways (Peter et al. 2014; Pailan and Saha 2015; Fang et al. 2016):

1. By monitoring gradual decrease in the amount of parent organophosphate compound in the growth medium (due to microbial metabolism) with respect to time
2. By monitoring gradual increase in the amount of hydrolytic intermediates followed by their subsequent decrease, indicating their utilization and metabolic conversion

As case study, for example, for metabolic study of parathion, the decrease in the residual amount of parathion in microbial culture inoculated test growth flask can be compared with blank (i.e., where no microbial inoculants are added) with respect to time as shown in Fig. 14.4a.

Another way of monitoring the metabolism is by quantifying the amounts of major hydrolytic intermediates produced as a result of the degradation of parent compound. As evident from Fig. 14.4b, by studying the fate of four major hydrolysis intermediates of parent organophosphate compound, parathion, one can conclude that the bacterial culture in the question can metabolically utilize parent organophosphate compound with concomitant formation of the first intermediate (4-nitrophenol, which accumulates in culture medium initially) followed by its gradual utilization (as its amount decreases) and then by formation of other intermediates (*p*-benzoquinone, hydroquinone, and benzenetriol). The temporal trend of the graph indicated the utilization of all the intermediates (as they decrease gradually).

Quantification of organophosphate compounds and its other hydrolytic intermediates can be carried out by HPLC technique. As evident from Fig. 14.5, compounds can be identified by comparing retention time of the test samples to that for authentic standards (from a well-known source like Sigma Aldrich). For quantification, specific peak area as well as height of the analytes from the test sample (extracted at different time intervals) is compared to that of the standard (for which standard curves are generated).

14.11 Identifying the Intermediate Compounds Produced Due to Metabolic Breakdown of Organophosphate

The reliable techniques to detect and identify different hydrolytic intermedites of organophosphate compound (e.g. parathion) are TLC, GC screening, GC-MS and LC-MS/MS followed by NIST (National Institute of Standard Technology) library

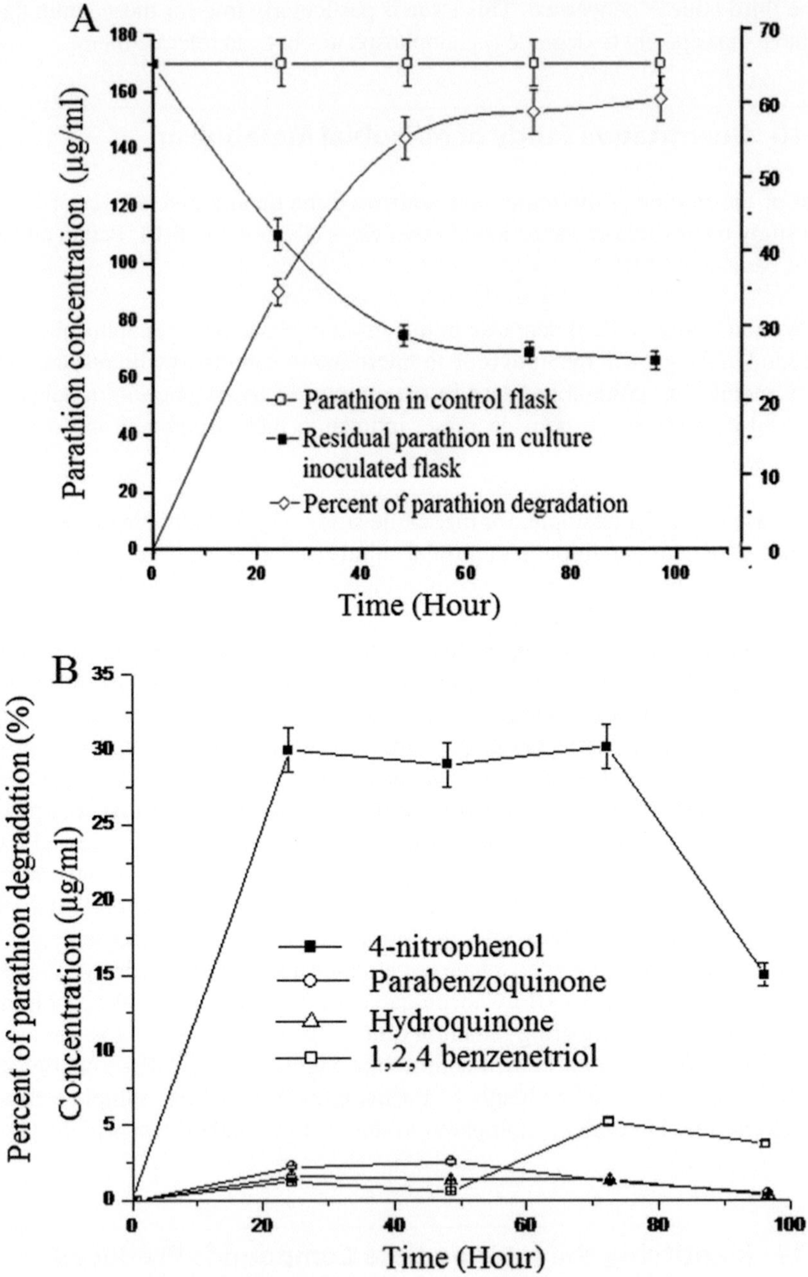

Fig. 14.4 Parathion degradation profile of BUR11. (**a**) Parathion degradation profile by the strain BUR11 and (**b**) fate of intermediates during parathion degradation by the strain BUR11 (Pailan and Saha 2015)

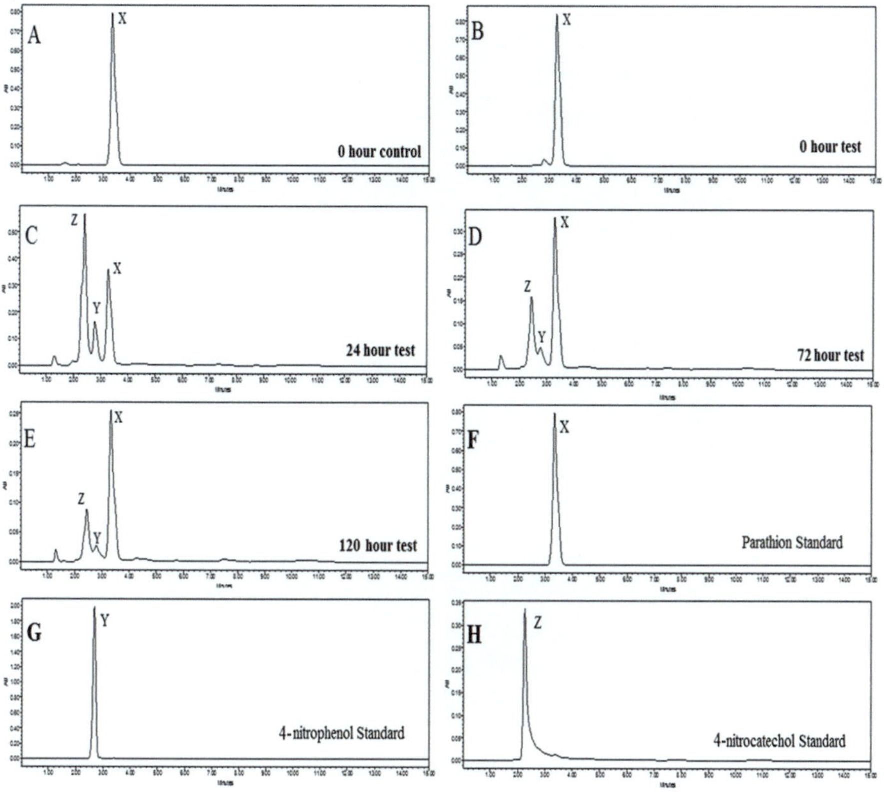

Fig. 14.5 Parathion degradation by a bacterial strain. The elution profile of each sample is shown as individual chromatograms. 0 h control sample (**a**), 0 h test sample (**b**), 24 h test sample (**c**), 72 h test sample (**d**), 120 h test sample (**e**), and elution profiles of standards (**f**, **g**, **h**). X-, Y-, and Z-labeled peak denotes parathion, 4-NP, and 4-NC, respectively

search. For the preliminary identification of hydrolytic intermediates during organophosphate (e.g., parathion) degradation, TLC is performed. Compounds were identified (Fig. 14.6) by comparing R_f value of the test samples to that for authentic standards (from Sigma Aldrich).

Through GC screening and library match, also the preliminary idea of hydrolytic intermediates can be obtained. However, for confirmed results, separation by GC-MS and/or LC-MS/MS techniques followed by the identification of intermediate compounds by comparing their mass spectrum profiles to that of the NIST library are universally accepted (Fig. 14.7)

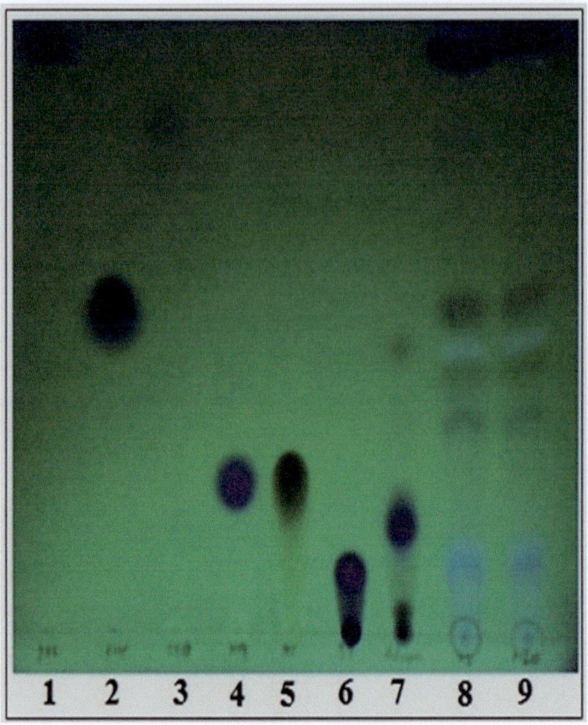

Fig. 14.6 Identification of metabolites of parathion degradation by TLC. Authentic standards **1**, parathion; **2**, 4-NP; **3**, PBQ; **4**, HQ; **5**, 4-NC; **6**, BT; **7**, 4-AP. And **8** and **9** correspond to 72 and 120 h extract of parathion-grown culture, indicating the detection of 4-NP, PBQ, HQ, and BT during the course of degradation (Pailan and Saha, 2015)

14.12 Factors That Affect Organophosphate Degradation

Several factors have been reported to affect the process of organophosphate degradation (both in soil and in laboratory batch cultures). These are as follows:

14.12.1 Substrate Concentration

Biodegradation of a particular pollutant depends upon the concentration of pollutant occurring in the contaminated site. Usually, a concentration which is too high may be toxic for the microbes, while lower concentration may not be sufficient to induce the microbial enzyme system involved in the degradation process (Block et al. 1993; Morra 1996). It has been reported that with the increasing concentration of organophosphate pesticides, there is a decrease in the microbial population (Shan et al. 2006). A dosage of 4 l/hac of chlorpyrifos was recorded to be inhibitory to the total soil microbial population (Pandey and Singh 2004). The average half-life of chlorpyrifos was reported to be increased with the increasing chlorpyrifos concentration of the soil (Hua et al. 2009).

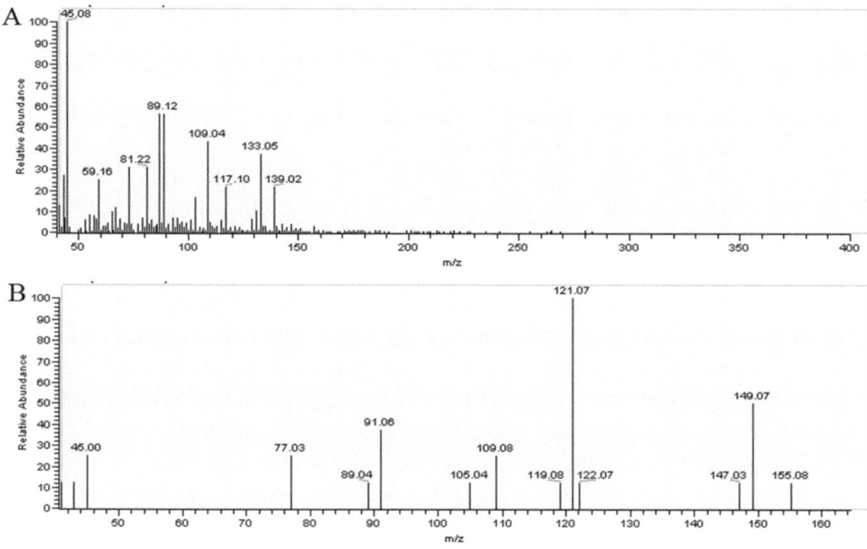

Fig. 14.7 GC-MS spectra obtained from the bacterial culture extract of parathion-grown broth culture. (**a**) 4-NP and (**b**) 4-NC were found as major compounds as hydrolysis products). The compounds were identified and confirmed from the NIST library

14.12.2 pH

It is one of the most important factors for the degradation of organophosphate compounds in soil and other habitats. Majority of the organophosphate pesticides are subject to base catalyzed hydrolysis at higher alkaline pH, around 8 (Greenhalgh et al. 1980). The degradation of chlorpyrifos was reported to be slow in acid soil (pH 4.7) and high in alkaline soil (pH 7.7–8.4), by Singh et al. 2003. Biodegradation of chlorpyrifos by *Bacillus laterosporus* DSP was reported to be enhanced by increasing the pH from 7 to 9 (Wang et al. 2006; Zhang et al. 2012b). A study of the effect of pH on biodegradation of malathion and dimethoate by *Pseudomonas frederiksbergensis* indicated decrease in half-life (almost by twofold) with increasing pH from neutral to pH 8 (Al-Qurainy and Abdel-Megeed 2009). For fungal culture, *Fusarium* sp. LK, biodegradation of chlorpyrifos was reported in the range of pH 6.5–9 (Wang et al. 2005).

14.12.3 Inoculum Size

The population of microorganisms involved in degradation is also reported to be an important factor. Inoculum size ranging from 10^6 to 10^8 cells/g of soil was suggested to be sufficient for bioremediation of pesticides from their contaminated sites (Comeau et al. 1993). Biodegradation of fenamiphos and chlorpyrifos was reported to be influenced by inoculum size, while no degradation of chlorpyrifos by *Enterobacter* sp. was recorded below an inoculum concentration of 103 cells/g of soil. When soil was supplied with less than 105 cells/g of soil, no biodegradation of fenamiphos was recorded (Singh and Walker 2006).

14.12.4 Bioavailability/Solubility

For proper biodegradation, it is very essential that the pollutant be available/made available to the degrading microorganism(s). In general, many organophosphate compounds have less water solubility, and this factor has been reported to be responsible for its decreased degradation (Alexander 1999). Many hydrophobic organophosphate pesticides become entrapped in the nanopores of the organic matter of the soil and thus are not available for biodegradation at all. Addition of suitable material that solubilizes the pollutant or selection of biosurfactant-producing microorganisms has been reported to make these hydrophobic molecules available for biodegradation. The biosurfactants desorb the hydrophobic chemicals so as to make them available for degradation (Aronstein et al. 1991; Brown and Jaffe 2006; Zhu and Zhou 2008).

Biosurfactants are anionic or neutral (some are cationic) rhamnolipids, glycolipids, lipopeptides, phospholipids, fatty acids, particulate compounds, etc. which are of microbial origin and are used for solubilization of hydrophobic pollutants, with the aim of making it bioavailable and more suitable for biodegradation (Monteiro et al. 2007).

14.13 Chemotaxis and Metabolism of Organophosphate Insecticides

The movement of bacteria either toward or away from a chemical gradient is called bacterial chemotaxis. Chemotaxis is a natural phenomenon and is reported from diverse groups of bacteria. A chemical compound that affects the bacterium's movement is called the chemoeffector (stimulant). Chemicals that attract bacteria are called chemoattractants, and chemicals that repel them are called chemorepellents. Chemotaxis can be classified into two types, namely, metabolism dependent and metabolism independent (Pandey and Jain 2002; Baker et al. 2005). Till date several assays have been developed to check the chemotactic activity of a bacterium. These are swarm plate assay, drop plate assay, agarose-plug assay, etc. (Bhushan et al. 2000; Samanta et al. 2000; Pandey et al. 2002; Bhushan et al. 2004). As far as chemotaxis to pesticides/insecticides are concerned, survey of literature revealed reports pertaining only to two bacteria, namely, *Pseudomonas* sp. strain ADP (Liu and Parales 2009) and *Ralstonia eutropha* JMP134 (Hawkins and Harwood 2002). Both of them are reported to exhibit chemotaxis-mediated biodegradation of atrazine and 2,4-dichlorophenoxyacetate herbicides, respectively. There is hardly any literature on chemotaxis of bacteria toward organophosphate compounds except by *Pseudomonas* sp. strain WBC-3 (Zhang et al. 2008) and by *Pseudomonas putida* DLL-1. However, the latter publication is only available in Chinese language, and its English version is currently not available (Wen et al. 2007).

Recently, *Pseudomonas* sp. strain BUR11 was reported to exhibit positive chemotaxis toward two OP compounds, namely, parathion and chlorpyrifos (as well as their degraded intermediate products 4-NP, 4-AP, and TCP). Through a series of plate-based qualitative assays (drop plate & swarm plate) and quantitative assay, the chemotactic response was confirmed for the strain BUR11 (Figs. 14.8 and 14.9). However, the authors could not conclude whether this chemotactic response was metabolism dependent or independent. The study concluded on the importance of genetic analyses for better understanding of this chemotactic process; nevertheless, this was one of the unique confirmed reports of chemotactic response of bacterium toward organophosphate compounds in recent times (Pailan and Saha 2015).

14.14 Discovery of Organophosphate-Degrading Enzyme

Organophosphate-degrading enzyme was first described by Mazur in 1946 when he discovered the hydrolysis of diisopropyl fluorophosphate (DFP) by enzymes found in rabbit and human tissue extracts (Mazur 1946). For the first time, DFPase and sarinase enzymes were found to degrade organophosphate compounds. Later, the DFPase activities of several bacterial isolates for organophosphate degradation were described by Attaway et al. (1987). In 1992, the Nomenclature Committee of the International Union of Biochemistry and Molecular Biology listed them in the category of phosphoric triester hydrolases. These enzymes were further categorized into two subgroups based on their substrate specificities. The first subgroup is the organophosphorus hydrolases (also referred to as paraoxonase and phosphotriesterase; PTE) that prefer the substrates paraoxon and P-esters, which have P–O and P–S bond. The second subgroup is diisopropyl fluorophosphates (also including

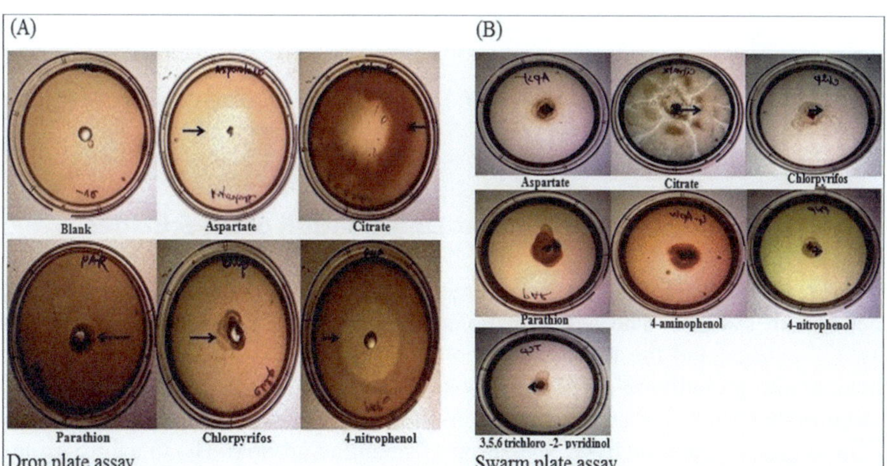

Fig. 14.8 (**a**) Drop plate assay and (**b**) swarm plate assay. Qualitative chemotactic response of BUR11 toward parathion, chlorpyrifos, 4-NP, 4-AP, and TCP (Pailan and Saha 2015)

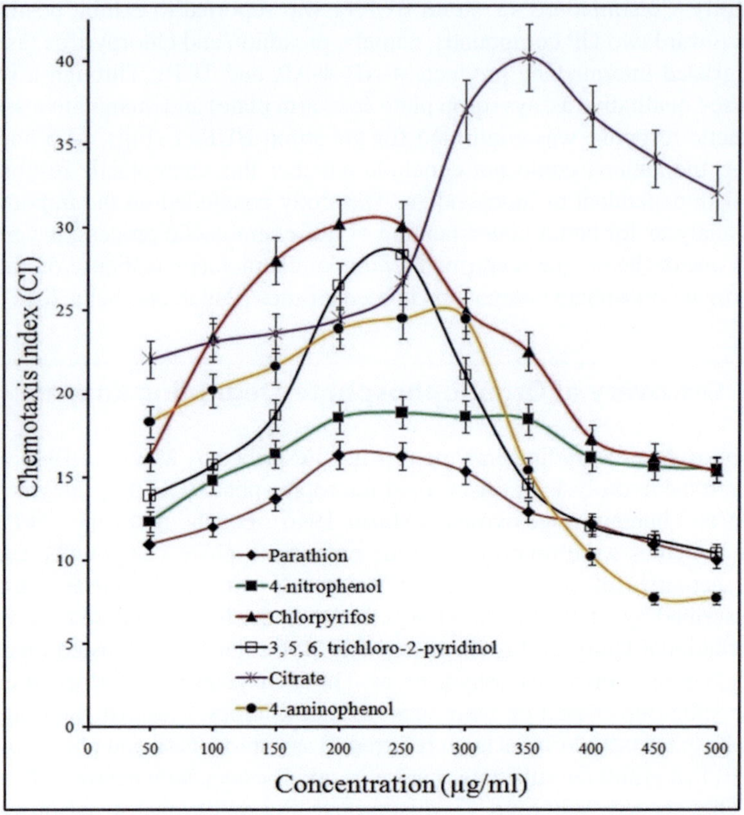

Fig. 14.9 Quantitative capillary assay. Quantitation of the chemotactic response and determination of the optimal response concentration for BUR11 chemotaxis toward different test compounds using capillary assays (Pailan and Saha 2015)

organophosphorus acid anhydrolase, *OPAA*), which are most active against organophosphate compounds with P–F or P–CN bonds (Cheng and DeFrank 2000).

14.14.1 Mechanism of Enzymatic Degradation of Insecticides

In case of insecticide degradation, three main enzymes are involved under two metabolism systems. The first metabolism system includes enzymes like hydrolases, esterases, and the mixed function oxidases (MFO), and the second system includes the glutathione *S*-transferases (GST) system (Li et al. 2007a). Several enzymes that catalyze metabolic reactions including hydrolysis, oxidation, addition of an oxygen to a double bond, oxidation of an amino group (NH_2) to a nitro group, addition of a hydroxyl group to a benzene ring, dehalogenation, reduction of a nitro group (NO_2) to an amino group, replacement of a sulfur with an oxygen, metabolism of side chains, and ring cleavage are required to degrade toxic insecticide into nontoxic intermediates (Ramakrishnan et al. 2011).

In most of the microorganisms, insecticides can be metabolized by a three-phase process. In phase I metabolism, the initial properties of parent compounds are transformed through oxidation, reduction, and hydrolysis to produce a more water-soluble and usually a less toxic product than parent. The second phase (phase II) involves conjugation of a pesticide or insecticide metabolite to a sugar or amino acid, which increases the water solubility and reduces toxicity, compared to the parent pesticide/insoluble metabolite. The third phase (phase III) involves conversion of phase II metabolites into secondary conjugates, which are also nontoxic. To carry out these processes, microorganisms like fungi and bacteria produce several intracellular or extra cellular enzymes including hydrolytic enzymes, peroxidases, oxygenases, etc. to accomplish the complete mineralization of toxic insecticides (Van Eerd et al. 2003).

14.14.2 Enzymes and Gene(s) Involved in Organophosphate Compounds Degradation

The organophosphate compounds are tri-esters of phosphates and their derivatives. Therefore, the most common enzyme that might be involved in their degradation is the esterase. Esterases are also categorized as hydrolases [enzyme that hydrolyzes a broad range of aliphatic, aromatic esters and organophosphates, Park and Kamble (2001)]. Various types of hydrolases involved in the degradation of organophosphate insecticides are as follows:

14.14.2.1 Phosphotriesterase (PTE)

Till date, the most well-addressed and discussed organophosphate-degrading enzyme is phosphotriesterases (PTE; Theriot and Grunden 2011). It is a metalloenzyme that hydrolyzes a variety of toxic organophosphate compounds (mainly those that act as nerve agents). PTE was first isolated from *Pseudomonas diminuta* MG (Serdar et al. 1982) and *Flavobacterium* sp. (Sethunathan and Yoshida 1973). This enzyme shows a highly catalytic activity toward various organophosphate insecticides. The PTE was further subcategorized into three groups on the basis of insecticide it acted upon (i.e., based on substrate). These are:

A. Organophosphorus hydrolase (OPH)
B. Methyl parathion hydrolase (MPH)
C. Organophosphorus acid anhydrolase (*OPAA*)

These three are encoded by *opd*, *mpd*, and *opaA* genes, respectively.

A. Organophosphorus Hydrolase (OPH)
   Many of the enzymes known to hydrolyze organophosphorus esters are referred as organophosphorus hydrolase [OPH; EC 3.1.8.1]. OPH is the most widely studied bacterial enzyme in OP degradation, exhibiting high catalytic activity and wide range of organophosphate substrate specificity (oxon and thion) (Yang et al. 2006; Ortiz-Hernandez and Sanchez-Salinas 2010). It is a zinc-containing

homodimeric membrane protein reported from *Flavobacterium* sp. strain ATCC 27551 and *Pseudomonas diminuta* MG (Sethunathan and Yoshida 1973; Serdar et al. 1982). It can hydrolyze organophosphate compounds at a rate approaching the diffusion limits (Horne et al. 2002a). The gene (*opd*) coding for OPH enzymes has been reported to be plasmid borne. The first *opd* gene (within a 66kb plasmid, pCMS1) was reported from *Pseudomonas diminuta* (Sethunathan and Yoshida 1973; Serdar et al. 1982; Mulbry et al. 1986; Singh and Walker 2006). Similar *opd* gene has been identified from various *Pseudomonas* strain by using Southern hybridization analysis (Chaudhry et al. 1988). *Flavobacterium* sp. strain ATCC 27551 and *Pseudomonas diminuta* MG contain identical *opd* genes as well as the OPHs purified from these have identical or very similar in amino acid sequences (Serdar et al. 1982; Mulbry and Karns 1989; Siddavattam et al. 2003), but it is not clear how this has occurred as the genes are on very different plasmids (Harper et al. 1988). Omburo et al. 1992 isolated an *opd* gene encoding a 40 kDa homodimer parathion hydrolase, containing divalent zinc ions as a cofactor. Horne et al. (2002a) suggested that PTE is a 384-amino-acid protein with a molecular mass of approximately 35 kDa when it is cleaved from its signal peptide. The two native Zn^{2+} ions of this enzyme can be substituted with either Co^{2+}, Ni^{2+}, Cd^{2+}, or Mn^{2+} with/without the restoration of catalytic activity. Recent findings have shown that two metal atoms are closely associated and the water molecule that attacks the phosphoryl center is bound directly to the binuclear metal center (Benning et al. 1995; Vanhooke et al. 1996).

B. Methyl Parathion Hydrolase (MPH)

Singh (2009) reported that MPH is present in several phylogenetically unrelated bacteria and is active against several organophosphate compounds but has a narrower substrate range than OPH. The crystal structure of the MPH (which is a member of the β-lactamase superfamily) from *Pseudomonas* sp. WBC-3 has been solved by Dong et al. (2005). MPH is a dimer in which each subunit has a mixed-hybrid, binuclear zinc center. MPH is not similar to any other PTEs, even though several PTEs can degrade methyl parathion. The MPH is coded by *mpd* gene. Molecular studies and phylogenetic analyses confirmed that *mpd* genes have evolved separately from *opd* genes. Unlike *opd* genes, most of the known *mpd* genes have been isolated from one country (China), indicating that the environment has an influence on *mpd* evolution (Singh 2009). Whole-genome sequence analysis also suggests that *mpd* and β-lactamase gene homologues are present in other bacteria, such as *Methylibium petroleiphilum* (locus tag NC 008825), *Azoarcus* sp. (locus tag AM 406670), *Leptothrix cholodnii* (locus tag CP 00001013), *Chromobacterium violaceum* (locus tag AE O16825), and *Sinorhizobium meliloti* 1021 (locus tag AE 006469). Interestingly, an AHL lactonase (*N*-acyl homoserine lactone) from *Bacillus thuringiensis* also belongs to the β-lactamase superfamily. AHL lactonase has some promiscuous PTE activities, so it is possible that OPH and MPH have evolved from different lactonase enzymes (Afriat et al. 2006).

C. Organophosphorus Acid Anhydrolase (OPAA)

Another organophosphate-degrading enzyme that has received considerable attention is *OPAA* [encoded by *opaA* (organophosphorus acid anhydrolase) gene], first isolated from halophilic species *Alteromonas undina* and *Alteromonas haloplanktis* (Cheng et al. 1993, 1999). This enzyme belongs to the dipeptidase family and does not share enzyme or gene-sequence homology either with OPH or MPH. This indicates that the organophosphate-degrading function of *OPAA* might have evolved from different progenitors (Singh 2009). *OPAA*s from the species of *Alteromonas* sp. JD6.5, *Alteromonas undina*, and *Alteromonas haloplanktis* are structurally and functionally similar to each other. They share a molecular weight between 50 and 60 kDa, having an optimum pH from 7.5 to 8.5 and temperature optima ranging from 40 °C to 55 °C, and require Mn^{2+} for their maximum catalytic activity (Cheng et al. 1997). *OPAA*s are highly active and more specific to OP nerve agents than OPHs. The amino acid sequence of *OPPA* of *Alteromonas* sp. JD 6.5 shares 49% and 31% similarity with dipeptidase or prolidase and aminopeptidase of *E. coli* (Theriot and Grunden 2011).

Since the property of organophosphate degradation is gene mediated, the same can be used to develop novel strains for in situ application purpose by genetic engineering process. In most of the cases, the genes are defined to be located either in plasmids or in chromosomes (Concepcio'n et al. 2012). In this way, many authors reported organophosphate degradation property using recombinant bacterial strains (Yang et al. 2005; Xu et al. 2007). Very recently, Farivar et al. (2017) reported construction of a recombinant organophosphate-degrading *Pseudomonas plecoglossicida* strains with *opd* gene from *Flavobacterium* sp. ATCC 27551 using the pUC57 plasmid.

A thorough list of organophosphate-degrading enzymes, genes, and source microorganisms from which the enzymes were isolated so far is summarized in Table 14.3.

14.14.2.2 Other Enzymes Involved in Insecticide Degradation

Survey of literature suggested some other enzymes having organophosphate-degrading activities. These are as follows:

14.14.2.2.1 Oxidoreductase

Oxidoreductases are a broad group of enzymes that carry out transfer of electrons from one molecule (the reductant or electron donor) to another (the oxidant or electron acceptor). Many of these enzymes require additional cofactors, to act as either electron donors, electron acceptors, or both. These enzymes have applications in bioremediation. There are the enzymes that catalyze an oxidation/reduction reaction by including the molecular oxygen (O_2) as electron acceptor. In these reactions, oxygen is reduced into water (H_2O) or hydrogen peroxide (H_2O_2). The oxidases are a subclass of the oxidoreductases. These enzymes not only catalyze oxidation reduction reaction of toxic compounds but also catalyze the oxidation reaction of various pesticides, insecticides, as well as herbicides (Scott et al. 2008).

Table 14.3 List of organophosphate-degrading enzymes, genes, and their source microorganisms

Organisms	Encoding genes (accession no.)	Degrading enzyme	References
Pseudomonas diminuta	*opd* (M29593)	OPH	Serdar et al. (1982)
Flavobacterium sp.	*opd* (M22863)	OPH	Harper et al. (1988)
Pseudomonas diminuta MG,	*opd* (M20392)	Phosphodiesterase	McDaniel et al. (1988)
Flavobacterium sp. strain ATCC 27551	*opd* (M29593)	Parathion hydrolase gene	Mulbry and Karns (1989)
Flavobacterium sp. ATCC27551	*opd* (AJ421424) (M20392)	OPH	Mulbry and Karns (1989)
Escherichia coli, Bacillus cereus	ND	Phosphonatase	Chen et al. (1990)
Nocardia sp.	*adpB*	ADPase	Mulbry (1992)
Mycobacterium sp. or *Nocardia* sp. strain B-1	*opaA* (AAA25371)	–	Mulbry (1992)
Pseudomonas spp.	*glpA and B*	C-P lyase	Penaloza-Vazquez et al. (1995)
Burkholderia caryophylli	*pehA*	PEH	Dotson et al. (1996)
Alteromonas sp. JD6.5	*opaA*	OPAA	Cheng et al. (1996)
Alteromonas undina *Alteromonas haloplanktis* ATCC 23821	*opaA* (U29240) Prolidase gene (U56398)	OPAA-2 OPAA	Cheng et al. (1996, 1997)
Nocardioides sp. strain C190	*trzN*	*s*-triazine hydrolase	Mulbry et al. (2002)
Burkholderia sp. strain NF100	*opd/mpd*	Fenitrothion-hydrolyzing enzyme	Hayatsu et al. (2000)
Plesiomonas sp. M6	*mpd* (AF338729)	MPH	Zhongli et al. (2001)
Moraxella sp.	*oph*	OPH	Shimazu et al. (2001)
Agrobacterium radiobacter	*opdA* (AY043245)	OPDA	Horne et al. (2002a)
Pseudomonas monteilii	*hocA*	HOCA (hydrolysis of coroxon)	Horne et al. (2002b)
Flavobacterium balustinum	*opd* (AJ426431)	Parathion hydrolase	Siddavattam et al. (2003)
Flavobacterium sp. ATCC 27551	*opd* (AJ421424)	–	Siddavattam et al. (2003)
Delftia acidovorans	*pdeA gene* (AF548455)	Phosphodiesterase (PdeA)	Tehara and Keasling (2003)
Escherichia coli	*pepA*	AMPP (aminopeptidase P)	Jao et al. (2004)

(continued)

Table 14.3 (continued)

Organisms	Encoding genes (accession no.)	Degrading enzyme	References
Pseudomonas pseudoalcaligenes	*ophc2* (AJ605330)	OPHC2	Ningfeng et al. (2004)
Pseudomonas sp. WBC-3	*mpd* (AY251554)	MPH	Liu et al. (2005)
Brucella melitensis mp-7 (AY331581)	*mpd* (AY627039)	MPH	Zhang et al. (2005, 2006a, b)
Achromobacter xylosoxidans mp-2 (AY331576)	*mpd* (AY627034)	MPH	Zhang et al. (2005, 2006a, b)
Pseudaminobacter sp. mp-1 (AY331575) strain no. AF072542	*mpd* (AY627033)	MPH	Zhang et al. (2005, 2006a, b)
Pseudaminobacter salicylatoxidans (AY331575), strain no AF072542	*mpd* (AY627033)	MPH	Zhang et al. (2005)
Ochrobactrum tritici mp-3, mp-4, mp-5, mp-6 (AY331577, AY331578, AY331579, AY331580), strain no. AF508089	*mpd* (AY627035, AY627036, AY627037, AY627038)	MPH	Zhang et al. (2005)
Burkholderia sp. FDS-1 (AY550913)	*mpd2/opd* (DQ173274, AY646835)	MPH	Zhang et al. (2006a, b)
Stenotrophomonas sp. strain YC-1 (DQ537219)	*mpd* (DQ677027)	MPH	Yang et al. (2006)
Burkholderia sp. NF100	*fedA, fedB*	Fenitrothion hydrolase gene (OPH)	Tago et al. (2006)
Flavobacterium sp. MTCC 2495	*mpd* (AY766084)	OPH	Kumar et al. (2006)
Pseudomonas putida DLL-1	*mpd*	MPH	Liu et al. (2005)
Pseudomonas pseudoalcaligenes	*ophc2*	OPH	Chu et al. (2006)
Sphingomonas sp. DSP-2 (AY994060)	*mpd* (DQ356953)	MPH	Li et al. (2007a, b)
Sphingomonas sp. CDS-1	*mpd*	MPH	Jiang et al. (2007)
Burkholderia sp. JBA3	*ophB* (EF495210)	OPH	Taesung et al. (2007)
Arthrobacter sp. L1	*mpd* (EF055988)	MPH	Li et al. (2008a, b)
Pseudomonas sp. (DSP-1, DSP-3), *Sphingomonas* sp. DSP-2, *Stenotrophomonas* sp. DSP-4	*mpd*	MPH	Li et al. (2008a, b)
Pseudomonas stutzeri strain HS-D36	*mpd*	MPH	Wang et al. (2008)

(continued)

Table 14.3 (continued)

Organisms	Encoding genes (accession no.)	Degrading enzyme	References
Pseudomonas stutzeri strain HS-D36	*mpd* (EF515812)	MPH	Guo et al. (2009)
Ochrobactrum sp. Yw18	*mpd* (DQ843607)	MPH	Singh (2009)
Ochrobactrum sp. M231	*mpd* (EU596456)		Tian et al. (2010)
Stenotrophomonas sp. SMSP-1 (EU312979)	*ophc2* (EU651813)	OPHC2	Shen et al. (2010a, b)
Lactobacillus brevis (WCP902)	*opd* B	–	Islam et al. (2010)
Pseudomonas sp.	Carboxyl esterase gene	Carboxyl esterase	Goda et al. (2010)
Sphingomonas sp. JK1	*opd* (EU709764)	OPH	Kumar and D'Souza (2010)
Burkholderia cepacia	*mpd B* (DQ001540)	MPH	Ekkhunnatham et al. (2012)
Bacillus pumilus W1	*opd A*	OPH	Ali et al. (2012)
Stenotrophomonas maltophilia MHF ENV20	*mpd*	OPH	Dubey and Fulekar (2012)
Kocuria sp.	*opd*	OPH	Neti and Zakkula (2013)
Pseudomonas sp. strain YF-5	*mpd*	MPH	Liu et al. (2014)
Sphingomonas sp. strain TDK1 and *Sphingobium* sp. strain TCM1		Haloalkylphosphorus hydrolases (TDK-HAD, TCM -HAD)	Abe et al. (2014)
Pseudomonas sp. BF1-3 (KJ849233)	*ophB*	OphB	Barman et al. (2014)
Acinetobacter sp.	*AbOPH* gene	OPH	Chen et al. (2015)
Sphingomonas sp. DC-6	*dmhA*	Amidohydrolase (DmhA)	Chen et al. (2016)
Reports from fungi			
Pleurotus ostreatus Chaetomium thermophilum		Laccase	Amitai et al. (1998)
Aspergillus niger	*opd*	A-OPH	Liu et al. (2001)
Penicillium lilacinum	*opd*	OPH	Liu et al. (2004)
Cladosporium cladosporioides Hu-01	–	CHP (chlorpyrifos hydrolase)	Gao et al. (2012)

14.14.2.2.2 Mixed Function Oxidase (MFO)

In the reaction catalyzed by the MFO (EC 1.14.14.1), an atom of one molecule of oxygen is incorporated into the substrate, while the other is reduced to water. For this reason, the MFO requires nicotinamide-adenine dinucleotide phosphate (NADPH) and O_2 for its operation. It is an enzyme system comprising of two enzymes, cytochrome P450 and NADPH-cytochrome P450 reductase; both are membrane proteins. They are also known as cytochrome P-450-dependent monooxygenase or P450 system. The genes encoding the different isozymes comprise a superfamily of over 200 genes grouped into 36 families based on their sequence similarity. Cytochrome P450 enzymes are active in the metabolism of a wide variety of xenobiotics (Khaled et al. 2012). The cytochrome P450 family is a large, well-characterized group of monooxygenase enzymes that have long been recognized for their potential in many industrial processes, particularly due to their ability to oxidize or hydroxylate substrates in an enantiospecific manner using molecular oxygen (Urlacher et al. 2004). Many cytochrome P450 enzymes have a broad substrate range and have been shown to catalyze biochemically recalcitrant reactions such as the oxidation or hydroxylation of nonactivated carbon atoms. These properties are ideal for the remediation of environmentally persistent pesticide residues. Over 200 subfamilies of P450 enzymes have been found across various prokaryotes and eukaryotes. MFOs metabolize a wide range of compounds such as OPs, carbamates, pyrethroids, DDT, inhibitors of the chitin synthesis, juvenile hormone mimics, etc. (Alzahrani 2009).

14.14.2.2.3 Glutathione *S*-Transferase (GST)

The GSTs (EC 2.5.1.18) are a group of enzymes that catalyze the conjugation of hydrophobic components with reduced glutathione. In this reaction, the thiol group of glutathione reacts with an electrophilic place in the target compound to form a conjugate which can be metabolized or excreted. GSTs are involved in many cellular physiological activities such as detoxification of endogenous and xenobiotic compounds, intracellular transport, biosynthesis of hormones, and protection against oxidative stress (Sheehan et al. 2001; Hayes et al. 2005; Oakley 2005). Broadly, GSTs are divided into four major families: (a) cytosolic GSTs, (b) mitochondrial GSTs, (c) microsomal GSTs, and (d) bacterial fosfomycin resistance proteins (Armstrong 1997; Hayes et al. 2005). A very recent report by Longkumar et al. (2014) revealed that GST was involved in dealkylation of methyl parathion (OP compound) by a bacterial strain *Acinetobacter baumannii* DS002. Unlike in other organophosphate-degrading bacterial strains, in the genome of *Acinetobacter baumannii* DS002, there is no conserved gene encoding an organophosphate-degrading enzyme. The absence of such *opd* gene and the induction of a GST-like protein in the presence of organophosphate insecticides suggested the existence of a novel organophosphate-degrading pathway in *Acinetobacter baumannii* DS002. Longkumar and his colleagues also discovered the existence of multiple *gst* genes in *Acinetobacter baumannii* DS002 and observed the expression of these *gst* genes and involvement of resulting GST enzyme in dealkylation of methyl parathion that eventually reduces toxicity of the parent compound (Longkumar et al. 2014).

14.15 Role of OPH in Phosphate Acquisition from Organophosphate Compounds

Among the PTEs, OPH (a metalloenzyme requiring Zn as cofactor) is the most well-studied and characterized enzyme as far as structural and catalytic properties are concerned (Omburo et al. 1992; Kuo and Raushel 1994). It is best studied from *Brevundimonas diminuta* (recently reclassified as *Sphingopyxis wildii*) (Parthasarathy et al. 2017a). It located in the periplasmic space as multi-protein complexes, and it interacts with several systems like phosphate-specific transport (Pst) system, ABC transporters, and efflux pump AcrZ/TolC. It is reported to anchor to periplasmic face of the inner membrane through a diacylglycerol linked to the invariant cysteine residue. This enzyme also contains a signal peptide with a conserved cysteine residue at the junction of its cleavage site. The signal peptide contains a characteristic Tat motif which is common for proteins that are translocated across the inner membrane in a prefolded state (Parthasarathy et al. 2016). Based on bioinformatic analyses, the c-terminal of OPH has been predicted to be in the cytoplasmic side (Parthasarathy et al. 2017a, b). Apart from triesterase activity, this enzyme has also been shown to possess lactonase activity. Due to that, OPH has been hypothesized to have evolved from lactonases (whose function s for quorum quenching) for the uptake of phosphate from the surrounding environment (Afriat-Jurnou et al. 2012).

It seems PTEs located in the periplasmic space converts organic organophosphate (that enters into the periplasmic space through after crossing the outer membrane) into phosphodiesters which ultimately gets converted into inorganic phosphate by the combined action of phosphodiesterase and phosphatase. OPH has been postulated to be involved in phosphate acquisition from organophosphate compounds through its interaction with components of the outer membrane (such as ABC-type transporters, TolC, etc.) known to be involved in phosphate transport in bacterial cells (Parthasarathy et al. 2016). Although these studies provide some idea toward the utilization of phosphotriesterases as the sole source of phosphate (at least in *Sphingopyxis wildii*), there is a huge lacuna as far as the transport processes operate in this organism.

14.16 Concluding Remarks

In spite of advances in cultivation methods, the total number of culturable microbes recoverable from any environmental niche is very low compared to what exists naturally. The wealth of information, regarding organophosphate metabolism that we have gained from existing diversity, is only the tip of the iceberg as we are far from knowing the exact boundaries of microbial diversity on earth. Moreover, a lot more studies has to be carried out with anaerobic microbes and their metabolic studies with respect to organophosphate compounds. Compared to aerobic metabolism, during anaerobic process more substrates are needed to be metabolized to provide equitable amount of ATP, and prospect of cleaning xenobiotic substrate is more through anaerobic degradation than aerobic degradation. Thus, more systematic studies for exploration of organophosphate biodegradation by anaerobic microorganisms should be made. This will not only increase our bio-resource in terms of novel microbes, gene, and enzyme pool for biotechnological aspect of the environmental cleanup process but also may lead to complete understanding of the overall degradation process and their links with other ecological processes on our planet earth. Microbial metabolism of organophosphate compounds in the environment is a complex, less understood process that depends upon the community diversity of the microflora residing in the habitat, energy, and nutrient flow as well as stress response metabolism of microbes. Unfortunately due to lack of our understanding toward holistic system wide understanding of complex interaction between degrading microbes, their genes, enzymes and multivariate environmental factors along with the complex microbial community (de Lorenzo 2008). Very recently, in order to understand the relationships in holistic manner, metagenomic approach was undertaken and it has shown promising results (Jeffries et al. 2018). The results of such approach highlighted the value holistic system-wide metagenomic approaches as a tool to predict microbial degradation in the context of the ecology of contaminated habitats. As pointed earlier, understanding the adaptation strategies taken by a microorganism to tolerate organophosphate toxicity and maintain cellular homeostasis will help us to understand the metabolism in a better way. Another huge lacuna is the process of signaling which facilitates the degradation process. Future studies will incorporate similar approaches to enrich our understanding the relationship.

Acknowledgments Authors are grateful to SERB, New Delhi, for providing fund to carry out work on organophosphate degradation and to the University of Burdwan, Burdwan, West Bengal.

References

Abe K, Yoshida S, Suzuki Y, Mori J et al (2014) Haloalkylphosphorus hydrolases purified from *Sphingomonas* sp. strain TDK1 and *Sphingobium* sp. strain TCM1. Appl Environ Microbiol 80:5866–5873

Abhilash PC, Singh N (2009) Pesticide use and application: an Indian scenario. J Hazard Mater 165:1–12

Abraham J, Silambarasan S (2013) Biodegradation of chlorpyrifos and its hydrolyzing metabolite 3,5,6-trichloro-2-pyridinol by *Sphingobacterium* sp. JAS3. Process Biochem 48:1559–1564

Afriat L, Roodveldt C, Manco G, Tawfik DS (2006) The latent promiscuity of newly identified microbial lactonases is linked to a recently diverged phosphotriesterase. Biochemist 45:13677–13686

Afriat-Jurnou L, Jackson CJ, Tawfik DS (2012) Reconstructing a missing link in the evolution of a recently diverged phosphotriesterase by active-site loop remodeling. Biochemistry 51:6047–6055

Ajaz M, Rasool SA, Sherwani SK, Ali TA (2012) High profile chlorpyrifos degrading *Pseudomonas putida* MAS-1 from indigenous soil: gas chromatographic analysis and molecular characterization. Int J Basic Med Sci Pharma 2:58–61

Aktar MW, Paramasivam M et al (2009) Impact assessment of pesticide residues in fish of Ganga river around Kolkata in West Bengal. Environ Monit Assess 157:97–104

Alexander M (1999) Biodegradation and bioremediation. Academic Press, London

Ali M, Naqvi TA, Kanwal M et al (2012) Detection of the OP degrading gene *opd*A in the newly isolated bacterial strain *Bacillus pumilus* W1. Ann Microbiol 62:233–239
Al-Qurainy F, Abdel-Megeed A (2009) Phytoremediation and detoxification of two organophosphorous pesticides residues in Riyadh area. World Appl Sci J 6:987–998
Alzahrani AM (2009) Insects cytochrome P450 enzymes: evolution, functions and methods of analysis. Global J Mol Sci 4:167–179
Amitai G, Adani R et al (1998) Oxidative biodegradation of phosphorothiolates by fungal laccase. FEBS Lett 438:195–200
Anwar S, Liaquat F, Khan QM, Khalid ZM, Iqbal S (2009) Biodegradation of chlorpyrifos and its hydrolysis product 3,5,6-trichloro-2-pyridinol by *Bacillus pumilus* strain C2A1. J Hazard Mater 168:400–405
Armstrong RN (1997) Structure, catalytic mechanism, and evolution of the glutathione transferases. Chem Res Toxicol 10:2–18
Aronstein BN, Calvillo YM, Alexander M (1991) Effect of surfactants at low concentrations on the desorption and biodegradation of sorbed aromatic compounds in soil. Environ Sci Technol 25:1728–1731
Arora PK, Sasikala C et al (2012) Degradation of chlorinated nitroaromatic compounds. Appl Microbiol Biotechnol 93:2265–2277
Attaway H, Nelson JO et al (1987) Bacterial detoxification of diisopropyl fluorophosphate. Appl Environ Microbiol 53:1685–1689
Awad NS, Sabit HH et al (2011) Isolation, characterization and fingerprinting of some chlorpyrifos-degrading bacterial strains isolated from Egyptian pesticides-polluted soils. Afr J Microbiol Res 5:2855–2862
Baker MD, Wolanin PM, Stock JB (2005) Signal transduction in bacterial chemotaxis. Bio Essays 28:9–22
Balamurugan K, Ramakrishnan M et al (2010) Biodegradation of methyl parathion and monochrotophos by *Pseudomonas aeruginosa* and *Trichoderma viridae*. Asian J Sci Technol 6:123–126
Barman DN, Haque MA et al (2014) Cloning and expression of *ophB* gene encoding organophosphorus hydrolase from endophytic *Pseudomonas* sp. BF1-3 degrades organophosphorus pesticide chlorpyrifos. Ecotoxicol Environ Saf 108:135–141
Barton JW, Kuritz T et al (2004) Reductive transformation of methyl parathion by the cyanobacterium *Anabaena* sp. strain PCC7120. Appl Microbiol Biotechnol 65:330–335
Begum SS, Arundhati A (2016) A study of bioremediation of methyl parathion in vitro using potential *Pseudomonas sp.* isolated from agricultural Soil, Visakhapatnam, India. Int J Curr Microbiol App Sci 5:464–474
Bending GD, Friloux M, Walker A (2002) Degradation of contrasting pesticides by white rot fungi and its relationship with ligninolytic potential. FEMS Microbiol Lett 212:59–63
Benning MM, Kuo JM et al (1995) Three dimensional structure of the binuclear metal center of phosphotriesterase. Biochemist 34:7973–7978
Berg G, Grube M et al (2014) The plant microbiome and its importance for plant and human health. Front Microbiol 5:491. https://doi.org/10.3389/fmicb.2014.00491
Bhagobaty RK, Malik A (2008) Utilization of chlorpyrifos as a sole source of carbon by bacteria isolated from wastewater irrigated agricultural soils in an industrial area of western Uttar Pradesh, India. Res J Microbiol 3:293–307
Bhushan B, Samanta SK et al (2000) Chemotaxis and biodegradation of 3-methyl-4-nitrophenol by *Ralstonia* sp. SJ98. Biochem Biophys Res Commun 275:129–133
Bhushan B, Halasz A et al (2004) Chemotaxis-mediated biodegradation of cyclic nitramine explosives RDX, HMX, and CL-20 by *Clostridium* sp. EDB2. Biochem Biophys Res Commun 316:816–821
Block R, Stroo H, Swett GH (1993) Bioremediation- why doesn't it work sometimes? Chem Eng Prog 89:44–50
Briceño G, Fuentes MS et al (2012) Chlorpyrifos biodegradation and 3,5,6-trichloro-2-pyridinol production by actinobacteria isolated from soil. Int Biodeterior Biodegrad 73:1–7

Brown D, Jaffe PR (2006) Effects of nonionic surfactants on the cell surface hydrophobicity and Apparent Hamaker constant of a *Sphingomonas* sp. Environ Sci Technol 40:195–201
Bumpus JA, Kakar SN, Coleman RD (1993) Fungal degradation of organophosphorus insecticides. Appl Biochem Biotechnol 39:715–726
Chao Y, Zhu Y et al (2008) Development of an autofluorescent whole-cell biocatalyst by displaying dual functional moieties on *Escherichia coli* cell surfaces and construction of a co-culture with OP-mineralizing activity. Appl Environ Microbiol 74:7733–7739
Chaudhry GR, Ali AN, Wheeler WB (1988) Isolation of a methyl parathion-degrading *Pseudomonas* sp. that possesses DNA homologous to the *opd* gene from a *Flavobacterium* sp. Appl Environ Microbiol 54:288–293
Chen CM, Ye QZ et al (1990) Molecular biology of carbon-phosphorus bond cleavage. Cloning and sequencing of the *phn* (*psiD*) genes involved in alkylphosphonate uptake and C-P lyase activity in *Escherichia coli* B. J Biol Chem 265:4461–4471
Chen S, Yang L et al (2011) Biodegradation of fenvalerate and 3-phenoxybenzoic acid by a novel *Stenotrohomonas* sp. strain ZS-S-01 and its use in boremediation of contaminated soils. Appl Microbiol Biotechnol 90:755–767
Chen S, Liu C et al (2012) Biodegradation of Chlorpyrifos and its hydrolysis product 3,5,6-trichloro-2-pyridinol by a new fungal strain *Cladosporium cladosporioides* Hu-01. PLoS One 7:e47205
Chen L-Z, Li Y-L, Yu Y-L (2014) Soil bacterial and fungal community successions under the stress of chlorpyrifos application and molecular characterization of chlorpyrifos-degrading isolates using ERIC-PCR. J Zhejiang Univ-Sci B (Biomed & Biotechnol) 15:322–332
Chen J, Luo X-J et al (2015) Marked enhancement of *Acinetobacter* sp. organophosphorus hydrolase activity by a single residue substitution Ile211Ala. Bioresour Bioprocess 2:39
Chen Q, Chen K et al (2016) A novel amidohydrolase (*DmhA*) from *Sphingomonas* sp. that can hydrolyze the organophosphorus pesticide dimethoate to dimethoate carboxylic acid and methylamine. Biotechnol Lett 38:703–710
Cheng TC, DeFrank JJ (2000) Hydrolysis of organophosphorus compounds by bacterial prolidases. In: Zwanenburg B et al (eds) Enzymes in action green solutions for chemical problems, vol 33. Kluwer Academic, Dordrecht, pp 243–261
Cheng T-C, Harvey SP, Stroup AN (1993) Purification and properties of a highly active organophosphorus acid anhydrolase from *Alteromonas undina*. Appl Environ Microbiol 59:3138–3140
Cheng T-C, Haevey SP, Chen GL (1996) Cloning and expression of a gene encoding a bacterial enzyme for decontamination of organophosphorus nerve agents and nucleotide sequence of the enzyme. Appl Environ Microbiol 62:1636–1641
Cheng T, Liu L et al (1997) Nucleotide sequence of a gene encoding an organophosphorus nerve agent degrading enzyme from *Alteromonas haloplanktis*. J Ind Microbiol Biotechnol 18:49–55
Cheng T-C, DeFrank JJ, Rastogi VK (1999) *Alteromonas* prolidase for organophosphorus G-agent decontamination. Chem Biol Interact 120:455–462
Cho CM-H, Mulchandani A, Chen W (2004) Altering the substrate specificity of organophosphorus hydrolase for enhanced hydrolysis of chlorpyrifos. Appl Environ Microbiol 70:4681–4685
Chu X-Y, Wu N-F et al (2006) Expression of organophosphorus hydrolase OPHC2 in *Pichia pastoris*: Purification and characterization. Protein Expr Purif 49:9–14
Comeau Y, Greer CW, Samson R (1993) Role of inoculum preparation and density on the bioremediation of 2,4-D-contaminated soil by bioaugmentation. Appl Microbiol Biotechnol 38:681–687
Concepcio'n C-F, Danta'n-Gonza'lez E et al (2012) Isolation of the *opdE* gene that encodes for a new hydrolase of *Enterobacter* sp. capable of degrading organophosphorus pesticides. Biodegradation 23:387–397
Cycon' M, Wojcik M et al (2011) Biodgradation kinetics of the benzimidazole fungicide thiophanate methyl by bacteria isolated from loamy sand soil. Biodegradation 22:573–583
Daughton CG, Hsieh DP (1977) Parathion utilization by bacterial symbionts in a chemostat. Appl Environ Microbiol 34:175–184

de Lorenzo V (2008) Systems biology approaches to bioremediation. Curr Opin Biotechnol 19:579–589

Deng S, Chen Y et al (2015) Rapid biodegradation of organophosphorus pesticides by *Stenotrophomonas* sp. G1. J Hazard Mater 297:17–24

Dong Y-J, Bartlam M et al (2005) Crystal structure of methyl parathion hydrolase from *Pseudomonas* sp. WBC-3. J Mol Biol 353:655–663

Dotson SB, Smith CE et al (1996) Identification, characterization and cloning of a phosphonate monoester hydrolase from *Burkholderia caryophilli* PG2982. J Biol Chem 271:25754–25761

Dubey KK, Fulekar MH (2012) Chlorpyrifos bioremediation in Pennisetum rhizosphere by a novel potential degrader *Stenotrophomonas maltophilia* MHF ENV20. World J Microbiol Biotechnol 28:1715–1725

Dumas DP, Caldwell SR et al (1989) Purification and properties of the phosphotriesterase from *Pseudomonas diminuta*. J Biol Chem 264:19659–19665

Ekkhunnatham A, Jongsareejit B et al (2012) Purification and characterization of methyl parathion hydrolase from *Burkholderia cepacia* capable of degrading OP insecticides. World J Microbiol Biotechnol 28:1739–1746

Fang H, Xiang YQ et al (2008) Fungal degradation of chlorpyrifos by *Verticillium* sp. DSP in pure cultures and its use in bioremediation of contaminated soil and pakchoi. Int Biodeterior Biodegrad 61:294–303

Fang L-C, Chen Y-F et al (2016) Complete genome sequence of a novel chlorpyrifos degrading bacterium, *Cupriavidus nantongensis* X1. J Biotechnol 227:1–2

Fang-Yao L, Ming-zhang H et al (2007) Biodegradation of methyl parathion by *Acinetobacter radioresistens* USTB-04. J Environ Sci 19:1257–1260

Farhan M, Khan AU et al (2012) Biodegradation of Chlorpyrifos using indigenous *Pseudomonas* sp. isolated from industrial drain. Pak J Nutr 11:1183–1189

Farivar TN, Peymani A et al (2017) Biodegradation of paraoxon as an OP pesticide with *Pseudomonas plecoglossicida* transfected by *opd* gene. Biotech Health Sci. https://doi.org/10.17795/bhs-45055

Fodale R, Pasquale CD et al (2010) Isolation of organophosphorus-degrading bacteria from agricultural mediterranean soils. Fresenius Environ Bull 19:2396–2403

Fulekar MH, Geetha M (2008) Bioremediation of Chlorpyrifos by *Pseudomonas aeruginosa* using scale up technique. J Appl Biosci 12:657–660

Gallo MA, Lawryk NJ (1991) Organic phosphorus pesticides. In: Hayes WJ, Laws ER (eds) Handbook of pesticide toxicology. Academic Press, San Diego, pp 917–1123

Gao Y, Chen S et al (2012) Purification and characterization of a novel chlorpyrifos hydrolase from *Cladosporium cladosporioides* Hu-01. PLoS One 6:e38137

Ghanem I, Orfi M, Shamma M (2007) Biodegradation of Chlorpyrifos by *Klebsiella* sp. isolated from an activated sludge sample of waste water treatment plant in damascus. Folia Microbiol 52:423–427

Ghosh PG, Sawant NA et al (2010) Microbial biodegradation of OP pesticides. Int J Biotech Biochem 6:871–876

Goda SK, Elsayed IE et al (2010) Screening for and isolation and identification of malathion-degrading bacteria: cloning and sequencing a gene that potentially encodes the malathion-degrading enzyme, carboxylesterase in soil bacteria. Biodegradation 21:903–913

Greenhalgh R, Dhawan KL, Weinberger P (1980) Hydrolysis of Fenitrothion in model and natural aquatic systems. J Agrlc Food Chem 28:102–105

Guha A, Kumari B, Roy MK (1997) Possible involvement of plasmid in degradation of malathion and chlorpyrifos by *Micrococcus* sp. Folia Microbiol 42:574–576

Guo Z, Yuan Y et al (2009) Function analysis of OP pesticides hydrolase from *Pseudomonas stutzeri* HS-D36. doi: https://doi.org/10.1109/ICBBE.2009.5162870.

Gupta PK (2004) Pesticide exposure-Indian scene. Toxicology 198:83–90

Hao J, Liu J, Sun M (2014) Identification of a marine *Bacillus* strain C5 and parathion-methyl degradation characteristics of the extracellular esterase B1. Bio Med Research Int 2014:863094. https://doi.org/10.1155/2014/863094

Harish R, Supreeth M, Chauhan JB (2013) Biodegradation of OP pesticide by soil fungi. Advanced Bio Tech 12:04–08

Harper LL, McDaniel CS et al (1988) Dissimilar plasmids isolated from *Pseudomonas diminuta* MG and a *Flavobacterium* sp. (ATCC 27551) contains identical opd genes. Appl Environ Microbiol 54:2586–2589

Hawkins AC, Harwood CS (2002) Chemotaxis of *Ralstonia eutropha* JMP134 (pJ4) to the herbicide 2,4-dichlorophenoxyacetate. Appl Environ Microbiol 68:968–972

Hayatsu M, Hirano M, Tokuda S (2000) Involvement of two plasmids in fenitrothion degradation by *Burkholderia* sp. strain NF100. Appl Environ Microbiol 66:1737–1740

Hayes JD, Flanagan JU, Jowsey IR (2005) Glutathione transferases. Annu Rev Pharmacol Toxicol 45:51–88

Horne I, Sutherland TD et al (2002a) Identification of an *opd* (OP degradation) gene in an *Agrobacterium* isolate. Appl Environ Microbiol 68:3371–3376

Horne I, Sutherland TD et al (2002b) Cloning and expression of the phosphotriesterase gene *hocA* from *Pseudomonas monteilii* C11. Microbiologica 148:2687–2695

Hua F, Yunlong Y et al (2009) Degradation of chlorpyrifos in laboratory soil and its impact on soil microbial functional diversity. J Environ Sci 21:380–386

Hussaini SZ, Shaker M, Iqbal MA (2013) Isolation of bacterial for degradation of selected pesticides. Adv Biores 4:82–85

Ifediegwu MC, Agu KC et al (2015) Isolation, growth and identification of chlorpyrifos degrading bacteria from agricultural soil in Anambra State, Nigeria. Univ J Microbiol Res 3:46–52

Islam SMA, Math RK, Cho KM (2010) Organophosphorus hydrolase (*OpdB*) of *Lactobacillus brevis* WCP902 from Kimchi is able to degrade organophosphorus pesticides. J Agric Food Chem 58:5380–5386

Iyer R, Iken B, Tamez T (2011) Isolation, molecular and biochemical identification of Paraoxon-metabolizing *Pseudomonas* Species. J Bioremed Biodegra 2:132

Iyer R, Stepanov VG, Iken B (2013) Isolation and molecular characterization of a novel *Pseudomonas putida* strain capable of degrading OP and aromatic compounds. Adv Biol Chem 3:564–578

Jabeen H, Iqbal S, Anwar S (2015) Biodegradation of chlorpyrifos and 3, 5, 6-trichloro-2-pyridinol by a novel rhizobial strain *Mesorhizobium* sp. HN3. Water Environ J 29:151–160

Jao S-C, Huang L-F et al (2004) Hydrolysis of OP triesters by *Escherichia coli* aminopeptidase P. J Mol Catalys B: Enz 27:7–12

Jayasri Y, Naidu MD, Malllkarjuna M (2015) Biodegradation of the chlorpyrifos pesticide by bacteria isolated from groundnut agricultural soils in Kadapa Basin. The Ecoscan 9:143–146

Jeffries TC, Rayu S et al (2018) Metagenomic functional potential predicts degradation rates of a model organophosphorus xenobiotic in pesticide contaminated soils. Front Microbiol 9:147. https://doi.org/10.3389/fmicb.2018.00147

Jiang J, Zhang R et al (2007) Parameters controlling the gene-targeting frequency at the *Sphingomonas* species rrn site and expression of the methyl parathion hydrolase gene. J Appl Microbiol 102:1578–1585

John EM, Sreekumar J, Jisha MS (2016) Optimization of chlorpyrifos degradation by assembled bacterial consortium using response surface methodology. Soil Sediment Contam: An Int J 25:668–682. https://doi.org/10.1080/15320383.2016.1190684

Jones AS, Hastings FL (1981) Soil microbe studies. In: Hastings FL, Coster JE (eds) Field and laboratory evaluations of insecticides for southern pine beetle control, vol 21. Southern Forest Experiment Station, Forest Service, SE, USDA, pp 13–14

Kadiyala V, Spain JC (1998) A two component monooxygenase catalyzes both the hydroxylation of p-nitrophenol and the oxidative release of nitrite from 4-nitrocatechol in *Bacillus sphaericus* JS905. Appl Environ Microbiol 64:2479–2484

Kanekar PP, Bhadbhade BJ, Deshpande NM (2004) Biodegradation of OP pesticides. Proc Indian Natri Sci Acad B70:57–70

Kannan K, Tanabe S, Giesy JP, Tatsukawa R (1997) Organochlorine pesticides and polychlorinated biphenyls in food stuffs from Asian and oceanic countries. Rev Environ Contam Toxicol 152:1–55

Karolin KP, Meenakumari KS, Subha P (2015) Isolation and characterization of novel chlorpyrifos degrading fungus *Isaria Farinosa*. J Chem Chem Eng 9:403–407

KaviKarunya S, Reetha D (2012) Biological degradation of chlorpyrifos and monocrotophos by bacterial isolates. Int J Pharm Biol Arch 3:685–691

Kawahara K, Tanaka A et al (2010) Reclassification of a parathione degrading *Flavobacterium* sp. ATCC 27551 as *Sphingobium fuliginis*. J Gen Appl Microbiol 56:249–255

Keprasertsupa C, Suchart Upatham ES et al (2001) Degradation of methyl parathion in an aqueous medium by soil bacteria. Sci Asia 27:261–270

Khaled A, Miia T et al (2012) Metabolism of pesticides by human cytochrome P450 enzymes. In: Vitro-a survey, insecticides-advances in integrated pest management. *InTech*. Available at: http://cdn.intechopen.com/pdfs/25674/InTechMetabolism_of_pesticides_b_human_cytochrome_p450_enzymes_in_vitro_a_survey.pdf

Kim J-R, Ahn Y-J (2009) Identification and characterization of chlorpyrifos-methyl and 3,5,6-trichloro-2-pyridinol degrading *Burkholderia* sp. strain KR100. Biodegradation 20:487–497

Kim T, Ahn J-H et al (2007) Cloning and expression of a parathion hydrolase gene from a soil bacterium, *Burkholderia* sp. JBA3. J Microbiol Biotechnol 17:1890–1893

Kim CH, Choi JS et al (2013) Biodegradation of chlorpyrifos (CP) by a newly isolated *Naxibacter* sp. strain CY6 and its ability to degrade CP in soil. Korean J Microbiol 49:83–89

Kitagawa W, Kimura N, Kamagata Y (2004) A novel p-Nitrophenol degradation gene cluster from a Gram-Positive bacterium, *Rhodococcus opacus* SAO101. J Bacteriol 186:4894–4902

Kulshrestha G, Kumari A (2011) Fungal degradation of chlorpyrifos by *Acremonium* sp. strain (GFRC-1) isolated from a laboratory-enriched red agricultural soil. Biol Fertil Soils 47:219–225

Kumar S (2011a) Isolation, characterization and growth response study of chlorpyrifos degrading bacteria from cultivated soil. Int J Adv Engineer Technol 2:199–203

Kumar S (2011b) Bioremediation of chlorpyrifos by bacteria isolated from the cultivated soils. Int J Pharm Bio Sci 2:359–366

Kumar J, D'Souza SF (2010) An optical microbial biosensor for detection of methyl parathion using *Sphingomonas* sp. immobilized on micro plate as a reusable biocomponent. Biosens Bioelectron 26:1292–1296

Kumar J, Jha SK, D'Souza SF (2006) Optical microbial biosensor for detection of methyl parathion pesticide using *Flavobacterium* sp. whole cells adsorbed on glass fiber filters as disposable biocomponent. Biosens Bioelectron 21:2100–2105

Kumar SV, Fareedullah M et al (2010) Current review on Organophosphorous poisoning. Archives Appl Sci Res 2:199–215

Kuo JM, Raushel FM (1994) Identification of the histidine ligands to the binuclear metal center of phosphotriesterase by site-directed mutagenesis. Biochemistry 33:4265–4272

Lakshmi A (1993) Pesticides in India: risk assessment to aquatic ecosystems. Sci Total Environ 134:243–253

Lakshmi CV, Kumar M, Khanna S (2009) Biodegradation of chlorpyrifos in soil by enriched cultures. Curr Microbol 58:35–38

Latifi AM, Khodi S et al (2012) Isolation and characterization of five chlorpyrifos degrading bacteria. Afr J Biotechnol 11:3140–3146

Li X, Schuler MA, Berenbaum MR (2007a) Molecular mechanisms of metabolic resistance to synthetic and natural xenobiotics. Annu Rev Entomol 52:231–253

Li X, He J, Li S (2007b) Isolation of chlorpyrifos degrading bacterium, *Sphingomonas* sp. strain Dsp-2, and cloning of the *mpd* gene. Res Microbiol 158:143–149

Li Y, Li W, Zhang C, Yan Y (2008a) Isolation, characterization of methyl-parathion degrading strain L1 and cloning of the *mpd* gene. Biotechnol Bull:2008-06

Li XH, Jiang JD et al (2008b) Diversity of chlorpyrifos degrading bacteria isolated from chlorpyrifos-contaminated samples. Int Biodeterior Biodegrad 62:331–335

Liu X, Parales RE (2009) Bacterial chemotaxis to atrazine and related *s*-triazines. Appl Environ Microbiol 75:5481–5488

Liu Y-H, Chung Y-C, Xiong Y (2001) Purification and characterization of a dimethoate-degrading enzyme of *Aspergillus niger* ZHY256, isolated from sewage. Appl Environ Microbiol 67:3746–3749

Liu X, You M et al (2003) Isolation of chlorpyrifos-degrading *Aspergillus* sp. Y and measurement of degradation efficiency. Chinese J Appl Environ Biol 9:78–80

Liu Y-H, Liu Y et al (2004) Purification and characterization of a novel organophosphorus pesticide hydrolase from *Penicillium lilacinum BP303*. Enz Microbial Technol 34:297–303

Liu H, Zhang J-J et al (2005) Plasmid-borne catabolism of methyl parathion and p-nitrophenol in *Pseudomonas* sp. strain WBC-3. Biochem Biophys Res Commun 334:1107–1114

Liu Z, Chen X et al (2012) Bacterial degradation of Chlorpyrifos by *Bacillus cereus*. Adv Mater Res 356–360:676–680

Liu Z, Xie J et al (2014) Isolation of an organophosphorus-degrading strain *Pseudomonas* sp. strain YF-5 and cloning of *mpd* gene from this strain. J Pure Appl Microbiol 8:587–591

Longkumar T, Parthasarathy S et al (2014) OxyR dependent expression of a novel glutathione S-tranferase (*Abgst01*) gene in *Acinetobacter baumannii* DS002 and its role in biotransformation of OP insecticides. Microbiologica 160:102–112

Lu P, Li Q et al (2013) Biodegradation of chlorpyrifos and 3,5,6-trichloro-2-pyridinol by *Cupriavidus* sp. DT-1. Bioresour Technol 127:337–342

Madhuri RJ, Rangaswamy V (2009) Biodegradation of selected insecticides by *Bacillus* and *Pseudomonas* sps in ground nut fields. Toxicol Int 16:127–132

Marinho G, Rodrigues K et al (2011) Glucose effect on degradation kinetics of methyl parathion by filamentous fungi species *Aspergillus niger* AN400. Eng Sanit Ambient 16:225–230

Mastumura F, Boush GM (1968) Degradation of insecticides by a soil fungus, *Trichoderma viridae*. J Econ Entomol 61:610–612

Maya K, Singh RS et al (2011) Kinetic analysis reveals bacterial efficacy for biodegradation of chlorpyrifos and its hydrolyzing metabolite TCP. Process Biochem 46:2130–2136

Maya K, Upadhyay SN et al (2012) Degradation kinetics of chlorpyrifos and 3,5,6-trichloro-2-pyridinol (TCP) by fungal communities. Bioresour Technol 126:216–223

Mazur A (1946) An enzyme in animal tissue capable of hydrolyzing the phosphorous-fluorine bond of alkyl fluorophosphates. J Biol Chem 164:271–289

McDaniel CS, Harper LL, Wild JR (1988) Cloning and sequencing of a plasmid-borne gene (*opd*) encoding a phosphotriesterase. J Bacteriol 170:2306–2311

Min-Kyeong C, Kim K-D et al (2009) Genetic and phenotypic diversity of parathion-degrading bacteria isolated from rice paddy soils. J Microbiol Biotechnol 19:1679–1687

Mishra A (2015) Microbial degradation of methyl parathion by a soil bacterial isolate and consortium. Int J Res Studies Biosci:15–19

Misra D, Bhuyan S et al (1992) Accelerated degradation of methyl parathion, parathion and fenitrothion by suspensions from methyl parathion and *p*-nitrophenol-treated soils. Soil Biol Biochem 24:1035–1042

Monteiro SA, Sassaki GL et al (2007) Molecular and structural characterization of the biosurfactant produced by Pseudomonas aeruginosa DAUPE 614. Chem Phys Lipids 147:1–13

Morra MJ (1996) Bioremediation in soil: Influence of soil properties on organic contaminants and bacteria. In: Crawford RL, Crawford DL (eds) Bioremediation: principles and application. Cambridge University Press, Cambridge, pp 35–60

Mulbry WW (1992) The aryldialkylphosphatase-encoding gene *adpB* from *Nocardia* sp. strain B-l: cloning, sequencing and expression in *Escherichia coli*. Gene 121:149–153

Mulbry WW, Karns JS (1989) Parathion hydrolase specified by the *Flavobacterium opd* gene: Relationship between the gene and protein. J Bacteriol 171:6740–6746

Mulbry WW, Karns JS et al (1986) Identification of a plasmid-borne parathion hydrolase gene from *Flavobacterium* sp. by southern hybridization with *opd* from *Pseudomonas diminuta*. Appl Environ Microbiol 51:926–930

Mulbry WW, Zhu H et al (2002) The triazine hydrolase gene trzN from *Nocardioides* sp. strain C190: Cloning and construction of gene-specific primers. FEMS Microbiol Lett 206:75–79
Munnecke DM, Hsieh DPH (1974) Microbial decontamination of parathion and *p*-nitrophenol in aqueous media. Appl Microbiol 28:212–217
Munnecke DM, Hsieh DPM (1976) Pathways of microbial metabolism of parathion. Appl Environ Microbiol 31:63–69
Nelson LM (1982) Biologically-induced hydrolysis of parathion in soil: isolation of hydrolyzing bacteria. Soil Biol Biochem 14:219–222
Neti N, Zakkula V (2013) Analysis of chlorpyrifos degradation by *Kocuria* sp. using GC and FTIR. Curr Biotica 6:466–472
Ningfeng W, Minijie D et al (2004) Isolation, purification and characterization of a new organophosphorous hydrolase OPH2. Chin Sci Bull 49:268–272
Oakley AJ (2005) Glutathione transferases: new functions. Curr Opin Struct Biol 15:716–723
Omar SA (1998) Availability of phosphorus and sulfur of insecticide origin by fungi. Biodegradation 9:327–336
Omburo GA, Kuo JM et al (1992) Characterization of the zinc binding site of bacterial phosphotriesterase. J Biol Chem 267:13278–13283
Ortiz-Hernandez ML, Sanchez-Salinas E (2010) Biodegradation of the OP pesticide tetrachlorvinphos by bacteria isolated from agricultural soils in Mexico. Rev Int Contam Ambient 26:27–38
Pailan S, Saha P (2015) Chemotaxis and degradation of OP compound by a novel moderately thermo-halo tolerant *Pseudomonas* sp. strain BUR11: evidence for possible existence of two pathways for degradation. PeerJ 3:e1378. https://doi.org/10.7717/peerj.1378
Pailan S, Gupta D et al (2015) Degradation of OP insecticide by a novel *Bacillus aryabhattai* strain SanPS1, isolated from soil of agricultural field in Burdwan, West Bengal, India. Int Biodeterior Biodegrad 103:191–195
Pailan S, Sengupta K et al (2016) Evidence of biodegradation of Chlorpyrifos by a newly isolated heavy metal tolerant bacterium *Acinetobacter* sp. strain $MemCl_4$. Environ Earth Sci 75:1019. https://doi.org/10.1007/s12665-016-5834-8
Pakala SB, Gorla P et al (2007) Biodegradation of methyl parathion and *p*-nitrophenol: evidence for the presence of a p-nitrophenol 2-hydroxylase in a Gram-negative *Serratia* sp. strain DS001. Appl Microbiol Biotechnol 73:1452–1462
Pandey G, Jain RK (2002) Bacterial chemotaxis toward environmental pollutants: role in bioremediation. Appl Environ Microbiol 68:5789–5795
Pandey S, Singh DK (2004) Total bacterial and fungal population after chlorpyrifos and quinalphos treatments in groundnut (Arachis hypogaea L.) soil. Chemosphere 55:197–205
Pandey G, Chauhan A et al (2002) Chemotaxis of a *Ralstonia* sp. SJ98 toward co-metabolizable nitroaromatic compounds. Biochem Biophys Res Commun 299:404–409
Park N-J, Kamble ST (2001) Decapitation impact effect of topically applied chlorpyrifos on acetylcholinesterase and general esterases in susceptible and resistant German Cockroaches (Dictyoptera: Blattellidae). J Econ Entomol 94:499–505
Parthasarathy S, Parapatla H, Nandavaram A et al (2016) OP Hydrolase is a lipoprotein and interacts with pi-specific transport system to facilitate growth of *Brevundimonas diminuta* Using op insecticide as source of phosphate. J Biol Chem 291:7774–7785
Parthasarathy S, Azam S, Lakshman Sagar A et al (2017a) Genome-guided insights reveal OP-degrading *Brevundimonas diminuta* as *Sphingopyxis wildii* and define its versatile metabolic capabilities and environmental adaptations. Genome Biol Evol 9:77–81
Parthasarathy S, Parapatla H, Siddavattam D (2017b) Topological analysis of the lipoprotein OP hydrolase from *Sphingopyxis wildii* reveals a periplasmic localization. FEMS Microbiol Lett 364:fnx187. https://doi.org/10.1093/femsle/fnx187
Penaloza-Vazquez A, Mena GL et al (1995) Cloning and sequencing of the genes involved in glyphosate utilization by *Pseudomonas pseudomallei*. Appl Environ Microbiol 61:538–543
Peter JK, Masih H et al (2014) OP pesticide (methyl parathion) degrading bacteria isolated from rhizospheric soil of selected plants and optimization of growth conditions for degradation. Int J Res 1

Pino N, Peñuela G (2011) Simultaneous degradation of the pesticides methyl parathion and chlorpyrifos by an isolated bacterial consortium from a contaminated site. Int Biodeterior Biodegrad 65:827–831

Pino NJ, Dominguez MC, Penuela GA (2011) Isolation of a selected microbial consortium capable of degrading methyl parathion and *p*-nitrophenol from a contaminated soil site. J Environ Sci Health 46:173–180

Puri SN, Murthy KS, Sharma OP (2013) Pest problems in India-current status. Ind J Plant Protec 27:20–31

Qiu X-H, Bai W-Q et al (2006) Isolation and characterization of a bacterial strain of the genus *Ochrobactrum* with methyl parathion mineralizing activity. J Appl Microbiol 101:986–994

Ramakrishnan B, Megharaj M et al (2011) Mixtures of environmental pollutants: effects on microorganisms and their activities in soils. Rev Environ Contam Toxicol 211:63–120

Ramanathan MP, Lalithakumari D (1996) Methyl parathion degradation by *Pseudomonas* sp. A3 immobilized in sodium alginate beads. World J Microbiol Biotechnol 12:107–108

Ramanathan MP, Lalithakumari D (1999) Complete Mineralization of Methyl parathion by *Pseudomonas* sp. A3. Appl Biochem Biotechnol 80:1–12

Rani NL, Lalithakumari D (1994) Degradation of methyl parathion by *Pseudomonas putida*. Can J Microbiol 4:1000–1004

Rani MS, Lakshmi KV et al (2008) Isolation and characterization of a chlorpyrifos degrading bacterium from agricultural soil and its growth response. Afr J Microbiol Res 2:026–031

Rao AV, Sethunathan N (1974) Degradation of parathion by *Penicillium waksmani* isolated from flooded acid sulphate soil. Arch Microbiol 97:203–208

Rayu S, Nielsen UN, Nazaries L, Singh BK (2018) Isolation and molecular characterization of novel chlorpyrifos and 3,5,6- trichloro-2-pyridinol-degrading bacteria from sugarcane farm soils. Front Microbiol 8:518. https://doi.org/10.3389/fmicb.2017.00518

Rodrigues GN, Alvarenga N et al (2016) Biotransformation of methyl-parathion by marine-derived fungi isolated from ascidian *Didemnum ligulum*. Biocatalys Agri Biotechnol 7:24–30

Romeh AA, Hendawi MY (2014) Bioremediation of certain organophosphorus pesticides by two biofertilizers, Paenibacillus (*Bacillus*) *polymyxa* (Prazmowski) and *Azospirillum lipoferum* (Beijerinck). J Agric Sci Technol 16:265–276

Rosenberg A, Alexander M (1979) Microbial cleavage of various organophosphorus insecticides. Appl Environ Microbiol 37:886–891

Sabdono A (2007) Biodegradation of chlorpyrifos by a marine bacterium *Bacillus firmus* strain BY6 associated with branching coral *Acropora* sp. J Coastal Develop 10:115–123

Sabdono A, Radjasa OK (2008) Phylogenetic diversity of OP pesticides-degrading coral bacteria from mild-West coast of Indonesia. Biotechnology 7:694–701

Samanta SK, Bhushan B et al (2000) Chemotaxis of a *Ralstonia* sp. SJ98 toward different nitroaromatic compounds and their degradation. Biochem Biophys Res Commun 269:117–123

Sasikala C, Jiwal S et al (2012) Biodegradation of chlorpyrifos by bacterial consortium isolated from agriculture soil. World J Microbiol Biotechnol 28:1301–1308

Savitha K, Raman DNS (2012) Isolation, identification, resistance profile and growth kinetics of chlorpyrifos resistant bacteria from agricultural soil of Bangalore. Res Biotechnol 3:08–13

Scott C, Pandey G et al (2008) The enzymatic basis for pesticide bioremediation. Indian J Microbiol 48:65–79

Serdar CM, Gibson DT et al (1982) Plasmid involvement in parathion hydrolysis by *Pseudomonas diminuta*. Appl Environ Microbiol 44:246–249

Sethunathan N, Yoshida T (1973) A *Flavobacterium* sp. that degrades diazinon and parathion. Can J Microbiol 19:873–875

Shan M, Fang H et al (2006) Effect of chlorpyrifos on soil microbial populations and enzyme activities. J Environ Sci 18:4–5

Sharma J, Gupta KC, Goel AK (2013) Isolation and identification of potential methyl parathion degrading bacteria from Gwalior arable soil. Int J Pharm Bio Sci 4:192–202

Sharmila M, Ramanand K, Sethunathan N (1989) Effect of yeast extract on the degradation of organophosphorus insecticides by soil enrichment and bacterial cultures. Can J Microbiol 35:1105–1110

Sheehan D, Meade G et al (2001) Structure, function and evolution of glutathione transferases: implications for classification of non mammalian members of an ancient enzyme superfamily. Biochem J 360:1–16

Shen YJ, Hong YF et al (2007) Isolation, identification and characteristics of a phoxim-degrading bacterium XSP-1. Huan Jing Ke Xue 28:2833–2837

Shen Y-J, Lu P, Mei H et al (2010a) Isolation of a methyl parathion-degrading strain *Stenotrophomonas* sp. SMSP-1 and cloning of the *ophc2* gene. Biodegradation 21:785–792

Shen Y-J, Lu P et al (2010b) Isolation of a methyl parathion-degrading strain *Stenotrophomonas* sp. SMSP-1 and cloning of the *ophc2* gene. Biodegradation 21:785–792

Shimazu M, Mulchandani A, Chen W (2001) Simultaneous degradation of organophosphorous pesticides and *p*-nitrophenol by genetically engineered *Moraxella* sp. with surface expressed organophosphorous hydrolase. Biotechnol Bioeng 4:318–324

Siddaramappa R, Rajaram KP, Sethunathan N (1973) Degradation of Parathion by bacteria isolated from flooded soil. Appl Microbiol 26:846–849

Siddavattam D, Khajamohiddin S et al (2003) Transposon-like organization of the plasmid-borne OP degradation (*opd*) gene cluster found in *Flavobacterium* sp. Appl Environ Microbiol 69:2533–2539

Silambarasan S, Abraham J (2013) Ecofriendly method for bioremediation of chlorpyrifos from agricultural soil by novel fungus *Aspergillus terreus* JAS1. Water Air Soil Pollut 224:1369

Singh BK (2009) Organophosphorous-degrading bacteria: ecology and industrial applications. Nat Rev Microbiol 7:156–164

Singh BK, Walker A (2006) Microbial degradation of organophosphorous compounds. FEMS Microbiol Rev 30:428–471

Singh BK, Walker A et al (2003) Effect of soil pH on the biodegradation of chlorpyrifos and isolation of a chlorpyrifos-degrading bacterium. Appl Environ Microbiol 69:5198–5206

Singh BK, Walker A et al (2004) Biodegradation of chlorpyrifos by *Enterobacter* strain B-14 and its use in bioremediation of contaminated soils. Appl Environ Microbiol 70:4855–4863

Singh DP, Khattar JIS et al (2011) Chlorpyrifos degradation by the cyanobacterium *Synechocystis* sp. strain PUPCCC 64. Environ Sci Pollut Res 18:1351–1359

Somara S, Siddavattam D (1995) Plasmid mediated OP pesticide degradation by *Flavobacterium balustinum*. Biochem Mol Biol Int 36:627–631

Spain JC, Gibson DT (1991) Pathway for Biodegradation of p-Nitrophenol in a *Moraxella* sp. Appl Environ Microbiol 57:812–819

Sreenivasulu C, Aparna Y (2001) Bioremediation of methyl Parathion by free and immobilized cells of *Bacillus* sp. isolated from soil. Bull Environ Contam Toxicol 67:98–105

Sukirtha TH, Usharani MV (2013) Production and qualitative analysis of biosurfactant and biodegradation of the OP by *Nocardia mediterranei*. J Bioremed Biodeg 4:198. https://doi.org/10.4172/2155-6199.1000198

Taesung K, Ahn J-H et al (2007) Cloning and expression of a parathion hydrolase gene from a soil bacterium, *Burkholderia* sp. JBA3. J Microbiol Biotechnol 17:1890–1893

Tago K, Sekiya E et al (2006) Diversity of fenitrothion- degrading bacteria in soil from distant geographical areas. Microbes Environ 21:58–64

Tehara SK, Keasling JD (2003) Gene cloning, purification, and characterization of a phosphodiesterase from *Delftia acidovorans*. Appl Environ Microbiol 69:504–508

Thengodkar RRM, Sivakami S (2010) Degradation of chlorpyrifos by an alkaline phosphatase from the cyanobacterium Spirulina platensis. Biodegradation 21:637–644

Theriot CM, Grunden AM (2011) Hydrolysis of OP compounds by microbial enzymes. Appl Microbiol Biotechnol 89:35–43

Tian J, Wang P et al (2010) Enhanced thermo stability of methyl parathion hydrolase from *Ochrobactrum* sp. M231 by rational engineering of a glycine to proline mutation. FEBS J 277:4901–4908

Tomlin C (2000) The pesticide manual. BCPC, Surrey, UK

Urlacher VB, Lutz-Wahl S, Schmid RD (2004) Microbial P450 enzymes in biotechnology. Appl Microbiol Biotechnol 64:317–325

Van Eerd LL, Hoagland RE et al (2003) Pesticidemetabolism in plants and microorganisms. Weed Sci 51:472–495

Vanhooke JL, Benning MM et al (1996) Three dimensional structure of the zinc-containing phosphotriesterase with the bound substrate analog diethyl 4-methylbenzylphosphonate. Biochemist 35:6020–6025

Vijayalakshmi P, Usha MS (2012) Optimization of Chlorpyrifos degradation by *Pseudomonas putida*. J Chem Pharm Res 4:2532–2539

Walker AW, Keasling JD (2002) Metabolic engineering of *Pseudomonas putida* for the utilization of parathion as a carbon and energy source. Biotechnol Bioeng 78:15–721

Wang JH, Zhu LS et al (2005) Degrading characters of 3 chlorpyrifos degrading fungus. Chinese J Appl Environ Biol 11:211–214

Wang X, Chu X et al (2006) Degradation characteristics and functions of chlorpyrifos degradation bacterium *Bacillus laterosporus* DSP. Acta Pedol Sin 43:648–654

Wang B, Li X et al (2008) Cloning and expression of the *mpd* gene from a newly isolated methyl-parathion-degrading strain of bacteria. Acta Sci Circumst 28:1969–1975

Wang S, Zhang C, Yan Y (2012) Biodegradation of methyl parathion and *p*-nitrophenol by a newly isolated *Agrobacterium* sp. strain Yw12. Biodegradation 23:107–116

Wang D, Xue Q et al (2015) Isolation and characterization of a highly efficient chlorpyrifos degrading strain of *Cupriavidus taiwanensis* from sludge. J Basic Microbiol 55:229–235

Wen Y, Jiang J-D et al (2007) Effect of mutation of chemotaxis signal transduction gene cheA in *Pseudomonas putida* DLL-1 on its chemotaxis and methyl parathion biodegradation. Acta Microbiol Sin 47:471–476

Whitman WB, David CC, William W (1998) Prokaryotes: the unseen majority. Proc Natl Acad Sci U S A 95:6578–6583

Wu S et al (2015) Isolation and characterization of a novel native *Bacillus thuringiensis* strain BRC-HZM2 capable of degrading chlorpyrifos. J Basic Microbiol 55:389–397

Xie H, Zhu L et al (2010) Immobilization of an enzyme from a *Fusarium* fungus WZ-I for chlorpyrifos degradation. J Environ Sci 22:930–1935

Xu G, Li Y et al (2007) Mineralization of chlorpyrifos by co-culture of *Serratia* and *Trichosporon* spp. Biotechnol Lett 29:1469–1473

Xu GM, Zheng W et al (2008) Biodegradation of chlorpyrifos and 3,5,6-trichloro-2-pyridinol by a newly isolated *Paracoccus* sp. TRP. Int Biodeterior Biodegrad 62:51–56

Yang L, Zhao YH et al (2005) Isolation and characterization of a chlorpyrifos and 3,5,6-trichloro-2-pyridinol degrading bacterium. FEMS Microbiol Lett 251:67–73

Yang C, Liu N et al (2006) Cloning of *mpd* gene from a chlorpyrifos-degrading bacterium and use of this strain in bioremediation of contaminated soil. FEMS Microbiol Lett 265:118–125

Yang C, Dong M et al (2007) Reductive transformation of parathion and methyl parathion by *Bacillus* sp. Biotechnol Lett 29:487–493

Yu YL, Fang H et al (2006) Characterization of a fungal strain capable of degrading chlorpyrifos and its use in detoxification of the insecticide on vegetables. Biodegradation 17:487–494

Zhang R, Cui Z et al (2005) Diversity of organophosphorus pesticide degrading bacteria in a polluted soil and conservation of their organophosphorous hydrolase genes. Can J Microbiol 51:337–343

Zhang R, Cui Z et al (2006a) Cloning of the organophosphorus pesticide hydrolase gene clusters of seven degradative bacteria isolated from a methyl parathion contaminated site and evidence of their horizontal gene transfer. Biodegradation 17:465–472

Zhang Z, Hong Q et al (2006b) Isolation of fenitrothion-degrading strain *Burkholderia* sp. FDS-1 and cloning of *mpd* gene. Biodegradation 17:275–283

Zhang J, Xin Y et al (2008) Metabolism-independent chemotaxis of *Pseudomonas* sp. strain WBC-3 toward aromatic compounds. J Environ Sci 20:1238–1242

Zhang JJ, Liu H et al (2009) Identification and characterization of catabolic para-Nitrophenol 4-Monooxygenase and para- Benzoquinone reductase from *Pseudomonas* sp. Strain WBC-3. J Bacteriol 191:2703–2710

Zhang Q, Wang BC et al (2012a) Plasmid-mediated bioaugmentation for the degradation of chlorpyrifos in soil. J Hazard Mater 221-222:178–184

Zhang S, Sun W et al (2012b) Identification of the para-nitrophenol catabolic pathway, and characterization of three enzymes involved in the hydroquinone pathway, in *Pseudomonas* sp. 1-7. BMC Microbiol 12:27

Zhang R, Xu X, Chen W, Huang Q (2016) Genetically engineered *Pseudomonas putida* X3 strain and its potential ability to bioremediate soil microcosms contaminated with methyl parathion and cadmium. Appl Microbiol Biotechnol 100:1987–1997

Zhongli C, Shunnpeng L, Guoping F (2001) Isolation of Methyl parathion degrading strain M6 and cloning of the methyl parathion hydrolase gene. Appl Environ Microbiol 67:4922–4925

Zhu L, Zhou W (2008) Partitioning of polycyclic aromatic hydrocarbons to solid-sorbed nonionic surfactants. Environ Pollut 152:130–137

Zhu J, Zhao Y, Qiu J (2010) Isolation and application of chlorpyrifos-degrading *Bacillus licheniformis*-ZHU-1. Afr J Microbiol Res 4:2410–2413

MIX
Papier aus verantwortungsvollen Quellen
Paper from responsible sources
FSC® C105338
FSC
www.fsc.org

If you have any concerns about our products,
you can contact us on
ProductSafety@springernature.com

In case Publisher is established outside the EU,
the EU authorized representative is:
Springer Nature Customer Service Center GmbH
Europaplatz 3, 69115 Heidelberg, Germany

Printed by Libri Plureos GmbH
in Hamburg, Germany